Bailey's Industrial Oil and Fat Products

BAILEY'S INDUSTRIAL OIL AND FAT PRODUCTS

Volume 2
Fourth Edition

Edited by

DANIEL SWERN

Fels Research Institute and Temple University

Authors

ROBERT R. ALLEN
MARVIN W. FORMO
R. G. KRISHNAMURTHY
G. N. McDERMOTT
FRANK A. NORRIS
NORMAN O. V. SONNTAG

1807 1982

A Wiley-Interscience Publication

JOHN WILEY & SONS
New York · Chichester · Brisbane · Toronto · Singapore

Library of Congress Cataloging in Publication Data:

Bailey, Alton Edward, 1907–1953.
  Bailey's Industrial oil and fat products.

  "A Wiley-Interscience publication."
  Includes bibliographical references and indexes.
  1. Oils and fats.   I. Swern, Daniel, 1916–   .
II Formo, Marvin W.   III. Title.   IV. Title: Industrial
oil and fat products.
TP670.B28   1979   665   78-31275
ISBN 0-471-83957-4 (v.1)   AACR1
ISBN 0-471-83958-2 (v.2)

Printed in the United States of America

10 9 8 7 6 5 4 3 2 1

Authors

ROBERT R. ALLEN
Exploratory and Applied Research
Anderson Clayton Foods
Richardson, Texas

Chapter 1

NORMAN O. V. SONNTAG
Red Oak, Texas

Chapters 2, 7

FRANK A. NORRIS
Kraftco Corporation
Research and Development
Glenview, Illinois

Chapters 3, 4

R. G. KRISHNAMURTHY
Kraftco Corporation
Research and Development
Glenview, Illinois

Chapter 5

MARVIN W. FORMO
Cargill Research Department
Minneapolis, Minnesota

Chapter 6

G. N. McDERMOTT
The Procter & Gamble Company
Cincinnati, Ohio

Chapter 8

v

Preface

Volume 2 of the 4th edition of *Industrial Oil and Fat Products* is a departure from prior revisions as it is not solely an updating of subjects from earlier editions. Chapters 7 (Analytical Methods) and 8 (Environmental Aspects) are entirely new. Their inclusion is a reflection of the growing importance of these two subjects in modern oil and fat technology. Increasing concern for environmental aspects by consumers, regulatory agencies, and manufacturers makes Chapter 8 exceedingly timely even though it describes a subject undergoing rapid change.

Chapter 1 (Hydrogenation) updates a subject that is possibly the most important unit process in oil and fat technology. The subject is extremely complex and not well understood in spite of all the past and ongoing research in the field. Only the more important aspects of hydrogenation could be included in this edition owing to space limitations. Chapter 2 (Fat Splitting, Esterification, and Interesterification) describes the techniques used to obtain some of the most important chemical raw materials derived from oils and fats. Chapters 3 (Extraction of Fats and Oils) and 4 (Refining and Bleaching) describe how oils and fats are obtained from their widely diverse natural sources and then processed to be suitable for the multitude of industrial and edible applications of oils and fats. Chapter 5 (Cooking Oils, Salad Oils, and Salad Dressings) describes in some detail selected important commercial uses for oils and fats in edible products that continue to show healthy growth. Chapter 6 (Miscellaneous Oil and Fat Products) describes numerous applications of fats and oils predominantly in industrial areas.

Again, it is a distinct pleasure to thank my secretary, Dorothy A. Wyszynski, for the multitude of tasks she undertook (voluntarily) in helping to edit and organize the manuscripts for this volume. My wife, Ann R. Swern, was invaluable in checking galley proofs and preparing the voluminous index. I also thank James L. Smith, Editor, for patience and varied assistance. And, finally, to all the contributors who labored mightily and withstood my constant prodding with good humor and grace, I extend my heartfelt thanks and continued friendship.

DANIEL SWERN

Philadelphia, Pennsylvania
August 1981

Contents

1
Hydrogenation

Introduction (1)

Hydrogenation of oils and fats is the largest single chemical reaction in the fatty oil processing industry. Simply stated, hydrogenation is the addition of hydrogen to the ethylenic linkages or double bonds by reaction with hydrogen in the presence of a metal catalyst.

However, the hydrogenation reaction is complicated by the simultaneous isomerization, both positional and geometrical, of the unsaturated bonds. Also, vegetable oils are glycerides of fatty acids that contain one, two, three, or more unsaturated bonds in each fatty acid. Since each double bond may be isomerized or hydrogenated at different rates depending on its position or environment in the molecule, the overall reaction is quite complex. Also, since oils are composed of triglyceride esters, the position of the fatty acid on the glycerol determines the physical properties of the molecule. Thus the partial hydrogenation of soybean oil would result in the production of a minimum of 30 different fatty acids, the *cis* and *trans* forms of partially hydrogenated linolenic, linoleic, oleic acids, so there would be at least 4000 different triglycerides possible.

It is apparent that the hydrogenation of vegetable oils is extremely complex. However, processors have learned how to control the reaction to produce many products for the market.

It is difficult to overemphasize the importance of the hydrogenation process in modern oil and fat technology. It is employed on a vast scale in the soap, industrial oil, and edible-fat industries; for converting liquid oils to hard or plastic fats; for converting soft fats to firmer products; and for improving the resistance of fats and oils to deterioration through oxidation or flavor reversion. It has made a major contribution to the present high degree of interchangeability among a wide variety of fats and oils. The most obvious result of the introduction of hydrogenation on a wide scale has been the establishment of liquid oils, such as cottonseed, soy-

1

bean, and other vegetable oils, and some marine oils, as adequate substitutes for originally more expensive meat fats.

Also, it has had a wide effect on the worldwide agricultural economy since it made possible the use of highly unsaturated vegetable oils obtained as a byproduct from seeds used as feed for animals. For example, the phenomenal world production of soybeans (1,891,676,000 bushels in 1974) came only after the ability to utilize soybean oil by hydrogenation was obtained. Thus the cost of the protein part of the bean was lower, and this has resulted in the production of more animals, principally cattle, since a low-cost source of high protein feed is available.

1.1   HISTORICAL

The modern hydrogenation process originated in the classical research of Sabatier and Senderens (2) carried out during 1897–1905, which demonstrated the feasibility of effecting hydrogenation of unsaturated organic materials in a simple apparatus and without an undue occurrence of side reactions, using nickel or other relatively inexpensive metal as a catalyst.

Actually, Sabatier's experiments encompassed hydrogenation in the vapor phase only; hence his technique was not directly applicable to relatively nonvolatile triglycerides. A process for the liquid-phase hydrogenation of fatty oils was patented by Normann (3) in 1903. Title to the Normann patent passed to the British firm of Joseph Crossfield and Sons, and hydrogenation is said to have been employed on a limited scale in the treatment of whale oil in England in 1906 or earlier (4). Potentially, however, the greatest use for the process lay in the United States, where a vast production of cottonseed oil awaited technical developments that would permit its conversion to the plastic edible fat demanded by American tradition and custom.

In 1909 the American rights to the Crossfield patents were acquired by the Procter and Gamble Company, which placed its hydrogenated cottonseed oil shortening, Crisco, on the market in 1911. Promotion and marketing of the new product were prosecuted with vigor and with success sufficient to arouse the strong interest of other American processors. Later a court decision invalidated the Burchenal patent (5), under whose broad claims the Procter and Gamble shortenings were then manufactured, and the way was cleared for the manufacture of comparable products by other firms. At present hydrogenation is employed by virtually every American producer of shortening or margarine oils and by most margarine manufacturers elsewhere in the world, as well as by many processors of nonedible oils and fats.

1.2 GENERAL

Primarily, hydrogenation is a means of converting liquid oils to semisolid, plastic fats suitable for shortening or margarine manufacture. However, it also accomplishes various other desirable purposes, including enhancement of the stability and the improvement of the color of the fat.

The reaction requires a catalyst; the catalyst employed in commercial hydrogenation consists basically of nickel, although catalysts such as copper chromite are now being used. Also, minor amounts of copper, alumina, zirconium, and other substances may be incorporated with the nickel for their "promoter" action. There has been some use of massive catalysts for continuous hydrogenation, but by far the greatest amount of hydrogenation is carried out with powder catalysts composed of the metal or metals in a finely divided form, prepared by special methods, and often supported on a highly porous, inert, refractory material, such as diatomaceous earth. The catalyst is suspended in the oil during hydrogenation and at the conclusion is removed by filtration. Although catalysts decrease in activity with repeated use, in most cases inactivation is slow, and a single charge of catalyst may be used a number of times.

For hydrogenation to take place, gaseous hydrogen, liquid oil, and the solid catalyst must be brought together at a suitable temperature. In ordinary practice it may be assumed that the hydrogen is first caused to dissolve in the oil, with the hydrogen-laden oil then brought into contact with the catalyst by mechanical means. In the usual type of equipment, reaction is brought about by agitating the suspension of catalyst and oil in a closed vessel in an atmosphere of hydrogen. Agitation of the catalyst–oil mixture serves the double purpose of promoting solution of hydrogen in the oil and continuously renewing the oil at the catalyst surface. The solubility of hydrogen and other gases in oil increases linearly with temperature and, of course, pressure.

The rate of hydrogenation under the conditions just outlined depends on the temperature, the nature of the oil, the activity of the catalyst, the concentration of the catalyst, and the rate at which hydrogen and unsaturated oil molecules are supplied together to the active catalyst surface. The composition and the character of the hydrogenated product may vary according to the positions of the double bonds, which are hydrogenated, as well as certain isomerizing influences accompanying the reaction, and are highly dependent on the conditions of hydrogenation.

Although only the hydrogenation of glycerides has been mentioned here thus far, the process is equally applicable to fatty acids, nonglyceride esters, and other unsaturated fatty acid derivatives.

The hydrogenation of fatty materials is accomplished to produce a more oxidatively stable product and/or change a normally liquid oil to a semi-solid or solid fat with melting characteristics designed for a particular product. Thus there are a myriad of products produced, and the possible products from the hydrogenation of the different fatty materials is almost infinite. Thus to prepare a consistent product with the desired charac-teristics, the reaction parameters must be controlled.

However, the effects of the many reaction conditions and character-istics of the hydrogenation are more easily understood after a consider-ation of the nature of the reaction.

Chemistry of Hydrogenation

2.1   MECHANISM

The basic hydrogenation of an unsaturated carbon–carbon double bond appears to be very simple but is extremely complex:

$$-CH{=}CH- + H_2 \xrightarrow{\text{catalyst}} -CH_2 - CH_2-$$

As this reaction shows, hydrogenation can take place only when the three reactants have been brought together—the unsaturated oil, a liquid, the catalyst (which is a solid), and hydrogen gas. Thus the physical mech-anism of bringing the reactants together has been devised without un-derstanding what happened when the reactants were together in the cor-rect structure to cause reaction.

The three phases of the system—gas, liquid, and solid—are brought together in a heated stirred reactor with hydrogen available under pressure in the headspace of the reactor. The hydrogen must be dissolved in the liquid–solid phase before reaction can occur since the dissolved hydrogen is the only hydrogen available for reaction. The hydrogen may then diffuse through the liquid to the solid catalytic surface. In general, at least one of the reactants must be chemisorbed on the surface of the catalyst. How-ever, the reaction between unsaturated hydrocarbons and hydrogen pro-ceeds by way of surface organometallic intermediates.

In general, a heterogenous reaction involves the following series of steps: (a) diffusion of reactant(s) to catalyst surface; (b) adsorption; (c) surface reaction; (d) desorption; and (e) diffusion of product(s) from ca-talyst surface. Albright (6) has amplified these steps to show the transfer, adsorption, desorption, hydrogenation, and isomerization steps.

Each unsaturated group of the fatty acid chain can transfer back and forth between the main body of the oil and the bulk surface of the catalyst. These unsaturated groups can be adsorbed on the catalyst surface. Each adsorbed unsaturated group can react with a hydrogen atom to form an unstable complex that is a partially hydrogenated double bond. Some of the complexes may react with another hydrogen atom to complete the saturation of the double bond. If the complex does not react with another atom of hydrogen, a hydrogen is removed from the adsorbed molecule and the "new" unsaturated bond is desorbed. Both the saturated and the unsaturated bonds are desorbed from the catalyst surface and diffused into the main body of the oil. Thus not only are some of the bonds saturated, but some may also be isomerized to new positions or new geometric forms.

A similar series of steps occurs when one of the double bonds of a polyunsaturated fatty chain is hydrogenated. Isomerization reactions also occur in these cases, and at least part of the double bonds are isomerized to new positions. If a methylene-interrupted diene is reacted on the catalyst surface, the double bonds may be conjugated before saturation of one of the bonds. Also, the conjugated diene may be desorbed from the catalyst surface into the main body of oil before being readsorbed and partially saturated.

If the mixture to be hydrogenated contains both monoenes and dienes and polyenes, there may be competition between the different unsaturated systems for the catalyst surface. Thus the dienes may be preferentially adsorbed from the oil to the catalyst surface and partially isomerized and/ or hydrogenated to a monoene and then desorbed to diffuse to the main body of the oil. The di- and polyenes are preferentially adsorbed until their concentration in the oil is very low, and the monoenes then may be adsorbed and reacted.

Since the oils that are hydrogenated are composed of a mixture of fatty acids, the selectivity of the reaction is very important.

2.2 SELECTIVITY

The term "selectivity" as used in the industry had two meanings as applied to the hydrogenation reaction and products. The term originally was defined by Richardson et al. (7) as the conversion of linoleic acid to a monoene, compared to the conversion of the monoene to stearic acid. This was also known as *chemical selectivity* since it compared the rates of chemical reactions.

Another type of selectivity was applied to catalysts. If a catalyst was "selective," it produced an oil of softer consistency or lower melting point at a given iodine number (value) (IV). Since this could not be made a very quantitative term, this definition was at best somewhat vague and the catalyst one called "selective" might be labeled something else by others. Since neither type of "selectivity" could be measured with any precision, the term was used only for comparisons.

In 1949 Bailey (8) demonstrated that the following model could be used to measure the relative reaction rate constants for each hydrogenation step during the batch hydrogenation of linseed, soybean, and cottonseed oil:

$$\text{Linolenic} \underset{\text{isolinoleic}}{\overset{\text{linoleic}}{\rightleftarrows}} \longrightarrow \text{oleic} \longrightarrow \text{stearic}$$

Using this model, Bailey considered each reaction to be first order and irreversible and thus developed the kinetic equations to represent the concentration of each acid group as a function of time. Relative reaction rate constants were calculated for several runs, and the ratio of the reaction rate constant for linoleic to oleic, divided by the reaction rate constant for oleic to stearic, was the selectivity (ratio) of the reaction. If the ratio was 31 or above, the hydrogenation was selective and below 7.5, nonselective. However, as the calculation of the reaction rate constants was very laborious, little use was made of the quantitative measure.

A simpler reaction sequence was proposed by Albright in 1965 (9). Since the triene (linolenic acid) produces several different dienes (isolinoleic acids) when one double bond is hydrogenated and since there would be little difference in the rates of hydrogenation of the mixture of dienes, these were included in one term. Also, since the addition of 2 moles of hydrogen to linolenic acid to produce oleic acid directly has not been shown to occur, the shunt was eliminated from the model, and since geometric and positional isomers that are formed were believed to have almost the same reactivity, these were not included in the model. Thus the model is simplified to

$$\text{Linolenic} \xrightarrow{K_1} \text{linoleic} \xrightarrow{K_2} \text{oleic} \xrightarrow{K_3} \text{stearic}$$

The first-order, irreversible kinetic equations for the hydrogenation of unsaturated fatty acids are

$$Ll = Ll_0 e^{-K_1 t} \tag{1}$$

$$L = Ll_0 \left(\frac{K_1}{K_2 - K_1} \right) (e^{-K_1 t} - e^{-K_2 t}) + L_0 e^{-K_2 t} \tag{2}$$

$$Ol = Ll_0 \left(\frac{K_1}{K_2 - K_1}\right)\left(\frac{K_2}{K_3 - K_1}\right)(e^{-K_1 t} - e^{-K_3 t})$$

$$- Ll_0 \left(\frac{K_1}{K_2 - K_1}\right)\left(\frac{K_2}{K_3 - K_2}\right)e^{-K_2 t} - e^{-K_3 t}$$

$$+ L_0 \left(\frac{K_2}{K_3 - K_2}\right)(e^{-K_2 t} - e^{-K_3 t}) + Ol_0 e^{-K_3 t} \qquad (3)$$

where Ll, L, and Ol are the mole percentages of the linolenic, linoleic, and oleic groups, respectively, at time t (since mole and weight percentages are almost equal, weight percentages were used); and where K_1, K_2, and K_3 are the reaction rate constants, and t is time.

The equations were programmed and solved by a digital computer for the hydrogenation of cottonseed, soybean, peanut, corn, and linseed oils. Pseudo time values were used and K_2 was assigned a value of 1. Dissociation values K_1 and K_3 were varied to cover ratios of K_2/K_3 of 2–50. The gain in stearic acid was plotted against the starting and ending linoleic, L/L_0, at each selectivity ratio SR to give a family of lines. Figure 1.1 shows the plot for the hydrogenation of soybean oil. Thus from these plots and the starting and ending composition of the partially hydrogenated oil the selectivity ratio SR is found for the reaction. Thus, the term SR or selectivity ratio has replaced the term "selectivity."

Also, the SR may be calculated from the expression

$$SR = \frac{100}{S - S_0} \frac{1.0}{a\, \exp[b(L)L_0]c\, \exp[-d(L)/L_0]}$$

where for soybean oil hydrogenation the constants are

$$a = 1.260$$
$$b = 2.065$$
$$c = 0.771$$
$$d = 2.299$$

This equation may be programmed into a programmable calculator for calculation of the SR.

The SR may be calculated from experimental data by use of the kinetic equations solved by a digital computer (10).

Equation 1 is solved for K_1 directly from the decrease in linolenic acid, but since equations 2 and 3 have no formal solutions for K_2 and K_3, an

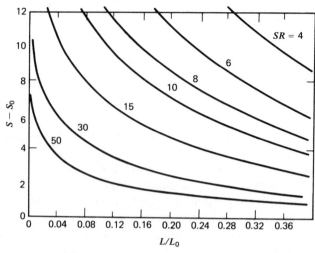

Figure 1.1   Calculation of selectivity ratio from fatty acid composition data.

iterative program is used. The computer approximates K_2 by initializing K_2 as a fraction of K_1. The value of L is then calculated and compared to the experimental L value. If the calculated L is not within 0.1% or less of the experimental L, then K_2 is incremented until the two values of L agree. Then K_3 is calculated by the same converging process. If experimental time t values are used, the calculated reaction rate constants are the experimental reaction rate constants. The ratios of the reaction rates $K_2/K_3 = SR$, and K_1/K_2 is the linolenic SR. This type of program may be used to calculate the results of the testing of commercial hydrogenation catalysts (11).

In the course of a practical hydrogenation the SR is not entirely constant, but if assumed to be constant, it is a good description of the compositional changes (12).

If $SR = 0$ All molecules react straight through to stearic acid. The use of platinum catalysts at low temperature and high pressure approximates this reaction.

If $SR = 1$ Equal reaction of oleic and linoleic acid.

If $SR = 2$ Equal reaction of each double bond. Linoleic acid reacts twice as fast as oleic acid.

If $SR = 50$ Linoleic acid reacts 50 times faster than oleic acid. A good nickel catalyst may produce this SR.

If $SR \geqslant 50$ All linoleic acid reacts before any oleic acid is hydrogenated.

Most commercial catalysts under the conditions used for commercial hydrogenations (10–50 psig, 150–225°C) have an SR of 30–90.

To illustrate how the SR may be used, Table 1.1 shows the results of hydrogenation under specific conditions of cottonseed oil (CSO) to 75 IV using three different catalysts. The fatty acid composition of the three products is different as is the SR; also, the solids curve for the three products was different. The one with the most stearic, lowest selectivity had the highest and slightly flatter solids curve; thus the SR agrees with the old definition of selectivity. The most selective is the reaction that gives an oil with the lowest melting point for a given IV. The ability to measure the SR with some precision has permitted much better control of hydrogenation and catalysts to produce much more consistent products.

Since soybean oil is the major oil in use in the United States, the hydrogenation of the linolenic acid with the least possible hydrogenation of linoleic acid is of great importance. Therefore, to measure the ability of catalysts to catalyze these reactions, the linolenic acid selectivity is important. This is the ratio of the reaction rate of linolenic acid compared to the reaction of linoleic acid, or K_1/K_2.

Figure 1.2 is a graph for the estimation of the linolenic acid (*Ln*) SR from the starting and ending composition of oils that contain linolenic acid. Allen (13) has plotted the ratios so that the *Ln* SR of fats of very different compositions may be estimated. Schmidt (14) has also presented a graphical method for the calculation of *Ln* SR. The computer program for the calculation of SR will furnish the *Ln* SR if the starting and ending fatty acid composition contains linolenic acid.

Although the SR and the *Ln* SR are very useful for the evaluation of

Table 1.1   Hydrogenation of CSO (400°F, 20 psig) to 75 IV

Catalyst	1	2	3
Palmitic	21.8	21.8	21.8
Stearic	3.6	4.0	4.8
Monoene	62.3	61.8	61.4
Diene	11.6	11.7	11.3
trans	37.8	35.7	36.6
SR	60	50	32

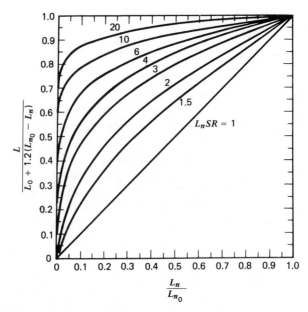

Figure 1.2 Calculation of linolenic acid selectivity ratio (*SR*) from fatty acid composition data.

catalysts and hydrogenation conditions, the calculated values are not constant. Since the calculation is based on a first-order reaction, the *SR* value may vary over the course of a hydrogenation if the reaction does not exactly obey pseudo-first-order kinetics. Since heterogeneous catalytic hydrogenations may display other-order reaction kinetics under certain conditions, the *SR* may change over the course of the reaction.

The very simple model for the reaction ignores all the isomers that are formed during hydrogenation. Linolenic acid may form diene isomers with the two double bonds widely separated. These isomers will hydrogenate like a monoene but will be analyzed as a diene. Also, the numerous positional and geometric monoene and diene isomers probably do not hydrogenate at the same rate. Therefore, the overall reaction rate "constants" may be changing during the reaction since they are the average of all the rate constants for hydrogenation of all the different isomers that are formed during the reaction.

The effect of cumulative poisons on the catalyst is not considered in the model. Catalyst poisons may affect only part of the reaction, so the *SR* may be changed by the poisons in the oil or the gas (15). Thus the *SR* of a catalyst would be different when used to hydrogenate different batches of oil.

2.3   REACTION ORDER AND REACTION RATES

In view of the preceding observations, it can hardly be expected that any definite order can be assigned to the reaction as a whole. On the other hand, it would be somewhat remarkable if the reaction rate bore no relation to the extent of unsaturation in the oil. Actually, under most conditions, hydrogenation will approach the character of a unimolecular reaction, where the rate of hydrogenation at any instant is roughly proportional to the unsaturation of the oil. However, the character of the reaction is markedly influenced by various conditions of hydrogenation.

A number of typical hydrogenation curves of cottonseed oil are shown in Figure 1.3, with the logarithm of the iodine number of the oil plotted against hydrogenation time. When so plotted, a true unimolecular or first-order reaction should yield a straight line, as in curve *B*. Curves similar to curve *B* are often obtained under average conditions of pressure, agitation, and catalyst concentration at moderate or low temperatures, that is, below about 300°F. At higher temperatures the shape of the hydrogenation curve resembles that of curve *C* since an increase in temperature accelerates the first stages of hydrogenation to a relatively greater extent than the latter stages; that is, it accelerates the conversion of linoleic to oleic acid to a greater extent than the conversion of oleic to stearic acid. In the case of curve *A*, hydrogenation more nearly approaches a linear rate. This type of curve is often obtained in the hydrogenation of relatively saturated oils, such as tallow, and is also sometimes observed in hydrogenation at a low pressure with a high concentration of catalyst, where

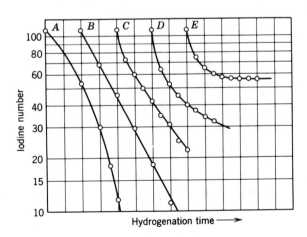

Figure 1.3   Typical cottonseed oil hydrogenation curves.

the rate of hydrogenation is determined by the rate of solution of hydrogen in the oil.

Curve D is characteristic of very high temperature hydrogenation, of hydrogenation with a very low concentration of catalyst, or of hydrogenation under conditions producing progressive slow poisoning of the catalyst during the reaction. Curve E represents a hydrogenation carried out with a self-poisoned nickel sulfate catalyst that was almost completely ineffective during the later stages of the reactions. Similar curves are obtained under conditions producing rapid catalyst poisoning.

As stated earlier, the simple reaction sequence

$$\text{Linolenic} \xrightarrow{K_1} \text{linoleic} \xrightarrow{K_2} \text{oleic} \xrightarrow{K_3} \text{stearic acid}$$

is only an approximation of the sum of many reactions. However, the reaction rate constants for the first-order reaction may be calculated. Figure 1.4 shows the experimental points compared to the calculated composition of hydrogenated soybean oil by using the reaction rate constants calculated from the experimental data by use of a digital computer. The agreement is quite good for the reaction.

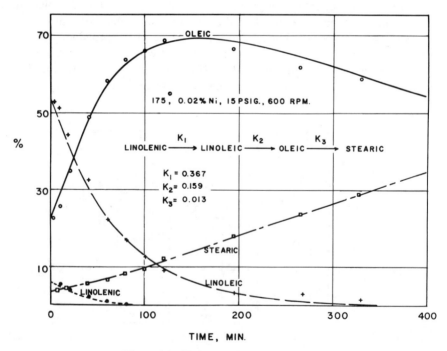

Figure 1.4   Hydrogenation of soybean oil.

From the calculated rate constants, the linolenic acid is hydrogenated 2.3 times as fast as the linoleic acid (K_1/K_2) and the linoleic acid reacts 12.5 times faster than oleic acid (K_2/K_3).

2.4 ISOMERIZATION

During the hydrogenation of fats the double bond may be saturated or isomerized while it is adsorbed on the catalyst surface. Both positional and geometric isomers are formed and are very important to the production of partially hydrogenated fats.

Since the *trans* form of octadecenoate has a higher melting point [trielaidin, melting point (mp) 42°C] than the *cis* form (mp 4.9°C), but lower than the saturated (tristearin, mp 73.1°C), the level of *trans* double bonds and saturates determines the solids content of fats at various temperatures. The saturated octadecanoate determines the solids at high temperatures, and the *trans* form performs the same function at the lower temperatures.

Both geometrical and positional isomers are formed on the catalyst surface by reaction with a hydrogen atom, basically by way of the Horiute–Polanyi mechanism.

As the double bond is adsorbed on the catalyst surface, the first reaction is with a hydrogen atom. Since this leaves a very active intermediate, another hydrogen atom may add to the adjacent position, with the molecule desorbed as a saturated one. However, if there is not a hydrogen atom available, a hydrogen may be removed from a chain carbon atom by the catalyst. Since the hydrogens on either side of the "active center" are activated, either may be removed. If the original hydrogen is removed, the original double bond is reformed and the molecule is desorbed. However, if the other hydrogen is removed, the double bond is shifted one position from the original position. Which of the two hydrogens on the carbon are removed determines whether the new double bond will be *cis* or *trans*. Also, the double bond in the original position maybe converted to *trans*. The double bonds in the new position may also be shifted. As hydrogenation proceeds, the isomerized double bonds tend to be shifted further and further along the chain, and the *trans* isomer content will increase until the monoenes are saturated (16).

The methylene-interrupted system of double bonds found in linoleic, linolenic, and other acids may also undergo an isomerization step. However, the hydrogens on the methylene group between the double bonds are quite labile. As the pentadiene approaches the catalyst, one of the hydrogens may be removed by the catalyst, thus causing a shift of one double bond to the conjugated position and a hydrogen adds to the carbon at the end of the conjugated system. When the bond shifts it may be *cis* or *trans* (*trans* predominates). Evidently, the conjugated system is tightly chemisorbed to the catalyst surface since it is hydrogenated to a monoene very rapidly and then desorbed. The remaining double bond may be *cis* or *trans* and may be shifted one position from the original position (17). Thus from *cis,cis*-9,12-octadecadienoate, hydrogenation produces the 9, 10, 11, and 12 *cis* and *trans* monoenes. Most of the *trans* will be in the 10 and 11 positions.

Effects of Process Conditions

The four main reaction parameters are temperature, pressure, agitation, and catalyst concentration. Of course, the type of oil and the type of catalyst also determine the product produced by hydrogenation. Changes in the reaction parameters, using the same type of oil and catalyst, are made according to the type of product desired. Although these reaction parameters are all interrelated, they are discussed singly for better understanding.

3.1   EFFECT OF TEMPERATURE

Hydrogenation, like other chemical reactions, is accelerated by an increase in temperature. The effect of temperature on the reaction rate is somewhat less than in ordinary reactions but is variable, as shown in Figure 1.5. At high rates of agitation the rate, shown as IV drops per minute, steadily increases, but at low agitation the effect of temperature slowly decreases as the temperature increases, thus indicating that several factors are affecting the reaction rate.

Also, the formation of *trans* unsaturation is affected by temperature. As the temperature of reaction is increased, the formation of *trans* unsaturation increases almost linearly. However, as shown in Figure 1.6, the pressure also affects isomer formation. Temperature also increases the SR value, and the increase is almost linear with temperature; pressure also affects the SR.

The observed effect of temperature on the hydrogenation is the sum of all the effects on the many steps of the reaction. An increase in temperature increases the solubility of hydrogen in the oil (18). Also, the higher temperature will lower the viscosity of the oil, thus increasing the agitation, and the hydrogen may diffuse from the bubble through the surface to the oil phase. Higher temperatures cause a faster reaction on the catalyst surface, so with increased agitation and pressure, hydrogen is

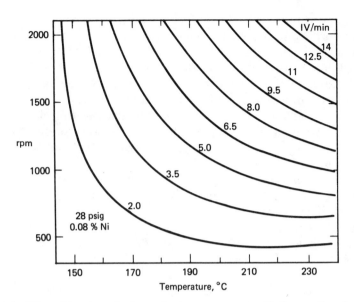

Figure 1.5 Effect of reaction agitation and temperature on rate of hydrogenation of soybean oil.

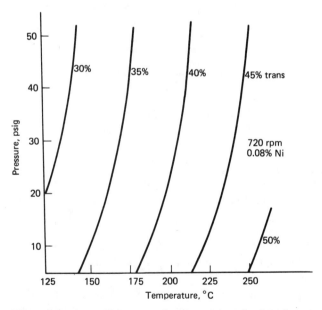

Figure 1.6 Effect of reaction temperature and pressure on production of trans unsaturation in 80-IV soybean oil.

kept supplied to the catalyst surface for the saturation. However, if the temperature alone is increased, although more hydrogen is supplied to the catalyst surface, the reaction is very rapid and the hydrogen on the catalyst may be partially depleted. This would account for the increased isomerization at higher temperatures. There is not enough hydrogen on the catalyst surface to complete the saturation, so the catalyst takes back a hydrogen and a geometrically isomeric double bond results. The same reasoning would apply to the *SR*. The diene would tend to be hydrogenated much faster than the monoene because the depleted supply of hydrogen on the surface would cause more diene conjugation with subsequent fast saturation of one bond. However, the rate of increase of the *SR* with temperature would tend to decrease since the supply of hydrogen on the catalyst would become more and more depleted as the temperature increased. Although the reaction rates all increase, the available hydrogen atoms would decrease, and the overall effect would be a decrease in the rate of increase with temperature.

Heat of Reaction.   Hydrogenation is an exothermic reaction, and the hydrogenation of the usual vegetable oil results in an increase in temperature of the oil of 1.6–1.7°C for each unit decrease in iodine value (19). Rogers and Siddiqui (20) determined the heats of hydrogenation (ΔH_h) of the methyl esters of the common unsaturated fatty acids (Table 1.2).

Table 1.2   Heats of hydrogenation
of methyl esters

Methyl Ester	ΔH_h(kcal/mole)
Palmitoleate	− 29.30 ± .24
Palmitelaidate	− 32.43 ± .60
Oleate	− 29.14 ± .26
Elaidate	− 28.29 ± .15
Linoleate	− 58.60 ± .39
Linoelaidate	− 55.70 ± .13
Linolenate	− 85.40 ± .58

Thus the heat of hydrogenation of fatty esters is not very different from that of other aliphatic compounds in the liquid phase (28–29 kcal per double bond).

This heat of hydrogenation is used by industrial operators to supply heat to the reactor. The reaction mass is heated to some minimum temperature and the hydrogen admitted. The heat of reaction is used to heat the reactants to some maximum temperature where the temperature is controlled.

3.2   EFFECT OF PRESSURE

Most industrial hydrogenations of oils are performed under hydrogen pressures of 10–60 psig. Although this is a very limited range of pressures, changes in pressure have a profound effect on the products produced.

The solubility of hydrogen in vegetable oils may be approximated as a linear function of temperatures and may be written (21)

$$\text{l stp/kg oil} = (47.04 + 0.294t) \times 10^{-3}p$$

where t is temperature in degrees Celsius and p is pressure in atmospheres between 1 and 10 atm. Thus the amount of hydrogen in the oil doubles if the pressure is doubled. For example, at 200°C and 30 psig the solubility is 0.216 l stp/kg, and at 60 psig the solubility is 0.432 l stp/kg.

Figure 1.7 shows the effect of pressure on the rate of hydrogenation

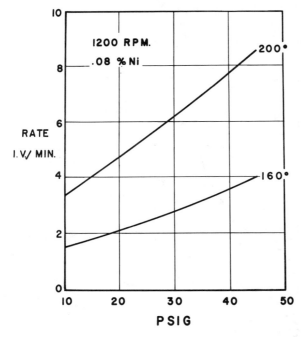

Figure 1.7   Effect of reaction pressure and temperature on rate of hydrogenation of soybean oil.

of soybean oil at two different temperatures. If the pressure is doubled, the rate approximately doubles at both temperatures. However, this reaction was very well agitated.

The effect of pressure on isomerization is limited, and the rate of increase is less at higher pressures. At low pressures the hydrogen dissolved in the oil does not cover the catalyst surface, but at high pressures, especially at low temperatures, an increase in pressure does not change the formation of isomers because the catalyst surface is already covered and the pressure is sufficiently high to supply all the hydrogen needed to increase the rate of saturation, but the rate of isomerization does not change.

The effect of pressure on the SR, shown in Figure 1.8, changes less at the higher pressures for the same reason. At the higher pressures the SR rate of increase is less than at the lower pressures. This also results from the concentration of dissolved hydrogen on the catalyst surface that is available for reaction.

3.3   EFFECTS OF AGITATION

The heterogenous hydrogenation of oils involves not only several consecutive and simultaneous chemical reactions, but also physical steps of mass transfer of the gas and liquid to and from the solid catalyst surface.

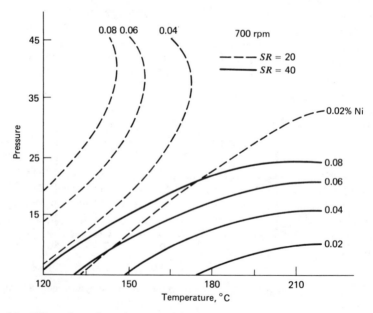

Figure 1.8   Effect of reaction pressure on selectivity ratio of hydrogenation of soybean oil.

Therefore, the reaction mass must be agitated. The agitation must accomplish the distribution of heat or cooling for temperature control, and it must keep the solid catalyst suspended throughout the mass for uniformity of reaction.

Industrial hydrogenators use the system of bubbling the hydrogen into the liquid at the bottom of the tank. The hydrogen available for reaction comes partly from the fraction of the hydrogen bubbles that are adsorbed during their passage through the oil and from the gas in the headspace that is stirred back into the oil. Agitation is provided by a central shaft with one or more blades that may be flat blades or of the turbine type. Also, the position of the blade on the shaft is important (22, 23). The speed of the agitators and the tendency to cavitate affect the amount of hydrogen in the oil. There may be flat plate baffles to break the flow. The configuration, the number, and the placement of the coils for heating and cooling are also important in agitation, as well as the size and the quantity of hydrogen bubbles going through the oil. Thus the "agitation" of the reaction mass is the sum of many factors that can be only estimated.

The main function of agitation is to supply dissolved hydrogen to the catalyst surface. Therefore, as shown in Figure 1.5, at low temperatures of hydrogenation the change of hydrogenation rate is less at higher agitation speeds. Thus at slow reaction rates the agitation at higher speeds is sufficient to provide almost enough hydrogen to the catalyst, and increased agitation does not change the hydrogen supply. However, at high hydrogenation temperatures there is a rapid change of hydrogenation rate with change in agitation, so the supply of hydrogen limits the rate of hydrogenation.

Agitation has a great effect on the *SR* of the reaction (Figure 1.9). The higher the agitation, the lower the *SR* since the selectivity drops if the catalyst is supplied with sufficient hydrogen. Also, isomerization is decreased with an increase in agitation (24).

Since the agitation of pilot plant or laboratory hydrogenators may be changed quite easily, this offers a method to match the operations of a small reactor to a plant size converter so that the same products will be obtained. In practice, the small equipment is operated under identical conditions as the large equipment and the products analyzed. The *SR*, the *trans* unsaturation, and the rate of reaction of the two products will indicate in which direction the agitation of the small units needs to be changed. Several test hydrogenations may be necessary to find the optimum hydrogenation conditions.

Converters with different internal structures of bracing, coils, baffles, and so on may be equated by changes in speed of the agitator shaft or type and placement of the agitator blades until the converters produce the same products under the same conditions.

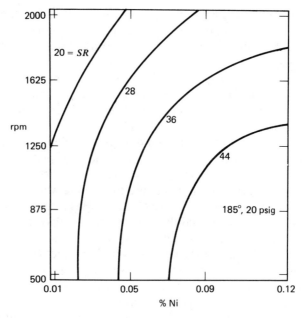

Figure 1.9 Effect of reaction agitation and catalyst on selectivity ratio (*SR*) of hydrogenation of soybean oil.

3.4   CATALYST CONCENTRATION

The catalyst concentration may be varied over a wide range, but economic considerations dictate the use of minimum catalyst consistent with rapid reaction.

As shown in Figure 1.10, increasing the amount of nickel catalyst for hydrogenation of oils offers decreasing returns. At low level of catalyst, the increase of catalyst concentration causes a corresponding increase in the rate of hydrogenation. However, if more and more catalyst is used, a point is finally reached at which no further increase in rate is observed. Also, increasing the nickel catalyst decreases the formation of *trans* unsaturation, but only slightly. As shown in Figure 1.6, change in agitation produces a much greater change in the formation of *trans* unsaturation than a change in catalyst concentration. Increasing the catalyst only slightly decreases the selectivity of the reaction.

Thus in hydrogenation under the usual conditions obtained during commercial operations, conditions other than catalyst concentration are limiting.

If the rate of agitation is sufficiently high, resistance to mass transfer is eliminated. Eldib and Albright (25) observed that the rate of hydrogenation of cottonseed oil varied directly with the temperature, the absolute pressure, and the catalyst concentration.

Type of Catalyst on Selectivity.   The porous structure of catalysts is believed to affect the rate and the selectivity of the reaction (26). If the spaces are relatively wide (≥25 Å), the liquid composition in the pores differs little from the bulk composition. The exchange of the partially hydrogenated materials from the pores to the bulk is rapid, and no buildup of hydrogenated material in the pores is allowed. However, in pores that have medium width (20–25 Å) the products of hydrogenation build up inside the pores, and if the pores are long, significant concentration gradients are set up along the pores. The material in the pores runs ahead of that in the liquid, and thus more saturated species are produced and the selectivity decreases. In the extreme case the triglycerides entering the pores will be converted to tristearin, and thus the selectivity will be 0. The hydrogen concentration also has a bearing on the pore transport effects (27). At high hydrogen concentrations, high pressure, intense stirring, low temperature, and low catalyst level, the hydrogenation rate in

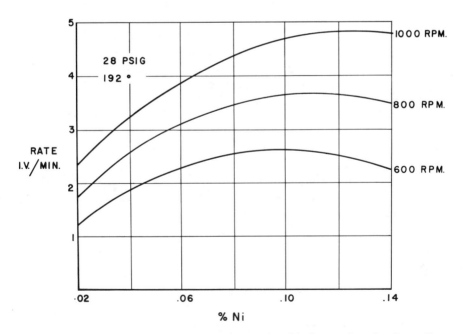

Figure 1.10   Effect of catalyst concentration on rate of hydrogenation of soybean oil.

a pore is increased and the saturated triglyceride concentration increases with a depression of the selectivity.

The selectivity of nickel catalysts for the hydrogenation of linolenic acid over linoleic (*Ln SR*) has been shown to be about 2–2.3. Changes in process conditions do not change this ratio, so evidently the diene and the triene are hydrogenated by the same mechanism. The presence of the third double bond would double the chance of hydrogenation of one double bond in the triene compared to the diene. However, when copper chromite catalysts are used to hydrogenate soybean oil, linolenic selectivity ratios of 8–12 are found (28–31).

The copper chromite catalyst operates somewhat differently from nickel to achieve the high *Ln SR*. Evidently, the copper chromite catalyst causes conjugation of the linolenate triene on the catalyst surface. Since conjugated trienes are much more reactive than linolenate and hydrogenate some 200 times faster, they do not accumulate in the products. They hydrogenate to a conjugated diene before they desorb from the catalyst surface. The conjugated dienes are further reduced to monoenes. The monoenes are not reduced to saturates by the catalyst since hydrogenation with this catalyst must involve two or more double bonds.

This catalyst also has the ability to cause extensive isomerization of a conjugated system. Nickel, platinum, or palladium do not have this property. The isomerization may extend to the terminal position since considerable monoene with a terminal double bond may be found in the hydrogenated product. Also, some conjugated diene may remain in the oil, particularly if low temperatures are used (32). The isomerization reactions proceed at a greater rate than does the hydrogenation of the dienes to the monoene.

The type of catalyst as well as the process conditions determine the product obtained; *trans* unsaturates are also formed by the copper chromite catalyst and may approach 70% of the monoenes formed by hydrogenation. This equilibrium point is also obtained by nickel catalysts.

Table 1.3 is a summary of the effect of process conditions of hydrogenation on the rate of reaction, the selectivity, and the formation of *trans* unsaturation or isomerization.

Effects of Substrate

4.1   EFFECT OF POSITION OF DOUBLE BOND ON REACTIVITY

In general, it is recognized that hydrogenation rates differ with the structure of the fatty esters. Since hydrogenation of oils produces many mon-

Table 1.3 Summary of process conditions in hydrogenation

Increase in	Rate	SR	Isomerization
		Results	
Temperature	+ + + +	+ + + +	+ + + +
Pressure	+ + +	– – –	– – –
Agitation	+ + + +	– – – –	– – – –
Catalyst concentration	+ +	–	–

oene isomers, the competitive hydrogenation rates of the positional monoene isomers have been studied.

Allen (33) showed that the *cis*-6-, *cis*-9-, and *cis*-12-octadecenoate esters hydrogenated at the same rate. However, if the double bonds were isomerized extensively, the double bonds nearer the carboxyl end tended to be hydrogenated at a slighter faster rate than those near the terminal end of the molecule. The nickel or palladium catalyst used did not cause a shift of the double bond into the 17 position, the terminal end.

Similar results were obtained by Scholfield et al. (34), but they found that the *cis*-15-octadecenoate was hydrogenated faster than the *cis*-9. A terminal double bond in the 17 position was shown to hydrogenate over 90 times faster than a *cis*-9. The *trans* isomers hydrogenated more slowly than the *cis*.

A study of the competitive rates of hydrogenation of isomeric methyl octadecadienoates (35) showed that the *cis*-9, *cis*-12-octadecadienoate (linoleate) is reduced the most rapidly of all the dienes studied. The relative rates of the positional isomers tend to decrease with increasing number of methylene groups between the double bonds, except when one of the double bonds is in the more reactive 15 position. The *trans,trans*-diene is hydrogenated at a slower rate than *cis,cis*-linoleate.

Hydrogenation of Oils with Conjugated Double Bonds. On hydrogenating methyl *cis,cis*-10:11, 12:13-octadecadienoate, Allen (36) found that when 1 mole of hydrogen was added, 1,2, 1,4, and 3,4 addition had taken place with equal ease to produce an equimolar mixture of *cis*-10:11, *trans*-11:12, and *cis*-12:13 octadecenoates. There was no shift of the double bond away from the conjugated system. It was believed that all the *trans* isomer formed was in the 11:12 position. Similarly, Scholfield et al. (37) observed that when conjugated linoleic acid was hydrogenated as the sodium soap, 1,2, 1,4, and 3,4 addition took place with equal ease. With *trans,trans*-conjugated linoleic acid, they found that 1,4 addition exceeded 1,2 and 3,4 addition and that all the reduced acids were *trans*.

Koritala et al. (38) found that conjugation of the two double bonds in a diene increases the reactivity some 10–18 times over the pentadiene system of linoleate.

The hydrogenation of a conjugated triene system has received considerable attention. Lemon (39) hydrogenated tung oil in a laboratory apparatus under conditions similar to those commonly obtained in commercial practice and observed very little development of diene conjugation. Similarly, Hilditch and Pathak (40) concluded from observations made during the hydrogenation of eleostearate that the primary reaction consisted almost entirely of the simultaneous addition of four atoms of hydrogen to form monoethenoid esters.

Allen and Kiess (41), on the other hand, have produced evidence to show that the first mole of hydrogen adds to a conjugated triene system in all possible positions. Their data indicate a step-by-step addition of 2 moles of hydrogen to produce equimolar amounts of 9:10-, 10:11-, 11:12-, 12:13-, and 13:14-octadecenoates.

Koritala et al. (38) showed that hydrogenation of the conjugated triene system of eleostearate resulted in 50–90% reduction of the triene to a monoene without going through the diene stage. With catalysts, except copper chromite, the conjugated triene is held on the catalyst until a monoene is formed that is then desorbed by the catalyst. However, a copper chromite catalyst permits the reduction of a conjugated triene to a conjugated diene.

Some of the conflicting conclusions may be due in part to the different conditions of hydrogenation used; it is more likely that it reflects the more highly refined analytical techniques available to and used by the later workers, but none of the results should be accepted without reservation.

4.2   SELECTIVITY WITH RESPECT TO DIFFERENT CLASSES OF GLYCERIDE

The selectivity of hydrogenation of triglycerides depends on two major factors: (1) the type of unsaturated fatty acids in the molecule and (2) the number of unsaturated acids per triglyceride. As would be anticipated from the preceding discussion, those triglycerides containing natural or conjugated linoleic or linolenic acids hydrogenate most rapidly. However, when these acids are depleted and the only remaining unsaturation is in monoethenoid acids or diethenoid acids with widely spaced double bonds, the rate of hydrogenation apparently approaches a probability pattern; that is, the more unsaturated acids per molecule, the greater the probability of that molecule undergoing hydrogenation. The data of Hilditch and Jones (42) on the hydrogenated cottonseed oil and olive oil, although old, are of interest. As can be seen in Table 1.4, the rate of hydrogenation of the diunsaturated glycerides is about double that of the monounsaturated in both instances.

Effect of the Unsaturated Acyl Position in Triglycerides on the Hydrogenation Rate. Several reports (43–46) suggest that there is no relation between the position of the unsaturated acyl group on the glyceride molecule and its hydrogenation rate. However, Drozdowski (47) showed that the unsaturated groups in positions 1 and 3 hydrogenated slightly faster than did the unsaturated acyl group in position 2. The influence of the conditions of hydrogenation was not stated. These results were obtained by interpretation of the results of positional analyses by using pancreatic lipase to liberate the fatty acids in positions 1 and 3.

4.3   HYDROGENATION OF MONOESTERS AND FREE FATTY ACIDS

The early work of Hilditch and Moore (48) has shown that mixed monoesters of oleic and linoleic acids hydrogenate with the same selectivity as do the mixed triglycerides of natural oils, but that much less selectivity is evident in corresponding mixtures of the free fatty acids. The comparative results of these investigators on cottonseed oil, ethyl esters of the oil, and the mixed free acids are shown in Figure 1.11. The lessened

Table 1.4   Composition of olive oil and cottonseed oil hydrogenated to different degrees of unsaturation (42)

Iodine Number	Saturated Acids (%)	Trisaturated Glycerides (%)	Disaturated Monounsaturated Glycerides (%)	Monosaturated Diunsaturated Glycerides (%)
		Hydrogenated Olive Oil		
37.6	58.1	25.5	23.3	51.2
31.7	65.1	30.8	33.7	35.5
20.1	78.7	46.5	43.1	10.4
11.9	87.8	67.5	28.4	4.1
		Hydrogenated Cottonseed Oil		
42.8	52.2	15.9	24.8	59.3
29.8	67.1	33.5	33.3	33.2
20.1	78.1	44.0	46.3	9.7
13.2	86.0	65.3	27.4	7.3

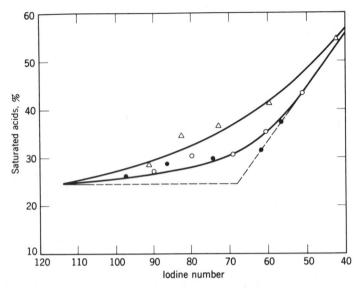

Figure 1.11 Production of saturated acids in the hydrogenation at 180°C of cottonseed oil (open circles); ethyl esters of cottonseed oil fatty acids (closed circles); and free fatty acids of cottonseed oil (triangles) (48).

selectivity in free acids is explainable on the basis of the strong polarity of the free carboxyl groups, which become competitive with active methylene groups in the process of adsorption on the catalyst and thus diminish the difference existing between oleic and linoleic acids in the combined form (49).

Feuge et al. (50) have observed that the rate of hydrogenation of methyl oleate is about triple that of triolein, with the latter attaining a lower equilibrium ratio of *trans* to *cis* isomers.

The presence of a considerable quantity of free fatty acids in an oil reduces the rate of hydrogenation. According to Pelley (51), even saturated acids can markedly reduce the reaction rate; however, the free acids in an oil hydrogenate preferentially to the glycerides. This somewhat paradoxical effect appears reasonable if it is assumed that the carboxyl groups are adsorbed more readily by the catalyst than are unsaturated groups but are desorbed less readily.

Johnston et al. (52) have studied the hydrogenation of monoesters and other model compounds using the most modern separation and analytical techniques. Thus the hydrogenation of methyl linolenate at 140°C with 0.5% nickel and 1.1 moles of hydrogen at atmospheric pressure, followed by separation of reaction products by means of countercurrent distribution, gas liquid chromatography, and low-temperature crystallization, yielded one triene, four dienes, and nine monoenes; *trans* isomers and *cis,cis*-nonconjugated dienes were also determined.

Studies on the kinetics of hydrogenation of methyl linolenate suggested that there is evidence for a direct triene-to-monoene shunt. Kinetics of the consecutive reactions involved were also described (52, 53).

Using ultraviolet spectrophotometry and various conditions of hydrogenation, Willard and Martinez (54) found large differences in the rate of hydrogenation of the three double bonds of linolenic acid. Initial attack on the 12:13 double bond is apparently favored.

Hydrogenation in Practice

5.1 HYDROGENATION EQUIPMENT

Almost all the oil hydrogenated commercially is processed in batch equipment. Although the development and adoption of continuous processes have been the general rule in other oil processing operations, continuous equipment has not become popular for hydrogenation. This fact may be attributed to considerable mechanical difficulties in achieving the close control and high selectivity of hydrogenation required by modern processors and the diversity of hardening operations in most plants, which require greater flexibility than is readily obtained in a continuous process. Continuous hydrogenation processes are under continual investigation.

The principal object in designing a hydrogenation reactor is to obtain good mixing of the hydrogen and oil; mixing of the catalyst–oil mass is of minor importance. Hence an efficient hydrogenator is essentially an efficient and easily controlled gas absorber. In the early days of the hydrogenation process, favor was divided between the so-called Wibuschewitsch system, in which the oil is sprayed into an atmosphere of hydrogen, and the Normann system, in which hydrogen is bubbled into a body of the oil. In the United States, at least, the relatively elaborate Wilbuschewitsch system is now seldom used, and modern development of the Normann system has been in the direction of simplifying rather than elaborating the apparatus.

The hydrogenation plants now operating commercially in this country are essentially divided between "recirculation" systems, in which agitation and dispersion of hydrogen within the oil are achieved by continuously recycling hydrogen in large volumes through the reactor, and the newer "dead-end" systems in which the reactor is supplied only with as much hydrogen as is absorbed, and where dispersion of the hydrogen is assisted by mechanical agitation.

Batch Equipment. Recirculation System.   The general design and lay-
out of a plant for hydrogenation by the recirculation of hydrogen is shown
in Figure 1.12. The reactor proper is a tall cylindrical vessel, usually built
to hold 20,000–60,000 lb of oil, containing heating coils and a distribution
device in the bottom for breaking up the injected hydrogen into many

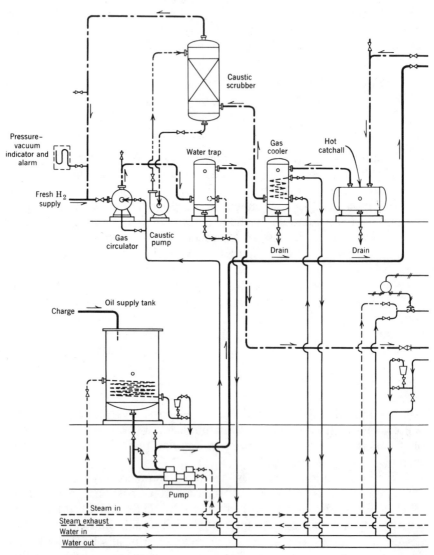

Figure 1.12   Hydrogenation plant employing hydrogen recirculation. (Courtesy of the Foster
Wheeler Corporation.)

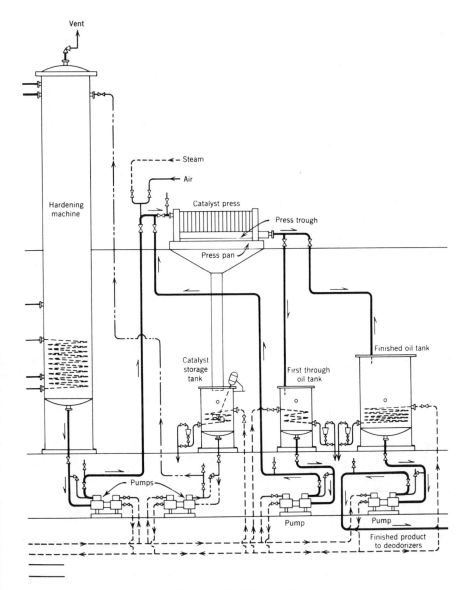

Figure 1.12 (*Continued*)

small streams. It provides a blower or compressor of large capacity that continuously withdraws hydrogen from the headspace of the reactor, forces it through a purification train, and sparges it back into the oil. Undue buildup of nitrogen or other nonremovable impurities in the hydrogen is avoided by intermittent venting to the atmosphere. With steam–hydrocarbon hydrogen of good quality, venting of 3–4% of the total hydrogen is adequate (55).

Although hydrogen is sometimes supplied to a recirculating reactor from high-pressure storage tanks, it is more usual to store the hydrogen in a large low-pressure holder. The recirculating blower may be designed to build up a moderate pressure on the reactor, but more commonly the reaction is carried out at a pressure only slightly above that of the atmosphere. There are considerable differences of opinion as to the benefits of the hydrogen purification train, and this portion of the apparatus varies considerably in design. The earlier plants subjected the recirculating gas to water washings, caustic scrubbing, and, in some cases, even to contact with activated carbon. In using the relatively pure hydrogen supplied by modern electrolytic, steam–hydrocarbon, or steam–iron plants, there is no real advantage in elaborate purification, and some processors employing this system omit all steps except water washing and retain this step only as a convenient means of cooling the exit hydrogen before it reaches the recirculating blower.

In the operation of the recirculating system the reactor is almost always kept filled with hydrogen under holder pressure at all times; that is, it floats on the low-pressure holder, except when hydrogenation is actually in process. When the catalyst is added with the oil charge, as is usual, hydrogenation takes place during the heating period, and thereafter as long as is required to reduce the oil to the desired iodine number; water is admitted to the cooling coils to carry away heat of reaction and maintain the temperature within proper limits. At the conclusion of the reaction, hydrogenation is stopped by shutting down the blower, and the finished charge is pumped out through an external cooler to a filter press where the catalyst is removed, and thence to "postbleaching" and to storage.

Practice varies in the handling of the catalyst in both recirculating and dead-end systems. A small catalyst mix tank is usually provided for receiving the catalyst press cake and resuspending it in oil for reuse, or for melting and slurrying fresh catalyst. In some plants the identity of each charge of catalyst is preserved as it is used in successive runs. In others, the catalyst slurry from each finished batch of oil is pumped into holding tanks and mixed with other similar catalyst, with different tanks being provided for "first-run" catalyst, "second-run" catalyst, and so on.

Batch Equipment. Dead-End System. Modern dead-end hydrogenators, or converters as they are more commonly called, are closed, vertical, cylindrical steel vessels, of about 10,000–60,000-lb capacity, designed for full vacuum and 100–150 lb of working pressure (Figure 1.13). Coils within the vessel are provided to take either steam for heating or water for cooling. A vertical shaft carrying a motor-driven agitator is inserted through a stuffing box or a seal in the top head. As these reactors require less headspace above the oil charge, they are generally somewhat shorter in relation to their diameter than the reactors used with the recirculating system.

Only special agitators, designed to give efficient hydrogen dispersion, are satisfactory for dead-end converters. A widely used agitator at present has two or more turbine-type impellers of a special shrouded design, with the top impeller being placed relatively near the surface of the oil and provided with a suction sleeve that projects into the headspace and induces hydrogen from the headspace back into the oil. A newer but highly efficient agitator (56) employs impellers consisting of a series of flat paddles set vertically around the periphery of a flat horizontal disk. Vertical banks of tubes around the inner wall of the vessel serve the double purpose of heating or cooling the charge and providing baffles to limit swirling of the oil and improve hydrogen dispersion. In converters equipped with this type of agitator, induction of hydrogen from the headspace occurs through a vortex extending down to the top impeller.

An important accessory of the dead-end plant is a steam ejector system capable of quickly evacuating the empty converter and maintaining a vacuum of about 28 in. Hydrogen is supplied to the converter from high-pressure storage tanks, through a reducing valve set at a constant operating pressure, and is admitted to the converter only when hydrogenation is actually in progress.

In operation, the dead-end converter is kept under vacuum while the charge of oil and catalyst is being pumped in, and also during the heating period, to deaerate and dry the oil before it reaches a high temperature and also to prevent reaction during heating and thus avoid hydrogenation over a range of temperatures. When steam in the heating coils has carried the batch to the required operating temperature, the vacuum line is closed and hydrogen is admitted to the vessel and allowed to build up to the desired pressure.

At the end of the hydrogenation period the flow of hydrogen is stopped; the hydrogen in the headspace, with accumulated impurities, is blown off to the atmosphere; vacuum is again placed on the vessel; and the charge is cooled to about 160–190°F by water in the coils, with the agitator run-

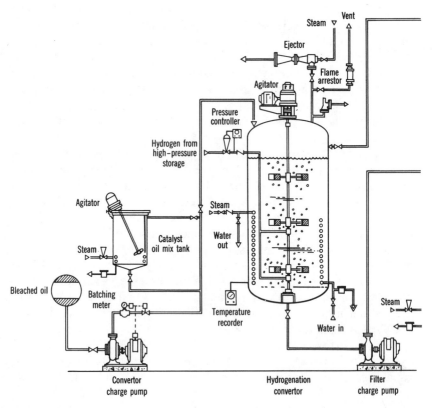

Figure 1.13 Hydrogenation plant, dead-end type. (Courtesy of the Votator Division, Chemetron Corporation.)

ning. Then the vacuum is broken, with the first cloudy oil through the press being diverted back to the converter. A vent line to the atmosphere should be installed between double block valves in the hydrogen line to the converter, and this should be opened except when the converter is under pressure, to prevent accidental leakage of hydrogen into the vessel.

Some processors operate mechanically agitated converters in such a manner as to combine certain typical features of the dead-end and the recirculating systems. For example, external oil coolers are used, the converters are not always evacuated, and a small flow of hydrogen is sometimes bled from the converter headspace back to the hydrogen plant for recompression, or a small blower is installed to give limited external hydrogen recirculation.

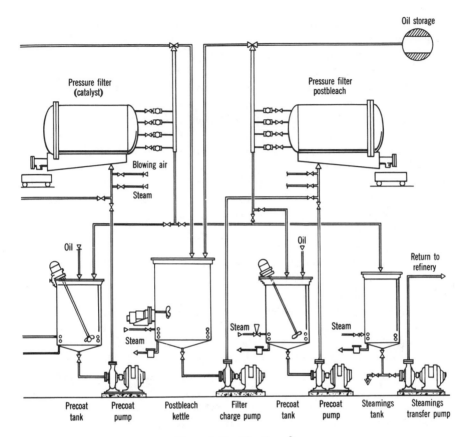

Figure 1.13 (*Continued*)

Hydrogen losses in the operation of the dead-end system depend on the amount of hydrogen discharged from the headspace of the converter at the conclusion of the run and on the care exercised in preventing hydrogen leakage. In well-designed and well-operated plants using pure hydrogen and producing shortening and margarine oil stocks from cottonseed and soybean oil, losses do not exceed 3–5% of the hydrogen reacted.

The claimed advantages of dead-end hydrogenation with evacuation of the converter, in comparison with hydrogenation by gas recirculation, are (*a*) prevention of oxidation and hydrolysis of the oil, through deaeration and dehydration of the charge, (*b*) more positive control of the reaction and consequent improved product uniformity (the entire reaction is carried out at a definite and constant temperature, and the amount of hydrogen absorbed by the oils is easily ascertained from the pressure drop on the hydrogen supply tank), (c) greater latitude in selectivity and decreased dependence on the temperature as a factor determining selectivity (as the operating pressure can be widely varied), and (d) simpler, less expensive, and more easily maintained equipment.

A system that permits more rapid turnover of the equipment has been described by Bentz (57).

In a batch system the oil must be cooled before catalyst filtration to prevent exposure of the hot oil to oxygen; the oil is cooled in the converter. Thus the converter cannot be used for the next batch until the cooling–filtration cycle is complete. To make more efficient use of the equipment, the Bentz system employs a hydrogenator on top of another tank so that when the hydrogenation is completed, the batch is dropped rapidly through the line in the bottom of the hydrogenator into the evacuated holding tank for cooling and filtration. The hydrogenation vessel is then ready to receive another batch of oil, which is heated and hydrogenated while the previous batch is cooled and filtered. Equipment for batch sizes up to 60,000 lb is in use.

For many years, proponents of the two batch systems, dead-end or recirculation, have argued the relative merits of each. However, only dead-end batch systems have been built in the past few years. There is considerable heat loss when the hot recirculation gases are cooled and then reheated to reaction temperature during the recirculation process. Also, the mechanical agitator of the dead-end reactor is more efficient for converting energy to high shear agitation than is possible in the recirculation system, and the dead-end system can be operated over a wider range of pressures than the recirculation process.

5.2   CONTINUOUS HYDROGENATION

Since refining, water washing, bleaching, deodorization, and packaging are continuous or semicontinuous in the modern production of edible oils, the continuous hydrogenation of fats would be desirable. This would enable the manufacturer to make the most economical use of space, labor, and energy.

Like any continuous system, the optimum value of continuous hydrogenation is realized when it is used to produce large quantities of a single base stock. The major problem with continuous operation is the off-specification material that is produced when switching base stocks. Since most manufacturers produce a variety of base stocks, this seems to be the major deterrent for use. Also, the modern batch system allows operation by one person, so there are no savings in labor costs by a continuous process.

However, several continuous processes have been designed and are now in use.

Mills et al. (58 a–c) described a continuous process for the production of hydrogenated cottonseed and soybean oils. Refined and bleached oil is pumped from a storage tank and heated almost to reaction temperature in a preheater. The heated oil is mixed with a catalyst slurry to obtain 0.02–0.1% nickel in the oil. The oil–catalyst slurry is joined by the inlet hydrogen and flows concurrently to a vertical reactor. The reactor is essentially a series of compartments with stationary baffles in each compartment. Each compartment is stirred by an agitator on a common shaft. A horizontal baffle with very small clearance around the agitator shaft closes each compartment; thus there is little or no back mixing between compartments. The oil–catalyst–hydrogen mixture passes through a pressure-reducing valve and the excess hydrogen flashes off. The hydrogenated oil–catalyst mixture is cooled in a heat exchanger and filtered. Pressures of 30–90 psig of hydrogen are used and the reactor temperature is controlled by a heat exchanger to the jacket of the reactors. The oil is hydrogenated in 2.5–10 min. Thus the rate of reaction may be as high as 25–30 IV reduction per minute. This reaction rate cannot be achieved in a batch reactor.

Another continuous process has been described by Schmidt (59). The oil and the catalyst are mixed and go to a reactor that is a column divided by a series of plates into a number of reaction chambers. Hydrogen is metered into the column and flows concurrently with the oil and the catalyst. Only enough hydrogen is metered into the reactor to obtain the desired reduction, and any excess hydrogen increases the pressure in the reactor. This increases the rate of reaction until the pressure falls. Since the gas volume/liquid ratio is very low, the IV of the product can be controlled to within 1.5 IV units. Back mixing is prevented by the flow of hydrogen through the apertures between the reaction chambers. No agitators are used since the flow of gas is sufficient to keep the catalyst in suspension. As the hydrogenated oil flows from the reactor through a pressure-reducing valve, the hydrogen is separated and pumped back into the reactor with the desired quantity of make-up hydrogen. The catalyst is filtered from the cooled oil, and part of the used catalyst is returned to the catalyst make-up tank to be mixed with fresh catalyst. Pressures of 30–60 psi with temperatures of 180–200°C are used in the reactor. The plants for neutral oil hydrogenation, which may have capacities of 50–200 tons/24 hours are made of mild steel and those for fatty acid hydrogenation, of chrome–nickel–molybdenum steel.

Table 1.5 shows the consumption of services per metric ton of neutral

Table 1.5   Consumption of Services per Metric Ton of Neutral Oil

Steam 60 psig	100 kg
Cooling water at 20°C	4 m³
Power	13 kWh
Hydrogen (99.5% +)	72 m³ stp
Catalyst (Ni)	0.2–0.5 kg
Operators	1 person per shift

oil based on a reduction of the iodine value of 60 (60). Thus the cost per ton of oil may be calculated depending on the size of the plant and local conditions.

Another process for continuous hydrogenation is described by Coombes et al. (61). The heated oil plus catalyst is pumped through a pipeline reactor and hydrogen is introduced at spaced intervals over the length of the reactor. The hydrogen is introduced in a manner to provide highly turbulent two-phase flow of the bubble type. Only enough hydrogen is introduced to provide the desired reduction in iodine value. The velocities of the materials in the reactor are sufficient to keep the catalyst suspended and prevent back mixing in the reactor tube. The hydrogenated oil goes through a back pressure valve, and the released hydrogen may be recirculated. The hot oil is cooled by an economizer that heats the incoming oil. Catalyst filtration completes the process.

The method described by Kehse (62, 63) employs a reactor that consists of a series of perforated plates so that the heated mixture of catalyst and oil flows along a tortuous path along the upper face of the plate and then onto the upper face of the next lower plate while simultaneously passing a stream of hydrogen through the perforations of the plates in an upward direction. The quantity and the pressure of the hydrogen are sufficient to prevent downward flow of the oil through the perforations. Since the oil is flowing in a stream across the plates and from plate to plate, back mixing is prevented and the hydrogenated oil is homogenous. The temperatures and pressures may be varied to produce any desired reduction. A discussion of the design considerations for a reactor with baffles has been presented (64, 65).

Stationary catalysts or fixed-bed reactors have the advantage of no catalyst filtration, thus simplifying the equipment in the process. One major disadvantage is the gradual poisoning of the catalyst bed by the feed and incoming hydrogen, so the product from the reactor gradually changes. Eventually the catalyst bed must be reactivated or changed.

Mukherjee (66) has described a fixed-bed reactor that employs a pelleted catalyst. The vegetable oil is pumped through a preheater and mixed with the hydrogen in the top of the reactor, thus preventing back mixing in the reactor. A back pressure valve maintains the desired hydrogen pressure in the reactor. Various pelleted catalysts are used to obtain different products. Also, hexane solutions of oil are hydrogenated.

A unique catalyst for continuous hydrogenation of vegetable oil has been described by Kirsch (67, 68). An unsaturated oil is rapidly and continuously stirred with suspended particles of a macroporous SiO_2–Ni catalyst while hydrogen is passed through the mixture from the bottom and drawn off through the top. The agitation is designed so that the catalyst particles remain in the center of the charge while the hydrogenated oil is drawn off through a filter at the top while fresh oil is added at the bottom. A temperature of 272°F and atmospheric pressure produced a hydrogenated soybean oil that was suitable for a margarine base stock. By the use of this catalyst, 28,000 lb of oil may be hydrogenated per pound of nickel, whereas in the batch process, 1 lb of nickel would treat only 5000 lb of oil.

In spite of the vast amount of work on the development of continuous hydrogenation processes, they are used to a limited extent in industrial hydrogenation of lipid materials.

5.3   REMOVAL OF METALS FROM HYDROGENATED OILS

Hydrogenated oils, and particularly those prepared with a wet-reduced catalyst, or copper chromite, contain traces of nickel or copper in the form of nickel or copper soaps or colloidal metal that should be removed from edible products. Nickel removal is usually accomplished by a so-called postbleaching step, in which the filtered oil from the converter, at a temperature of about 180°F, is treated with 0.1–0.2% of bleaching earth, preferably of the activated type, and filtered. A very small amount of phosphoric acid or other metal scavenger is sometimes added in the bleaching step. For the adsorption and removal of colloidal nickel, Borkowski and Schille (69) recommend the use of activated carbon, which is added to the oil along with the catalyst, prior to hydrogenation. Carbon equal to 10–20 times the amount of nickel in the catalyst is reported to yield a metal-free filtered oil and also to improve the color of the oil.

The presence of traces of copper in oils is particularly damaging because of the powerful prooxidant effect of copper. Oils hydrogenated in the presence of copper chromite catalysts may contain 0.2–4.0 ppm copper after filtration. The amount of copper may be related to the free fatty acid content of the oil. Postrefining or postbleaching with clay or carbon does not remove enough copper to render the oil stable. Treatment with citric acid solution or treatment with a cation exchange resin followed by citric acid reduces the copper content to less than 0.03 ppm, and the oil is stable (70).

Another method of catalyst removal is the electrofilter that is reported to remove finely divided (<200 μ) solids from dielectric fluids on the basis of particle behavior in a gradient electrical field (71).

The separator system is relatively simple and is a cylindrical steel shell packed with pieces of ceramic or glass beads. A high-voltage electrode is centered in the medium. During the filtration cycle the electrodes are energized and a gradient electrical field is developed across the ceramic or glass beads. As the oil containig solids is pumped through the filter, the particles are polarized and captured at bead contact points by an induced dipole. The particles may be removed from the filter by deenergizing the electrodes and back flushing with oil. In this way catalyst may be removed from hydrogenated oil and then pumped back into the reactor without exposure to air.

6.1   CONJUGATED HYDROGENATION

Conjugated hydrogenation or catalytic transfer hydrogenation refers to the process of hydrogenation when the source of hydrogen is an organic molecule and a catalyst is used to effect a transfer of hydrogen to the unsaturated molecule.

The hydrogen donor can be any compound whose oxidation potential is sufficiently low so that the hydrogen transfer can occur under mild conditions; thus chosen compounds for donors are usually hydroaromatics, unsaturated terpenes, and alcohols. Although cyclohexene, because of its ready availability and high degree of reactivity, is the preferred donor, most of the work with fat hydrogenation has employed alcohols as the donor.

The process has not attained commercial technical importance, probably because of the comparative cost of gaseous hydrogen and donor hydrogen and the cost of removing the hydrogen donors; some work has shown possibilities.

The hydrogenation of peanut oil and safflower oil with the use of nickel catalyst and alcohols as hydrogen donors was reported by Basu and Chakrabarty (74).

With ethyl alcohol at 225°C, 145 psig, 1% Ni, peanut oil was reduced from 94 to 71.3 IV in 6 hours. Little saturated acids were formed. However, if isopropyl alcohol was used as the donor under the same conditions, safflower oil was reduced from 144.4 to 68.8 IV in 0.25 hour. This illustrates the easy removal of the hydrogen of a secondary alcohol to give a ketone. Acetone is produced when isopropyl alcohol is used as a donor. This subject has been reviewed recently (75).

6.2 CHARACTERISTICS OF HYDROGENATED FATS (OTHER THAN MELTING POINT OR CONSISTENCY)

The characteristic of fats most directly affected by hydrogenation is the iodine number, which decreases in direct proportion to the amount of hydrogen absorbed. Characteristics that do not depend on unsaturation, such as saponification number, hydroxyl number, Reichert–Meissl number, and content of unsaponifiable matter, are substantially unchanged by hydrogenation.

Certain nonglyceride constituents of fats and oils are reduced during hydrogenation, including carotenoid pigments and probably also unsaturated hydrocarbons. The hydrogenation of carotenoids usually causes a marked reduction in the color of vegetable oils. Unbleached palm oil, which is deep orange–red, is no darker than other vegetable oils after it is hydrogenated. The color of bleached cottonseed oil, measured on the Lovibond scale, is often decreased as much as 50%, and that of soybean oil is usually reduced even more.

Concurrently with the reduction of carotenoids, the vitamin A activity of oils is destroyed. However, by operating at low temperatures (e.g., 40–60°C) and high pressures (200–1000 psi) it is reported (76) that fish oils can be partially hydrogenated without material loss of vitamin potency.

The free fatty acid content of dry oils is not measurably affected by hydrogenation. However, water, from either the oil or hydrogen, will cause an increase in the free fatty acids by hydrolysis. Therefore, all the reactants need to be dry.

Hydrogenation, if extensive, will destroy the natural flavor and odor of any oil or fat, producing in its stead a distinctive, rather unpleasant "hydrogenation odor" that must be removed from edible fats by steam deodorization. This odor is attributed by Mielck (77) to ethers of higher and lower alcohols, possibly in conjunction with acetyl carbinol resulting from hydrogenation of the diglycerides that are present in traces in all oils.

Silveira et al. (78) made a clear distinction between the flavors produced by the hydrogenation of oils and the characteristic hardening flavors that develop in hardened linseed and soybean oils during storage. He found the latter flavor to be derived from saturated aldehydes and ketones (79). However, Keppler attributed the flavor to the formation of 6-*cis* and 6-*trans*-nonenal that could come from the oxidation of 9,15 and/or 8,15 octadecadienoic acids that are formed during the hydrogenation of linolenic acid (80,81). Later work by Yasuda et al. (82) indicated that the hydrogenation flavor is derived from 2-*trans*-6-*trans*-octadienal.

Thus it is believed that the flavor comes from the isomeric unsaturated acids produced by hydrogenation, but the particular compound (s) has (have) not been identified with certainty.

The stability of oils as measured by accelerated oxidation tests is progressively increased as the oils are hydrogenated. The relationship between stability and the composition of hydrogenated oils is discussed elsewhere.

Hydrogenation reduces the refractive index of oils. The relationship between iodine number and refractive index depends largely on the average molecular weight of the glycerides. This relationship is very nearly the same for many fats and oils (Figure 1.14), although rapeseed and other erucic acid oils, which are high in molecular weight, and coconut oil and other lauric acid oils, which are low molecular weight, deviate quite markedly from the average. The correlation between iodine number and refractive index is not precise for any given variety of oil, but the refractive index will indicate the iodine number of hydrogenated oil within one or two units or even less for a single lot of oil hydrogenated under a fixed set of conditions (Figure 1.15). Since refractive index measurements are easily and quickly made, they are often used to monitor and control the hydrogenation. Determination of iodine number by a rapid method, such as that of Hoffman and Green (83), can be conducted almost as quickly and, of course, gives a closer indication of the progress of hydrogenation.

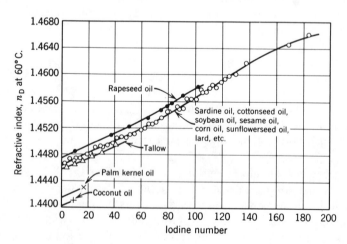

Figure 1.14 Iodine number versus refractive index for various hydrogenated oils.

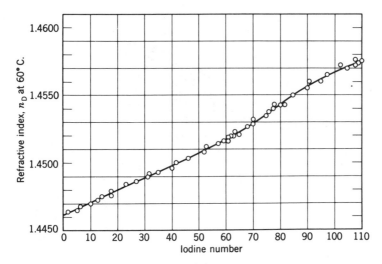

Figure 1.15   Iodine number–refractive index relationship in hydrogenated cottonseed oil.

Effect of Hydrogenation on Melting Point, Consistency, and Solid Fat Index (by Dilatometry).   During the first stages, of hydrogenation the consistency of an oil and its characteristics related to consistency, such as melting point, softening point, congeal point, and solid fat index, depend a great deal on the conditions of hydrogenation and the characteristics of the catalyst. For example, if the consistency is measured by the Feuge–Bailey micropenetration method (84), the micropenetration at 22.5°C (72.5°F) of cottonseed oil hydrogenated to an iodine number of 65 may be as low as about 25 or as high as about 125, according to the quantities of saturated and isooleic acids in the sample. The melting point varies according to the saturated fatty acid content and hence depends on reaction selectivity. At an iodine number of 67, Williams (85) recorded melting points for hydrogenated cottonseed oil varying from 35.9 to 53.1°C, as the temperature of hydrogenation was decreased from 180 to 130°C.

In the later stages of hydrogenation the percentage of saturated acids in the oil becomes a function simply of the iodine number of the oil, and the percentage of isooleic acid is not greatly different under different conditions of hydrogenation, so that melting points and other variables are quite predictable from the iodine numbers. Typical curves of iodine numbers versus titer or melting point for various highly hydrogenated oils are shown in Figures 1.16 and 1.17, respectively. The titer of hydrogen-

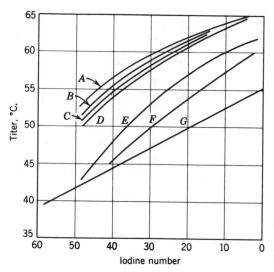

Figure 1.16 Iodine number versus titer for various hydrogenated oils: (*A*) sunflower seed; (*B*) corn; (*C*) soybean; (*D*) sesame; (*E*) cottonseed; (*F*) tallow; (*G*) sardine.

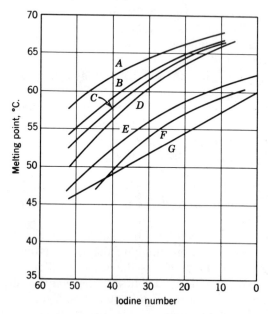

Figure 1.17 Iodine number versus melting point (clear point in closed capillary tube) for various hydrogenated oils: (*A*) sunflowerseed; (*B*) corn; (*C*) soybean; (*D*) sesame; (*E*) cottonseed; (*F*) tallow; (*G*) sardine.

ated cottonseed oil exhibits the peculiarity of passing through a minimum at an iodine number between 80 and 85.

The typical alteration of consistency with hydrogenation is illustrated in the case of different oils by the micropenetration curves in Figures 1.18 and 1.19. Similarly, the influence of degree of hydrogenation on solid fat index curves is shown in Figures 1.20 and 1.21.

The lack of selectivity with respect to the hydrogenation of glycerides, quite as much as the equally unavoidable formation of *trans* isomers, renders it impossible to produce hydrogenated fats or oils which are equivalent to some of the natural fats. Natural fats tend to be softer in consistency and lower in melting point than is possible in hydrogenated oils of the same iodine number. Olive oil, for example, has an iodine number of about 80 but remains liquid down to quite low temperatures. Cottonseed and soybean oil hydrogenated to this iodine number are lardlike in consistency at ordinary atmospheric temperature. Consequently, lard substitutes can be prepared by the hydrogenation of vegetable oils to iodine numbers in the same range as those of lard (Figure 1.22).

Methods for Hydrogenation Control.   Even though different batches of oil are hydrogenated under uniform conditions to precisely the same iodine number, slight variations inevitably occur in their composition and

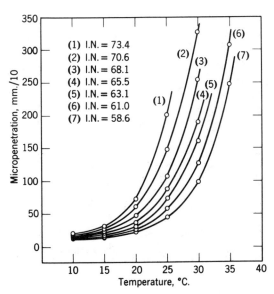

Figure 1.18 Typical micropenetration curves of cottonseed oil hydrogenated to different degrees (86).

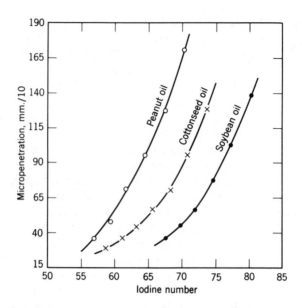

Figure 1.19 Typical variations in the consistency of different oils with hydrogenation; micropenetration at 22.5°C (mm/10) versus iodine number of the oils (84).

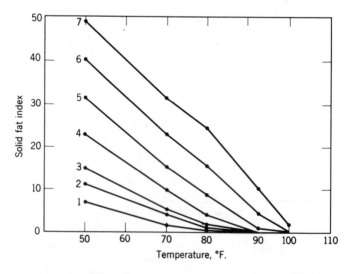

Figure 1.20 Typical solid fat index curves for soybean oil hydrogenated to different iodine numbers: (1) 99.3; (2) 90.5; (3) 88.4; (4) 82.5; (5) 78.0; (6) 72.7; (7) 70.0.

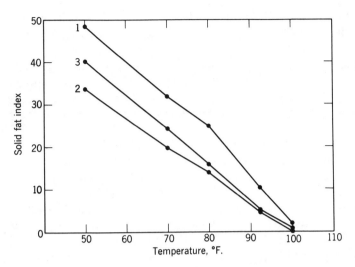

Figure 1.21 Typical solid fat index curves for different oils hydrogenated to similar iodine numbers: (1) soybean oil, 70.0; (2) cottonseed oil, 69.0; (3) peanut oil, 72.1.

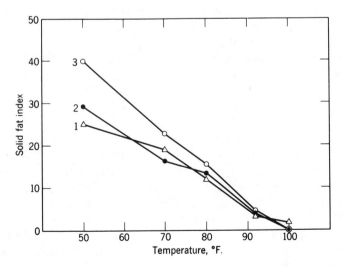

Figure 1.22 Solid fat index curves for lard and for a cottonseed oil and a soybean oil hydrogenated to lardlike consistency: (1) lard, IN 67.3; (2) hydrogenated cottonséed oil, IN 75.4; (3) hydrogenated soybean oil, IN 72.7.

characteristics. Hence it is necessary to control the hydrogenation of edible oils according to their physical characteristics, rather than strictly according to the quantity of hydrogen absorbed. In practice it is customary to adjust the conditions of hydrogenation so as to obtain the proper balance between the formation of saturated and isooleic acids and then carry hydrogenation of the oil far enough to bring all batches to a uniform consistency at ordinary atmospheric temperatures, that is, in the range of about 70–80°F.

The problem of hydrogenation control is less difficult if more than one hydrogenated stock is blended to produce the product in question and if a number of converter batches can be blended in making up each lot. In the manufacture of some stocks it is almost essential to interrupt hydrogenation at some point short of the endpoint and check the consistency of the oil by analysis before finishing the batch. However, after the product endpoint is established for a given lot of oil and catalyst, the hydrogenation of subsequent batches can usually be satisfactorily controlled by hydrogen absorption or refractive index alone, inasmuch as irregularities can usually be traced to variations in the oil or the catalyst.

Of the various analytical methods for judging the consistency of hydrogenated fats, the solid fat index (SFI), determined by dilatometry, has become the most widely used, even though it is only empirical. The American Oil Chemists' Society (AOCS) has adopted a method (86) (Cd 10-57) in which solid fat indices at 10, 21.1, 26.7, 33.3, and 37.8°C are used to characterize shortenings and margarine oils. For the purpose of hydrogenation control, however, the determination of all five points is often found to be too time consuming. Consequently, most manufacturers have adopted one-, two- or three-point modifications of the official procedure to be used for control purposes. The modified procedures seldom yield precisely the same values as those obtained by the AOCS method, but they do bear a sufficiently constant relationship to make them satisfactory for practical use.

A major advantage of the SFI measurement is that it can be used to predict the consistency of an oil over a wide temperature range. Consequently, an SFI curve can be used to characterize the plastic range of a shortening or the firmness of a margarine oil at 50°F and its liquefaction at mouth temperature. It is not, however, a direct measure of consistency. Two different oils of the same SFI may have considerably different consistencies, depending on the inherent crystallizing characteristics of the oils and the way they are handled in chilling and tempering. For any specific oil or blend of oils handled in a standardized manner, it is possible to predict with reasonable accuracy the consistency of either finished margarine or shortening from the SFI curves of the oils, once sufficient experience has been obtained to establish the relationship for this oil or blend of oils. If either the oil or the subsequent handling is changed, a new relationship must be established by trial and error. A pair of typical SFI curves is shown in Figure 1.23, comparing a shortening with a margarine oil.

Another method that has been used in hydrogenation control is the setting or congeal point (86). The details of this method have never become standardized and vary from one laboratory to another, with corresponding variations in the congeal point itself. Some laboratories cool a molten fat sample under carefully controlled conditions until a "cloud point," corresponding to a standard degree of turbidity, is reached, with the latter noted either visually or photoelectrically.

Again, congeal points are not direct measures of the consistency of the fat, but rather are characteristics that may be correlated with the consistency at 70–75°F, provided that the selectivity of hydrogenation is known.

Hard oil, stearine, or "flakes" is usually hydrogenated to a definite titer (or clear capillary tube melting point). In highly hydrogenated products the relationship between titer and melting point is dependable; hence hydrogenation may be satisfactorily controlled by the refractive index or the iodine number.

In following closely the course of hydrogenation by any of the preceding methods the best procedure is to maintain a running plot of hydrogenation time against refractive index as a general guide to the course of reaction. During the first stages this plot may be made quite roughly, being esti-

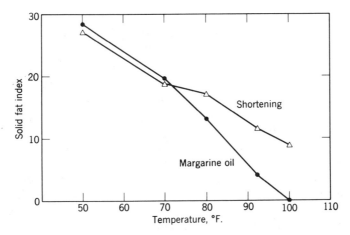

Figure 1.23   Solid fat index curves for a typical shortening and margarine oil.

mated from refractometer readings or from the drop in pressure of the hydrogen tank supplying the converter.

Micropenetration or other critical characteristics should be determined at the same time as the solid fat index. Then from the relationship between the two (by consulting families of typical curves established from previous runs) it is easily possible to determine by extrapolation how much further the iodine number should be dropped to bring the critical characteristic to the desired value. The proper drop in iodine number can be brought about by extending the hydrogenation time according to indications of the time–iodine number curve or by continuing the reaction until the required amount of hydrogen (calculated from gauge, i.e., psig, readings) is absorbed.

Naturally, the more closely the end of the reaction is approached before a check is made, that is, the less extrapolation required, the more precisely the endpoint will be determined. In very careful work it may be necessary to make repeated checks on more than one sample. If the endpoint is thus approached very cautiously, or in any case if reaction is relatively rapid, the reaction will have to be stopped while the oil samples are being analyzed.

The content of *trans* isomers has an effect on the physical properties of hydrogenated oils (87). The relationship between *trans* content and congeal point, Wiley melting point, and SFI has been reported by Stingley and Wyobel (87).

For control in the blending of hydrogenated stocks to produce margarine oils or shortening, dilatometric methods are in most common use.

6.3   HYDROGENATION OF HARD OILS OR STEARINE

Hydrogenated stearines, or hard oils, are principally manufactured for use in making blended-type shortenings, for imparting high-temperature body to all-hydrogenated shortenings, or for stiffening soft lards. They may also be desired as raw materials for the production of stearic acid, however, or for a variety of other purposes. The terms "vegetable stearine" and "animal stearine" are applied respectively to vegetable and animal oils or fats that have been hydrogenated to a brittle consistency and an iodine value usually less than about 20 and often in the range of 5–10. The stiffening capacity of a hard oil is very closely measured by its titer. Vegetable stearine was originally prepared as a substitute for oleostearine, which has a titer of about 50°C and must be mixed with cottonseed oil in the ratio of about 20 to 80 parts to produce a shortening of the proper consistency. For blending purposes, it is now customary to hydrogenate vegetable oils to a titer of 58°C or above; not more than 10–15% of such a hard oil is required in a blended shortening.

Some oils can be hardened to considerably higher titers than others (Figure 1.16); soybean oil, sunflower oil, corn oil, or sesame oil may be hydrogenated to a titer as high as 65°C. It is not convenient to use these oils in a highly hardened form for blended-shortening manufacture, however, because they solidify in undesirable polymorphic forms if reduced to too low an iodine number. This causes trouble in the solidification and the subsequent handling of the shortening. This tendency, as well as the high titer, is undoubtedly a result of the predominance of C_{18} acids in these oils, thus causing the completely hardened product to consist largely of tristearin. Cottonseed oil, palm oil, tallow, or fish oils contain a sufficient proportion of acids higher and lower than C_{18} to ensure that little tristearin will be formed and, at 58–60°C titer, will solidify in the more desirable polymorphic forms.

Hydrogenation to produce stearine is the least critical of all hydrogenation operations; such considerations as selectivity and isooleic acid suppression are naturally unimportant in the manufacture of this product. If a catalyst is to be used only for hard oil production, if may be selected purely on the basis of its activity. To make the reaction as rapid as possible, the highest possible pressure is usually used, and the temperature is often allowed to rise as high as 400–425°F. At high temperatures and pressures, the rate at which hydrogenation can be conducted will depend on the capacity of the cooling coils in the hydrogenator for carrying away the heat of the exothermic reaction. Ordinarily about 3–4 hours will be required to reduce cottonseed oil to an iodine number of 10 and a titer of 60°C at a pressure of 60 psig, a maximum temperature of 400°F, and a catalyst concentration equivalent to 0.03–0.04% of fresh, active nickel. If only new catalyst is used, the best results will be obtained by reserving a portion of the catalyst to be added toward the end of the hydrogenation; for example, 0.02–0.04% may be added to the hydrogenator at the beginning of the run, and 0.01–0.02% may be added after the iodine number is reduced to about 30 or 40. Catalysts that have been partially inactivated in the manufacture of other products may be used in making hard oil, although some catalysts may produce excessive amounts of free fatty acids in the oil if used in extremely high concentrations.

6.4 HYDROGENATION OF SHORTENING STOCKS

In the shortening trade in the United States, so-called all-hydrogenated shortenings constitute a class of products that are manufactured to rigid standards of color, flavor, odor, stability, consistency, and performance. In general, they are hydrogenated in such a manner as to obtain the lowest iodine number and the highest stability that is consistent with the proper consistency and performance characteristics. The proper consistency at 70–80°F, together with reasonably good body at higher and lower temperatures, can be obtained with iodine numbers in the range of 75–80 in the case of hydrogenated stock for general purpose, cake and icing shortenings, and 65–75 for uses where stability is more important than a wide range of plasticity. These base stocks are blended with varying amounts (up to about 10%) of the hard stocks described in the preceding section to obtain the best plastic range.

Iodine numbers such as those listed in the preceding paragraph are obtained by adjusting the hydrogenation conditions so as to minimize the formation of saturated acids and *trans* isomers. This adjustment is obviously rather delicate, since conditions leading to high degrees of selectivity normally tend to cause increased formation of *trans* isomers. Some hydrogenators conduct the operation in successive steps at two different temperatures, employing a low temperature in the first stage to minimize *trans* isomer formation and a high temperature in the second stage to reduce the percentage of linoleic acid.

The analytical data contained in Table 1.6 are more or less representative of the characteristics of all-hydrogenated standard and high-stability shortenings.

Although all-hydrogenated and blended shortenings formerly comprised two entirely distinct types, changes in shortening manufacture in the last decade have made classification on this basis essentially without

Table 1.6   *Analyses of typical all-hydrogenated vegetable shortenings*

Analyses	Standard	High-Stability
Hard stock added	10%	5%
Iodine number	73.2	69.2
Melting point (FAC)	119°F	109°F
Linoleic acid[a]	6.4%	1.5%
Solid fat index at		
10°C	27.1	43.8
21.1°C	18.4	27.5
26.7°C	16.9	22.0
33.3°C	11.8	11.3
37.8°C	8.7	4.7

[a] By ultraviolet absorption method.

practical meaning from the point of view of utility. Numerous high-stability and standard shortenings of excellent quality have been manufactured from all-hydrogenated meat fats, all-hydrogenated blends of meat fats and vegetable oils, and blends of meat fats and hydrogenated vegetable oils. In most cases these are hydrogenated in such a way as to obtain maximum selectivity.

6.5   HYDROGENATION OF MARGARINE OILS

The hydrogenation of margarine oils, like that of all-hydrogenated shortenings, is a highly critical process, although the characteristics desired in the finished products are somewhat different. Shortenings are preferably as soft and plastic at the lower temperatures as is feasible and at the same time have some body at a temperature in the neighborhood of 98°F, thus necessitating an SFI curve that is as nearly horizontal as possible, having a minimum rate of change in solid glycerides for each unit change in temperature. Although margarine must also be as plastic as possible at lower temperatures (e.g., refrigerator temperature), it must have sufficient solid glycerides at 70–80°F to allow it to be formed and packed in the customary prints and should be able to hold its shape and not separate oil for a reasonable period of time at 80–90°F, and it must melt completely at the temperature of the human body, so as not to be "gummy" in the mouth. Consequently, the SFI curve for margarine oil will have a considerably higher rate of change per unit of temperature change than shortenings. Preferably, the curve should be somewhat convex, with a lesser rate of change at low temperatures than at the higher ones (see Figure 1.23).

The desired characteristics in margarine oil are best obtained by blending two or more stocks hydrogenated to different degrees. Commonly, a major proportion of cottonseed or soybean oil hydrogenated to an iodine number in the range of 80–90 is blended with a minor proportion of an oil hardened to an iodine number of about 55–65. Hydrogenation conditions are selected to produce minimum "solids" at the lower temperatures of the SFI curve. Analytical data for a typical blended margarine oil and its two components are shown in Table 1.7.

6.6   HYDROGENATION OF HARD BUTTER SUBSTITUTES

In the manufacture from liquid oils of hard butter substitutes suitable for confectionery coating and so on, the combination of relative hardness and nongreasiness at 70–80°F with substantially complete melting at body temperature is best obtained by the most selective hydrogenation to pro-

Table 1.7   Analytical data for a typical margarine oil formulated from 80 parts of a softer component and 20 parts of a harder component

Analyses	Margarine Oil Blend	Softer Component	Harder Component
Iodine number	79	86	55
Melting point (FAC)	100°F	86°F	111°F
Solid fat index at			
10°C	27	18	66
21.1°C	16	7	59
26.7°C	12	2	57
33.3°C	3.5	0	43
37.8°C	0	0	27

duce an SFI curve with the greatest possible rate of change for each unit change in temperature, especially at higher temperatures. Peanut oil is considered a particularly suitable oil for making this type of product.

Ziels and Schmidt (88) have patented a method of hydrogenating oils to produce a hard butter substitute. This method involves the use of a special catalyst that has been deliberately sulfur poisoned. In one experiment cited a fat quite similar in physical characteristics to cocoa butter was prepared from peanut oil by hydrogenating with 0.5% nickel at 338°F and 20 lb of pressure to an iodine number of 68.9. The melting point was 38.3°C and the content of solids at 97°F was only 3.2%, but the fat was quite hard up to about 80–85°F; the solids content at 68°F was 62.8% (as compared to about 64% for cocoa butter). The isooleic acid content of the product was 53.5%.

Theory of Catalysis

7.1   GENERAL CONSIDERATIONS

Definition of a Catalyst.   A catalyst, according to the classical definition of Ostwald, is a substance that alters the rate of a chemical reaction without affecting the energy factors of the reaction or being consumed in the reaction. Properly speaking, therefore, a catalyst cannot initiate but can only accelerate a reaction. In numerous instances, however, including the hydrogenation of fats and oils, reaction is so imperceptibly slow in the absence of a catalyst that the latter must be considered an essential element of the reacting system.

The permanent or self-regenerative nature of catalysts is very important. Although in the course of their action catalysts may enter into temporary combination with the reactants, such combinations are unstable and invariably are broken down at the completion of the reaction, to yield the catalyst in an unchanged form. Thus the catalyst enters into the reaction over and over again, and a relatively small amount may be capable of transforming very large amounts of material. Ordinarily, the concentration of nickel employed as a catalyst for the hydrogenation of fats does not exceed a few hundredths of 1% of the weight of the fat.

The fact that a catalyst cannot initiate reactions does not mean that the introduction of a catalyst into a reacting system may not influence the composition of the reaction products, or that different catalysts may not yield different products. In many cases reaction follows a number of alternative courses; hence the composition of the final products will depend on the relative rate of the various alternative reactions. Where a number of different reactions occur together, the addition of a catalyst to the system may accelerate some of them to a far greater extent than others. Furthermore, one catalyst may differ from another in its relative effect on the different reaction rates. The hydrogenation of fats and oils furnishes examples of such specificity of catalyst action. Thus the addition of 1 mole of hydrogen to a linoleic acid chain in a glyceride molecule may yield either normal oleic acid or isometric forms of this acid. Some nickel catalysts are much more inclined than others to produce the isomeric forms.

Heterogeneous Catalysis.   The type of catalysis that is the most important in industry, and the type that operates in fat hydrogenation, is heterogeneous catalysis. By definition, a heterogeneous system is one in which the catalyst and the reactants exist in different physical states. Heterogeneous catalysis is to be somewhat sharply distinguished from homogeneous catalysis, in which the catalyst and the reactants comprise a single phase. In homogeneous catalysis the catalyst functions in the form of individual molecules, which are uniformly distributed throughout the reacting system. Thus the question of catalyst physical structure or of surface phenomena does not enter. On the other hand, in heterogeneous catalysis it is the catalyst surface that performs the catalytic function; hence the nature of the surface is of extreme importance. A catalyst operating in a homogeneous system is defined simply in terms of its chemical constitution and its concentration in the system. With all other factors controlled, the effect of a homogeneous catalyst of definite composition is exactly predictable on the basis of its concentration. If the catalyst is an immiscible solid, however, its behavior will depend not only on its chemical composition, but also to a very large degree on both the nature and the extent of its surface. The fact that the characteristics of a solid catalyst are determined so largely by the submicroscopic character of its surface renders the study and and the control of such catalysts very difficult. Apparently similar catalysts may differ enormously in their activity and specific action.

In heterogeneous catalysis it is now generally assumed that reaction proceeds through the formation of unstable intermediate compounds or adsorption complexes, in which the catalyst is temporarily combined with one or more of the reactants. If such compounds exist, it is probable that in most cases they are not definite chemical combinations but consist merely of strongly bound molecules of the reactant that are held to the catalyst surface by secondary valence forces or by π complexing. In any event, it is essential that they be unstable, that is, capable of being either decomposed or desorbed, to permit reaction to proceed according to the following scheme:

$$\text{Catalyst + reactants} \xrightarrow{\text{adsorption}} \text{catalyst–reactant complex}$$
$$\xrightarrow{\text{desorption}} \text{reaction products + regenerated catalyst}$$

Catalysis in Relation to Energy of Activation. Chemical reactions, catalyzed or uncatalyzed, do not occur instantaneously principally because of a pattern of molecular distribution of energy that ensures that at any instant only a few molecules of the reacting substances will be at a sufficiently high energy level. The critical energy for a specific reaction, known as the energy of activation, is represented graphically in Figure 1.24 as the height of a potential barrier opposing the reaction.

Modern views relative to the energy factor in catalysis have been reviewed by Berkman et al. (89). More recently, Grosse (90) has presented

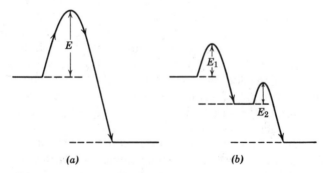

(a)   (b)

Figure 1.24 Graphical representation of the activation energy factor in (a) uncatalyzed and (b) catalyzed reactions. [After Grosse (90).]

a simplified treatment of a catalytic action from the standpoint of energy relationships, from which the following is taken.

The rate of a chemical reaction k is determined by the integrated Arrhenius equation

$$k = a\, e^{-E_{act}/RT}$$

where T is the absolute temperature, a is a factor related to the concentration of the reactants, and E_{act} represents the activation energy. Because of the exponential character of this equation, a relatively slight change in activation energy will have a large effect on the reaction rate. Thus, for example, if the activation energy at 300°K is 50,000 calories, a 10% lowering of this energy requirement will increase the reaction rate 4400 times; even a 1% lowering will increase it 2.3 times.

Catalysts increase the reaction rate through their influence on the activation energy. A catalyst breaks the reaction up into two successive steps: the combination of catalyst and reactants to form an unstable intermediate compound and the breakdown of this compound to yield a new product and the free catalyst. This has the effect of permitting the energy barrier to be surmounted in two small steps, rather than one large one (Figure 1.24). In other words, two reactions with relatively low activation energies are substituted for a single reaction with a high activation energy. The alteration caused in the reaction rate is measured by the difference between the activation energy of the uncatalyzed reaction and that of the slower of the two steps of the catalyzed reaction. In view of the exponential relation outlined previously, it is readily apparent that the catalyzed reaction may proceed at a rate that very greatly exceeds that of the uncatalyzed reaction.

7.2   THEORY OF CATALYST STRUCTURE

Since heterogeneous catalysis is a surface phenomenon, an essential requirement in an active catalyst is a highly extended surface. With all other factors being equal, the smaller the individual catalyst particles, the more active the catalyst will be.

In spite of the obvious relationship of surface area to catalyst activity, it does not follow that the activity is determined solely by the magnitude of the surface area. The latter may be made very large without the catalyst necessarily being very active. In fact, metallic nickel dispersed to the colloidal state may be virtually devoid of catalytic activity. All evidence indicates that the activity of a hydrogenation catalyst is due to a certain degree and kind of heterogeneity in the catalyst surface. The development of this heterogeneity will not occur under all conditions but must be achieved by special methods of catalyst preparation.

The various phenomena associated with heterogeneous catalysis are best explained on the basis of the "active spots" theory of Taylor (91). This theory assumes that the metal atoms on the surface of the catalyst possess varying degrees of unsaturation, according to the extent to which they are elevated above the general catalyst surface, or otherwise released from the mutually restraining influence of their neighboring atoms. The relatively few metal atoms that are thus highly unsaturated are the ones capable of entering into temporary combination with the hydrogen and the unsaturated oil, thereby furthering the hydrogenation reaction. Each unsaturated atom or concentration of unsaturated atoms constitutes an active spot or center. The catalytic activity of each unsaturated atom corresponds to the extent of its unsaturation (Figure 1.25).

An alternative theory of catalyst structure envisions the active portions of a catalyst as areas where the normal crystal lattice of the metal is slightly expanded to fit more exactly the dimensional requirements for two-point adsorption on either side of a double bond. In the case of nickel, the normal interatomic spacing of 2.47 Å is actually a little smaller than the theoretical optimum. Expansion may occur as metallic nickel is produced by the reduction of a nickel compound.

Hydrogenation catalysts are seldom prepared directly from massive nickel but are made by first combining the nickel with other elements, as in nickel oxide, nickel hydroxide, nickel carbonate, nickel formate, and nickel–aluminum alloy, and then reducing the resulting compound to regain the nickel in metallic form. The efficacy of this procedure, in producing active nickel atoms relatively free from restraint by neighboring atoms, is evident from the schematic representation of catalyst reduction shown in Figure 1.26.

There is considerable evidence that the hydrogenation of an ethylenic

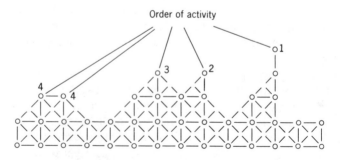

Figure 1.25 Schematic representation of a catalyst surface.

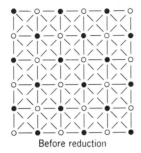 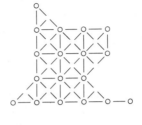

Before reduction   After reduction

Figure 1.26 Schematic representation of effect of reduction on a nickel catalyst. White circle, nickel atom; black circle, atom of oxygen, aluminum, and so on.

compound must be preceded by two-point adsorption of the carbon atoms on either side of the double bond (92). This requirement would impose certain dimensional limitations on the space lattice of any catalytically active metal. Actually, the metals that are all effective in the hydrogenation of double bonds (nickel, cobalt, iron, copper, platinum, and palladium) all have interatomic spacings close to that (2.73 Å) calculated as optimum for such two-point adsorption.

7.3 MISCELLANEOUS CHARACTERISTICS OF CATALYSTS

Catalysts are quite sensitive to heat and may be inactivated by temperatures much below the fusion point of massive nickel. This inactivation is apparently the result of a sintering process, which causes the active, projecting nickel atoms to assume more stable positions on the catalyst surface. Their sensitivity to heat makes their reduction to metallic form somewhat critical with respect to temperature since there is usually a small interval between the temperature at which reduction becomes rapid and that at which sintering begins. Catalysts precipitated in the form of nickel hydroxide or carbonate on kieselguhr must usually be reduced below about 1000°F.

Freshly reduced catalysts that have not been in contact with oil are highly pyrophoric as a result of the reactive nature of their unsaturated nickel atoms and the fact that they retain much adsorbed hydrogen at the end of the reduction period. If exposed to air at this stage of their preparation and allowed to oxidize, they are completely inactivated. Their pyrophoric properties and tendency to become inactivated through oxidation disappear, however, after the catalyst surface has become coated with oil. This behavior of catalysts is, of course, readily explainable on the basis of the theory of active spots or unsaturated nickel atoms.

Hydrogenation catalysts of the precipitated type are almost invariably supported on diatomaceous earth or other porous refractory material, and other catalysts, such as those prepared by the treatment of nickel alloys, are mixed with diatomaceous earth before use. Supporting a catalyst in this manner so greatly increases its activity and otherwise modifies its behavior that there can be little doubt that the support contributes largely to the catalyst structure. The support appears to protect the catalyst from sintering since supported catalysts may be successfully reduced at somewhat higher temperatures than the corresponding unsupported catalysts. It also appears reasonable to suppose that the presence of the support influences the dispersion of the catalyst particles and provides more numerous points for the attachment of unsaturated nickel atoms. In addition, there is the possibility that silica or other material in contact with nickel may produce active nickel atoms and hence that activity is centered at the boundaries between the nickel and the support. The latter viewpoint is quite in accord with the common observation that the activity of a metallic catalyst may be enhanced or "promoted" by the presence of another metal or metallic oxide that is not in itself a catalyst for the reaction.

7.4   CATALYST POISONING

Since the activity of a catalyst depends on the presence of a relatively few metallic atoms of an unusually high degree of reactivity, it is to be expected that these atoms will display a marked avidity for many substances other than hydrogen or glycerides if such substances are present as impurities in the reacting system. Furthermore, if the impurities are of such a nature that they are not readily desorbed by the catalyst but are held fast to the active atoms, they will gradually concentrate on the catalyst surface, saturating the active atoms and rendering the catalyst inactive. Substances that are thus able to cause catalysts to become inactive are termed catalyst "poisons." They may be troublesome when present even in traces in the reactants since the amount of catalyst is always small in relation to the amount of reactants, and the active portion of the catalyst is, in turn, small in comparison with the total catalyst.

Catalyst poisoning may be irreversible, leading to permanent inactivation of the catalyst, or it may be reversible if under certain conditions the poison can be removed and the original activity of the catalyst restored. Both reversible and irreversible catalyst poisoning are encountered in the hydrogenation of fats and oils.

Gaseous Poisons.   Among the worst poisons for nickel catalysts are the gaseous sulfur compounds: hydrogen sulfide, carbon disulfide, sulfur dioxide, carbon oxysulfide, and so on. These compounds are of considerable concern in practical hydrogenation since they may occur as impurities in crude hydrogen prepared by the steam–iron, water gas–catalytic, or hydrocarbon reforming processes. They are rapidly adsorbed by nickel catalysts and poison the catalysts irreversibly.

Experience with the sulfur poisoning of commercial nickel catalysts is in general accord with the active-spot theory of catalyst structure. The amount of sulfur required to poison a catalyst is in direct proportion to the activity of the catalyst. In the case of good catalysts, however, the amount of sulfur required for complete poisoning is rather larger than is to be expected from the assumption that the active nickel comprises a relatively small part of the total surface. Although inactive catalysts may be poisoned by 0.5–1.0 grams of sulfur per 100 grams of nickel, active catalysts are not rendered substantially inoperative until they have absorbed 3.0–5.0 grams of sulfur per 100 grams of nickel. If it is assumed that the sulfur is combined stoichiometrically with the nickel to yield NiS, the amounts of sulfur mentioned previously correspond to about 5.5–9.0% of the total nickel in the catalyst. When a single catalyst is progressively poisoned by the addition of successive portions of a sulfur-containing compound, the activity of the catalyst after each addition corresponds closely to the residue of active nickel, as calculated from the sulfur added and the sulfur required for complete poisoning (Table 1.8).

Experiments with the three sulfur compounds, hydrogen sulfide, carbon disulfide, and sulfur dioxide, indicate that the amount of sulfur required

Table 1.8   Activity versus sulfur content of a poisoned catalyst

Approximate active nickel content of catalyst	6.5%
Experiment 1	
Calculated percentage of total nickel combined with sulfur (as NiS)	0.9%
Residual active nickel	5.6%
Hydrogenation time required to reduce the iodine number of cottonseed oil from 107 to 25	30 min
Experiment 2	
Calculated percentage of total nickel combined with sulfur	2.9%
Residual active nickel	3.6%
Hydrogenation time, as previously	48 min
Experiment 3	
Calculated percentage of total nickel combined with sulfur	6.0%
Residual active nickel	0.5%
Hydrogenation time as previously	280 min

to inactivate a given catalyst is substantially the same regardless of the compound from which it is derived. The avidity of catalysts for the sulfur in the three compounds varies somewhat, however, and increases in the order the compounds are named.

Besides the sulfur compounds mentioned previously, the catalyst poison most likely to cause trouble in the hydrogenation of fats and oils is carbon monoxide, which is also present in small amounts in unpurified steam–iron and steam–hydrocarbon hydrogen. Carbon monoxide is adsorbed more slowly than are the sulfur compounds and poisons the catalyst reversibly. It may be removed and the catalyst restored to its original activity by interrupting the hydrogenation and continuing to agitate the catalyst–oil mixture under a reasonably good vacuum (about 25 in Hg or better) for a short time.

The poisoning effect of carbon monoxide is highly dependent on the conditions, and particularly the temperature, of hydrogenation. If hydrogenation is conducted in the neighborhood of 400°F, carbon monoxide in the amount of 0.5% in the hydrogen will slow the operation to a barely noticeable degree. At 350°F its effect will be much more noticeable, and at 300°F it will not be possible to hydrogenate with hydrogen containing as little as 0.1% carbon monoxide without frequent evacuation of the hydrogenator. At a temperature of 200°F it is virtually impossible to hydrogenate oils with gas containing more than a few thousandths of 1% of carbon monoxide. The poisoning effect of carbon monoxide is relatively more troublesome at low hydrogen pressures.

Most other gases that may be present as impurities in hydrogen, including carbon dioxide, nitrogen, and methane, are not catalyst poisons, although in a hydrogenation apparatus of the "dead-end" type they will, of course, slow the reaction if allowed to accumulate in the headspace of the hydrogenator and will dilute the hydrogen. Small amounts of water vapor in the hydrogen appear to have no poisoning effect but cause the formation of free fatty acids.

Poisons in the Oil.   The matter of catalyst poisoning through impurities in the oil has been discussed to some extent previously in connection with the effect of the catalyst concentration on the course of hydrogenation. Not a great deal is known about the natural oil and fat impurities that may function as catalyst poisons. Free fatty acids in small concentrations have little effect on the activity of the catalyst. The carotenoid pigments likewise appear to be devoid of poisoning tendencies since there is no correlation between the color of an oil and the readiness with which it may be hydrogenated. There are catalyst poisons (probably phosphatides) in most crude oils, however, for these oils are more easily hydrogenated after refining. In the case of some oils, for example, crude fish oil, treatment of the crude oil with a liberal quantity of active bleaching clay is as effective in removing catalyst poisons as is alkali refining.

Sodium and other alkali soaps are very pronounced catalyst poisons. Soap sufficient to seriously poison the catalyst will seldom be found in vegetable oils that have been well bleached after alkali refining, but lard, edible tallow, or other light-colored animal fats that require little or no bleaching are often soapy enough after refining to cause trouble in hydrogenation. Soap is quite effectively adsorbed by used catalysts that have become largely inactivated; hence in hydrogenating alkali-refined lard and so on it is sometimes expedient to give the fat a pretreatment with old catalyst before fresh catalyst is added.

A wide variety of substances were tested for their effect on nickel catalyst by Ueno (93), who reported the following to be poisons: soaps of potassium, sodium, lithium, magnesium, barium, beryllium, iron, chromium, zinc, cadmium, lead, mercury, bismuth, tin, uranium, and gold; copper hydroxide; ammonium molybdate; boric, arsenious, and hydrochloric acids; glycolic, lactic, hydroxystearic, oxalic, succinic, fumaric, malic, citric, and tartaric acids; sodium taurocholate; iron, zinc, lead, mercury, sulfur, tellurium, selenium, and red phosphorus; proteins, blood albumin, blood fibrin, gelatin, glycerol, lecithin, sucrose, dextrose, mannitol, starch, morphine, strychnine, amygdalin, potassium cyanide, zinc oxide, and aluminum silicate.

The following were found to be without poisoning effect: soaps of calcium, strontium, aluminum, cerium, nickel, manganese, copper, silver, vanadium, thorium, and platinum; nickel acetate, butyrate, stearate, lactate, oxalate, and succinate; and tungstic acid, fatty acids, nucleic acid, nickel, tin, zirconium, aluminum, copper, hemoglobin, cholesterol, squalene, and glycogen.

The effective life of a catalyst probably depends on the extent to which it adsorbs poisons from the oil. Some slight content of poisonous substances appears to be unavoidable, even in the most carefully refined oils. Even if the latter consisted of absolutely pure glycerides, it is quite possible that some degree of catalyst poisoning might occur through the formation and the adsorption of oxidation products or other degradation products of the oil. The literature on catalyst poisoning—particularly with respect to poisons in the oil—has been reviewed in detail by Bodman et al. (94).

Carbon monoxide is often observed to accumulate in closed hydrogenation systems, and it is generally considered to be formed by thermal decomposition of the oil. However, this compound is also a trace impurity in much commercial hydrogen, and there appears to be some doubt that there is any appreciable evolution of carbon monoxide from oil, except at quite elevated operating temperatures.

Drozdowski and Zajac (95) found that poisons or "inhibitors" increased the induction period of the hydrogenation as well as decreased the catalyst activity.

Effect of Poisoning on Catalyst Characteristics Other Than Activity. Partial poisoning of a catalyst can affect characteristics other than its activity. Sulfur-poisoned catalysts, in particular, produce larger quantities of isooleic acids in hydrogenated oils. Catalysts that have become partially inactivated in use produce hydrogenated fats that are higher in isooleic acids at a given saturated acid level or higher in saturated acids at a given isooleic level than the fats obtained with fresh catalysts.

7.5   WET-REDUCED CATALYSTS

One method of preparing nickel catalysts that has long been popular comprises of first converting the nickel to the easily decomposable salt of an organic acid and then decomposing the salt by heat. The latter operation, in which the nickel is reduced to the metallic form, is carried out while the catalyst is suspended in oil; hence catalysts of this type are termed "wet-reduced." In the preparation of wet-reduced catalysts it is essential that the nickel salt be reducible at a temperature below that at which thermal decomposition of the oil becomes pronounced. Various nickel salts have been proposed for this purpose, but the one that has found perhaps the greatest practical use is the formate, prepared by treating precipitated nickel hydroxide or nickel carbonate with formic acid.

Nickel formate begins to decompose at about 300°F by the following reaction:

$$Ni(OOCH)_2 2H_2O = Ni + 2CO_2 + H_2 + 2H_2O$$

Decomposition becomes rapid above about 375°F. Theoretically, 100 lb of nickel formate yield 31.8 lb of metallic nickel.

In practice, the reduction of nickel formate is carried out in specially designed closed vessels equipped with agitators, means for heating to a high temperature, and accurate temperature control. The oil in which the catalyst is suspended may be of any type. The ratio of oil to nickel formate in the charge is not critical; the use of 2–4 parts of oil to 1 part of formate is common practice. Ordinarily the rate of heating is as rapid as the design of the equipment permits. A maximum temperature in the neighborhood of 475°F is usually employed (96, 97).

Reduction is continued for an arbitrary time, which may be upward of 1 hour, or until laboratory tests reveal that the product has attained the maximum activity. During the reduction period, and also during the cooling period following reduction, a slow current of hydrogen is bubbled through the oil. Since the reduction is thermal in nature and involves no combination of hydrogen with oxygen in the catalyst, the chief function of the hydrogen is merely to sweep decomposition products out of the oil. As the catalyst becomes active hydrogenation of the oil begins, and at the end of the reduction period it is usually completely hardened. Hugel (97) recommends that the removal of decomposition products be accomplished by maintaining a vacuum on the reducer, rather than by using a current of hydrogen.

If the catalyst is to be used in the plant in which it is prepared, the batch is cooled to about 200°F; kieselguhr equivalent to one to four times the weight of the nickel is added; the original oil, which will have suffered some thermal decomposition, is filtered from the catalyst; and fresh oil is substituted in sufficient amount to make a fairly thick paste or slurry. Usually the fresh oil is a liquid oil of the same kind that is to be hydrogenated.

To obtain a product that is more easily stored, shipped, and handled, manufacturers usually suspend the catalyst in a highly hardened oil that may be conveniently cast into blocks, granulated, or formed into flakes over a chill roll (98). Nickel formate catalysts generally have good activity and desirable characteristics with respect to selectivity and the formation of isooleic acids. Since the catalyst is activated while in contact with oil, the somewhat troublesome operation of transferring the reduced catalyst to oil without access to air is avoided.

The chief disadvantage of wet-reduced catalysts is that they usually contain nickel particles of colloidal or near-colloidal dimensions, which are very difficult to filter from the oil after hydrogenation is completed. The formation of colloidal nickel may be minimized by careful attention to the reduction temperature and time, but even under the best conditions wet-reduced catalysts are relatively difficult to filter cleanly from the oil. Oils hydrogenated with wet-reduced catalysts must usually be treated with bleaching earth or otherwise refined to ensure complete removal of the catalyst, whereas good dry-reduced catalysts can in most cases be removed completely by simple filtration.

7.6 DRY-REDUCED CATALYSTS

Dry-reduced catalysts are prepared by precipitating nickel hydroxide or nickel carbonate on diatomaceous earth or other refractory support, drying and grinding the precipitate, and reducing the resultant powder at a high temperature with a current of hydrogen. The preparation of dry-reduced catalysts differs fundamentally from that of wet-reduced catalysts in that the activity of the former is determined primarily in the step of precipitation whereas the activity of the latter depends principally on the conditions of reduction. Dry-reduced catalysts are, of course, finally activated by reduction, but this latter operation is relatively simple and straightforward in comparison to the highly critical process of precipitation.

Catalysts by Conventional Precipitation.   Since nickel sulfate is the cheapest and most readily available nickel salt, it sometimes serves as the starting material for dry-reduced catalysts. However, nickel catalysts precipitated directly from the sulfate have the property of producing excessive amounts of isooleic acids unless subjected to very prolonged washing for the removal of sulfates; hence nitrate is a preferable material for most purposes. Nickel nitrate may be prepared directly by the action of nitric acid on the metal or by precipitating nickel sulfate solution with an alkali and dissolving the precipitate in nitric acid.

Alkalies used for precipitating the catalyst include sodium hydroxide, ammonia, sodium carbonate, and sodium bicarbonate; the latter is perhaps the most generally suitable. Diatomaceous earth equivalent to one to two times the weight of nickel is suspended in the solution of nickel salt at the beginning of the precipitation to provide a support for the catalyst.

The activity of a catalyst of this type depends to an overwhelming degree on such operating details as the temperature at which precipitation takes place, the rate at which the nickel salt and alkali solutions are mixed, the excess of alkali employed, and the boiling to which the suspended precipitate is subjected. Suitable conditions for the preparation of an active catalyst must be established by trial and error, and once they are established it will be found that even minor deviations may affect the activity of the catalyst greatly. In fact, even the most careful standardization of the manufacturing procedure scarcely suffices to eliminate all variations in the catalyst. The electrolytic method of precipitation, to be described later, is recommended for the production of the most uniform and active catalysts.

The following methods of precipitation yield active catalysts from nickel sulfate and nickel nitrate, respectively.

Method No. 1.   First, 240 lb of $NiSO_46H_2O$ is dissolved in 1100 gal of distilled water, and the solution is brought to boiling. To the solution is added 60 lb of kieselguhr. A fresh solution of 145 lb of $NaHCO_3$ in 300 gal of cold distilled water is made up separately, and this is added uniformly over a period of 1.0–1.5 hours to the $NiSO_4$ solution while the latter is kept continuously and vigorously boiling. There are then added 60 lb of additional kieselguhr, and the boiling is continued for 0.5–1.0 hour longer. At the end of this time the solution should be slightly alkaline to phenolphthalein. The precipitated catalyst is filtered, resuspended in 1000 gal of distilled water, boiled for a short time, and refiltered. This last operation is repeated once or twice, and the catalyst is then filtered, dried, and ground. It should dry to the form of a friable cake, easily disintegrated in the fingers to a fine powder of high specific volume. This catalyst requires about 8 hours of reduction at 900–950°F and contains about 25% nickel in the unreduced form.

Method No. 2. First, $Ni(NO_3)_2$ equivalent to 27 lb of nickel is dissolved in 500 gal of distilled water, and 25 lb of kieselguhr is added. Fifty pounds of 16 Bé sodium hydroxide and 20 lb of aluminum powder are mixed, and after foaming has subsided, 60 lb of sodium bicarbonate is added and the solution is made up to 300 gal with distilled water. The latter solution is then added to the $Ni(NO_3)_2$ solution over a period of 1 hour, while the temperature of the latter is raised from about 150°F to boiling. It is unnecessary for the aluminum to become completely dissolved during this operation. After precipitation is complete, the precipitated catalyst is filtered, resuspended in water, again filtered and resuspended and is then filtered, dried and ground. This catalyst contains about 22% nickel before it is reduced and requires a temperature of about 1000°F for reduction.

Dry Reduction of Catalysts. The apparatus commonly used for the dry reduction of catalysts consists of a horizontal steel drum with conical ends, which may be slowly revolved on its longitudinal axis. The drum is placed inside an insulated housing equipped with gas burners, which serves as a furnace. Hydrogen inlet and outlet pipes are provided to permit a current of hydrogen to be passed over the surface of the catalyst in the drum. After reduction is complete, which is usually a matter of several hours, the apparatus is cooled, the hydrogen flow is shut off, and the hydrogen inside the drum is displaced with a current of carbon dioxide. A tailpipe of large diameter attached to one of the conical ends of the drum is then uncapped, and the drum is tilted so that the reduced catalyst will discharge by gravity below the surface of oil contained in a small tank or drum. The discharge of reduced catalyst must be slow and accompanied by intensive stirring of the oil to prevent unwetted catalyst from floating to the surface of the oil and becoming oxidized and inactivated on contact with the air. The slurry of oil and catalyst resulting from this series of operations will usually contain 10–20% nickel and may, of course, be added directly to charges of the same oil in the hydrogenators.

7.7   Nickel Alloy or Raney Catalysts (99, 100)

A novel catalyst devised by Raney has been used quite extensively, not only for the hydrogenation of fats and oils, but also for various other hydrogenation processes, both in the laboratory and in commercial operation.

In the manufacture of Raney catalyst, nickel is first alloyed with aluminum, and the most common ratio of the two metals is about 1:1 by weight. The alloy is friable and easily reduced to a fairly fine powder on cooling. The catalyst is marketed in the form of such a powder.

The catalyst is prepared for use by the following procedure. In an open digestion tank is placed a considerable excess of strong (ca. 20%) sodium hydroxide solution. To this solution the desired amount of powdered alloy is slowly added, with the rate of addition limited by the extent of foaming. Much heat is generated by the reaction between sodium hydroxide and aluminum, so that heating of the solution in unnecessary during the addition of the powder. Proper disposition must, of course, be made of the hydrogen liberated during the reaction. After the alkali solution and powder have been combined, the contents of the tank are heated to 245–250°F and held at this temperature for 2–3 hours. At the end of this time the catalyst will be in the form of a sludge consisting almost entirely of metallic nickel, with the aluminum in solution in the form of sodium aluminate. The sodium aluminate and excess alkali are then removed by repeated washing with cold water, followed in each case by decantation of the wash water from the settled catalyst sludge. The washing must be very thorough, as even traces of alkali left in the catalyst will form a quantity of soap sufficient to poison it effectively when it is placed in contact with oil.

After washing is completed, the last wash water is drawn off as cleanly as possible, and the residual sludge is covered with oil. Heat is then applied, and vacuum, if it is available, is used to dry the sludge and replace the protective water with oil. The catalyst is highly pyrophoric; hence it would become inactivated if dried without protection from the air. The finished catalyst is in the form of relatively large particles, and it occasions little trouble in filtration. However, kieselguhr is usually added to the oil–nickel sludge after the latter is dried. Drying of the catalyst is preferably carried out in a tank separate from that used for digestion; otherwise, it is very troublesome to remove all traces of the added oil before the next batch of catalyst is digested. Any trace of oil in the digestion tank forms soaps that are highly poisonous to the catalyst.

Raney nickel is somewhat low in selectivity and thus has found limited use for hydrogenation of edible oils. However, it is used to hydrogenate fatty acids.

7.8 PROMOTION OF NICKEL CATALYSTS

A "promoter" is a metal or other substance that enhances the activity of a catalyst without being a catalyst for the reaction in question. The promotion of catalysts is much practiced in fields of catalysis other than the hydrogenation of fats, and the employment of catalysts of two, three, four, or even more components is by no means unusual. In such complex systems the action of some of the components may more properly be called "synergistic" than "promotive," since usually more than one component will possess catalytic activity alone. Promotion in its proper sense is, however, very common.

A number of mechanisms have been proposed to explain the phenomenon of promotion. It has been supposed that the promoter acts as a secondary catalyst, accelerating the formation or decomposition of intermediate compounds; assists in the adsorption of the reactants; or protects the catalyst from poisons. With catalysts for the hydrogenation of fats, however, it appears more reasonable to assume that the function of a promoter is simply structural and that it permits the development of larger numbers of active centers on the catalyst surface.

Promoters play a major role in the manufacture of fat hydrogenation catalysts, since catalysts of satisfactory selectivity can be prepared with their aid.

Metals referred to in the patent literature as useful promoters, either as such or in the form of oxides, are chromium, cobalt, thorium, zirconium, copper (101), titanium (102), and silver.

7.9 METALS OTHER THAN NICKEL AS CATALYSTS

Nickel–silver catalysts have recently been reported (103) to have very high (5–10) linolenic selectivity and low isomerization. Raw materials for these catalysts are water-soluble salts of nickel and silver, such as the acetates or the nitrates. Ammoniacal solutions of the salts are prepared and the carrier such as kieselguhr, alumina, and so on is impregnated with the solutions. The nickel is impregnated first, and then the silver. The nickel/silver ratio may be 100:10–12. The moist material is dried, calcined, and then reduced at 220–290°F in a stream of hydrogen for 2 hours, cooled, and immediately stirred into refined soybean oil under hydrogen. The catalyst is diluted for hydrogenation of soybean oil.

Copper chromite catalyst, because of its high (8–13) linolenic selectivity is used for the hydrogenation of the linolenic acid in soybean oil. One type of catalyst is manufactured by the method described by Lazier and Arnold (104). An aqueous solution of barium nitrate and cupric nitrate trihydrate is stirred during addition of a solution of ammonium chromate, prepared from ammonium dichromate and aqueous ammonia. The reddish brown precipitate of copper, barium, and ammonium chromate is washed, dried, and decomposed by heating in a muffle furnace at 350–450°F. The ignition residue is pulverized, washed with 10% acetic acid, dried, and ground to a fine black powder.

Copper on silica gel (105) or molecular sieves (106) also show high linolenic selectivity. To an aqueous solution of copper nitrate, silica gel or molecular sieve is added, and after the copper has been adsorbed, the solid is removed by filtration, dried, and calcined at 350°F for 2 hours. The catalysts are used for the hydrogenation of soybean oil.

Although palladium has not been used for commercial hydrogenations of fats and oils, it is a powerful catalyst that will operate at a temperature below that used for nickel catalyst and has been recommended for preparation of base stock oils for margarine and shortening (107). Although palladium is expensive, it is some 80–120 times more active than nickel, so only a few parts per million of catalyst need be used, and it may be reused several times (108). Palladium chloride solution is mixed with absorptive carbon and the metal reduced with alkaline formaldehyde. The catalyst is washed and dried. Usually, 5 or 1% palladium on carbon is prepared (109).

7.10   CATALYST TEST

All catalysts, or at least some batches as received by the user, should be tested for activity, isomerization activity, selectivity ratio (*SR*), and filterability.

The activity, the selectivity, and the isomerization ability may be determined by using the catalyst in a standardized small-scale hydrogenation using a "standard" oil and gas. The hydrogenator should have good temperature control as both activity and selectivity are affected by temperature.

One method (110) to test nickel catalysts makes use of a Paar medium-pressure apparatus with one blade stirrer at 600 rpm. A 200-gram batch of refined, bleached, and deodorized soybean oil and 0.04% Ni (from the test catalyst) are heated to 185°F under vacuum and then hydrogen is admitted to 20 psig. Samples are removed periodically until about 80 IV is reached. From the time, the fatty acid composition, and the iodine value of the sample, the activity, expressed as IV drops per minute, and the *SR* and the *trans* unsaturation at a specific IV may be calculated. These values are compared to previous tests of the same type of catalyst.

The filtration test to determine the presence of colloidal nickel in a catalyst consists of mixing a small sample of catalyst with carbon tetrachloride, followed by filtration of the sample and determination of the transmittance of the filtrate by a spectrophometer. The carbon tetrachloride suspension of catalyst is filtered through a 0.65 -μ filter, enabling particles of less than 0.65 μ to pass through and be determined as "colloidal" nickel.

If catalysts are purchased on the specifications of activity, selectivity, and isomerization ability and filterability, consistent catalysts and thus more consistent results of hydrogenation will be achieved.

The AOCS has a standard method (Ca 17-76) that describes the determination of hydrogenation catalytic activity. Also, this society is working on a standard method for determination of the selectivity of catalysts.

Hydrogen Production and Purification

8.1   MEASUREMENT AND PROPERTIES OF HYDROGEN

It is customary to measure commercial hydrogen in terms of standard cubic feet (scf), defined as cubic feet of dry hydrogen at 60°F and normal atmospheric pressure of 760 mm Hg or 14.70 psia. A standard cubic foot differs from a cubic foot measured at the standard laboratory temperature and pressure of 0°C and 760 mm, with 1 ft^3 at standard temperature and pressure (stp) equivalent to 1.057 scf. One scf of hydrogen weighs 0.00532 lb.

Hydrogen is highly flammable and readily forms explosive mixtures with oxygen or air. Mixtures of hydrogen and air are flammable when their hydrogen content by volume is greater than 4.00 but less than 74.2%. However, because of its extreme lightness, hydrogen admitted into the air rises rapidly and disperses, and explosive hazard from ordinary minor leakage within any adequately ventilated structure is not great.

8.2   ELECTROLYTIC PRODUCTION OF HYDROGEN

One of the oldest commercial processes for the manufacture of hydrogen and one still used to some extent is the electrolytic process, whercin an aqueous electrolyte is decomposed by means of a direct current to yield gaseous hydrogen and oxygen.

The essential parts of an electrolytic hydrogen plant are a motor–generator combination, rectifiers or other source of direct current, a number of electrolytic cells for the actual generation of hydrogen, a hydrogen holder for receiving hydrogen from the cells, a hydrogen compressor, and high-pressure storage tanks for receiving the compressed gas. Normally, oxygen is a byproduct of the process, and equipment must also be provided for receiving, compressing, and bottling the oxygen. A considerable supply of distilled water is required for replenishment of water consumed.

Two types of cell are used. The so-called unipolar cell consists essentially of alternate metal cathodes and anodes, which usually present not more than about 20 ft^2 of surface each, suspended close together in a rectangular tank containing the electrolyte. For the latter, a 25–30% solution of sodium or potassium hydroxide is used. Usually each cell contains only a few electrodes (in some cases only three), and hence the individual cells are relatively deep and narrow and limited in production capacity to 20–200 ft^3/hour each at the customary current density of about 75 amperes/ft^2. The desired capacity of the complete plant is obtained by setting up the proper number of unit cells; several hundred are employed in some of the larger plants. The electrodes within each cell are usually connected in parallel; the different cells are connected in a sufficient number of series to give proper operating characteristics at the voltage supplied by the rectifiers or the generator.

Diaphragms of asbestos cloth or other liquid-permeable material placed between the electrodes divert the evolved hydrogen and oxygen separately without intermixing to superimposed collecting chambers. The temperature of the electrolyte is allowed to rise to a maximum of about 170°F; if the design of the cell is such as to permit a greater rise, the cells are cooled by external circulation of water. The cells may be of either the open or the closed type, but in each the gas is collected at substantially atmospheric pressure.

Bipolar cells are sometimes referred to as cells of the *filter press* type since they resemble a plate and frame press in appearance and construction. In these cells each electrode, corresponding to the plate of a filter press, does double duty, serving on one side as the anode for one cell and on the other as the cathode for the adjoining cell. Passage of the current is from electrode to electrode in series from one end of the assembly to the other, thus eliminating electrical connections within each cell. This arrangement permits closer spacing of the electrodes and higher current densities than can be used in unipolar cells and hence a generally more compact apparatus. There are accompanying disadvantages, however, in their greater operation and maintenance difficulties. Cells of this type have not been popular in the United States. The two types of cell are similar with respect to efficiency of hydrogen production in terms of current utilization.

Electrolytic hydrogen from the better plants usually has a purity of at least 99.8%, where the impurities consist principally of oxygen. As such hydrogen is quite devoid of catalyst poisons, it requires no purification before use.

The design and to some extent the operation of an electrolytic hydrogen plant generally involves a compromise between fixed charges determined by the investment in the plant and operating charges for electric current. Considerable latitude is allowable in the operating conditions; the capacity of a plant of given size and cost may be increased by increasing the amperage applied to each cell. If the capacity is increased in this manner, however, there is at the same time a decrease in the efficiency of the plant in terms of cubic feet of hydrogen produced per kilowatt hour of current consumed.

Since the passage of a definite quantity of current will always produce the same amount of hydrogen regardless of the type of cell, the efficiency of the latter is to be measured in terms of the voltage required for its operation. Current amounting to 1 ampere hour liberates 0.01482 ft^3 of hydrogen measured at 0°C and 760 mm of pressure; this is equivalent to 0.0157 scf. Therefore, the cubic feet of hydrogen produced per kilowatt hour of direct current is equivalent to 15.7 divided by the voltage. In the more efficient cells, the voltage required is about 2.0–2.3 volts, so that the production per kilowatt hour is about 6.8–7.8 ft^3. Stated in another way, about 128–148 kWh of current is required for each 1000 ft^3 of hydrogen produced. There is some loss (usually about 10%) in converting from alternating current to direct current; hence the overall power consumption is in the neighborhood of 145–165 kWh per 1000 ft^3.

According to Taylor (111), substantially greater economy in the way of current consumption is hardly to be expected, regardless of possible future improvements in cell design. The theoretical decomposition potential of water, in the absence of overvoltage, is 1.23 volts, and with any form of technically feasible electrodes the overvoltage increases this potential to not less than 1.5 volts. The necessity of providing a further potential to overcome the resistance of electrolyte and diaphragms further increases the minimum practical voltage to about 2 volts.

In the past the electrolytic process has been favored by many hydrogenators of fatty oils because of the high degree of purity of the hydrogen. However, this consideration is now of much less importance with the advent of efficient purification processes for hydrogen made by other methods. The capital investment required for an electrolytic plant is relatively high, and much floor space is required. Further, except where very cheap power is available or the byproduct oxygen is unusually valuable, the operating costs are usually higher than for other methods. It possesses the marked advantage, however, of being equally efficient in large and small installations and of lending itself to ready expansion in production capacity. Where a capacity of less than about 2500 ft^3/hour is required or where cheap hydroelectric power is available, it still may be the preferred process.

8.3   STEAM–IRON PROCESS

A considerable part of the total hydrogen produced for fat hardening is made by the steam–iron process. In principle, this process is simple, involving merely the alternate oxidation and reduction of a hot iron ore mass. Reduction is accomplished by means of the hydrogen and carbon monoxide in ordinary water gas, made by blowing hot coke with steam, or reformer gas made by reacting steam with natural gas or other light hydrocarbon. In the oxidation cycle the ore mass is blown with superheated steam, and the oxygen of the latter combines with the previously reduced ore and the hydrogen is set free. The exact changes in the oxygen content of the iron mass in practical operation have not been fully established but are commonly supposed to be as follows:

Reduction:

$$Fe_3O_4 + H_2 \rightarrow 3FeO + H_2O \qquad (1)$$
$$3FeO + 3H_2 \rightarrow 3Fe + 3H_2O \qquad (2)$$
$$Fe_3O_4 + CO \rightarrow 3FeO + CO_2 \qquad (3)$$
$$3FeO + 3CO \rightarrow 3Fe + 3CO_2 \qquad (4)$$

Steaming:

$$3Fe + 3H_2O \rightarrow 3\,FeO + 3H_2 \qquad (5)$$
$$3FeO + H_2O \rightarrow Fe_3O_4 + H_2 \qquad (6)$$

There is also some interaction of carbon monoxide and steam during the reduction cycle to produce carbon dioxide and hydrogen:

$$CO + H_2O \rightarrow CO_2 + H_2 \qquad (7)$$

Thus carbon monoxide may function either directly as a reducing agent or indirectly through the production of hydrogen by reaction 7.

In addition to reaction 7, there are side reactions that do not contribute to hydrogen production but are mentioned because they affect the purity of the hydrogen. At the high temperatures prevailing in the steam–iron generator, reduced iron catalyzes the reaction

$$2CO \rightarrow CO_2 + C \tag{8}$$

Because of this reaction there is always a slight deposit of carbon on the ore mass at the end of the reduction period. During the subsequent period of steaming any carbon thus deposited reacts with the steam to produce both carbon dioxide and carbon monoxide as impurities in the hydrogen:

$$C + H_2O \rightarrow CO + H_2 \tag{9}$$
$$C + 2H_2O \rightarrow CO_2 + 2H_2 \tag{10}$$

Water gas ordinarily contains about 0.2% of hydrogen sulfide or other gaseous sulfur compounds. During the reducing phase these combine with the iron contact mass to form iron sulfide. The iron sulfide is in turn reactive with steam, and thus a considerable part of the sulfur eventually finds its way into the hydrogen in the form of hydrogen sulfide and traces of organic sulfur compounds.

In practice, carbon and sulfur will be deposited on the contact mass more rapidly than they are removed. Hence it is customary to follow each steaming period with a short period of aeration, during which air is blown through the ore to burn off carbon and sulfur accumulations. Failure to remove these accumulations results in virtually complete inactivation of the contact mass (112).

An additional source of impurities in steam–iron hydrogen is the reducing gas itself. After each reducing period, the generator is purged with steam for a few seconds before collection of hydrogen is begun, but this is insufficient to eliminate all water gas from the system. Prolonged purging is, of course, wasteful of hydrogen.

The composition of unpurified steam–iron hydrogen varies somewhat according to the type of generator and its mode of operation but is ordinarily approximately as shown in Table 1.9. Most of the impurities are removed in subsequent operations.

Table 1.9   Composition of unpurified steam–iron hydrogen

Carbon dioxide	0.5–1.0%	Hydrogen sulfide	0.05–0.15%
Carbon monoxide	0.2–0.5%	Organic sulfur	0.1–0.5 grain per 100 ft^3
Oxygen	0.0–0.1%		
Nitrogen	0.3–1.0%	Hydrogen	98.0–99%

A hydrogen generator very commonly used in the United States is of the Bosch or Bamag type. This generator, which is shown in Figure 1.27, consists of a single tall cylinder, lined with firebrick and insulating brick. The space within the generator is divided into two superimposed sections. In the lower section is a grate supporting the iron ore contact mass; the upper section is filled with a checkerwork of firebrick that serves as a steam superheater. Connections are provided at the extreme top of the generator for air and steam and between the contact mass and the superheater for air. In a large line leading from the bottom of the generator is a cross, with quick-opening valves from the latter leading to a water–gas inlet line, a hydrogen outlet line, and a purge line. All valves, including those mentioned previously and the stack valve, are interlocked and operated from an elevated platform near the top of the generator. In the larger modern plants the valves are automatically opened and closed by a timing device.

The generator is operated at about 1500°F and as continuously as is feasible, since considerable time is required to warm up a cold unit.

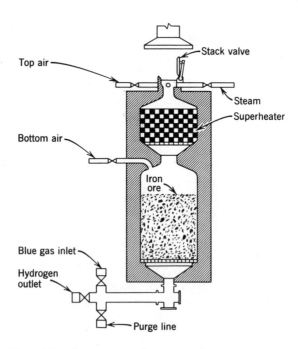

Figure 1.27   Steam–iron hydrogen generator, Bamag type.

Operating Cycle.   The complete operating cycle comprises of the steps described in the following paragraph.

REDUCING.   Reducing gas is drawn from a holder and delivered to the bottom of the generator by a centrifugal blower designed to produce a pressure of 20–30 in of water. It passes up through the hot contact mass, reducing the latter, and at the bottom of the superheater meets a current of bottom air. Combustion of the partially spent gas with this air in the superheater stores up heat in the latter for the subsequent steaming period.

PURGING.   The reducing gas is shut off and hydrogen production is begun by passing steam through the superheater and then down through the contact mass. The first portions of gas are discharged through the purge line to the atmosphere to sweep out of the system the reducing gas remaining there at the end of the reducing period.

STEAMING.   The purge valve is closed, the hydrogen outlet valve is simultaneously opened, and steaming is continued. The hydrogen produced is collected.

AIR BLOWING.   Top air is blown through the generator from top to bottom to burn off accumulations of carbon and sulfur on the contact mass.

REDUCING.   Reducing gas is again delivered to the generator, and so on.

Steam–Iron Hydrogen Production: General.   The times allowed for the different portions of the cycle are determined by considerations of economy. There is, of course, a steady diminution in the rate of hydrogen production as steaming is continued and a similar diminution in the rate of iron oxide reduction during the reducing period. The amount of hydrogen produced during the average period of steaming is governed by the amount of reduction accomplished in the corresponding reduction period. The reduction and steaming periods are thus approximately equal in length and may together be in the neighborhood of 10–15 min. Both purging and burning-off periods are much shorter (e.g., 10–20 sec each).

The largest single item of expense in making steam–iron hydrogen with water gas is coke for the manufacture of the reducing gas. Water gas manufacture is a fairly well standardized process, and the coke requirements of different plants do not vary greatly in terms of the volume of water gas produced. However, the efficiency of hydrogen generators widely varies with generator design and mode of operation. In general, practice in steam–iron hydrogenation generation has improved greatly within the past 25 years. Formerly, a consumption of reducing gas amounting to 2.5–3.0 times the volume of hydrogen produced was considered good practice. In modern plants water gas/hydrogen ratios as low as 1.7–1 or even 1.5–1 are obtained. The latter ratio corresponds to a coke consumption of about 45–50 lb per 1000 ft^3 of hydrogen. More recently, reducing gas made by reforming natural gas with steam has been substituted for water gas in a number of steam–iron plants, with a considerable gain in operating economy.

Since reactions 1 to 8 are reversible, with an equilibrium point depending on the temperature of reaction as well as the concentration of reactants, all the factors mentioned previously will be affected by the temperature at which the generator is operated, the composition of the reducing gas, and the reduction–steaming cycle adopted. A detailed discussion of the equilibria involved in the manufacture of steam–iron hydrogen is furnished by Taylor (111), who particularly points out the importance of working principally on the Fe_3O_4–FeO cycle (reactions 1, 3, and 6) rather than the FeO–Fe cycle (reactions 2, 4, and 5), and at a relatively high temperature, to obtain maximum economy in the consumption of blue gas. Actually, the reactions occurring in a steam–iron hydrogen generator are undoubtedly more complex than as represented by Taylor or the preceding equations because of the tendency of metallic iron and the various iron oxides to form a continuous series of solid solutions (113) comprising materials containing both less oxygen than FeO and more oxygen than Fe_3O_4. In practice, the utilization of both steam and blue gas is more efficient than can be predicted from Taylor's data.

Although the overall reaction is theoretically exothermic, radiation losses from the generator and the necessity for heating excess quantities of steam and reducing gas require that a certain proportion of the reducing gas be burned to maintain the temperature of the generator. Radiation losses are greatly reduced in single-unit generators, however, and particularly in those that are large in comparison with the older multiretort units.

The type of iron ore employed has an important bearing on the efficiency of steam–iron generators. Ore from only a few localities has the requisite properties of high reactivity and sufficient ruggedness to withstand prolonged use without disintegration.

The purity of steam–iron hydrogen is much less a problem now than formerly since methods are available for efficiently removing all the impurities mentioned previously, with the exception of nitrogen. In the better-operated plants, hydrogen is now quite commonly produced with a purity in excess of 99.5%; the impurities consist of nitrogen and methane, both of which are inert toward hydrogenation catalysts. The individual purification processes are each discussed in later sections.

In addition to the hydrogen generator proper, equipment required for a complete steam–iron plant includes a reducing gas generator, gas holders for reducing both gas and hydrogen, a water scrubber and a cooler for the crude hydrogen obtained from the generator, air and blue gas blowers, a hydrogen compressor (if the gas is to be stored under high pressure), hydrogen storage tanks or a holder, and equipment for purifying the hydrogen.

Efficient operation of a steam–iron hydrogen plant requires metering and accounting of steam, coke, reducing gas, and hydrogen; frequent analysis of both reducing gas and hydrogen; and constant and expert attention to all operating details. Checking of the hydrogen purity is much facilitated by the ready adaptability of continuous indicators or recorders of the thermal conductivity type.

8.4 STEAM–HYDROCARBON PROCESS

The steam–hydrocarbon or hydrocarbon reforming process for hydrogen manufacture has been developed to a high degree of perfection in the United States and has been adopted by a number of oil and fat processors because of its economy and simplicity of operation, its flexibility, and the high degree of purity of the hydrogen (Figure 1.28). The latter is limited practically only by the nitrogen content of the hydrocarbon used as a raw material. The following analysis (99), recorded for the product of a plant using liquid propane, illustrates the degree of purity obtainable under favorable conditions:

Carbon dioxide	0.001%
Carbon monoxide	0.001%
Oxygen	0.005%
Methane	0.018%
Nitrogen	0.007%
Hydrogen	99.968%

Either commercial propane or natural gas of low nitrogen content usually serves as the raw process material, although other light hydrocarbons may be used. The hydrocarbons must be free of sulfur compounds. If the latter consist only of mercaptans, as is usually the case in natural gas or propane derived from natural gas, they may be removed by scrubbing

Figure 1.28   Steam–hydrocarbon hydrogen plant. (Courtesy of the Girdler Corporation.)

with caustic soda solution. If carbon oxysulfide or certain other organic sulfur compounds common to refinery propane are present, they must be converted to hydrogen sulfide before scrubbing is carried out by passing the hydrocarbon vapors over a catalyst consisting of bauxite or other metallic oxide at about 700°F.

Metered amounts of steam and the desulfurized hydrocarbon vapor are mixed and passed through furnace tubes containing a catalyst maintained at about 1500°F. A furnace of the type described by Shapleigh (114) is employed. It consists of one or more long alloy steel tubes supported vertically within refractory walls and heated by burners supplied with natural or manufactured gas, fuel oil, or the same stock of propane as used for the feed. Flow of the mixture is downward, countercurrent to the hot flue gases. Space velocities of 600 volumes of hydrocarbon gas per hour per apparent unit volume of catalyst are attained.

Reaction takes place according to the following typical equation, which applies to propane:

$$C_3H_8 + 3H_2O \rightleftharpoons 3\,CO + 7H_2$$

A portion of the carbon monoxide formed undergoes further reaction (the so-called water gas shift reaction):

$$CO + H_2O \rightleftharpoons CO_2 + H_2$$

The mixture of hydrogen, carbon monoxide, and carbon dioxide issuing from the reforming furnace is mixed with steam to cool it to about 700°F, and the mixture is passed through a bed of another catalyst. Here the water gas shift reaction mentioned in the preceding paragraph converts 90–95% of the carbon monoxide to carbon dioxide and hydrogen. The exit gases are cooled to about 100°F, after which the carbon dioxide is removed by scrubbing with monoethanolamine solution as described in the succeeding section on hydrogen purification. The scrubbed gas contains about 1% of carbon monoxide, together with negligible amounts of methane, nitrogen, and so on. Two additional stages of carbon monoxide conversion, identical with that described previously, reduce the carbon monoxide content to about 0.001%. As an alternative to the third stage of conversion, the last small proportion of carbon monoxide may be eliminated by methanation.

The water gas shift reaction, unlike the reforming reaction, is exothermic; hence there is some increase in temperature of the gases in the first carbon monoxide convertor. This is desirable, however, as the capacity of the unit is increased at some slight expense to the completeness of conversion. Heat exchangers may be installed to permit heat transfer to the cold inlet gases from the hot exit gases of each convertor following the first.

The following utilities and material requirements are given (115) for the hydrocarbon reforming process per 1000 scf of hydrogen produced:

Propane, process	2.75 gal
or natural gas	250 ft³
Fuel equivalent	350,000 Btu
Steam	380 lb
Cooling water	1800 gal
Electric power	2 kWh
Chemicals, catalyst	2 cents

With each 100 ft³ of hydrogen, 300 ft³ (35 lb) of pure carbon dioxide is produced as a byproduct. These figures apply to the manufacture of hydrogen of high purity, without the use of gas–gas heat exchangers.

In addition to the features enumerated in the preceding paragraphs, a considerable advantage of the steam–hydrocarbon process is its great flexibility. If, for any reason, the hydrogenating plant becomes unable to take the full capacity of the hydrogen generating plant, the output of the latter can be readily reduced by as much as 75% without significant decrease of operating economy.

Steam–hydrocarbon plants may be built with any desired capacity, but they are generally considered most economical in sizes of about 5000 ft^3/ hour or greater.

8.5   HYDROGEN PRODUCTION WITH SELECTIVE ADSORPTION PURIFICATION

Since hydrogen gas is lost during hydrogenations by venting to remove gaseous impurities from the headspace, the demand for very pure hydrogen has increased. Therefore, the new plants for conversion of hydrocarbons to hydrogen have added purification by adsorption (116).

The process description can be followed by referring to the flow diagram shown in Figure 1.29. The process steps for the production of hydrogen may be summarized as follows:

1. Feedstock hydrogenation.
2. Feedstock desulfurization.
3. Steam–hydrocarbon reforming.
4. Carbon monoxide shift conversion.
5. Gas cooling.
6. Hydrogen purification by fixed-bed adsorption.
7. Off-gas recovery as fuel.
8. Steam generation.

Feedstock Hydrogenation.   The hydrogen plant can be designed for any mixture of hydrocarbon feedstocks ranging from natural gas to naphtha petroleum fractions.

The feedstock usually contains sulfur compounds that are a poison to both reforming and shift conversion catalysts and thus must be removed prior to reforming. A typical first step in removing the sulfur compounds is to convert them to hydrogen sulfide. This is accomplished by hydrotreating.

A small volume of recycle hydrogen is mixed with the hydrocarbon. The mixture is heated to about 750°F by heat exchange with the reformer effluent in the preheat exchanger. It is then sent to the hydrotreater, where sulfur compounds are converted to hydrogen sulfide over a bed of cobalt–molybdenum catalyst. The hydrotreated feedstock then passes to the desulfurizers.

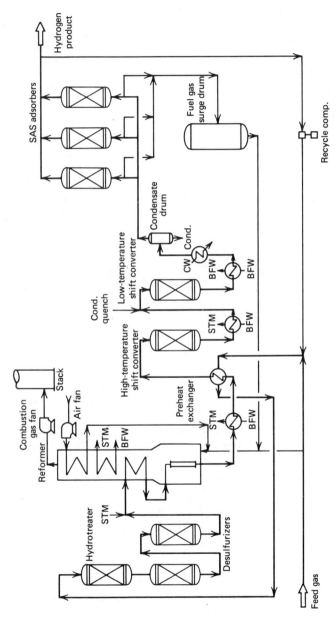

Figure 1.29 Hydrogen plant with selective adsorption purification (SAS). For naphtha feed, the preheat exchanger is replaced with a fired heater. The plant includes a deaerator, BFW pumps, and a steam drum.

81

Feedstock Desulfurization.   In the desulfurizers the sulfur is removed to less than 0.2 ppm by adsorption on zinc oxide catalyst. Desulfurization to this low level is required because even small traces of sulfur in the feedstock to the reformer will partially deactivate the reformer catalyst and will completely poison the low-temperature shift catalyst.

The desulfurizers consist of two beds. Each bed is in a separate vessel and normal flow is in series through the beds. This series arrangement reduces the total zinc oxide requirement since the front bed can be run to a higher degree of sulfur retention. The front bed becomes loaded long before sulfur breakthrough occurs in the trail bed. The piping–valving is arranged so that the front bed, when loaded, can be replaced while the trail bed remains onstream in "parallel flow." The trail bed then becomes the front bed and the fresh bed is switched to the trail bed. Under normal operating conditions, each bed is designed for a minimum life of 6 months before replacement.

Steam–Hydrocarbon Reforming.   The desulfurized feedstock is mixed with steam and heated to 950°F in the feed preheat coil located in the convection section of the reformer. A steam/carbon ratio of 3:0 is typically optimum for this type of plant. The mixture is then passed through catalyst–filled tubes in the reformer furnace. In the presence of nickel catalyst the feedstock reacts with steam to produce hydrogen, carbon oxides, and methane by the following processes:

$$C_nH_m \ (g) \ + \ H_2O \ (g) \ + \ heat \ \rightarrow \ nCO \ (g) \ + \ (m/2 \ + \ n) \ H_2 \ (g) \qquad (1)$$

$$CO \ (g) \ + \ H_2O \ (g) \ \rightarrow \ CO_2 \ (g) \ + \ H_2 \ (g) \ + \ heat \qquad (2)$$

Reaction 1 is the reforming reaction and reaction 2, the shift reaction. Both reactions produce hydrogen. Both reactions are limited by chemical equilibrium. The net reaction is endothermic, with the heat supplied by combustion in the radiant zone of the furnace. Typical outlet conditions of the reformer are 1550°F and 250 psig. The reforming reaction takes place under carefully controlled external firing conditions. The combustion gases leaving the reformer are used to preheat feed gas, to generate and superheat steam, and to preheat combustion air.

Carbon Monoxide Shift Conversion.   The process gas stream exiting the reformer is cooled to about 700°F by heat transfer in both the No. 1 process steam generator and in the preheat exchanger.

The process gas leaving the feed gas preheater enters the high-temperature shift converter. Shift conversion produces additional hydrogen by the following reaction:

$$CO\ (g)\ +\ H_2O\ (g) \rightarrow CO_2\ (g)\ +\ H_2\ (g)\ +\ heat$$

The reaction is limited by equilibrium, and the equilibrium is favored by low temperature.

The high-temperature shift converter contains an iron–chromium catalyst that converts approximately 60% of the incoming carbon monoxide by the shift reaction. The temperature of the gas leaving the high-temperature shift converter is approximately 830°F.

The process gas is then cooled to about 440°F by heat transfer in the No. 2 process steam generator and by condensate quench. The gas enters the low-temperature shift converter, where the total carbon monoxide conversion is further increased to almost 93% by a copper–zinc catalyst. The inlet temperature to this converter is controlled by varying the condensate quench flow rate. As in the first stage of shift conversion, there is a temperature rise across the catalyst bed. The exit temperature of the process gas is approximately 500°F.

Gas Cooling.   The shift effluent gas is cooled by heat exchange with deaerated boiler feedwater in the boiler feedwater heater and with deaerator feedwater in the deaerator feedwater heater. The gas is further cooled to 100°F by heat exchange with cooling water in the process gas cooler. Condensate is collected in the process condensate separator and is recirculated as quench water and deaerator feed. The process gas then flows to the hydrogen purification system.

Hydrogen Purification.   High-purity hydrogen is produced by selective adsorption of impurities. The unit operates on an adsorption-regeneration cycle to produce high-purity hydrogen.

ADSORPTION.   Each adsorber contains a fixed bed of activated alumina, activated carbon, and special molecular sieve. During the adsorption step the process gas flows through one of the adsorbers, *A, B,* or *C,* which selectively adsorbs the impurities from the hydrogen.

The adsorption step continues until the adsorbents are loaded with impurities. Then the process gas flow is switched to another adsorber and the spent bed is regenerated. Pure hydrogen product is delivered at 200 psig and 100°F.

REGENERATION.   During the regeneration step the impurities are removed
from the adsorbents. Regeneration is carried out as follows:

a. The adsorber is depressurized to eliminate a fraction of the impurities.
    The gas goes to the fuel gas surge drum for use as fuel.
b. The adsorbents are purged with hydrogen to eliminate the residual
    impurities.
c. The adsorber is repressurized to adsorption pressure with pure hy-
    drogen.

Offgas Recovery.   The offgas from the adsorber regeneration flows to
the fuel gas surge drum, from which it is sent on flow control to the
reformer burners to provide a major portion of the fuel requirements.

Steam Generation.   Steam generation satisfies the process steam requirements and produces a substantial amount of export steam (Table
1.10).

The steam system consists of a steam generation coil located in the
reformer convection section, two process steam generators, a steam
drum, a deaerator, a deaerator feedwater heater, a boiler feedwater
heater, and boiler feedwater pumps.

8.6   LIQUID HYDROGEN

With the advent of the space program, liquid hydrogen became available
for general use. Since liquid hydrogen is only about 1/850 the volume of
hydrogen gas, the transport and storage space requirements are much
less.

To liquefy hydrogen, the crude gas is purified to over 99.9% hydrogen

Table 1.10   Utility requirements

Item	Units/m scf H_2
Hydrocarbon feed	0.37 mm Btu LHV
Hydrocarbon fuel	0.10 mm Btu LHV
Electric power	0.50 kWh
Cooling water circulation (20°F rise)	100 gal
Treated boiler water	130 lb
Export steam	80 lb
Catalyst and adsorbents	4 cents

and liquefied by the same general principles used for air liquefication. One major modification is the use of liquid nitrogen to precool the hydrogen. Also, a catalyst is used to convert the normal 25% *para*-hydrogen, 75% *ortho*-hydrogen to 100% *para*-hydrogen. Since the equilibrium concentration of liquid hydrogen at its boiling point (20°K) is almost pure *para*-hydrogen, freshly made liquid normal hydrogen converts slowly to *para*-hydrogen. The heat released by this exothermic conversion causes vaporization of a large portion of the liquid hydrogen. Thus the gas is converted to *para*-hydrogen in the liquefying machine (117).

Liquid hydrogen is transparent and odorless and has a density of 4.4 lb/ft^3, which is about one-fourteenth that of water (118). It is not corrosive or reactive. The boiling point is -253°C, and at the critical point liquid hydrogen has a pressure of 118 psig.

Liquid hydrogen may be stored, preferably outside, in tanks constructed of 304 stainless steel inner containers inside a carbon steel outer casing (119). Space between the container and the casing is filled with powder-type insulation and then evacuated and sealed at the factory. Hydrogen gas at 250 psig is obtained from the tank.

Liquid hydrogen may be delivered by tank truck and thus is a convenient source of very pure hydrogen for the hydrogenation of fats and oils.

9.1   HYDROGENATION OF INEDIBLE FATS AND FATTY ACIDS

Fatty acids and inedible fats high in free fatty acids, when hydrogenated, are usually saturated as nearly as possible, although there is some partial hydrogenation of soft greases and marine oils for soapmaking. The hydrogenation of fatty acids is quite similar to the hydrogenation of vegetable oils to produce hard oils, except that temperatures above about 350°F and sometimes above 300°F are avoided, and high pressures are depended on to accelerate the reaction. The converters, which must be built of type 316 stainless steel or other alloy resistant to corrosion, are usually designed for a working pressure of 200–300 lb. Hydrogenation is relatively difficult; 0.2–0.5% of nickel is required, and the catalyst is largely inactivated with one usage.

The foregoing has dealt only with the hydrogenation of ethylenic linkages in fatty acid chains to produce more highly saturated acids, esters, or glycerides, and hydrogen for the reaction is supplied in the molecular, gaseous form. In addition to this common hydrogenation process, a number of other catalytic hydrogenation reactions are applicable to fats or their derivatives, and some of them are of considerable technical importance.

9.2   HYDROGENATION FOR PRODUCTION OF FATTY ALCOHOLS

Alcohols of the higher fatty acids, such as lauryl alcohol, myristyl alcohol, and palmityl alcohol, are in considerable demand as intermediates for the manufacture of detergents and wetting agents of the sodium alkyl sulfate type. These are prepared on a large scale by the catalytic hydrogenation of the corresponding acids, esters, or glycerides according to one of the following reactions:

In the case of the acids:

$$RCOOH + 2H_2 \rightarrow RCH_2OH + H_2O$$

In the case of the esters or glycerides:

$$RCOOR' + 2H_2 \rightarrow RCH_2OH + R'OH$$

Priority in the development of catalytic hydrogenation for the production of fatty alcohols is not clear. Apparently, processes were developed at about the same time in Germany (120–122) and in the United States (123, 124). The basic U.S. patents on the process were granted to Lazier and cover the hydrogenation of the fatty acids; their esters with methyl, ethyl, or other monohydric alcohols; and their glycerides. An important feature of the Lazier patents is the use of catalysts of the copper chromite type first disclosed by Adkins and Connor (125) and patented by Lazier (126). Ordinary nickel catalysts are reported to be effective in the reduction of carboxyl groups at the high temperatures and pressures employed in the process, but the resulting products consist largely of hydrocarbons rather than alcohols. The fatty acid salts of a number of metals, including copper, zinc, lead, manganese, cobalt, and mercury, are claimed as catalysts in the patent issued to Schrauth and Bottler (127).

Although fatty acids, esters, or glycerides may be used as the starting materials for the production of fatty alcohols, the glycerides are generally preferred because of convenience and cost. The principal disadvantage in the use of glycerides is that the glycerol produced as a byproduct of the process is not completely stable under the conditions of the reaction and hence cannot be recovered quantitatively. If esters of monohydric alcohols are hydrogenated, the alcohols may be largely recovered. The problem of alcohol recovery is avoided if fatty acids are employed as a starting material, but free acids are corrosive to the hydrogenation equipment and may enter into side reactions.

It has been observed by Guyer et al. (128) that fatty acids of low molecular weight are much more difficult to hydrogenate to alcohols than are the higher-molecular-weight acids. Thus, for example, in a series of comparable laboratory experiments covering the range of saturated acids from C_4 to C_{12}, lauric acid produced a 94% yield of alcohol with almost no ester or unchanged acid, whereas butyric acid produced but 10.9% alcohol, with 53.1% ester.

The hydrogenation is carried out by the batch method, at a temperature between 280 and 360°C (536 and 680°F) and at a pressure in the neighborhood of 3000 psi. The catalyst, consisting of copper chromite or copper chromite promoted with cadmium, cerium, and so on, is used in the powder form, and agitation of the catalyst and reacting material is accomplished by circulation of the hydrogen.

Relatively high yields of alcohols are obtained; thus, for example, Adkins and Folkers (123) obtained yields of alcohols from the esters as high as 97.5–98.5%.

Ordinarily the reaction is not selective with respect to the reduction of carboxyl groups; that is, unsaturated acids are hydrogenated to the corresponding saturated acids at the same time that hydrogenation of the carboxyl groups is accomplished. Under certain conditions, however, it may be reasonably selective. For example, in the laboratory Sauer and Adkins (128b), using a zinc chromite catalyst and a very high catalyst/fatty material ratio, were able to obtain a 68% yield of unsaturated alcohols from oleic acid esters. More recently, Pantulu and Achaya (128b) hydrogenolyzed oleic and ricinoleic acids at 220–250°C and 3500–4000 psi with copper–cadmium salt catalysts and lost only about 10% of the unsaturation.

Richter et al. (129) showed in the mixed copper–cadmium catalyst that the cadmium stabilized the copper sol and also was an intermediate in the hydrogenation of oleic acid to the unsaturated alcohol.

The use of alloy steel reactors is necessary in the hydrogenation of fatty materials to alcohols to avoid corrosion of the equipment and metallic contamination of the product.

A somewhat different method of operation is employed in the so-called lead soap process, which is covered by a series of patents issued to Richardson and Taylor (130). In this process the fatty acids are reacted with litharge to form lead soaps, and the latter are then treated with gaseous hydrogen, without the use of an additional catalyst, at a pressure of about 4000 psi and at a temperature in excess of 300°C (e.g., 340°C). Reaction occurs to produce alcohols in both the free and esterified forms, according to the equations

$$(RCOO)_2Pb + 5H_2 \rightarrow 2RCH_2OH + 2H_2O + Pb$$

and

$$(RCOO)_2Pb + 3H_2 \rightarrow RCOOCH_2R + 2H_2O + Pb$$

The lead is recovered in molten, metallic form. In a typical example given in one of the patents the reaction mixture from the treatment of lead oleate consisted of less than 1% oleic acid, about 21% oleyl oleate, and about 77% free oleyl alcohol. The product had an iodine value of 81, indicating that about 85% of the double bonds of the oleic acid had escaped saturation.

The lead soap process appears particularly desirable for the preparation of unsaturated fatty alcohols. Various other metals, notably cadmium, may be employed instead of lead as intermediates for forming the lead soaps, although they are less satisfactory. The process is adaptable to continuous operation.

9.3   FATTY ALCOHOLS BY SODIUM REDUCTION

The classical Bouveault–Blanc method for producing fatty alcohols from the corresponding fatty esters by sodium reduction may be considered a variation of the hydrogenation process wherein hydrogen for reduction of the carbonyl group is derived from an alcohol through the action of sodium. The product of reduction consists of a mixture of sodium alkoxides that is hydrolyzed to obtain the free fatty alcohols, the glycerol or other alcohol with which the fatty acids were originally esterified, and the regenerated reducing alcohol.

The reaction is complex (131), but the overall result may be represented by the following typical equations, applying to interaction between triglycerides and ethyl alcohol:

$$C_3H_5(OOCR)_3 + 12Na + 6C_2H_5OH \rightarrow 3RCH_2ONa$$
$$+ C_3H_5(ONa)_3 + 6C_2H_5ONa$$

and

$$RCH_2ONa + H_2O \rightarrow RCH_2OH + NaOH$$
$$C_3H_5(ONa)_3 + 3H_2O \rightarrow C_3H_5(OH)_3 + 3NaOH$$
$$C_2H_5ONa + H_2O \rightarrow C_2H_5OH + NaOH$$

In commercial practice one of the higher alcohols, for example, methylcyclohexanol or methylamyl alcohol, is used as the reducing alcohol rather than ethyl alcohol. They may be safely used at a higher temperature than ethyl alcohol, they are less likely to enter into side reactions and are easier to recover in an anhydrous form for reuse. A detailed description and discussion of an improved method employing theoretical quantities of reactants and an inert solvent (toluene and xylene) to increase the fluidity of the reaction mass has been published by Hansley (131). It is carried out at atmospheric pressure and a temperature of about 140°C.

Sodium reduction has been and is still used on a commercial scale for the production of fatty alcohols. It has the advantage over catalytic hydrogenation of effecting no change in the unsaturation of the fatty acid chain and hence is particularly desirable for the manufacture of products containing a high proportion of unsaturated alcohols. The design and operation of a commercial plant operating on hydrogenated coconut oil have been described by Kastens and Peddicord (132).

9.4   HYDROGENATION OF NITRILES FOR PRODUCTION OF FATTY AMINES

A hydrogenation process of some technical importance is the hydrogenation of nitriles derived from fatty acids to form amines with a nickel or cobalt catalyst. Ordinarily both primary and secondary amines are produced, according to the equations

$$
\underset{\text{Nitrile}}{RCN} + \underset{\text{Hydrogen}}{2H_2} = \underset{\text{Primary amine}}{RCH_2NH_2}
$$

$$
\underset{\text{Primary amine}}{RCH_2NH_2} + \underset{\text{Nitrile}}{RCN} + \underset{\text{Hydrogen}}{2H_2} = \underset{\text{Secondary amine}}{\overset{\displaystyle RCH_2}{\underset{\displaystyle RCH_2}{{\diagdown}\;NH\;{\diagup}}}} + \underset{\text{Ammonia}}{NH_3}
$$

A maximum production of primary amines is generally desired, and the secondary amines are merely undesirable byproducts. Production of secondary amines is inhibited by the addition of gaseous ammonia to the hydrogenator at the beginning of the reaction and the maintenance of a substantial pressure of ammonia relative to that of hydrogen as hydrogenation progresses (133). It is claimed by Young and Christensen (134) that further improvement in the yield of primary amines (e.g., ca. 50–80%) is obtained by the addition of water and caustic soda or other alkaline substance to the nitriles before the latter are hydrogenated. On the other hand, an increased yield of secondary amines can be obtained removing ammonia continuously from the reaction zone as the reaction proceeds (135).

According to Kenyon et al. (136), in commercial practice primary amines are obtained in yields of about 85% by hydrogenating with Raney nickel catalyst at 150°C (302°F) at 200 lb of pressure. For a maximum yield of secondary amines, higher temperatures and a different catalyst are employed. The use of a temperature of 285°F and a pressure of 250 lb is mentioned by Potts and McBride (137).

9.5   HYDROGENATION IN SOLVENTS

Hydrogenation in solution is of considerable interest. Solvent-extracted oils are becoming more prevalent and, since miscella refining has several obvious advantages, it should be possible to hydrogenate the oil in the extracting solvent. There have been many claims to the superiority of products from solvent hydrogenations; one apparent advantage is that the reaction may be conducted at lower temperatures. Other advantages are lower oil loss, easier filtering because of lower viscosity, and perhaps even greater selectivity.

Hydrogenation in the presence of solvents has been covered in two patents (138, 139). Koritala and Dutton (140) studied the effect of solvents on the comparative rates of hydrogenation of linolenate and linoleate. The linolenic selectivity, during the hydrogenation of linseed-safflower oil in the presence of polar solvents such as dimethyl formamide and 5% palladium on alumina catalyst ratio, was increased to 4 compared to 2 for the usual hydrogenation. Thus it appears polar solvents affect the course of hydrogenation.

Aids for Hydrogenation Calculations

One standard cubic foot of H_2 = 0.00532 lb; 1 liter = 0.085 gram.
To reduce 1000 lb of oil, one IV unit = 0.0795 lb of H_2.
To reduce 1000 lb of oil, one IV unit = 14.15 ft^3 of H_2 stp.
To reduce 1000 kg of oil, one IV unit = 0.0795 kg of H_2.
To reduce 1000 kg of oil, one IV unit = 883.3 liters of H_2 stp.
One pound of oil reduced by one IV unit produces 1.6–1.7 Btu.
One kilogram of oil reduced by one IV unit produces 888–943 gram calories.

References

1. General references: H. Adkins, *Reactions of Hydrogen with Organic Compounds Over Copper-Chromium Oxide and Nickel Catalysts*, University of Wisconsin Press, Madison, 1937; L. F. Albright, *J. Am. Oil Chemists' Soc.*, **40**, 16, 17, 26, 28, 29 (1963);

R. R. Allen, *J. Am. Oil Chemists' Soc.*, **37**, 521–523 (1960); **39**, 457–459 (1962); S. Berkman, J. C. Morrell, and G. Egloff, *Catalysis*, Reinhold, New York, 1940; J. W. Bodman, E. M. James, and S. J. Rini, in *Soybeans and Soybean Products*, K. S. Markley, Ed., Chapter 17, Interscience, New York, 1950; C. Ellis, *Hydrogenation of Organic Substances*, 3rd ed., Van Nostrand, New York, 1930; T. P. Hilditch and C. C. Hall, *Catalytic Processes in Applied Chemistry*, 2nd ed., Van Nostrand, New York, 1937; E. Hugel, in *Chemie und Technologie der Fette und Fettprodukte*, H. Schonfeld, Ed., Vol. 2, Springer, Vienna, 1937, pp. 135–214; H. W. Lohse, *Catalytic Chemistry*, Chemical Publishing, Brooklyn, N.Y., 1945; J. W. McCutcheon, *Can. Chem. Process Ind.*, **33**, 53–57 (1939); National Research Council, *Twelfth Report of the Committee on Contact Catalysis*, Wiley, New York, 1940; P. Sabatier, *Catalysis in Organic Chemistry*, translated by E. E. Reid, Van Nostrand, New York, 1922; H. S. Taylor, *Industrial Hydrogen*, Chemical Catalog Company (Reinhold), New York, 1921; H. I. Waterman, *Hydrogen of Fatty Oils*, Elsevier, New York, 1951; J. W. E. Conen, *J. Am. Oil Chemists' Soc.*, **53**, 382–389, 1976; J. W. E. Conen, *Chem. Ind.* (18), 709–722 (1978).

2. P. Sabatier, in *Catalysis in Organic Chemistry*, translated by E. E. Reid, Van Nostrand, New York, 1922.

3. W. Normann, *British Pat.* 1,515 (1903).

4. G. M. Weber and C. L. Alsberg, *The American Vegetable Shortening Industry*, Food Research Institute, Stanford University, 1934.

5. J. J. Burchenal, U.S. Pat. 1,135,351 (1915).

6. L. F. ALbright, *J. Am. Oil Chemists' Soc.*, **47**, 490–493 (1970).

7. A. S. Richardson, C. A. Knuth, and C. H. Milligan, *Ind. Eng. Chem.*, **16**, 519–522 (1924).

8. A. E. Bailey, *J. Am. Oil Chemists' Soc.*, **26**, 596–601 (1949).

9. L. F. Albright, *J. Am. Oil Chemists' Soc.*, **42**, 250–253 (1965).

10. R. O. Butterfield, H. J. Dutton, *J. Am. Oil Chemists' Soc.*, **44**, 549–550 (1967).

11. R. R. Allen, unpublished data.

12. J. W. E. Conen, *J. Am. Oil Chemists' Soc.*, **53**, 382–389 (1976).

13. R. R. Allen, *J. Am. Oil Chemists' Soc.*, **44**, 466–467 (1967).

14. H. J. Schmidt, *J. Am. Oil Chemists' Soc.*, **45**, 520–522 (1968).

15. R. R. Allen, *J. Am. Oil Chemists' Soc.*, **45**, 312A (1968).

16. R. R. Allen and A. A. Kiess, *J. Am. Oil Chemists' Soc.*, **32**, 400–405 (1955).

17. R. R. Allen and A. A. Kiess, *J. Am. Oil Chemists' Soc.*, **33**, 355–359 (1956).

18. J. Wisniak and L. F. Albright, *Ind. Eng. Chem.*, **53**, 375–380 (1961).

19. H. P. Kaufmann, *Studien auf dem Fettegebiet*, Verlag Chemie G.m.b.H., Berlin, 1935, pp. 234–251; see also C. F. Homboe, *Ber.*, **71**, 532–541 (1938).

20. D. W. Rogers, N. A. Siddiqui, *J. Phys. Chem.*, **79**, 574–577 (1975).

21. K. Andersson, M. Heil, L. L. Lowendahl, and N. H. Schoon, *J. Am. Oil Chemists' Soc.*, **51**, 171–173 (1974).

22. J. Wisniak and S. Stefanovic, *J. Am. Oil Chemists' Soc.*, **44**, 545–546 (1967).

23. J. Wisniak, S. Stefanovic, E. Rubin, Z. Hoffman, and Y. T. Almon, *J. Am. Oil Chemists' Soc.*, **48**, 379–383 (1971).

24. R. R. Allen and J. E. Covey, Jr., *J. Am. Oil Chemists' Soc.*, **47**, 494–496 (1970).

25. I. A. Eldib and L. F. Albright, *Ind. Eng. Chem.*, **49**, 825–831 (1957).

26. J. W. E. Conen, H. Boerma, B. G. Linsen, and B. DeVries, *Proc. 3rd Internat. Congr. Catal.*, Amsterdam, The Netherlands, **2**, 1387 (1964).

27. J. W. E. Conen, *J. Am. Oil Chemists' Soc.*, **53**, 382–389 (1976).

28. S. Koritala, *J. Am. Oil Chemists' Soc.*, **47**, 463–466 (1970).

29. S. Koritala and H. J. Dutton, *J. Am. Oil Chemists' Soc.*, **43**, 556–558 (1966).

30. G. R. List, C. D. Evans, R. E. Beal, L. T. Black, K. J. Moolton, and J. C. Cowan, *J. Am. Oil Chemists' Soc.*, **51**, 239–243 (1974).

31. S. Koritala, *J. Am. Oil Chemists' Soc*, **52**, 240–243 (1975).

32. E. Kirshner and E. R. Lowrey, *J. Am. Oil Chemists' Soc.*, **47**, 467–469 (1970).

33. R. R. Allen, *J. Am. Oil Chemists' Soc.*, **41**, 521–523 (1964).

34. C. R. Scholfield, T. L. Mounts, R. O. Butterfield, and H. J. Dutton, *J. Am. Oil Chemists' Soc.*, **48**, 237–239 (1971).

35. J. M. Snyder, C. R. Scholfield, T. L. Mounts, R. O. Butterfield, and H. J. Dutton, *J. Am. Oil Chemists' Soc.*, **52**, 244–247 (1975).

36. R. R. Allen, *J. Am. Oil Chemists' Soc.*, **33**, 301–304 (1956).

37. C. R. Scholfield, E. P. Jones, J. A. Stolp, and J. C. Cowan, *J. Am. Oil Chemists' Soc.*, **35**, 405–409 (1958).

38. S. Koritala, R. O. Butterfield, and H. J. Dutton, *J. Am. Oil Chemists' Soc.*, **50**, 317–320 (1973).

39. H. W. Lemon, *Can. J. Res.*, *F25*, 34–43 (1947).

40. T. P. Hilditch and S. P. Pathak, *Proc. Roy. Soc. (Lond.)*, *A198*, 323–337 (1949).

41. R. R. Allen and A. A. Kiess, *J. Am. Oil Chemists' Soc.*, **33**, 419–422 (1956).

42. T. P. Hilditch and E. C. Jones, *J. Chem. Soc.*, **1932**, 805–820.

43. W. J. Bushnell and T. P. Hilditch, *J. Chem. Soc.*, **1937**, 1767–1774.

44. F. H. Mattson and R. A. Volpenhein, *J. Am. Oil Chemists' Soc.*, **39**, 307–308 (1962).

45. H. V. Tumer, R. O. Feuge, T. L. Ward, and E. R. Cousins, *J. Am. Oil Chemists' Soc.*, **41**, 413–417 (1964).

46. K. Schilling, *Fette Seifen Anstrichm.*, **70**, 389–393 (1968).

47. B. Drozdowski, *J. Am. Oil Chemists' Soc.*, **54**, 600–603 (1977).

48. T. P. Hilditch and C. W. Moore, *J. Soc. Chem. Ind.*, **42**, 15–16T (1923).

49. T. P. Hilditch, *Nature*, **157**, 586 (1946).

50. R. O. Feuge, M. B. Pepper, R. T. O'Connor, and E. T. Field, *J. Am. Oil Chemists' Soc.*, **28**, 420–426 (1951).

51. R. G. Pelley, *J. Soc. Chem. Ind.*, **46**, 449–454 (1927).

52. A. E. Johnston, D. McMillan, H. J. Dutton, and J. C. Cowan, *J. Am. Oil Chemists' Soc.*, **39**, 273–276 (1962).

53. C. R. Scholfield, J. Nowakowska, and H. J. Dutton, *J. Am. Oil Chemists' Soc.*, **39**, 90–95 (1962).

54. J. G. Willard and M. L. Martinez, *J. Am. Oil Chemists' Soc.*, **38**, 282–286 (1961).

55. F. B. White and S. Faulkner, abstracts of papers, 23rd Fall Meeting, American Oil Chemists' Society, Chicago, 1949.

56. C. J. O'Boyle, *Ind. Eng. Chem.*, **42**, 1705–1714 (1950).

57. I. C. Bentz, U.S. Pat. 3,271,433 (1966).

58. (a) V. Mills, J. H. Sanders, and H. K. Hawley, U.S. Pat. 2,520,422 (1950); (b) U.S. Pat. 2,520,423 (1950); (c) U.S. Pat. 2,520,424 (1950).

59. H. J. Schmidt, *J. Am. Oil Chemists' Soc.*, **47**, 134–136 (1970).

60. Lurgi, *Continuous and Discontinuous Hydrogenation of Fatty Acids and Neutral Oils*, p. 9. Lurgi Gesellschaft Fiir Wirme und Chemotechnik MBH.

61. W. A. Coombs, R. A. Zavada, J. E. Hansen, W. A. Singleton, and R. R. King, U.S. Pat. 3,792,067 (1974).

62. W. Kehse, U.S. Pat. 3,634,471 (1972).

63. A. G. Pintsch Bamag, French Patent 1,454,722 (1966).

64. W. Kehse, *Fette Seifen Anstrichm.*, **65**, 217–221 (1963).

65. W. Kehse and K. Mechler, *Chem. Proc. Eng.*, **44**, 431–433 (1963).

66. K. D. Mukherjee, I. Kiewitt, and M. Kiewitt, *J. Am. Oil Chemists' Soc.*, **52**, 282–288 (1975).

67. W. F. Kirsch, J. D. Potts, and W. J. Sarouynin, Jr., Belgian Pat. 630,383 (1963); *Chem. Abstr.*, **60**, 13803 (1964).

68. W. F. Kirsch, U.S. Pat. 3,123,626 (1964).

69. C. J. Borkowski and J. L. Schille, U.S. Pat. 2,365,045 (1944).

70. R. E. Beal, K. J. Moulton, H. A. Moser, and L. T. Black, *J. Am. Oil Chemists' Soc.*, **46**, 498–500 (1969).

71. A. D. Franse and L. C. Waterman, U.S. Pat. 3,324,026 (1967).

72. G. R. Fritsche and L. W. Haniak, U.S. Pat. 3,928,158 (1977).

73. J. H. Crissman, G. R. Fritsche, F. B. Hamel, and L. W. Hilty, U.S. Pat. 4,059,498 (1978).

74. H. N. Basu and M. M. Chakrabarty, *J. Am. Oil Chemists' Soc.*, **43**, 119–121 (1966).

75. G. Brieger and T. J. Nestrick, *Chem. Rev.*, **74**, 567–580 (1974).

76. J. G. Blaso (to Natural Vitamins Corp.), U.S. Pats. 2,307,756 and 2,311,633 (1943); D. J. Hennessy (to Vitaminoil Laboratories, Inc.), U.S. Pat. 2,321,913 (1943); H. I. Waterman and J. A. van Dijk, U.S. Pat. 2,143,587 (1939).

77. H. Mielck, *Seifensieder-Ztg.*, **57**, 241–242 (1930).

78. A. Silveira, Y. Masuda, and S. S. Chang, *J. Am. Oil Chemists' Soc.*, **42**, 85–86 (1965).

79. S. S. Chang, Y. Masuda, B. D. Mookerjee, and A. Silveira, *J. Am. Oil Chemists' Soc.*, **40**, 721–725 (1963).

80. J. G. Keppler, J. A. Schols, W. H. Feenstra, and P. W. Meijboom, *J. Am. Oil Chemists' Soc.*, **42**, 246–249 (1965).

81. P. W. Meijboom and G. A. Jongenotter, *J. Am. Oil Chemists' Soc.*, **48**, 143 (1971).

82. K. Yasuda, R. J. Peterson, and S. S. Chang, *J. Am. Oil Chemists' Soc.*, **52**, 307–311, (1975).

83. H. P. Hoffman and C. E. Green, *Oil Soap*, **16**, 236–238 (1938).

84. R. O. Feuge and A. E. Bailey, *Oil Soap*, **12**, 78–84 (1944).

85. K. A. Williams, *J. Soc. Chem. Ind.*, **46**, 448–449 (1927).

86. *Official and Tentative Methods of the American Oil Chemists' Society*, Cd 10–57 (1962).

87. D. V. Stingley and R. J. Wyobel, *J. Am. Oil Chemists' Soc.*, **38**, 201–305 (1961).

88. N. W. Ziels and W. Schmidt, U.S. Pat. 2,468,799 (1949).

89. S. Berkman, J. C. Morrell, and G. Egloff, *Catalysis*, Reinhold, New York, 1940.

90. A. V. Grosse, *Ind. Eng. Chem.*, **35**, 762–767 (1943).

91. H. S. Taylor, *Proc. Roy. Soc. (Lond.)*, **A108**, 105–111 (1925).

92. G. H. Twigg and E. K. Rideal, Trans. Faraday Soc., **36**, 533–537 (1940); O. Beek, in *Advances in Catalysis*, Vol. 2, Academic Press, 1950, p. 151; R. J. Kokes and P. H. Emmett, *J. Am. Chem. Soc.*, **81**, 5032–5037 (1959).

93. S. Ueno, *J. Soc. Chem. Ind. Jpn.*, **21**, 898–939 (1918).

94. J. W. Bodman, E. M. James, and S. J. Rini, in *Soybeans and Soybean Products*, K. S. Markley, Ed., Interscience, New York, 1950, Chapter 17.

95. B. Drozdowski and M. Zajac, *J. Am. Oil Chemists' Soc.*, **54**, 595–599 (1977).

96. O. H. Wurster, *Ind. Eng. Chem.*, **32**, 1193–1199 (1940).

97. E. Hugel, in *Chemie und Technologie der Fette und Fettprodukte*, H. Schonfeld, Ed., Vol. 2, Springer, Vienna, 1937, p. 142.

98. M. L. Freed, U.S. Pat. 2,424,811 (1947).

99. M. Raney, U.S. Pats. 1,563,587 (1925), 1,628,190 (1927), and 1,915,473 (1933); *Ind. Eng. Chem.*, **32**, 1199–1203 (1940). See also R. Paul, *Bull. Soc. Chim. France*, **7**, 396–345 (1940). For improved catalysts, see H. V. Tumer, R. O. Feuge, and E. R. Cousins, *J. Am. Oil Chemists' Soc.*, **41**, 212–214 (1964).

100. A. E. Bailey, *Ind. Chem.*, **44**, 990–994 (1952); anon., *Chem. Week*, **93**, 57 (August 24, 1963).

101. V. Mills, J. H. Sanders, and H. K. Hawley, U.S. Pats. 2,520,422–2,520,424 (1950).

102. N. K. Nadirov, *Kinetika i Kataliz* **6**, 355–356 (1965); *Chem. Abstr.*, **63**, 6365i.

103. J. LeFebure and J. Baltes, *Fette, Seifen, Anstrichm.* **77**, 125–130 (1975).

104. W. A. Lazier and H. R. Arnold, *Org. Syn. Coll.*, **2**, 142, (1943).

105. S. Koritala, *J. Am. Oil Chemists' Soc.*, **49**, 83–84 (1972).

106. S. Koritala, *J. Am. Oil Chemists' Soc.*, **45**, 197–200 (1968).

107. M. Zajcew, *J. Am. Oil Chemists' Soc.*, **37**, 11–14 (1960).

108. M. Zajcew, *Engelhard Ind. Tech. Bull.*, **4** (4) 121–124 (1965).

109. R. Mozingo, *Org. Syn., Coll.*, **3**, 685 (1955).

110. R. R. Allen, *J. Am. Oil Chemists' Soc.*, (in press).

111. H. S. Taylor, *Industrial Hydrogen*, Chemical Catalog Company (Reinhold), New York, 1921.

112. S. Hurst, *Oil Soap* **16**, 29–35 (1939).

113. E. D. Eastman, *J. Am. Chem. Soc.*, **44**, 975–998 (1922).

114. J. H. Shapleigh, Re. 21,521 (1940) of U.S. Pat. 2,173,984 (1939).

115. *Hydrogen*, The Girdler Corporation, 1946.

116. Howe Baker Engineers, Tyler, Texas, private communication.

117. *Liquid Hydrogen*, Union Carbide, Form 9914-D, July 1974.

118. *Liquified Hydrogen Systems*, 1973, NFPA No. 50B.

119. *Liquid Hydrogen Tanks*, Union Carbide, TMH-9000-18000.

120. W. Normann, *Z. Angew. Chem.*, **44**, 714–717 (1931).

121. O. Schmidt, *Ber.*, **B64**, 2051–2053 (1931).

122. W. Schrauth, O. Schenck, and K. Stockdorn, *Ber.*, **B64**, 1314–1318 (1931).

123. H. Adkins and K. Folkers, *J. Am. Chem. Soc.*, **53**, 1095–1097 (1931).

124. W. A. Lazier, U.S. Pats. 1,839,974 (1932), 2,079,414 (1937), and 2,109,844 (1938).

125. H. Adkins and R. Connor, *J. Am. Chem. Soc.*, **53**, 1091–1095 (1931).

126. W. A. Lazier, U.S. Pats. 1,964,000 and 1,964,001 (1934).

127. W. Schrauth and T. Bottler, U.S. Pat. 2,023,383 (1935).

128. (a) A. Guyer, A. Bieler, and K. Jadberg, *Helv. Chim. Acta*, **30**, 39–43 (1947); (b) J. Sauer and H. Adkins, *J. Am. Chem. Soc.*, **58**, 1–3 (1937); (c) A. J. Pantulu and K. T. Achaya, unpublished results.

129. J. D. Richter, P. J. Vandenberg, *J. Am. Oil Chemists' Soc.*, **46**, 155–156 (1969).

130. A. S. Richardson and J. E. Taylor, U.S. Pats. 2,340,343–2,340,344, 2,340,687–2,340,691 (1955) and 2,375,495 (1945).

131. V. L. Hansley, *Ind. Eng. Chem.*, **39**, 55–62 (1947).

132. M. L. Kastens and H. Peddicord, *Ind. Eng. Chem.*, **41**, 438–446 (1949).

133. A. W. Ralston, *Oil Soap*, **17**, 89–91 (1940).

134. H. P. Young and C. W. Christensen, U.S. Pat. 2,287,219 (1942).

135. H. P. Young, U. S. Pat. 2,355,314 (1944).

136. R. L. Kenyon, D. V. Stingley, and H. P. Young, *Ind. Eng. Chem.*, **42**, 202–213 (1950).

137. R. H. Potts and G. W. McBride, *Chem. Eng.*, **57**, No. 2, 124–127 (1950).

138. C. H. Maryott, U.S. Pat. 1,097,456 (1914).

139. J. H. Sanders, U.S. Pat. 2,520,440 (1950).

140. S. Koritala and H. J. Dutton, *J. Am. Oil Chemists' Soc.*, **42**, 1150–1152 (1965).

2
Fat Splitting, Esterification, and Interesterification

The following are examples of similar reactions that are reversible: *fat splitting* (hydrolysis) (reaction 1a), in which fat or oil is hydrolyzed to yield free fatty acids and glycerol, and alkyl ester hydrolysis (reaction 1b); *esterification*, in which the free fatty acids are recombined with glycerol (reaction 2a), or with another alcohol (reaction 2b) (both with the elimination of water); and *interesterification*, in which a glyceride or other fatty acid ester is reacted with fatty acids (acidolysis) (reaction 3a), with 1 mole of alcohol (reaction 3b1), with 3 moles of alcohol (alcoholysis) (reaction 3b3), with glycerol (glycerolysis) (reaction 3c), with other triglyceride esters (bimolecularly) (reaction 3d), within itself (monomolecularly) (reaction 3e), or with the interchange of α-monoglycerides to β-monoglycerides (reaction 4). In the case of the two intramolecular reactions 3e and 4, it is rarely necessary or desirable to complete the reaction; bimolecular reactions are brought to completion only if one of the components is removed from the homogeneous reaction system. The different reactions involve similar groups and are, in general, responsive to the same catalysts. All involve migration of fatty acid radicals (Scheme 1).

Fat splitting has been extensively practiced for over a century for the production of fatty acids for soapmaking and the manufacture of candles. Esterification has been commercially important for perhaps 50 years. Interesterification is the most recent development of the three general reactions; only within the last 30 years has its potential been developed, particularly in the fields of shortenings, hard butters, and other products. Today all three reactions are of vital importance in the "tailor-making" of industrially important fat-derived products.

97

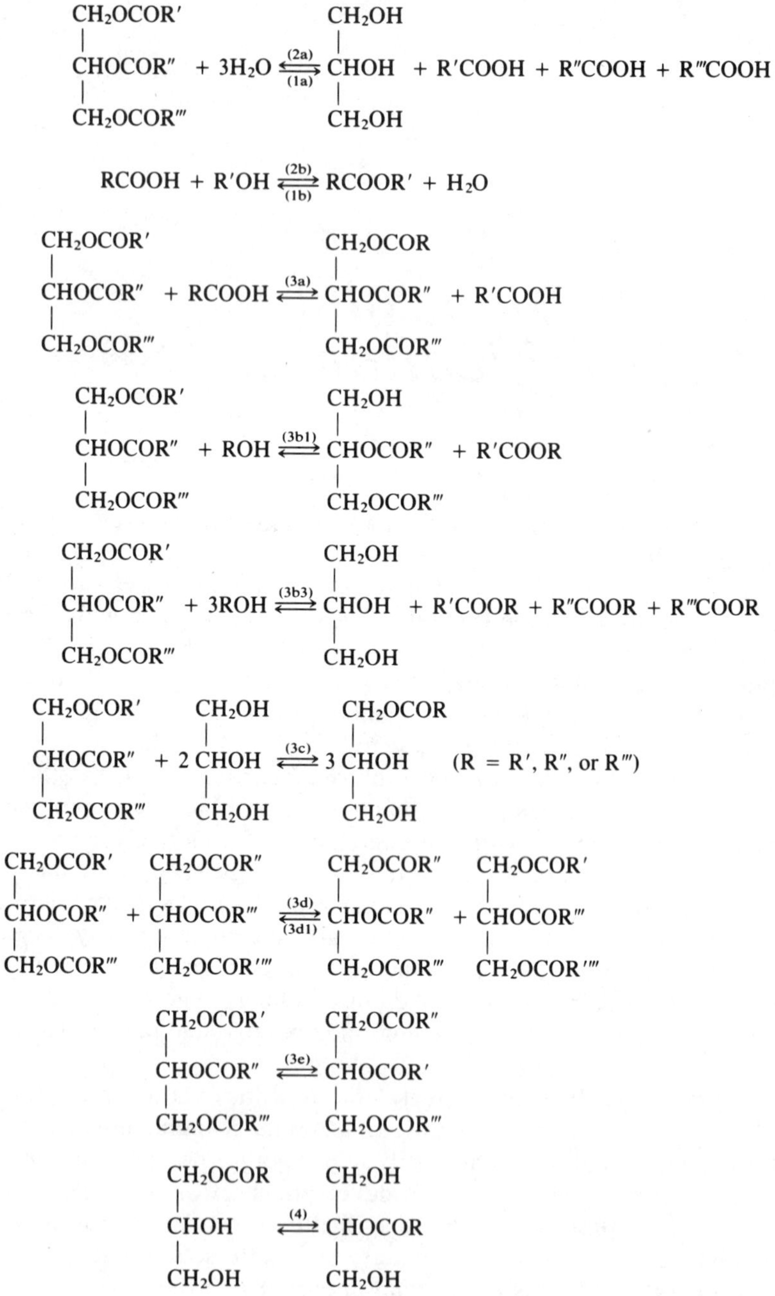

$$\begin{matrix} CH_2OCOR' \\ | \\ CHOCOR'' \\ | \\ CH_2OCOR''' \end{matrix} + 3H_2O \overset{(2a)}{\underset{(1a)}{\rightleftharpoons}} \begin{matrix} CH_2OH \\ | \\ CHOH \\ | \\ CH_2OH \end{matrix} + R'COOH + R''COOH + R'''COOH$$

$$RCOOH + R'OH \overset{(2b)}{\underset{(1b)}{\rightleftharpoons}} RCOOR' + H_2O$$

$$\begin{matrix} CH_2OCOR' \\ | \\ CHOCOR'' \\ | \\ CH_2OCOR''' \end{matrix} + RCOOH \overset{(3a)}{\rightleftharpoons} \begin{matrix} CH_2OCOR \\ | \\ CHOCOR'' \\ | \\ CH_2OCOR''' \end{matrix} + R'COOH$$

$$\begin{matrix} CH_2OCOR' \\ | \\ CHOCOR'' \\ | \\ CH_2OCOR''' \end{matrix} + ROH \overset{(3b1)}{\rightleftharpoons} \begin{matrix} CH_2OH \\ | \\ CHOCOR'' \\ | \\ CH_2OCOR''' \end{matrix} + R'COOR$$

$$\begin{matrix} CH_2OCOR' \\ | \\ CHOCOR'' \\ | \\ CH_2OCOR''' \end{matrix} + 3ROH \overset{(3b3)}{\rightleftharpoons} \begin{matrix} CH_2OH \\ | \\ CHOH \\ | \\ CH_2OH \end{matrix} + R'COOR + R''COOR + R'''COOR$$

$$\begin{matrix} CH_2OCOR' \\ | \\ CHOCOR'' \\ | \\ CH_2OCOR''' \end{matrix} + 2 \begin{matrix} CH_2OH \\ | \\ CHOH \\ | \\ CH_2OH \end{matrix} \overset{(3c)}{\rightleftharpoons} 3 \begin{matrix} CH_2OCOR \\ | \\ CHOH \\ | \\ CH_2OH \end{matrix} \quad (R = R', R'', \text{ or } R''')$$

$$\begin{matrix} CH_2OCOR' \\ | \\ CHOCOR'' \\ | \\ CH_2OCOR''' \end{matrix} + \begin{matrix} CH_2OCOR'' \\ | \\ CHOCOR''' \\ | \\ CH_2OCOR'''' \end{matrix} \overset{(3d)}{\underset{(3d1)}{\rightleftharpoons}} \begin{matrix} CH_2OCOR'' \\ | \\ CHOCOR'' \\ | \\ CH_2OCOR''' \end{matrix} + \begin{matrix} CH_2OCOR' \\ | \\ CHOCOR''' \\ | \\ CH_2OCOR'''' \end{matrix}$$

$$\begin{matrix} CH_2OCOR' \\ | \\ CHOCOR'' \\ | \\ CH_2OCOR''' \end{matrix} \overset{(3e)}{\rightleftharpoons} \begin{matrix} CH_2OCOR'' \\ | \\ CHOCOR' \\ | \\ CH_2OCOR''' \end{matrix}$$

$$\begin{matrix} CH_2OCOR \\ | \\ CHOH \\ | \\ CH_2OH \end{matrix} \overset{(4)}{\rightleftharpoons} \begin{matrix} CH_2OH \\ | \\ CHOCOR \\ | \\ CH_2OH \end{matrix}$$

Scheme 1

98

The various hydrolytic splitting methods are those customarily employed for the production of fatty acids from common fats and oils. One other method available for this purpose involves saponification of a fat or an oil, followed by removal of glycerol and acidulation of the soaps:

$$
\begin{array}{lll}
\mathrm{CH_2OCOR'} & \mathrm{CH_2OH} & \mathrm{R'COONa} & \mathrm{R'COOH} \\
| & | \\
\mathrm{CHOCOR''} \xrightarrow{\ 3\,\mathrm{NaOH}\ } \mathrm{CHOH} + \mathrm{R''COONa} \xrightarrow{\ 3\,\mathrm{HCl}\ } \mathrm{R''COOH} + 3\,\mathrm{NaCl} \\
| & | \\
\mathrm{CH_2OCOR'''} & \mathrm{CH_2OH} & \mathrm{R'''COONa} & \mathrm{R'''COOH}
\end{array}
$$

Certain heat-sensitive fatty acids are still produced in this way, and the method has general utility for special fatty acids. Cocoa butter, for example, can be more completely saponified than in usual conventional saponifications if 9.7% aqueous caustic soda is used at 71–100°C and if the acidulation is controlled at a pH of 6.6–6.8. The iodine number of the resulting fatty acids is said to be uniformly higher, frequently above 25 (1). Saponification/acidulation is subject to the frequent disadvantage of gel or emulsion formation during the acidulation step and large quantities of saltwater must be disposed of; consequently, hydrolytic splitting (Scheme 1, reaction 2a) is generally preferred.

Fat Splitting

2.1   GENERAL CONSIDERATIONS

Composition of Partially Split Fat.   The overall reaction of fat splitting (reaction 1a) occurs in three stages involving (*a*) the formation of two isomeric diglycerides; (*b*) the formation of α- and β-monoglycerides; and (*c*) the formation of glycerol. The reaction is quite complex (2–5). Hydrolysis is efficient when it is carried out under conditions of optimum homogeneity, that is, when water has the greatest solubility within the fatlike phases. Since water is increasingly soluble in diglycerides and monoglycerides and is also more soluble in the product fatty acids than in the triglycerides, fat splitting conditions are designed to afford a sufficiently fast hydrolysis rate to initiate the first phase of the reaction.

Partially split fat contains variable quantities of tri-, di-, and monoglycerides and glycerol depending primarily on the splitting conditions. Mono- and diglycerides corresponding to as much as 3.5% excess combined glycerol have been produced in fat partially split in an autoclave. Less mono- and diglycerides are produced by splitting in the Twitchell method and still less in fermentative splitting. No mono- and diglycerides have been detected in the partial saponification of a fat with strong alkali. Apparently, the rate of saponification of these intermediates is very rapid as a result of the intimate contact of these phases with aqueous caustic soda when emulsified with sodium soaps.

In the Twitchell splitting of coconut oil, Mueller and Holt (6) found that the proportions of mono-, di-, and triglycerides became constant early in the reaction at about 14, 33, and 42% by weight or 25, 33, and 42 mole %, respectively, (corresponding to about 18.3% combined glycerol) and did not change with subsequent change in the free glycerol concentration of the aqueous and fat phases (Figure 2.1).

In the uncatalyzed autoclave splitting of tallow and coconut oil, Mills and McClain (7) observed that with various degrees of splitting, the combined glycerol of the unsplit fat was constant at 18.5–20% for tallow and at 22–24% for coconut oil. They also found at equilibrium a constant ratio of combined glycerol to free glycerol in the fat phase (about 1.8:1 for tallow and 1.25:1 for coconut oil). Ester interchange among mono-, di-, and triglycerides may occur at higher temperatures. Actually, in the case of tallow the observed content of combined glycerol (18.5–20%) was not

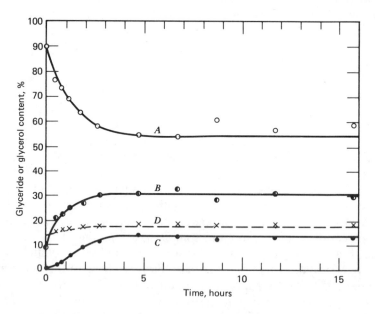

Figure 2.1 Composition of unsplit fat during the first boil in the Twitchell splitting of coconut oil: (*A*) triglycerides; (*B*) diglycerides; (*C*) monoglycerides; (*D*) combined glycerol content (6).

very different from that calculated for random distribution at a 1.8 : 1 ratio of combined to free glycerol (ca. 20.5%). In the case of coconut oil the observed content of combined glycerol was considerably less than that called for by random distribution, assuming equal average molecular weights for the free and combined fatty acids.

Mechanism of the Reaction.   It was first pointed out by Lascaray (8), and it is now generally accepted, that only a minor amount of splitting occurs at the water–fat interface and that fat splitting is essentially a homogeneous reaction that occurs through action of water dissolved in the fat phases. When plotted as degree of splitting versus time the reaction may be represented by a sinusoidal curve (see Figure 2.3). The reaction is characterized by an initial period when the reaction is slow and limited by the low solubility of water in the fat, a middle period during which reaction is relatively rapid, and a final period during which the rate diminishes as equilibrium between fatty acids and liberated glycerol is approached. Splitting is accelerated by increasing the temperature, largely because of the increased solubility of water in the fat phases.

All the industrial fat splitting methods have as their objectives the attainment of a high rate of hydrolysis, together with a high degree of completeness. This is achieved, more or less, by (*a*) the use of a large excess of water, (*b*) selection of the appropriate combination of temperature and pressure to optimize the solubility of water in the fat phases, with or without the use of suitable water-in-oil emulsifiers, and (*c*) removal of the byproduct glycerol.

Rate of the Reaction.   Splitting is accelerated by mineral acids, certain metal oxides, including particularly zinc and magnesium oxides (which form fat-soluble soaps), and sulfonic acids of the Twitchell type. Presumably, these function because they increase the solubility of water in the fat. During the greater part of the reaction they function in solution in the fat phase, although in the beginning they may also initiate hydrolysis through the promotion of emulsification. The water/fat ratio has an important bearing on the extent to which hydrolysis may be carried; hence in the last stage the rate of splitting increases with increase in the amount of water in the system. In the early stage, however, the reaction rate is not increased by the presence of large amounts of water.

The influence of temperatures in the range 150–220°C on the rate of splitting of fats is illustrated by Lascaray's data (8) for tallow treated in an autoclave with 60% water and 0.5% sodium hydroxide. The times indicated for the hydrolysis of 50% of the total fatty acids were as follows:

Temperature (°C)	Time (hours)
220	0.5
200	0.9
185	2.2
170	4.2
150	9.0

The rapid rate of splitting in the temperature range 225–280°C is illustrated by the work of Sturzenegger and Sturm (9), who split beef tallow and coconut and peanut oils in an autoclave using about 30 mole proportions of water at 30–70 atmospheres. The data are summarized in Table 2.1.

Maximum Splitting Obtainable. The fat splitting reaction is reversible, and a point of equilibrium between hydrolysis and reesterification will eventually be reached unless the liberated glycerol is removed from the sphere of reaction by some means. It is generally recognized that the completeness of hydrolysis depends on the concentration of glycerol in the fat phase. This concentration is directly proportional to the concentration in the water phase. Mills and McClain (7), splitting fats in an autoclave without a catalyst at 453°F (234°C) and 493°F (251°C), found the ratio of free glycerol in the fat phase to be 12.5:1 for tallow and 7.0:1 for coconut oil. The maximum degree of splitting is, therefore, a function of the concentration of the glycerol in the water; within the range of 80–100% hydrolysis the relationship between the two is substantially linear. Apparently, it is independent of temperature, concentration of water,

Table 2.1 *Influence of temperatures in the range 225–280°C on splitting time* (9)

Temperature (°C)	Time to Reach Equilibrium (min)			Half-Life Time (min)		
	Beef Tallow	Coconut Oil	Peanut Oil	Beef Tallow	Coconut Oil	Peanut Oil
225	156	158	156	62	58	50
240	82	85	85	34	33	25
260	47	46	53	18	23	16
280	34	33	33	8	10	10
260 (Catalyst 0.2% ZnO)	21	—	—	4	—	—

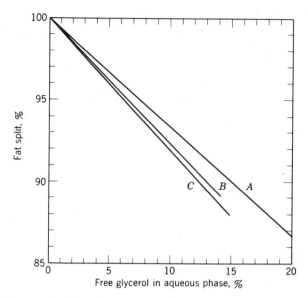

Figure 2.2 Maximum hydrolysis in fat splitting as a function of the glycerol content of the aqueous phase: (*A*) Twitchell splitting of palm kernel oil (10); (*B*) autoclave splitting of tallow (7); (*C*) autoclave splitting of coconut oil (7).

or the presence or absence of catalysts. The data of Mills and McClain (7), are shown in Figure 2.2 representing equilibrium in the uncatalyzed splitting of tallow and coconut oil, and that of Lascaray (10), representing the near-equilibrium conditions for the Twitchell splitting of palm kernel oil. Figure 2.3 illustrates the autoclave splitting of tallow with 60% water (8).

Miscellaneous Characteristics of the Reaction.   Kaufman and Keller (11) showed that the temperature independence of the point of equilibrium is an indication of a zero heat of reaction in fat splitting. A heat effect, however, is associated with the solution of water in the fat phase. In high-temperature splitting where a relatively large amount of water is dissolved, there may be a sufficient absorption of heat to cool the fat 50°F.

In fat splitting, unlike saponification, there is apparently no considerable difference in the reaction rates of saturated and unsaturated fatty acids. In enzymatic fat splitting at atmospheric pressures fatty acids of intermediate chain length, that is, C_{12} to C_{18}, are liberated most readily, with the ease of splitting decreasing as the chain becomes longer or shorter.

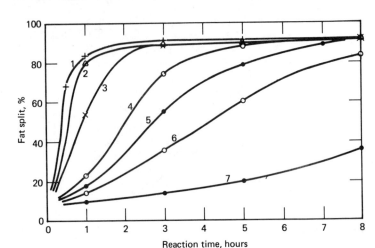

Figure 2.3   Autoclave splitting of tallow with 60% water: 1, 220°C (322 psig) with 0.5% NaOH as a catalyst; 2, 185°C (148 psig), catalyst 0.51% ZnO; 3, 200°C (210 psig), catalyst 0.5% NaOH; 4, 185°C (148 psig), catalyst 0.5% NaOH; 5, 185°C (148 psig), no catalyst; 6, 170°C (100 psig), catalyst 0.5% NaOH; and 7, 140°C (38 psig), catalyst 0.5% NaOH (8).

2.2   TWITCHELL PROCESS

The Twitchell process (2, 12) is no longer of significant importance in the United States, although it is still used throughout the world to split fats and oils in batch processes. Twitchell patented a reagent made from oleic acid and concentrated sulfuric acid (also ones from saturated fatty acids and concentrated sulfuric acid) with the capability of hydrolyzing fats and oils at atmospheric pressure on boiling with steam and water in open kettles. Quickly, Twitchell reagents, made by the condensation and ring sulfonation of an aromatic hydrocarbon with oleic acid, such as that from *m*-xylene, represented by the formula

$$CH_3(CH_2)_xCH(CH_2)_yCOOH$$

$$x + y = 15$$

were applied for this purpose. Later, alkylbenzenesulfonic acids and petroleum sulfonic acids came to be used (12a,b).

Sulfuric acid is generally used along with the reagent to depress the solubility of the reagent in the aqueous phase and increase its solubility in the fat phase where acid-catalyzed hydrolysis occurs. From a study of the distribution of the reagent between oil and water it has been shown that the degree of hydrolysis depends on the relative amount of the reagent in the oil phase (13). It was also found that sulfonic acids derived from commercial wetting agents of the alkylarylsulfonate sodium salt type were the best splitting reagents.

The Twitchell process has the advantage of requiring only relatively simple and inexpensive equipment. The operation is carried out in tanks at atmospheric pressure; wooden, lead-lined, or Monel metal tanks are employed, since the process requires some degree of acidity in the aqueous phase. Often the fatty stock to be subjected to Twitchell hydrolysis contains considerable quantities of albuminous material and other impurities. These may neutralize the catalyst and must be removed. It is more or less general practice to subject the fat charge to a period of boiling with dilute sulfuric acid for this purpose before splitting is started. Fat stocks of particularly poor quality are often acid refined with strong sulfuric acid before they are split.

The acid-washed fat is mixed with about 25–50% of its own weight of water and 0.75–1.25% of its weight of Twitchell reagent. If the fat has not been treated with acid, 0.1–0.2% of strong sulfuric acid is added. Splitting is carried out by boiling the mixture with open steam for a total of about 36–48 hours. The tanks are usually provided with covers to minimize contact of the charge with air and consequent darkening of the fatty acids. Discoloration, the long times required (need for extra boilings for overall high splitting yield to afford a more dilute glycerol more costly to concentrate), and high consumption of steam constitute the chief disadvantages of the Twitchell method.

The splitting is carried out in at least two, often three or even four, stages with the liquor or "sweet waters" containing the liberated glycerol drawn off at the end of each stage and replaced with freshwater, or more commonly with weaker sweet waters from a previous boil, with countercurrent operation maintained throughout a series of boils. Where two periods of boilings are employed, about 85–90% of the total splitting will take place in the first stage, and the remainder will take place in the second stage. Typical four-stage operation distributes splitting in the four stages in the respective proportions of about 60, 25, 10, and 5%. Smaller amounts of water are used if the operation is conducted in a number of stages. The degree of splitting obtained depends on the number of stages and the time allotted for the operation; ordinarily about 92–95% of the fat will be hydrolyzed. A higher degree of splitting requires an impracticably long time and produces excessively weak glycerol solutions.

Although an emulsion is produced during boiling, it generally breaks when steaming is discontinued. If a persistent emulsion is formed, a little sulfuric acid may be added to break it. The operation is finished by adding freshwater to the fatty acids and boiling to wash out the mineral acid remaining. Some operators add barium carbonate to neutralize the residual sulfuric acid.

A typical British Twitchell process hydrolysis of tallow has been described by Cox (14):

Twenty five tons of the fat, 10 tons of water, 200 lb. of sulfuric acid and about 1% "Kontackt" reagent (a sulfonaphthenic of petroleum origin) or equivalent proportion of other reagent are boiled with open steam until the split reaches 88% f.f.a. The emulsified mixture is allowed to settle into two phases and the lower aqueous layer is drawn off. Three tons of fresh water and 60 lb. of acid are added and the mixture is again boiled until 92% f.f.a. is attained. After settlement and removal of the water, a third boil with two tons of water and 40 lbs. of acid brings the split to 94–95% f.f.a. A final boil with one ton of fresh water serves to remove mineral acid from the split fatty acids prior to distillation. Settling usually requires about two hours.

The first three boils vary greatly in duration according to the quality of the fat being split and the proportion and type of the reagent used, but for neutral fats of good quality, i.e., those substantially free of impurities which render the reagent inactive, typical boiling times would be 20–24, 12 and 4 hours, respectively. Inferior grades of fat or smaller proportions of reagent necessitate longer boiling. The presence of appreciable proportions of free fatty acids in the raw material somewhat shortens the duration of the first boil.

Abroad, the possibility of modifying the Twitchell process to operation on a continuous basis has been considered (14). Hartman (15) pointed out that the relative velocities for Twitchell hydrolysis at 100 and 150°C are 0.03 and 1, respectively, and that "Kontackt-type" catalysts, but not old-type Twitchell reagents, could probably withstand and function effectively at 150°C (inferring higher than atmospheric pressure operation). However, the unavailability of satisfactory metals of construction capable of withstanding the severe acidic corrosion thus far restricts this possibility.

Kallyanpur et al. (15a) noted that whereas the Twitchell splitting of castor oil gave a yield of over 93%, splitting was accompanied by estolide formation with the average molecular weight of the estolides at equilibrium corresponding to that of a trimer. Estolides caused lower solubility of Twitchell catalysts in the fat phase; 45% of the catalysts were detected in the aqueous phases at equilibrium.

2.3   LOW-PRESSURE SPLITTING WITH CATALYST

Liquid water homogeneously dissolved in the fat phase is the primary requirement for industrial fat splitting; it is generally achieved satisfactorily by the concurrent use of high temperature and, when steam is used, high pressure. It is possible to avoid the use of costly high-pressure equipment and conduct hydrolysis of fats with superheated (gaseous) steam at relatively low pressures in the presence of a catalyst. A patent assigned to the Carad Corporation (16) claims that tallow at 200–280°C is 16% hydrolyzed at only 10–30-sec residence time through a specially designed pipe reactor with zinc oxide catalyst. A second pass of the oil phase through the system is said to afford 70% complete hydrolysis.

2.4   MEDIUM-PRESSURE AUTOCLAVE SPLITTING WITH CATALYST

The autoclave method of fat splitting is perhaps the oldest method employed on a commercial scale, and patents covering this process date back as far as 1854. Previously, this general method was termed "high-pressure splitting with catalyst." However, with the development of continuous, countercurrent fat splitting operating at 500–725 psig (pounds per square inch gauge), the more appropriate term "medium pressure" is applied to it today. Pressures in the range of 150–500 psig are employed in these batch operations. It is used variously for splitting high-grade stocks to produce light-colored fatty acids that do not require distillation, or for the splitting of heat-sensitive fats and oils that cannot tolerate the higher splitting temperatures utilized in high-pressure continuous countercurrent splitting. In addition to producing lighter-colored acids, the autoclave process is much more rapid than the Twitchell process. Examples of fats split by modifications of the autoclave batch process are coconut oil, babassu, and palm kernel oil.

The catalysts used in autoclave splitting are usually zinc, magnesium, and calcium oxides; zinc oxide is the most active and is universally preferred. About 2–4% of catalyst is used, often less than that if pressures employed are higher and in the 300–500-psig range. Color of the fatty acids is improved by the use of a small quantity of zinc dust.

The autoclaves are built in the form of tall cylinders that may be 4–6 ft in diameter and 20–40 ft in height and are commonly made of corrosion-resistant nickel–chrome alloys. The autoclaves are insulated and equipped with lines for the injection of steam but have no mechanical agitators.

In operation, the autoclave is charged with fat, catalyst, and water equivalent to about 30–60% of the weight of fat. Steam is blown through the mass to displace air in the headspace and dissolved air in the fat and water, the autoclave is then sealed, and steam is admitted to raise the internal pressure to at least 150 lb. The steam is injected at the bottom; condensation of steam within the vessel plus the venting of a small amount of steam maintains a sufficient steam flow to keep the charge agitated. A splitting of 95% or better can be achieved after about 6–10 hours. After the desired degree of splitting is obtained, the contents are blown into a settling tank, where the separated fatty acids are drawn off from the water–glycerol liquor. The fatty acids must then be treated with sulfuric acid or other mineral acid to decompose the soaps formed by the catalyst, after which they are washed free of mineral acid. Figure 2.3 shows the results of a series of tests carried out at different temperatures on tallow, using 60% water and 0.5% sodium hydroxide as catalyst and also shows the results of a corresponding test at 185°C, with 0.51% zinc oxide as catalyst, and a test at this temperature without a catalyst.

Autoclave fat splitting with Twitchell reagents is reported to be effective at temperatures below those required with lime, zinc oxide, and so on, but the necessity for a strongly acidic system acid so limits the materials from which the autoclave can be constructed as to render the method more or less impracticable.

2.5   Continuous, Uncatalyzed High-Pressure Countercurrent Splitting

Steadily developed since the first Ittner (17) and Mills (18) patents, the countercurrent, continuous, high-pressure process is the most efficient of current methods of fat hydrolysis. Without a doubt, it is the most inexpensive method for the large-scale production of saturated fatty acids, and for those unsaturated fatty acids with iodine numbers below about 120. The high temperatures and pressures used permit short reaction times, and full countercurrent oil and water flow produces high degrees of split.

The heart of the system is a tower 20–48 in in diameter, 60–80 ft high, of solid type 316 stainless steel or Inconel to withstand operating pressures of over 700 psi, and well insulated. The fat is introduced by means of a sparge ring 3 ft from the bottom, with a high-pressure feed pump. Water is introduced near the top of the column at a ratio of 40–50% of the weight of the fat. The fat rises through the hot glycerol–water collecting section at the bottom of the column and passes through the oil–water interface into the continuous phase, the oil layer in which the hydrolysis takes place. Direct injection of high-pressure steam quickly raises the temperature to 500°F (260°C). Pressure is maintained at 700–725 psi. Figure 2.4 is a schematic of the essential features of the Colgate–Emery fat splitting unit (19).

The continuous countercurrent high-pressure process splits fats and oils more efficiently than other processes and produces high degrees of split in a reaction time of only 2–3 hours, and very little discoloration of the fatty acids occurs. Because of the efficient internal heat exchange it affords very high steam economy. The principal objection to this process lies in the high equipment cost.

In large-volume industrial operation, continuous countercurrent high-pressure fat splitters are run for long periods of time with a minimum of stock changes. Startup and shutdowns are delicate, time consuming, and therefore costly. Long-range scheduling of this key equipment is required to optimize production of a variety of stocks. Although glycerol is usually not isolated with respect to fat source, except occasionally as in the case of hydrogenated fish oils, fatty acids must be isolated between stock changes. For example, it is preferred not to follow a split of Kosher vegetable stock with an animal fat; coconut or other lauric acid oils after tallow; and hydrogenated tallow (about 68% 18:0 content) after hydrogenated soya (about 89% 18:0 content). Cleanout maintenance depends on the cleanliness of the feedstocks and their pretreatment; it is usually

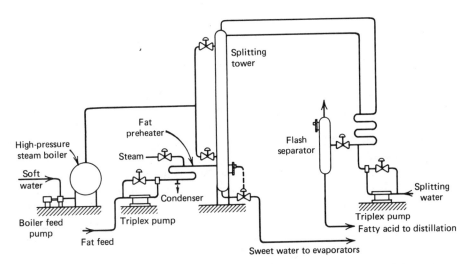

Figure 2.4   Flow sheet for continuous, high-pressure Colgate–Emery fat splitting unit.

required once every 12–18 months. Unplugging of screens and the removal of scale and deposits on walls and exposed internals is necessary. Monitoring of corrosion rate is mandatory, especially pitting of the Inconel liners, measurement of sparge holes, and overall wall thickness. Most industrial plants avoid unscheduled shutdowns by providing backup high-pressure pumps.

Continuous countercurrent high-pressure splitting is not applicable to all fats and oils. Although fat splitting at 700–770 psig and temperatures of 240–260°C is undoubtedly the fastest and cheapest method for splitting common fats, the high temperatures involved do not permit the splitting of sensitive triglycerides that contain conjugated double bonds, unconjugated systems capable of thermal conjugation, hydroxy-containing fats and oils that can be dehydrated [such as castor oil, which generates ricinoleic acid capable of interesterification and hydrogenated caster oil (19a)], or certain polyunsaturated vegetable or fish oils with iodine numbers of 130 and higher. The continuous countercurrent high-pressure splitting of alkyd resin-type soybean oil of iodine numbers of 140 and higher (and with a linolenic acid content of about 7–9%) begins to approach the temperature range where thermal polymerization interferes with ease of splitting and lowers the yield of split products. For these reasons methods involving splitting temperatures in the range of 200–240°C are also employed in the fatty acid industry. Batch splitting methods are used for the lower volume fatty acid specialty products.

2.6   ENZYMATIC FAT SPLITTING

Fat splitting through the use of lipolytic enzymes has been used for sensitive fats and oils (20), but it is not of any appreciable commercial use today. In the past, the enzyme preparation obtained from castor beans was largely employed. The beans were ground with water, and the ground mass was centrifuged to remove solid material. The resulting emulsion of water, oil, protein, and so on was then fermented at room temperature for 48 hours, after which it was ready for use.

Splitting directions for this type of hydrolysis are as follows: The splitting is carried out in open, conical-bottom, lead-lined tanks. The fat is mixed with 30–40% of its weight of water and about 6% of the ferment. Sufficient acetic acid or other acid is added to bring the pH of the system to about 5 and also manganese sulfate solution equivalent to 0.15–0.20% of the dry salt on the basis of the fat. Barium chloride is an activator of hydrolysis, whereas mercuric chloride, cupric chloride, silver nitrate, and copper sulfate are inhibitors (21). A temperature of 35°C is considered optimum, although reasonably rapid splitting is obtained at temperatures as low as 15°C and as high as 40°C. Outside this range of temperatures the activity of the enzyme falls off rapidly; hence the process is not applicable to fats of high melting point.

The mass in the tank is agitated to form an emulsion and then is held at the prescribed temperature for about 24–48 hours, with occasional stirring. As in other fat splitting processes, hydrolysis takes place rapidly at first and then more slowly, eventually reaching a point of equilibrium. About 90% splitting is reported to be optimum. After splitting has taken place, the emulsion is broken by heat and the addition of a little sulfuric acid, after which the fatty acids may be drawn off and washed and further processed in the usual way.

Enzymatic splitting is potentially useful for the splitting of highly unsaturated fatty acids such as those of fish oils. It has not been commercially developed, chiefly because of its incompleteness, sluggishness, and the general difficulty in handling it as a unit operation. Still, in the case of very sensitive fats and oils such as conjugated oils, enzymatic hydrolysis may be one of the few methods of generating fatty acids from triglycerides with minimal structural changes. While some castor oil fatty acids have been produced by this method, the technique is not generally used for this purpose today. Recourse may be made to gentle complete saponification and acidulation of the castor oil soaps if the fatty acids are desired; alternately, medium-pressure splitting at temperatures in the range of 160–200°C is more satisfactory. Conjugated unsaturated fatty acids sush as dehydrated castor oil fatty acids containing perhaps 32% of the 9,11-isomer require even more controlled and modulated conditions.

Desnuelle and Savary (22) discovered that pancreatic lipase selectively hydrolyzes triglycerides at the primary hydroxyls (1- and 3-positions, or α- and α'-) of the glyceride molecules largely before those at the central position (secondary hydroxyl or 2-position, or β-) are removed. The conditions can be adjusted so that the situation prevails exclusively, and the specificity is absolute (23, 24). This hydrolysis technique has been extensively used to determine triglyceride structure (25–27).

Another specific enzyme hydrolysis of potential value is that occurring with the lipase from the seed of *Vernonia anthelmintica*, which is capable of hydrolyzing the 2-position of trivernolin preferentially, while leaving the 1- and 3-positions essentially intact (28). On the other hand, pancreatic lipase hydrolysis of trivernolin proceeds preferentially at the 1- and 3-positions, as with other oils, indicating that the structure of this oil is probably not responsible for the 2-position specificity observed with the lipase from *Vernonia anthelmintica* (29).

2.7 GLYCEROL RECOVERY AND YIELDS

The theoretical amounts of glycerol recoverable from a series of common fats and oils are shown in Table 2.2, which includes figures representing practical plant yields of both 88% and anhydrous glycerol recoverable by existing conventional technology and corrected for the contents of unsaponifiable material and free fatty acids in the oils.

Much of the fat that is split industrially is poor grade and contains 10–50% of free fatty acids. Calculations of glycerol recoverable from such fats should be based on corrections for free fatty acids, mono- and diglyceride content, glycerol, nonsaponifiable matter, and moisture content (30).

Upgrading crude glycerol (''sweet water'') from splitting operations in the fatty acid industry requires the removal of dissolved salts, elimination of color and fat and oil impurities, concentration (evaporation of water), and/or distillation (31–34).

The sweet waters obtained by Twitchell hydrolysis of fats are ordinarily neutralized with lime, filtered to remove calcium sulfate, and concentrated to recover glycerol by the same methods that are used for processing spent lyes from soapmaking.

Autoclave-split or continuous process sweet waters are usually upgraded by (a) light lime treatment, (b) filtration, (c) evaporation concentration to 88–90%, and, probably, (d) distillation. Glycerol may also be upgraded by ion-exchange processing followed by evaporation concen-

Table 2.2 Theoretical and uncorrected glycerol contents of common fats and oils and practical plant yields obtained from them

Fat or Oil	Saponi- fication No.	Theoret- ical Molecu- lar Weight	Acid No.	Percent Glycerol by Weight		
				Theoret- ical Yield (%)	Practical Plant Yields	
					88%	100%
Rapeseed oil, high erucic	171.0	984.2	0.2	9.34	10.27	9.00
Soybean oil	192.7	873.3	0.4	10.53	11.57	10.15
Bleachable fancy tallow	198.0	850.0	3.6	10.82	11.70	10.27
Marine oil	196.3	857.4	2.0	10.73	11.69	10.26
Coconut oil	245.8	684.5	0.3	13.44	14.81	12.89

tration in which distillation may be eliminated (35), or, alternatively, glycerol may be extracted from aqueous solutions by ion exchange (36).

Esterification (37–39)

3.1   REESTERIFICATION OF FATTY ACIDS WITH GLYCEROL

The reesterification of fatty acids with glycerol was known as early as 1844 [e.g., Pelouze and Gelis (40) with butyric acid] and 1854 [e.g., Berthelot (41), who used sealed tube reactions at 200–270C]. The reaction (Scheme 1, Reaction 2a) is the reverse of fat splitting. It is carried out by reacting fatty acids and glycerol at an elevated temperature while maintaining a vacuum on the reaction vessel or using other means to continuously remove the water that is formed. A partial condenser must be inserted in the vacuum line to condense and return glycerol to the vessel while permitting the escape of the water vapor. To avoid metallic contamination of the product and to minimize corrosion by the hot fatty acids, industrial-scale esterifiers are constructed of molybdenum-stabilized stainless steel or Monel (see Section 3.4).

The reaction can be carried out either catalytically (39, 42–47) or in the absence of a catalyst (39, 42, 48, 49). Acid catalysts are frequently used (39, 43–47, 50), and various sulfonic acids appear to give the most rapid rate, especially if water is removed azeotropically during the reaction. Benzene- and toluenesulfonic acids are often used; the latter is claimed to be superior (45). Camphor β-sulfonic acid and naphthalene β-sulfonic acids are also among the best (43, 50). Acid catalysts frequently cause darkening of the products (43, 45, 46) and may initiate dehydration of the unreacted glycerol, with the undesirable formation of acrolein (46):

$$\begin{array}{ccc} CH_2{-}CH{-}CH_2 \\ | \quad\ | \quad\ | \\ OH \quad OH \quad OH \end{array} \xrightarrow{-2H_2O} \left[\begin{array}{c} CH_2{=}C{=}CH \\ | \\ OH \end{array} \right] \rightleftharpoons CH_2{=}CHCHO$$

Finely divided zinc or tin metal was employed by the Germans in their synthetic fatty acid esterification of glycerol intended for fat food use during World War II. Apparently some color was produced since a bleaching step was included in the process (51–54).

Feuge et al. (42) studied the catalyzed and uncatalyzed esterification of glycerol with peanut oil fatty acids and evaluated the catalytic performance of a whole series of metallic compounds in the triesterification reaction. Comparative data are included in Table 2.3.

Table 2.3   Relative activities of different inorganic esterification catalysts (42)[a]

Catalyst	FFA[b] (%)	Catalyst	FFA (%)
Control, no catalyst	16.1	NaOH	13.8
$AlCl_3 \cdot 6H_2O$	17.1	Ni (hydrogenation catalyst)[c]	16.4
Al_2O_3[c]	15.4	$NiCl_2 \cdot 2H_2O$	13.8
$CdCl_2 \cdot 2H_2O$	11.0	$PbCl_2$[c]	12.0
$FeCl_2 \cdot 6H_2O$	11.4	PbO	10.9
FeO[c]	13.9	$SbCl_3$	15.0
$HgCl_2$	15.0	$SnCl_2 \cdot 2H_2O$	2.8
$MgCl_2 \cdot 6H_2O$	13.7	$SnCl_2 \cdot 5H_2O$	2.4
MgO	13.6	SnO_2[c]	15.1
$MnCl_2 \cdot 4H_2O$	13.2	$ZnCl_2$	3.5
MnO_2	10.8	ZnO	11.8

[a] Peanut oil fatty acids reesterified at 200°C and 20-mm pressure with an equivalent quantity of glycerol, employing 0.000008 mole of the catalyst per gram of fatty acids. Free fatty acid content of the reaction mixtures at the end of 6 hours.
[b] Free fatty acid.
[c] Did not dissolve completely in the reaction mixture.

Esterification proceeds at a reasonable rate in a stirred flask under 20 mm pressure; that is, the free fatty acid content of the mixture is reduced to 3% in 6 hours if 0.0008 mole of tin chloride per 100 grams of fatty acids (0.18%) is used at 175°C (347°F), or if a similar molar proportion of zinc chloride (0.11%) is used at 200°C (392°F). Without a catalyst, equally rapid esterification can be obtained only above 250°C (482°F). Typical results are shown in Figures 2.5 through 2.7. Neither tin nor zinc chloride appears to be detrimental to the fat from the standpoint of causing polymerization, conjugation, or undue increase in color, and either can be completely removed by ordinary alkali refining. The catalyzed reesterification of glycerol with fatty acids to form triglycerides can be readily brought to a stage where no monoglycerides and relatively small proportions of diglycerides remain. To accomplish this, one of the better catalysts must be used, fatty acids must be present in 5–20% excess, and the reaction must be continued for 3–6 hours at 180–230°C while under subatmospheric pressure or while being stripped with an inert gas (44). In general, there is little difference in the acid-catalyzed esterification rates of fat-forming acids from butyric to stearic acid.

Recent developments in esterification catalysts have been directed toward innocuous (or easily removed), noncorrosive materials that are capable of inducing rapid esterification to light-colored products. Among

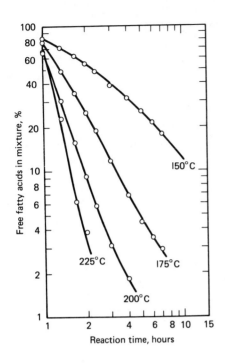

Figure 2.5 Reesterification of peanut oil fatty acids with an equivalent quantity of glycerol without a catalyst (42).

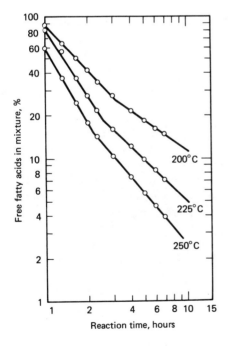

Figure 2.6 Reesterification of peanut oil fatty acids with an equivalent quantity of glycerol and 0.18% $SnCl_2 2H_2O$ as a catalyst (42).

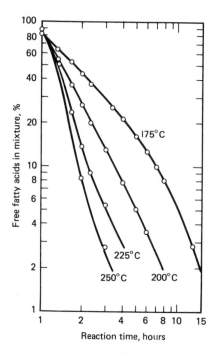

Figure 2.7 Reesterification of peanut oil fatty acids with an equivalent quantity of glycerol and 0.11% $ZnCl_2$ as a catalyst (42).

the numerous new catalysts suggested are hypophosphoric acid used alone or in conjunction with other acid catalysts (55), dibutyl tin oxide (56), titanium dichloride diacetate (57), and zinc dust [the last for triglycerides from fish oil fatty acids (58), alcohol-soluble edible triglycerides from high caprylic/capric fatty acids (59), or triglycerides from cottonseed soapstock-derived fatty acids (60)].

Continued interest in uncatalyzed esterification of glycerol for the production of edible triglyceride products is high since the use of any catalyst has several disadvantages. Many catalysts are inactivated above certain temperatures; several actually volatilize from the reaction medium, and others are especially sensitive to contaminants. In certain instances their removal from the ester product is a major problem that at best involves yield losses due to subsequent bleaching or refining operations.

It has been amply demonstrated that the reaction proceeds readily at high temperatures in the absence of catalyst provided the water is removed as formed. Yields of 95–98% of simple or mixed triglycerides have been claimed by heating equivalent quantities of fatty acids and glycerol at a reduced pressure of 2 cm when the temperature is raised slowly from 180 to 240°C over a 5–6-hour period (49, 61, 62). Similarly, it was observed that the esterification proceeds satisfactorily at atmospheric pressure if a stream of CO_2 gas is used to assist in the removal of water (62). Bellucci recommended a maximum temperature of 240°C (61), on the assumption that glycerides are unstable above that temperature (44). Darker-colored products obtained generally by Feuge et al. (42) may indeed be due to the higher temperatures employed by them.

The mechanism of the uncatalyzed direct esterification of glycerol has been studied by several workers (42, 44, 48). The reaction proceeds in two stages and follows second-order kinetics. With equimolar quantities of fatty acids and glycerol, rate differences appear to be related to relative solubilities of the acids in glycerol; the reaction rates are reported to be the same provided equivalent amounts of acids and glycerol are present (48, 61).

If partially esterified glycerides are desired, less than stoichiometric (<3:1) quantities of fatty acids to glycerol are employed. The crude products are variable, complex mixtures of unreacted fatty acids and glycerol, plus mono- (1- and 2-isomers), di-(1,2- and 1,3-isomers), and triglycerides. If these crude products are carefully refined to remove fatty acids but not monoglycerides by saponification, and salt- or freshwater-washed to remove glycerol, the products are mixtures of mono-, di-, and triglycerides. Table 2.4 shows the compositions of monoglyceride mixtures prepared

Table 2.4   Monoglycerides by direct esterification of glycerol with fatty acids (47)[a]

| | | | Acid-Free Product | | |
| | Fatty Acids Esterified (wt. %) | 1-Monoglyceride[b] (wt. %) | Glyceride Composition[c] (wt. %) | | |
Reactants			Mono-	Di-	Tri-
Glycerol + stearic acid	82.2	60.0	64.9	33.8	1.3
Glycerol + oleic acid	33.6	57.2	65.6	33.2	1.1
Glycerol + lauric acid	81.4	67.7	70.8	29.0	0.2

[a] Proportions: 0.63 mole of fatty acid to 1 mole of glycerol, 100°C, 0.4% p-toluenesulfonic acid catalyst, with acetonitrile as azeotroping agent; reaction time 6 hours.

[b] Periodic acid oxidation analysis.

[c] Silica gel chromatography separation.

by the direct esterification of glycerol (0.63 mole of fatty acid to 1 mole glycerol) with stearic, oleic, and lauric acids (*p*-toluenesulfonic acid catalyst) using acetonitrile as the solvent–azeotroping agent (47).

Feuge et al. (63) found that the direct esterification of glycerol is subject to other complex interactions. For example, 1-monostearin could be disproportionated in the presence of *p*-toluenesulfonic acid presumably by ester ester interchange to distearin, and the esterification of 1,3-distearin and oleic acid yielded 75% 1-oleodistearin and 25% 2-oleodistearin. Under controlled conditions, however, simple esterification of 1-monostearin at 80°C yielded as much as 72.3% diglycerides, and esterification of glycerol with stearic acid at 100°C yielded up to 70.1% monoglycerides (both calculated on a glyceride basis) (47).

3.2   ESTERIFICATION WITH OTHER POLYHYDROXY ALCOHOLS

Aside from glycerol, many di- and polyfunctional alcohols offer numerous possibilities for the preparation of useful fatty acid esters. These include ethylene glycol; propylene glycol (food emulsifiers); polyoxyethylene glycols (solvents, plasticizers, and industrial emulsifiers); trimethylolpropane, -ethane, and -butane (synthetic lubricants from saturated fatty acids); pentaerythritol (synthetic lubricants from saturated fatty acids, quick-drying protective coatings from unsaturated fatty acids), sorbitol, sorbitan and their ethoxylated derivatives (food emulsifiers), polyglycerol (viscous industrial emulsifiers), sucrose (cosmetic esters), mannitol, and starch.

In the esterification of ethylene glycol the uncatalyzed rate constants for a whole series of fatty acids (at ratios of 6 moles of glycol to 1–2 moles of fatty acid) at 150–180°C in diphenyl ether were found to be almost identical (64). Esterification catalysts, such as sulfuric acid, phosphorus pentoxide and zinc chloride, were superior to zinc oxide (65). The reaction of oleic acid in toluene could be catalyzed by ion-exchange resins for optimum yields of oleic acid monoester of 41–45% (66).

The kinetics of esterification of diethylene glycol with an equivalent proportion of caproic acid at 166°C were investigated by Flory (67), who observed that the reaction, after the initial stage, had the characteristics of a third-order reaction; that is, there was a linear relationship between the reaction time and the *square* of the concentration of carboxyl groups unreacted, rather than the first power, as in a bimolecular or second-order reaction. This is because the esterification, like that with monohydroxy alcohols, is catalyzed by hydrogen ions, so that the reaction rate is proportional to the product of the concentration of free hydroxyl groups multiplied by the square of the concentration of free carboxyl groups.

Similar characteristics are exhibited by the reaction of lauric acid with lauryl alcohol.

In the Soviet Union, diethylene glycol–mixed diesters have been prepared by the esterification of monoesters with representative synthetic fatty acids (odd and even carbon numbered fatty acids derived by catalytic oxidation of n-paraffinic hydrocarbons) in benzene at temperatures as low as 80–98°C with cationic-exchange resin catalysts; overall yields were 65–85.6% (68).

Optimum conditions for the esterification of propylene glycol to monostearate ester are reported by Chernysheva et al. (69) to be a 10:1 molar glycol/stearic acid ratio at 100–110°C with the use of an ion-exchange catalyst and periodic removal of water by distillation. The esterification was first order.

The esterification of 1,2-propanediol and 1,3-butanediol with stearic acid was investigated by Decouzon and Naudet (70). Table 2.5 shows the molar percentages of products obtained in the esterification of 1 mole of stearic acid with 1 mole of glycol. The ratio of acylation of the primary hydroxyl–secondary hydroxyl for 1,2-propylene glycol and 1,3-butylene glycol were found to be, respectively, 2.1 and 2.45.

There is industrial interest in the esterification of unsaturated fatty acids with pentaerythritol and other polyhydroxy alcohols containing more hydroxyl groups than glycerol, because the esters are useful in quick-drying protective coatings. Blagonravova and Lazarev (71) have reported that pentaerythritol is readily esterified with oleic or linoleic acids at 200–240°C (392–464°F) without a catalyst, and with no appreciable occurrence of oxidation or polymerization. With equivalent proportions of the reactants the reaction was found to be strictly bimolecular; the calculated energy of activation, 10,652 cal./mole, was close to that observed by others for the energy of activation in the esterification of fatty acids with glycerol or monohydroxy alcohols. The two unsaturated acids esterified with equal readiness.

Table 2.5   Molar percentages of components in the products obtained by esterification of equimolar 1,2-propanediol and 1,3-butanediol with stearic acid (70)

Diol Esterified	Reaction Product (Mole %)			
	Free Diol	Primary Monoester	Secondary Monoester	Diester
1,2-Propanediol	22.6	35.4	16.6	25.4
1,3-Butanediol	20.6	37.6	15.3	26.3

Burrell (72) has recommended that pentaerythritol esterification be carried out at 200–230°C (392–446°F) in a vessel equipped with an agitator designed to lift the heavy alcohol from the bottom of the reaction mixture. A 2–3% excess of the alcohol is used. Azeotropic distillation with a high-boiling hydrocarbon, for example, triisopropylbenzene, has been suggested as an alternative to operation under reduced pressure for the removal of water liberated in the reaction. Calcium naphthenate is a recommended catalyst. Litharge and sulfuric or aromatic sulfonic acid catalysts produce dark-colored products.

Konen et al. (73) found that a number of catalysts increase the initial reaction rate but have little or no effect on the time required to reach low acid values of 2–5 (1–2.5% free fatty acids). To avoid excessive heat bodying, they recommended a maximum reaction temperature of 450°F with 5% excess alcohol, and removal of water by stripping with an inert gas.

Mueller et al. (74) have reported that neither lead oxide nor stannous chloride has any appreciable effect on the rate at which tall oil reacts with glycerol, pentaerythritol, or sorbitol.

Recent experience (75) teaches that aluminum, nickel, and zinc alloys function well as esterification catalysts for polyols. Distilled C_{14-18} saturated and unsaturated fatty acids from cottonseed soapstocks were found to esterify completely mixed glycerol and trimethylolpropane after 1.5–2 hours at 195°C with the use of catalytic amounts of 50:30:20 AlNiZn alloy.

The formation of partial pentaerythritol esters is also possible. Those prepared from lauric, stearic, and oleic acids have been ethoxylated for further evaluation (76).

Fatty acid esters of polyglycerols are usually prepared by first polymerizing glycerol with an alkaline catalyst, such as NaOH or KOH, at temperatures of about 280°C for 6–12 hours to the desired polyglycerol mixture (as determined by the development of viscosity, the evolution of water or analysis for hydroxyl value), followed by esterification with the required amount and type of fatty acid without change of catalyst (77). The kinetics of the esterification of diglycerol with stearic acid have been studied (78). The reaction was second order, and the optimum rate was achieved with a catalyst level of 0.5%; higher catalyst levels had no effect on the rate. Baichwal and Lalla (79) recommend 1% NaOH as catalyst (wt./vol.) and the use of a nonoxidizing atmosphere (CO_2) for 8 hours; they prescribe a temperature of 270°C as optimum.

The esterification of fatty acids with the higher polyhydroxy alcohols, such as sorbitol or mannitol, is relatively involved—partly because of the ease with which the esters polymerize and partly because of the marked tendency of free hydroxyl groups of the alcohols to interact with the loss of water and the production of ethers. Konen et al. (73), using a ratio of 4 moles of mixed linseed oil fatty acids to 1 mole of sorbitol and operating at 450°F without a catalyst, were able to obtain an esterified product with an acid number of 22 in 8 hours and an acid number of 10 in 20 hours. The viscosity was 1.5–2.0 poises, indicating very little polymerization. It was found impossible to obtain a low acid number with higher ratios of fatty acids to alcohol.

Brandner et al. (80), from an extensive series of laboratory experiments, concluded that a molar fatty acid-sorbitol ratio of 4.5 to 1 was optimum for the production of varnish oils from linseed oil fatty acids. Calcium and barium acetates, carbonates, oxides, and hydroxides were found to be the most effective catalysts, 0.5% of mixed calcium and barium acetates (ratio 3 to 1) serving approximately to double the esterification rate and produce esters with acid values of 15–20 in 2–4 hours at 180–200°C (356–392°F), plus 8–12 hours at 250°C (482°F). Considerable bodying occurred at the higher temperature. In all cases it was impossible to attain complete esterification of the hydroxyl groups, regardless of the excess of fatty acids employed, and a check of the amount of water evolved indicated considerable ether formation. The reaction was complex but over an intermediate range of esterification had approximately the character of a reaction of the first order.

The ease with which a polyhydroxy alcohol can be esterified is determined largely by the relative proportions of primary and secodary hydroxyl groups, with the latter contributing to nonreactivity. Thus Konen et al. (73) rated dipentaerythritol, pentaerythritol, glycerol, and sorbitol in that order with respect to ease of esterification, whereas Burrell (72) observed that pentaerythritol is more readily esterified than erythritol, and that trimethylolpropane, pentaerythritol, and glycerol are to be placed in the order named.

Despite the fact that oil-soluble ester products prepared by the direct esterification of sucrose, that is, sucrose hexastearate, have desirable shortening additive properties (81), the direct esterication of sucrose is difficult because sugar caramelizes at about 105°C. Alcoholysis methods (Section 4.2) are usually resorted to for the preparation of sucrose and other carbohydrate esters. Starch may be esterified in xylene suspension with an acid catalyst.

Acetate, butyrate, and stearate esters of cornstarch are reported to be prepared by the use of the acid chloride, anhydride, or FFA at 50–75°C in the presence of a small quantity of an acidic catalyst like sulfuric acid. The products have some use as antistatic and optical additives for photographic film (82).

3.3   ESTERIFICATION WITH MONOHYDROXY ALCOHOLS

Esterification reactions of monohydroxy alcohols with fatty acids have been the subject of numerous investigations and reviews (37, 38, 83).

Mechanistically, the reaction occurs through elimination of the hydroxyl group of the acid with the hydrogen of the alcohol, based on investigations with isotopic oxygen in the esterification of benzoic acid with methanol (84):

$$C_6H_5C\underset{O^{16}H}{\overset{O^{16}}{\diagup}} + CH_3O^{18}H \rightarrow C_6H_5C\underset{O^{18}CH_3}{\overset{O^{16}}{\diagup}} + H_2O^{16}$$

In the 1968 publication (Part 5) of his compendium (38), Markley has pointed out the discrepancies in the velocity constants for the direct esterification of various fatty acids with methanol catalyzed by hydrochloric acid, as published in earlier editions of *Fatty Acids* as well as in the *International Critical Tables* (85). His corrections illustrate the fact that with the exception of the three lowest members, the direct esterification of saturated and unsaturated fatty acids in methanol are, irrespective of molecular weight, on the whole similar, provided the double bond is further removed than the 3,4-position. This is in general accord with the work of Sudborough and Gittins (86) and Hartman (48).

The reaction of fatty acids with methanol has probably been the subject of more investigation than any other type of direct esterification. Much of this attention arose because of the necessity to establish early in the development of the gas liquid chromatography (GLC) analytical technique for fatty acid assay a reliable series of laboratory methods for converting many types of fatty acid to methyl esters reasonably rapidly in quantitative yield without fractionation and without the need to remove the water of esterification. Furthermore, it was necessary that complete esterification occur without structural changes of any sort, particularly in the case of certain sensitive unsaturated fatty acids.

For analytical purposes, the assay of small samples permitted the use of up to and over 100 molar excesses of methanol without the need to remove water of esterification and still allowed the rapid attainment of complete reaction with acidic catalysts. Today, laboratory analytical conversions for all but the most extremely sensitive polyunsaturated fatty acids are available (87–92). Nevertheless, special precautions in the handling of certain types of volatile and water-soluble methyl esters are occasionally required.

Development work on the direct esterification of methanol with common fatty acids has included the continuous esterification of lauric acid with vapor phase feed of methanol using $KHSO_4$ as catalyst under reduced pressure (93) and a number of Russian reports dealing with the esterification of synthetic fatty acids (odd and even carbon numbered mixed acids from the catalytic oxidation of n-paraffinic hydrocarbons) with concentrated sulfuric acid as catalyst (94, 95) and uncatalyzed (96). For the C_{10-16} fraction, a 5:1 weight ratio of CH_3OH/fatty acids with 9% H_2SO_4 and two hours reflux appeared to be optimum (94). To avoid side reactions, 110°C was maximum for the acid-catalyzed reaction (95), and a pressure of 10 atmospheres and a weight ratio of 6.7:1 for CH_3OH/fatty acids was optimum for the uncatalyzed reaction (96).

Industrial methyl ester production is carried out from fats and oils by alkaline–catalyzed interesterification (methanolysis) (see Section 4.2) or by the direct esterification of inexpensive fatty acids. In the latter case the esterification can be carried out continuously with excess methanol at pressures of about 150 psig with sulfuric acid as catalyst; about 88–89% conversion to methyl esters is achieved *without the necessity for removal of the water*. Excellent yields of methyl esters can also be obtained by conventional batch esterification procedures in which many acidic catalysts are employed with molar excesses of 5–6:1 of methanol and fractional distillation to separate the water from the methanol. The products may be conveniently distilled if required.

The methyl esterification of acidulated vegetable soapstocks, such as that of cottonseed, soybean, and coconut, presents a unique problem in that these materials usually contain, in addition to fatty acids, 15–25% of neutral oils, and mono- and diglycerides. Attempts to prepare methyl esters with acidic catalysts and excess methanol affords rapid and essentially complete conversion of the free fatty acids but, unfortunately, slow alcoholysis of the oils, and mono- and diglycerides, even at temperatures of 100–200°C and pressures up to 300 psig, occurs. For this reason, a rapid and efficient conversion of acidulated soapstocks to methyl esters by a one-step acid catalyzed process is not yet practical (97). Actually, oils have been converted to methyl esters in the absence of catalysts at high pressures, for example, palm, castor, coconut and menhaden oils at 24:1 molar ratios of CH_3OH/oil at 175–300°C at about 800 psig. The conversions are slow and require 5.5–16 hours in each case (98).

On the other hand, while alkali-catalyzed alcoholysis of oils is reasonably rapid, alkali-catalyzed direct esterification of fatty acids is quite slow. For the efficient methyl esterification of acidulated soapstocks it is usually best to split the soapstocks by the Twitchell process, remove the glycerol, and directly esterify the resulting acids with Twitchell reagent catalysts; thus about 94% conversion to methyl esters can be readily achieved. Otherwise, the acid-catalyzed methyl esterification of unsplit soapstocks affords only 70–76% esterification of the fatty acid radicals. An alternate method consists in first esterifying the free fatty acid with the proper amount of glycerol (uncatalyzed) at temperatures in the range 210–230°C at 5–10 mm pressure, followed by interesterifying the triglycerides so produced with methanol using NaOH, KOH, or ZnO (99).

Fatty acids esterify far more rapidly than rosin acids with low-molecular-weight alcohols such as methyl, ethyl, propyl, and butyl. This difference is the basis for the analytical determination of fatty acids and rosin acids in tall oil. Mixed tall oil fatty acids are refluxed with methyl alcohol containing dry hydrogen chloride gas; the fatty acids esterify whereas the rosin acids are virtually unesterified. Differences in acid number before and after esterification permit calculation of both the fatty acids and the rosin acids (100–103). The McNicol method (100) has been largely displaced today by the Linder–Persson esterification procedure in which water is removed azeotropically (103).

Three other methods are available for the synthesis of monohydric alcohol esters of fatty acids:

1. Reaction of an alcohol and an acid anhydride to form an ester and an acid:

$$(\text{Reaction 5}) \ (R'CO)_2O + ROH \rightarrow R'COOR + R'COOH$$

2. Reaction of an alcohol and an acid chloride with elimination of HCl:

$$(\text{Reaction 6}) \ R'COCl + ROH \rightarrow R'COOR + HCl$$

3. Reaction of an alkyl halide and a metal salt of an organic acid with elimination of the metal halide:

$$(\text{Reaction 7}) \ R'COOM + RX \rightarrow R'COOR + MX$$

3.4   INDUSTRIAL PRODUCTION OF FATTY ACID ESTERS

Production equipment used for the manufacture of fatty acid esters is generally of the batch type and is as versatile as possible so that it can be used for a variety of products and operations with several catalysts. Figure 2.8 is a flowsheet for a versatile esterification unit and auxiliary equipment suited for the batch esterification of common fatty acids with alcohols, glycols, glycerol, or polyols, for azeotropic esterification, for

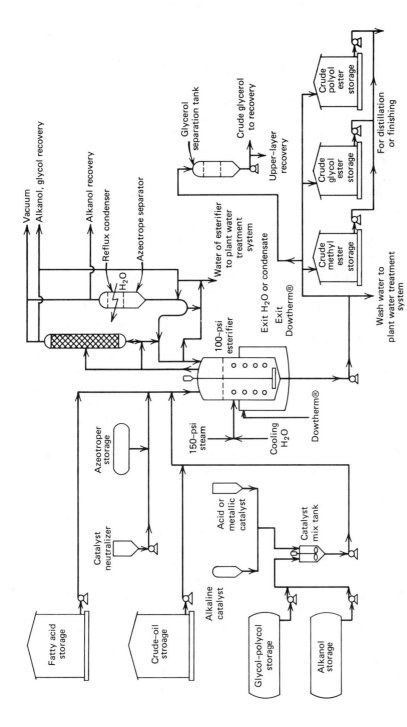

Figure 2.8 Flow sheet for versatile batch esterification plant suited for direct esterification of fatty acids with alkanols, low-molecular-weight glycols, glycerol, and other polyols, for azeotropic esterification, for heterogeneous ion-exchange resin-catalyzed esterification, and for certain fat and oil alcoholyses (methanolysis).

125

heterogeneous ion-exchange resin-catalyzed esterifications, and for certain fat and oil alcoholyses (methanolysis) (see Section 4.2).

Esterifiers used for the direct esterification of fatty acids generally are constructed of glass-lined metal (for low-temperature esterifications) or of acid-resistant materials of construction, such as Monel or, perhaps other recently developed corrosion resistant alloys like E-Brite, Incoloy 825 or Carpenter 20C$_6$ (for high-temperature esterifications). Stainless steel of the 316 type is entirely satisfactory for alcoholysis reactions although 305-type stainless steel or even ordinary carbon steel have been used for equipment restricted only to methanolysis. A compromise choice for a versatile unit would be one constructed from Monel, if highly acidic catalysts like sulfuric or sulfonic acids are replaced by noncorrosive catalysts, such as dibutyl tin oxide (56) and hypophosphorous acid (55).

The vessel contents are heated by internal coils with 150-psig (185°C) or 200-psig (200°C) steam; the coils are also used for circulating cooling water. For ultimate reaction temperatures at 280–300°C, such as during finishing-off stages of polyol esterifications, the unit is equipped with a Dowtherm® heating jacket. It is designed to be used up to 100 psig (for certain low-pressure ion-exchange resin-catalyzed esterifications) without removal of water, and it is also designed for vacuum operation down to 2 mm Hg in order that removal of water of esterification be facilitated and the excesses of hydroxylic or carboxylic components be readily removed by vacuum distillation. Basket-type meshed enclosures within the esterifier to hold heterogeneous esterification catalysts assure further versatility. Provision to monitor layer separation directly from the esterifier or separating tanks, including illuminated sight gauges below them, is required.   In practice, esterification of alcohols, glycols or glycerol is generally accomplished with a mild acid or metallic catalyst in the low boiling alcohol and included in the excess acid in glycerol triesterifications. Esterification procedures vary considerably according to the product desired and its quality. Use of elevated temperatures, azeotropic esterification, or of vacuum (if the boiling point of alcohol or glycol permits it) are employed to remove the water of esterification and to permit the reaction to proceed to completion. In the esterification of low-boiling glycols, such as ethylene and propylene glycols, it is customary to separate the evolved vapors through a packed fractionating column, to discard the water, and to return any separated glycol to the esterification unit. With glycerol or polyols such as pentaerythritol, finishing-off esterification temperatures as high as 260–300°C may be required. The excess of hydroxylic or carboxylic components remaining after esterification is complete is usually removed by distillation under reduced pressure after neutralization of the catalysts (acids with lime, alkalies, with phosphoric acid). Some operators use inert stripping gases such as nitrogen to facilitate the removal of low boiling components.

For process control in the batch esterification of fatty acids with excess alcohols, the determination of acid value (AOCS Te-1a-64) is generally employed; for the esterification of glycerol or polyols where excess fatty acid is used, the determination of hydroxyl value (AOCS Cd-13-60) is appropriate. Obviously, neutralization of either acidic or alkaline catalysts is required after esterification is complete if the ester is to be distilled; occasionally inert or innocuous catalysts are left in the ester, but most frequently it is desirable to remove them by neutralization and/or clay bleaching (which also improves the color of the crude ester) because catalysts are capable of reversing esterification. Tin, antimony, lead, and arsenic-containing catalysts are removed because of their toxicity. Catalyst removal may be monitored by analysis for certain trace metals by atomic absorption analytical methods, applicable to Sn, Hg, Ti, Pb, Cd, B, P, Ca, Mg, K, Na and several others.

Final product quality for fatty acid esters is generally defined by means of saponification number (AOCS Tl-1a-64), acid number, hydroxyl numbers, and fatty acid distribution, although the latter is rarely a purchase specification. Trace metal content, color (Gardner: AOCS Td-1a-64; APHA: AOCS Td-1b-64) and color stability (AOCS Td-3a-64), peroxide value (AOCS Cd-8-53) (particularly for edible products), and ash content (AOCS Tm-la) are other common specification criteria for fatty acid esters.

Interesterification (37, 38, 44, 104–108)

As used here, the term "interesterification" refers to reactions in which a fat or other material composed of fatty acid esters is caused to react with fatty acids, alcohols, or other esters with the interchange of fatty acid groups to produce a new ester. Thus, the reaction of an ester with an acid is called *acidolysis* (Scheme 1, reaction 3a), the reaction of an ester with an alcohol is called *alcoholysis* (Scheme 1, reactions 3b3 and 3c), and the reaction of one ester with another is termed *ester interchange* or *transesterification*.

4.1   INTERCHANGE BETWEEN A FAT AND FREE FATTY ACIDS (ACIDOLYSIS)

Except for sporadic interest acidolysis has attracted less general interest than either alcoholysis or ester interchange. Perhaps this is due to the supposition that these reactions are more sluggish and accompanied by more side reactions than alcoholysis, especially at elevated temperatures. Whatever the cause, the existing technical literature on acidolyis is relatively sparse.

Under the proper conditions, a mixture of fat and free fatty acids will interchange fatty acid radicals. Some of the reported data have concerned the introduction of acids of low molecular weight into a fat made up of higher fatty acids. Normann (109) was able to effect partial replacement of the fatty acids in triglycerides with butyric acid simply by heating the mixture without a catalyst. Schwartz (110) was able to react formic, acetic, or propionic acids with neutral coconut oil at 150–170°C in the presence of sulfuric acid as a catalyst, to produce a stable, low-melting plasticizer for pyroxylin (nitrocellulose).

Meade and Walder (110a,b) carried out the "acetolysis" of tallow and hydrogenated tallow by refluxing the fats for 24 hours in 98% acetic acid (3 vols./1 vol. fat) in the presence of 2 wt % toluenesulfonic acid catalyst. Under these conditions and in the presence of water, it is probable that the 18–18.5% of the triacetin, and the fatty acids and mixed acetoglycerides that were produced resulted from a series of successive hydrolyses followed by acetylations. In the case of tallow, the presence of some acetoxystearic acid could account for part of the approximately 20% loss of unsaturation observed.

Fatty acids of high molecular weight will also displace lower-molecular-weight acids, though perhaps less readily. In a noncatalyzed reaction of coconut oil with cottonseed oil fatty acids (111) for 2–3 hours at 260–300°C, and with removal of liberated lower-molecular-weight free fatty acids under reduced pressure, the saponification value was reduced from 258 to 245.

The method of Grün (112) is easier and accomplishes the desired result as a consequence of alcoholysis and esterification. The fat first reacts with glycerol to produce mono- and diglycerides; the free hydroxyl groups are then esterified with the free acid. The entire process can be conveniently conducted at 225–230°C in a stirred evacuated vessel in the presence of a catalyst, for example, 5% of metallic tin or zinc. The reactions of glycerolysis and esterification can be carried out either successively or concurrently. Gol'dberg et al. (113) used this procedure and an electroconductivity technique for control with oleic acid and refined sunflower oil without a catalyst. At 250°C, using a mixture of 50.7% oil, 14.3% oleic acid, and 26% glycerol, they were able to achieve a 20% weight replacement of the sunflower oil by oleic acid. Capric, stearic, and abietic acids could also be partially exchanged for the acyl groups of the oil. Whether acidolysis competes at all with esterification under the high temperature near-anhydrous conditions is problematical.

As in simple esterification, there should be free acid in excess in the reaction mixture at the conclusion of any of the preceding reactions, and the excess may be removed by either neutralization with alkali or distillation with steam under reduced pressure, if a neutral product is desired.

In all the foregoing processes of free acid interchange or acidolysis the operator has limited control over the composition of the final product. Because random interchange of fatty acid radicals occurs, any specific acid will distribute itself indiscriminately between the esterified and free acid portions. An improved method (114), whereby acids of high molecular weight may be incorporated in the esterified product preferentially, leaving the free acids predominantly those of lower molecular weight, involves operating at subatmospheric pressure in a vessel equipped with a fractionating column through which acids of lower molecular weight can be continuously removed. Zinc soaps are recommended catalysts.

The Eckey process is applicable either for introducing new acids into a fat or for altering the composition of a single fat by reduction of its content of lower-molecular-weight acids. In a typical example 300 parts of palm oil react with 350 parts of mixed palm oil fatty acids and 1.2 parts (0.18%) of zinc oxide at 280–290°C and 60-mm pressure for 2 hours while blowing with steam at the rate of 4.5% per hour. Acids amounting to about 34% of the charge, and consisting largely of palmitic acid (iodine number 2.5) are removed by distillation. The residue of glycerides, which contains 26% free fatty acids, is alkali refined, after which it has an iodine number of 75.8 and a cloud point of 59.3°F, as compared with an iodine number of 49.6 and a cloud point of 84.6°F for the original (refined) oil.

Chakrabarty and Talapatra (115) were able to exchange up to about 14% of the longer chain fatty acids of cottonseed oil, peanut oil, mahua oil (*Madhuca latifolia*), and palm oil with lauric acid using catalytic quantities of sulfuric acid or zinc, calcium, magnesium and aluminum oxides at $150 \pm 2°C$ for 3 hours. Sulfuric acid was found to be the best catalyst with 1 part of oil (by weight) and 1.2 parts of lauric acid for the displacement of high-molecular-weight fatty acids from an oil by low-molecular-weight fatty acids. Further, linoleic acid was shown to be preferentially displaced over oleic acid in an amount varying with its initial content in an oil with a corresponding increase in saturated acids content. Fatty acids above C_{18} were not displaced in the case of peanut oil. Chakrabarty and Kundu (116) also replaced 17% of the oleic acid in kusum oil (*Schleichera trijuga*) with lauric acid using sulfuric acid as catalyst and a 1:1 kusum oil/lauric acid ratio.

4.2 INTERCHANGE BETWEEN A FAT AND A FREE ALCOHOL (ALCOHOLYSIS)

The interchange of fatty acids between a fat and an alcohol, or alcoholysis, is analogous to the reaction of acidolysis and is the most important interesterification reaction from the practical point of view. Whereas in acidolysis there is an excess of carboxyl groups in the reaction mixture, to which available hydroxyl groups are redistributed, in alcoholysis there is an excess of hydroxyl groups, to which available carboxyl groups are redistributed. The redistribution, as in acidolysis, is wholly random unless one or more of the reaction products is removed from the reaction as it is formed. Alcoholysis of triglycerides, then, will normally produce not only free glycerol and triglycerides of altered or rearranged structure, but also mono- and diglycerides, as well as esters of the new alcohol, and also partial esters of this alcohol, if it happens to be polyhydroxy. In a homogeneous reaction mixture, with the proportions of the reactants known, the composition of the mixture at equilibrium can be calculated from probability considerations (117, 118). Often, however, the reaction is carried to substantially complete interchange of the two alcohols, by removing the originally esterified alcohol from the reaction zone as rapidly as it is set free.

The alcoholysis of fats is important commercially. It offers a means of preparing fatty esters other than glycerides and is often more suitable and convenient than the alternative method of splitting the fats and reesterifying with the particular alcohol.

Alcoholysis reactions have perhaps been studied and commercially developed to a greater extent than either of the other two interesterification reactions. As a consequence, two industrial applications, namely, the methanolysis of fats to methyl esters, and the glycerolysis of fats to mixed mono- and diglycerides have largescale industrial importance.

Aside from its industrial utility, methanolysis has been thoroughly investigated because of the necessity to calibrate and quantitate the GLC analytical technique for fatty acid distribution in fats and oils. A reliable series of laboratory methods has been developed for converting many types of fats and oils to methyl esters reasonably rapidly and directly in quantitative yield without fractionation and without the need to remove byproduct glycerol. (Although fatty acids can be used for GLC (119–123), methyl esters are preferred.) Furthermore, in order for the technique to be adaptable to the analytical methods, it is necessary that alcoholysis occur without structural changes in polyunsaturated fatty acid components.

For analytical purposes, the assay of small-sized samples frequently permitted the use of up to and over 100 molar excesses of methanol without the need to remove by product glycerol and still allowed the rapid completion of reaction with acidic or basic catalysts. Today, standard laboratory conversions for all but the most sensitive fats and oils are available (124–131). Nevertheless, special precautions in the handling of certain types of methyl ester are required, such as those from polyunsaturated oils containing five or more double bonds (132,133), low-molecular-weight methyl esters like methyl butyrate obtained from the methanolysis of butterfat (134), or, occasionally, those from olive oil (128,129).

Alcoholysis to Produce Monoesters.   The alcoholysis of a fat with a monohydroxy aliphatic alcohol of low molecular weight, such as methanol or ethanol, may be catalyzed by either acid or alkali, but the alkali-catalyzed reaction is generally superior in speed, completeness, and the relatively low temperature at which it can be effected.

Displacement of the glycerol in a fat by a low-molecular-weight alcohol, such as methyl or ethyl alcohol, is readily accomplished on a large scale by the method described by Bradshaw and Meuly (135). The process is remarkable not only for producing methyl or ethyl esters directly from the fat, without intervening hydrolysis, but also for taking place at low temperatures, and requiring no alloy steel or other special corrosion-resistant equipment. When methyl alcohol is used, the reaction with a hypothetical triglyceride is as follows:

$$CH_2O_2CC_{11}H_{23}$$
$$|$$
$$CHO_2CC_{13}H_{27} + 3CH_3OH \rightarrow C_3H_5(OH)_3 + 2C_{11}H_{23}CO_2CH_3 + C_{13}H_{27}CO_2CH_3$$
$$|\qquad\qquad\text{Methyl alcohol}\quad\text{Glycerol}\qquad\text{Methyl laurate}\qquad\text{Methyl myristate}$$
$$CH_2O_2CC_{11}H_{23}$$
$$\text{Fat}$$

In the past the reaction was carried out in open tanks constructed of ordinary carbon steel; today it is preferred to use sealed vessels (methanol vapors are highly toxic) of at least 305 stainless steel quality (for maximum equipment versatility), although ordinary carbon steel is satisfactory if the equipment is restricted to simple alcoholysis reactions (see Section 3.4). The fat must be clean, dry, and substantially neutral. It is heated to about 80°C (176°F), and to it is added commerical anhydrous (99.7%) methyl alcohol in which is dissolved 0.1–0.5% sodium or potassium hydroxide. The quantity of alcohol recommended is about 1.6 times that theoretically required for the reaction, although the alcohol may be reduced to as little as 1.2 times theoretical, if the operation is carried out in three steps. Alcohol amounting to more than 1.75 times the theoretical quantity does not materially accelerate the reaction and interferes with subsequent gravity separation of the glycerol.

After addition of the alcohol, the mixture is stirred for a few minutes and is then allowed to stand. The glycerol begins to separate almost immediately; since it is virtually anhydrous and much heavier than the other liquids, it readily settles to form a layer at the bottom of the tank. Conversion of the oil to methyl esters is usually 98% complete at the end of an hour.

The lower layer of glycerol contains not less than 90% of the glycerol originally present in the fat; the upper layer consists of the methyl esters, most of the unreacted alcohol and alkali, the remainder of the glycerol, and a small amount of soap. These impurities are removed from the esters by successive washes with small amounts of warm water.

The Bradshaw patent contemplates use of the methyl esters to make anhydrous soap by a continuous process. The esters are easily saponified by caustic soda or caustic potash at a low temperature, and the methyl alcohol is readily recovered for reuse. The process may also constitute a valuable means of obtaining monoesters for fractionation to produce "tailor-made" fats and oils. The methyl and ethyl esters of fatty acids are fluid, relatively stable, noncorrosive, and low-boiling derivatives, and in certain operations are preferred to the free fatty acids. Methyl esters are now much preferred over the ethyl esters for reasons of lower cost of manufacture and better pyrolytic stability during processes such as fractional distillation.

A series of patents (136) describes various continuous means for conducting the alcoholysis of fats with lower alcohols and other alcohols, and of separating and purifying the reaction products. In a review by Wright et al. (137) the proper conditions for alcoholysis of fats with methyl and ethyl alcohols have been considered in detail. The alkali-catalyzed alcoholysis method is completely successful only if the fat is almost neutral and the reaction mixture is substantially anhydrous. Failure to comply with either of these conditions causes soap formation, which leads to a loss of alkalinity and also the building up of a gel structure that prevents or retards separation and settling of the glycerol.

In ethanolysis difficulty may be anticipated if the free fatty acid content of the oil exceeds about 0.5%. In the reaction of 30 parts of ethyl alcohol, 100 parts of cottonseed oil, and 0.5% sodium hydroxide, the yield of glycerol is appreciably lowered by the presence of only 0.3% water in the reaction mixture. However, the effect of moisture may be partially compensated by use of additional alkali or alcohol. The water tolerance of this mixture is increased to 0.5–0.6% if the amount of catalyst is doubled or the amount of alcohol is increased to 40 parts, or to at least 0.8% if both catalyst and alcohol are so increased.

It was demonstrated by Wright et al. (137) that the rate of the overall reaction was limited principally by the time required for gravity separation of glycerol, since continuous centrifugal separation at 65°C with a holding time of only 5 min gave a fairly good yield of glycerol (about 85% of the theoretical). The claim of Bradshaw and Meuly (135) that less alcohol is required in incremental addition and separation of glycerol was confirmed in methanolysis, but not in ethanolysis, where this method caused gel formation. Acid catalysis of the reaction was found unsuitable, owing to the large excess of alcohol required. Although propyl alcohol was found to be less reactive than the lower alcohols, good results could be obtained by using a large excess.

The interesterification of peanut oil with ethanol has been studied in detail by Feuge and Gros (118), who found that the optimum temperature for the reaction is about 50°C (122°F). A higher yield of free glycerol is obtained at this temperature than at either 30°C (86°F) or 70°C (158°F).

Pathak and Bhatnagar (138) found that the methanolysis or butanolysis of castor oil could be conducted efficiently at relatively low temperatures. In the formation of crude methyl ricinoleate, a $\geqq 12:1$ molar ratio of methanol to castor oil at 30°C with 1% alkaline catalysts (based on oil) was effective; for crude butyl ricinoleate, a $6:1$ molar ratio of butanol to castor oil was sufficient.

A methanolysis reaction has been suggested as the second step in a patented two-step process (99) for the efficient conversion of acidic oils to methyl esters. The free fatty acids are first esterified uncatalyzed with added glycerol at 210–230°C at 5–10 mm and are then interesterified with methanol and alkaline catalysts to afford yields of totally available fatty acid components from the acidic oils of well over 90%.

The reverse of the preceding reactions in which methyl or other low-molecular-weight alcohol is displaced from its esters by glycerol to yield triglycerides and free alcohol is also possible, although slightly more difficult (see section on glycerolyis of methyl esters).

Glycolysis.   Some interesting glycolysis reactions have been reported with both fats and methyl esters using ethylene glycol and polyoxyethylene glycols. Burrell (72) evaluated the alcoholysis of soybean oil with ethylene glycol using several catalysts; of these calcium napthenate proved to be most effective. Celades and Paquot (139) found that the alcoholysis of methyl stearate (0.038 mole) with triethylene glycol, $HO(C_2H_4O_3)_3H$

(0.0609 mole) was complete in 15 min, as measured by the evolution of methanol at 100°C, if metallic sodium was added to the dry solution of reactants.

Glycerolysis of Fats (Manufacture of Mono- and Diglycerides).   About 80% of the reports in *Chemical Abstracts* on alcoholysis during 1962–1978 deal with fat glycerolysis. There are numerous commercial applications for the mixed mono- and diglyceride products of these reactions.

The manufacture of commercial mono- and diglyceride mixtures represents a special case of alcoholysis in which either fatty acids or a fat are caused to react with an excess of glycerol. It is carried out on a considerable scale as a preliminary process in the preparation of alkyd resins, to produce edible, oil-soluble, surface-active agents, and also for the production of detergents of the sulfated monoglyceride type. Mono-/diglyceride mixtures are useful intermediates for acetostearins (140), lactopalmitins and -stearins (141–143), diacetyltartaric acid esters (144, 145), succinylated monoglycerides (146–149), stearyl monoglyceride citrate (150), and ethoxylated monoglycerides (151–154) for use in the food industry (44, 155).

In the protective coating industry the reaction is usually carried out in the same kettles as are used for further reaction with resins, with the entire process being completed without cooling the initial reaction mixture or removing the alcoholysis catalyst. For edible use, mono- and diglycerides are somewhat more carefully prepared in special kettles, which are equipped with mechanical agitators and means for producing a vacuum. Ordinary carbon steel is not a suitable material for the kettles, as it contaminates the product excessively with iron soaps and produces a dark product; nickel, and 305 or 316 (preferably) stainless steels are satisfactory.

Figure 2.8 represents a flow sheet for versatile plant equipment suited for glycerolysis with certain modifications. The packed fractionating column is unnecessary. The internal coils used to heat the reactants with either 150- or 200-psig steam are also superfluous for heating and are used only to cool the finished products with cooling water. Usually, air-cooled vertical-pipe-type condensers are mounted on the esterifiers in order to condense and return glycerol to the unit or perhaps low-boiling monoglycerides (especially in the case of coconut oil glycerolysis). Provision to remove excess glycerol from the esterifier at the conclusion of the reaction by vacuum distillation is ordinarily provided. The lines for glycerol removal are constructed of large-diameter pipe to permit rapid removal of excess, thus minimizing product ester reversion at high temperatures. Agitators must have a high degree of torque to permit adequate mixing of the heavy immiscible glycerol with the fat early in the reaction during the heat-up stage.

To the fat charge is added about 25–40% of its own weight of glycerol and 0.05–0.20% of an alkaline catalyst, usually sodium hydroxide. Hydrated lime, Ca(OH)$_2$, optimally used at 0.06–0.10%, affords better colored products. The catalyst reacts with the fat to form soaps, which promote the reaction, at least in part, by increasing the solubility of glycerol in the fat phase (156). Heating of the charge is begun under reduced pressure, to remove dissolved air, and thereafter it is protected from oxidation by blanketing with an inert gas, such as nitrogen. (The previous practice of using hydrogen for blanketing is now being displaced as too hazardous.)

The temperature is carried to a maximum of 400–475°F, usually by Dowtherm heating, and reaction is complete in 1–4 hours. At the end of this time an equilibrium will have become established between the free glycerol in the mixture and the reaction products.

The reaction is conducted at a high temperature partly to hasten the reaction and partly to increase the miscibility of the reaction mixture with glycerol. As Hilditch and Rigg (43) pointed out some time ago, the extent to which reaction can be carried is limited by the comparative immiscibility of glycerides and free glycerol. Any glycerol forming a second liquid phase when equilibrium is reached cannot participate in the reaction. The maximum amount of glycerol reactable and miscible with highly hydrogenated cottonseed oil plus 0.1% sodium hydroxide (about 0.75% soap) has been determined at different temperatures by Feuge and Bailey (117) (Figure 2.9); presumably other fats of comparable molecular weight yield similar results, although with coconut oil, and so on, the solubility of glycerol is considerably greater. It will thus be seen that the proportion

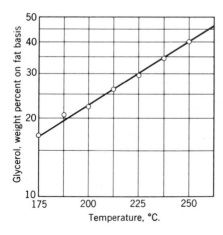

Figure 2.9 Maximum amount of glycerol reactable and miscible with highly hydrogenated cottonseed oil in the presence of soaps from 0.1% NaOH (fat basis) (117).

of glycerol that may be profitably used is dependent on the temperature, and varies from about 22.5% of the fat at 200°C (392°F) to 40% at 250°C (482°F). Larger percentages of glycerol may be used if a large amount of soap-forming catalyst is employed (157), or if the reaction is conducted in a high-boiling solvent in which glycerol and glycerides are mutually soluble. Phenol or cresols were recommended as solvents by Hilditch and Rigg (43). The use of dioxane has been patented by Richardson and Eckey (158).

Illustrative of the advantage of employing a solvent for glycerolysis is the work of Franzke et al. (159) on sunflower oil. Heating at 115–120°C for 3 hours using 10 parts of oil and 20 parts of glycerol with 0.3 parts of $NaHCO_3$ catalyst in excess pyridine affords, after removal of the excess glycerol on completion of the reaction, a 75% α-monoglyceride or 83% total monoglyceride yield; the same ratio of reactants at 200°C without a solvent yields 52% α-monoglyceride and 58% total monoglycerides. McKinney and Goldblatt (160) have described a process in which 78% monoglycerides were prepared by the glycerolysis of tung oil with sodium methoxide in pyridine solution, but Eaves et al. (161) could not duplicate these results. They obtained optimum yields of about 50% in laboratory-scale equipment and 32–45% yields in the pilot plant.

Several exhaustive studies have been carried out to determine the solubility of glycerol in fats at various temperatures (48, 162, 163), and, in general, it has been repeatedly stated that the degree of glycerolysis of fats is limited and determined by the miscibility of glycerol in the "fat." However, during glycerolysis the composition of the reaction mixture gradually changes as reaction progresses, that is, 1-monoglyceride content increases as triglyceride content diminishes, and the solubility of glycerol in the near-equilibrium mixture is, presumably, much higher than it is in the triglyceride starting material. Thus conclusions based on the solubility of glycerol in the starting materials are not valid when used to predict the optimum ratios of miscible reactants for any conversion to monoglycerides. The determination of the solubility of glycerol in fats without interchange catalysts present is not possible at temperatures above about 200°C since uncatalyzed interchange has already commenced at these temperatures. The assumption that glycerolysis is a strictly random distribution of acyl groups among all the "available" hydroxyl groups is not consistent with knowledge that the esterification rate (direct or interchange) of primary and secondary hydroxyl groups differs (164).

An illuminating examination of the intermolecular and intramolecular equilibria in fat glycerolysis by Brandner and Birkmeier (164) has revealed some interesting aspects of this complicated reaction: (a) the preference for primary hydroxyl acylation, as represented by appropriate equilibrium constants, is above 1; at 200°C about 2.3; and at room temperature, from 6 to 10; (*b*) there is a shift from β- to α-groups on cooling and at room temperature the equilibrium is only slowly attained in the solid state; (*c*) intermolecular rearrangement of acyl groups is largely arrested by cooling with hard monoglycerides, with or without neutralization of the catalyst; and (*d*) the equilibrium mixture of monoglycerides at 100°C is about 90% 1-monoglyceride.

Batch methods for the preparation of edible mono- and diglycerides are described in a number of patents, including those to Edeler and Richardson (165), Christensen (166), and others (167, 168). Continuous manufacture has been described by Arrowsmith and Ross (169), Birnbaum (170), Birnbaum and Lederer (171), Giddings and Davis (172), Alsop and Krems (173), and Chang and Wiedermann (174). Allen and Campbell (175), describe a continuous process applicable to soybean oil and operative at slight pressure (40 psi) reported to give α-monoglyceride product in 58.2% yield. The reaction between fat and glycerol will take place without a catalyst, but the rate is reasonably rapid only at very high temperatures (500–550°F).

Data for the rough estimation of the composition of commercial mono- and diglyceride preparations, based on hydrogenated cottonseed oil with an average molecular weight of 876.2, are given in Table 2.6. By reference to this table and Figure 2.9, it will be seen, for example, that at equilibrium, with a moderate amount of catalyst, the fat is miscible with about 30% of its own weight of added glycerol at 225°C (347°F) and that with this proportion of glycerol the glyceride portion of the reaction mixture will consist of about 58% (total) monoglycerides, 36% diglycerides, and 6% triglycerides by weight.

In the past, commercial preparations of so-called "60% monoglycerides" have been assumed to be all α-monoglycerides, easily analyzed by the conventional periodic acid method (176). From a direct analysis of these products for glycerol and α-monoglycerides it could be estimated that the composition conformed to 50–60% α-monoglycerides and 30–45% diglycerides. However, it is now well appreciated and confirmed by Brokaw et al. (177) that β-monoglycerides are also present, usually to the extent of 5–8% of the total monoglycerides. β-Monoglycerides have always been assumed to be functionally equivalent to α-monoglycerides in performance and application, but they do not respond to periodic acid (they are not vicinal glycols). They can be largely isomerized to α-monoglycerides for example, by perchloric acid isomerization (177a), and the entire monoglyceride content determined by periodic acid analysis, but this is not a feature of the standard method of analysis.

Theoretically, it should be possible on examination of the data in Table

Table 2.6 Theoretical composition of products prepared by reaction of hydrogenated cottonseed oil with glycerol (117)[a]

Percent of Total OH Groups Esterified	Parts Glycerol Reacted with 100 Parts Fat	Composition of Reaction Mixture (Mole %)				Composition of Fatty Product[b] (wt. %)			
		Tri-	Di-	Mono-	Free Glycerol	Mono-	Di-	Tri-	Combined Glycerol Content (%)
100	0	100.0	0	0	0	0	0	100.0	10.51
95	0.55	85.7375	13.5375	0.7125	0.0125	0.3	9.9	89.8	11.00
90	1.17	72.9	24.3	2.7	0.1	1.2	18.7	80.1	11.53
85	1.86	61.4125	32.5125	5.7375	0.3375	2.7	26.4	70.9	12.10
80	2.63	51.2	38.4	9.6	0.8	4.7	32.9	62.4	12.71
75	3.50	42.1875	42.1875	14.0625	1.5625	7.3	38.2	54.5	13.35
70	4.50	34.3	44.1	18.9	2.7	10.5	42.5	47.0	14.03
65	5.66	27.4625	44.3625	23.8875	4.2875	14.1	45.6	40.3	14.74
60	7.01	21.6	43.2	28.8	6.4	18.3	47.7	34.0	15.48
55	8.60	16.6375	40.8375	33.4125	9.1125	22.9	48.8	28.3	16.25
50	10.51	12.5	37.5	37.5	12.5	28.0	48.8	23.2	17.06
45	12.85	9.1125	33.4125	40.8375	16.6375	33.6	47.8	18.6	17.87
40	15.77	6.4	28.8	43.2	21.6	39.6	45.9	14.5	18.71
35	19.52	4.2875	23.8875	44.3625	27.4625	46.0	43.0	11.0	19.57
30	24.52	2.7	18.9	44.1	34.3	52.7	39.3	8.0	20.45
25	31.53	1.5625	14.0625	42.1875	42.1875	59.8	34.7	5.5	21.35
20	42.04	0.8	9.6	38.4	51.2	67.3	29.2	3.5	22.27
15	59.56	0.3375	5.7375	32.5125	61.4125	75.1	23.0	1.9	23.22
10	94.59	0.1	2.7	24.3	72.9	83.0	16.1	0.9	24.16
5	199.69	0.0125	0.7125	13.5375	85.7375	91.4	8.4	0.2	25.09
0	∞	0	0	0	100.00	100.0	0	0	26.05

[a] Calculated on basis of random distribution in a homogeneous reaction mixture. Average molecular weight of the fat, 876.2.
[b] Composition of reaction product freed of uncombined glycerol.

2.6 and Figure 2.9 to carry out the reaction at 250°C and achieve a product of 66% monoglyceride: 34% diglyceride content. This is never achieved in the conventional nonsolvent method of production. Failure to achieve this is probably due to the inapplicability of several of the basic assumptions on which the data in Table 2.6 were based, and also to the reversion of monoglycerides that occurs when reaction is complete but before catalyst neutralization or after it if neutralization is incomplete. During the cooling that precedes glycerol removal (by either washing or vacuum distillation), glycerol is less soluble in the fatty phase and separates into the lower layer. This shifts the equilibrium somewhat, and diglyceride and triglyceride are regenerated. At the highest temperatures that have been used, even complete catalyst neutralization does not guarantee that reversion will not occur. The catalyst neutralization and removal of glycerol are thus critical steps in production and must be carried out with care and control.

Gooding and Vahlteich (157) have reported that by using as a catalyst sufficient sodium bicarbonate to convert about 13% of the fat to soap they were able to produce from hydrogenated peanut oil a product containing an average of only 1.24–1.26 fatty acid radicals per molecule of combined glycerol. This corresponds to about 79–82% by weight of monoglycerides. Fats of even higher monoglyceride content may be produced by fractionating commercial monoglyceride preparations by molecular distillation or solvent crystallization. Eckey and Formo (178) have shown that by the "directed interesterification" technique involving concurrent ester interchange and crystallization, a fat such as cottonseed oil can be made to precipitate its saturated fatty acids in the form of nearly pure monoglycerides or diglycerides.

By molecular distillation, technical grades containing 40–50% monoglycerides will give distillates of 90–97% monoglycerides. Three-stage, centrifugal, short-path distillation produces practically instantaneous removal of monoglycerides with limited decomposition (195).

Monoglyceride preparations may be washed with concentrated solutions of sodium chloride, sodium sulfate, and so on for the removal of dissolved free glycerol, but the operation is difficult and not altogether satisfactory because of the high surface activity of the material and the ease with which emulsions are formed. Frequently free glycerol is removed by steam stripping under reduced pressure. Monick (179) points out that, based on 6% usage in emulsifier-type shortening, a maximum of 0.02% free glycerol is required in these shortenings to guarantee a smoke point of 340°F or better. This corresponds to a 0.30% maximum content in free glycerol for a monoglyceride emulsifier. He found that a two-stage extraction with 5% NaCl solution reduced free glycerol from

5.51–0.14%; furthermore, a continuous process for evaporation in a thin-film evaporator operated at 340°F, and 6 mm Hg absolute could be used to reduce the free glycerol content of commercial monoglycerides to below 0.30%.

The use of pressure in glycerolysis reactions appears to assist in the attainment of a more homogeneous reaction system. Castor oil glycerolysis is claimed (180) to produce 82% α-monoglyceride when reaction is conducted for 2 hours at 240°C with 800 grams of the oil and 2000 grams of glycerol if carbon dioxide, probably a coblanketing/catalytic agent, is used to 100 psig. Under similar conditions, coconut oil is said to afford 74.5% and peanut oil 73% α-monoglycerides.

The effect of small amounts of water (4–5% in the glycerol) on castor oil glycerolysis was studied by Dey et al. (181) at 250°C and 200 psig of CO_2. If the split acids were reesterified by depressurizing and heating for a short period of time, the effect of water was to increase the yield of monoglycerides.

Rheineck et al. (182) studied the optimum glycerolysis of linseed oil at a 4:1 weight ratio of oil to dry glycerol at four temperatures between 150° and 225°C using several catalysts, such as NaOH, litharge, and sodium glyceroxide. They analyzed the equilibrium products by TLC with respect to 1-mono-, 2-mono-, 1,2-di-, 1,3-di-, and triglycerides and determined the fatty acid distribution of each by GLC. A temperature of 200°C and 0.18% NaOH was about optimum, and equilibrium was attained in 2 hours. Sodium glyceroxide at 200°C was found to be just as effective as NaOH at 225°C. Generally, the attainment of the alcohol solubility point (a practical test used in the production of linseed oil monoglycerides for eventual alkyd resin manufacture) was at 35% 1-monoglyceride content, which is about 10% lower than the equilibrium concentration. Surprisingly, the authors found no particular difference between the fatty acid distribution in the raw material oil and the partial glycerides.

The controlled glycerolysis of tristearin may be used to produce high diglyceride of the *sym*-1,3-type, according to a patent assigned to the Glidden Company (183). Thus if 1 kg of tristearin is subjected to glycerolysis with only 52 grams of glycerol and 5 grams of CH_3ONa at 150°C for 2–3 hours and is then cooled to 80°C and the product recrystallized from tetrahydrofuran, the crystallized product consists of 13.3% triglyceride, 78.8% 1,3-diglyceride, and 5.2% monoglyceride. A recrystallization from isopropanol gives essentially pure 1,3-diglyceride.

Sperm whale oil can be transformed into a good quality "40% monoglyceride," according to Fujita et al. (184), if a sufficient glycerol/oil ratio is selected to afford a 35–44% α-monoglyceride product, using 1% K_2CO_3 at 200°C under 15 mm vacuum. Over 4 hours, 97% of the unsaponifiables are removed by distillation. Reports of other monoglyceride preparations include the glycerolysis of palm oil (185, 186) and investigations of a general nature (187, 188). Other monoglyceride processes include the use of glycidol (189) or 1,2-isopropylidene glycerol (190, 191) as intermediate reactants with fats.

The production of superglycerinated shortening and the behavior of mono- and diglycerides in the deodorization of fats is worthy of comment here. Addition of 6–8% mono- and diglycerides to shortening affords products known in the trade as *superglycerinated, emulsifier-type,* or *high-ratio* shortenings; these contain more combined glycerol than ordinary hydrogenated fats and oils. They allow the use of a higher ratio of sugar to flour and still permit the production of light and full-volumed cakes and are very popular in cakes, pastry, and pies, and are used substantially in the household as well as in commercial bakeries.

If mono- and diglycerides are added to the fat before deodorization, most of the excess combined glycerol will be driven off and mono- and diglycerides will be converted to triglycerides. If the mono- and diglycerides are to be incorporated into shortening to produce a superglycerinated product, the most satisfactory practice is to add them to the shortening while the latter is in the deodorizer, but near the end of the deodorization period, according to the method of Richardson and Eckey (192). If the mono- and diglycerides are prepared originally from a deodorized fat, they will require very little steam stripping to constitute an acceptable ingredient of an edible fat product.

Decomposition of the mono- and diglycerides is facilitated by the presence of catalysts. The experimental results in Table 2.7 are typical. They comprise the results of a series of tests in which hydrogenated cottonseed oil was treated with 22% of its weight of glycerol and 0.05% of sodium hydroxide, and then steam deodorized at 500°F for various lengths of

Table 2.7   Effect of steam deodorization at 500°F on decomposition of mono- and diglycerides (193)

Sample	Deodorization Time (hours)	Glycerol Content (%)	Excess Glycerol (%)
Original oil	—	11.25	—
Mono- and diglycerides, catalyst	2	14.85	3.60
removed	6	13.72	2.47
Mono- and diglycerides, catalyst not	2	11.65	0.40
removed	4	11.46	0.21
	7	11.30	0.05

time, with and without decomposition of the catalyst with strong phosphoric acid. In the strict absence of water or any foreign material acting as a catalyst, Ross et al. (193) found that pure monoglycerides could be heated to 200–230°C and distilled without substantial decomposition. Similarly, these workers found that anhydrous glycerol and fat, free of soap, did not react appreciably below decomposition temperatures. The reaction between ordinary commercial materials is apparently catalyzed by traces of water and of the residual soap always found in alkali-refined oils.

Feuge and Bailey (117) have shown that products partially decomposed by steam deodorization do not have the same composition as normal products of equivalent combined glycerol content. As shown in Figure 2.10, deodorization has the effect of increasing the diglyceride content at the expense of the monoglyceride and, more particularly, the triglyceride content of the mixture. Similar effects in deodorization are obtained with and without removal of the interesterification catalyst. In the manufacture of edible products it is important to remove the soaps formed by alkali or metal catalysts, since they have a deleterious effect on the keeping quality of the fat.

As oil-soluble surface-active materials, mono- and diglyceride preparations are usually judged on the basis of their effectiveness in lowering

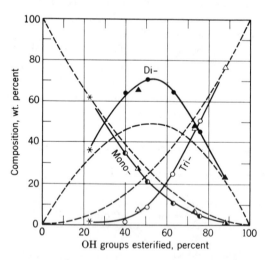

Figure 2.10 Composition of a mono- and diglyceride preparation after partial decomposition by steam deodorization (solid curves and plotted points represent compositions determined experimentally after deodorization at different temperatures, with and without destruction of the catalyst; broken curves represent compositions corresponding to random distribution) (117).

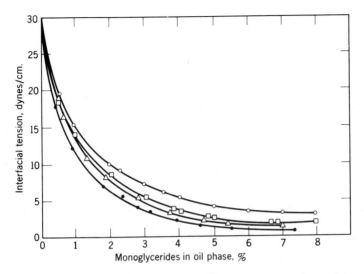

Figure 2.11 Interfacial tension against water of refined cottonseed oil containing varying percentages by weight of commercial distilled monoglycerides (ca. 93–95% pure). Curves from top to bottom of diagram are for monoglycerides of mixed cottonseed oil fatty acids, triple pressed stearic acid, palmitic acid, and lauric acid (195).

the interfacial tension of fats against water. The interfacial tension of a neutral oil such as cottonseed, soybean, or peanut oil against water will be approximately halved by the addition of 1% saturated monoglycerides and reduced to less than 1 dyne/cm by the addition of 6%. In commercial preparations it has been shown by Feuge (194) that lowering of the interfacial tension is caused almost wholly by the monoglycerides. Since common oils and fats have almost equal interfacial tensions (29.5–30.5 dynes/cm at 70°C), and also nearly identical interfacial tensions at equivalent monoglyceride concentrations, it follows that the monoglyceride content of a product may be estimated roughly from the very simple determination of its interfacial tension by the ring method, or in terms of "drop numbers."

The data of Kuhrt et al. (195) on the effect of different monoglycerides on the interfacial tension between refined cottonseed oil and water are shown in Figure 2.11. Additional data have been published by Feuge (194) and by Eckey and Formo (178).

Glycerolysis of Methyl Esters. The glycerolysis of methyl esters, which is the reverse reaction of the methanolysis of fats (Scheme 1, reaction 3b3), can be carried out by removing evolved methanol from the reaction, generally under reduced pressure.

Glycerolysis of methyl esters of lauric, palmitic, oleic, and linoleic acids has been studied in the Soviet Union (196) with a view to the synthesis of margarinelike oils. Cocoa bean fatty acid methyl esters have also been investigated for glyceride production (197).

Glycerolysis of methyl esters for monoglyceride production can be achieved if the proportions of reactants and conditions are adjusted. The methyl esters of safflower oil, tall oil, rice bran oil, sperm oil, and palmitic acid have been converted into 1-monoglycerides assaying 50.6–54.9% by Matsuyama et al. (198) by using 1 mole of methyl ester to 1.5–4.0 moles of glycerol with 0.1% KOH catalyst (on wt. of ester) at 215–220°C for 25–30 min and removing methanol by distillation. The relative yields from shorter-chain methyl esters, like methyl laurate and myristate, and from unsaturated acids, like oleic acid, are somewhat higher in monoglycerides than those from the longer-chain acids like palmitic and stearic acids under comparable conditions (199).

Essentially pure 1-monopalmitin was prepared by Matsuyama et al. (199a) from the 60.8% 1-monopalmitin produced by the glycerolysis of high-purity methyl palmitate. Careful distillation at about 150°C at 0.02 mm removed the excess glycerol, another distillation at 150–160°C at 0.008 mm gave a distillate, which after two recrystallizations from 1:1 hexane ether, gave 99.0% minimum purity 1-monopalmitin (assay by GLC of trimethylsilyl ether derivative).

Alcoholyses of Methyl Esters or Triglycerides with Tetrahydroxy-and Higher Alcohols.   The alcoholysis of fat with tetrahydroxy- or higher polyhydroxy alcohols has assumed industrial importance. The chief application is to convert drying oils to products of higher functionality and enhanced capacity for polymerization and quick and hard drying by reaction with higher polyols, such as pentaerythritol or sorbitol.

Burrell (72) found the alcoholysis of fats with pentaerythritol to be a convenient and economical method for preparing partial esters of mixed glycerides for further reaction with drying oil acids to form improved drying oils. With 1% calcium naphthenate as the catalyst the alcoholysis of refined linseed oil with pentaerythritol is complete in a few minutes at 230°C.

Wright et al. (200) have shown that the alcoholysis of glycerides with high-boiling alcohols, such as pentaerythritol, mannitol, sorbitol, methyl, and ethyl glycosides, and polyglycerols, can be carried to the point of substantial elimination of the glycerol. It is recommended that the oil be heated to 200–250°C with 1.0–1.5 equivalents of the alcohol plus 0.1–2.0% of a catalyst, such as lead salt or an alkaline compound. Glycerol is removed continuously by stripping with a current of steam or organic-solvent vapors under a reduced pressure of 40–300 mm. Analyses of a product so obtained were not given, but presumably it is a mixture of partial esters of the higher polyhydroxy alcohol.

Linseed oil (2320 parts) was heated by Mleziva and Hanzlik (201) in an atmosphere of CO_2 for 4 hours at 260°C while pentaerythritol (618 parts by weight) was added with stirring over a period of 2 hours; heating was continued for 9 additional hours. At the expiration of this time the reaction mixture had the following composition: pentaerythritol monoester 27%, pentaerythritol diester 53%, monoglyceride 15.9%, and pentaerythritol polyesters and triglycerides 4.1%.

Wright et al. (200) carried out experiments in which pentaerythritol was similarly used to displace methyl alcohol from the methyl esters of mixed linseed oil fatty acids. Products with acid values of 4–8 were obtained after 1–4 hours reaction at 225–280°C and atmospheric pressure with the use of 1% lead as a catalyst, added as naphthenate. Pentaerythritol, mannitol, sorbitol, and so on were successfully used for the alcoholysis of glycerides, with the liberated glycerol removed by distillation under reduced pressure through a partial condenser with the assistance of stripping steam. Methyl esters of fatty acids were also alcoholyzed with furfuryl alcohol and other higher monohydroxy alcohols.

Sucrosolysis.   The direct esterification of sucrose with fatty acids is presumably limited by the tendency of sucrose to caramelize if held at about 105°C; the same objection is said to prevail for the sucrosolysis of fats (202). Sucrosolysis of methyl esters appears to be the most satisfactory method for the preparation of sucrose esters, and much of the technical literature deals with innovations, with and without solvents and with reaction conditions designed for efficient conversions.

Sucrose monolaurate, prepared in 28% yield by sucrosolysis of methyl laurate in dimethylformamide, catalyzed by sodium methoxide, has been reported (203). It possessed a softening point of 90–91°C; $\alpha_D^{20°}$ + 42.5° (5% $CHCl_3$); saponification value 105 (theory, 106.4).

Lorand (203a) apparently first prepared sucrose monopalmitate by the acylation of sucrose with palmitic anhydride in monochloroacetic acid in the presence of magnesium perchlorate. However, Osipow et al. (202, 204) finally succeeded in efficiently preparing sucrose monoesters by the careful sucrosolysis of methyl esters in dimethylformamide or dimethylsulfoxide. The use of an immiscible organic solvent (xylene) to remove the sucrose monoester and to permit the sucrosolysis of methyl esters to proceed in a concentrated aqueous medium was found necessary to achieve good monoester yields.

A British patent (205) describes the sucrosolysis of methyl esters or esters of glycols, glycerol, pentaerythritol, and so on in the presence of a standard interchange catalyst in substituted morpholine or piperidine solvents. It is claimed that the process is applicable to the preparation of esters from nonreducing oligosaccharides like trehalose, glucoxylose, raffinose, melizitose, and gentianose.

During the late 1950s and early 60s two-phase processes for sucrosolysis of methyl esters were developed (206), including a kinetics study at atmospheric pressure. The sucrolysis was said to conform to a second-order reversible reaction.

Various innovations in the manufacture of sucrose esters with several solvents were carried out during 1960–1963. Solvents include dimethylformamide (206a–206e), dimethylsulfoxide (206f), or dimethylbenzylamine (206g). A British patent (206h) suggested the sucrosolysis of phenyl esters of fatty acids for the preparation of higher sucrose esters. Sucrose octalinoleates could be prepared by the nonsolvent sucrosolysis of phenyl linoleate with potassium carbonate at 110–120°C/0.1–0.3 mm by fractionally distilling phenol from the reaction medium.

In an effort to avoid the use of expensive and troublesome solvents and to permit the use of somewhat higher reaction temperatures than ordinarily employed, Osipow and Rosenblatt (207) suggested the utilization of a "microemulsion process" for the preparation of sucrose esters. Sucrose, methyl stearate, sodium stearate, and potassium carbonate suspended in propylene glycol were microdispersed until a transparent emulsion was formed. Using a starting molar ratio of 1.5:1:0.9 for sucrose, methyl stearate, and sodium stearate, the slow distillation of propylene glycol was continued until reaction completion; the product assayed 85% sucrose monostearate and 15% sucrose distearate.

Feuge et al. (208) were able to interesterify melted sucrose with esters like methyl "Carbitol" palmitate, monopalmitin, distearin, and 40% technical mono- and diglycerides (derived from hydrogenated cottonseed oil by glycerolysis) at 170–187°C without extensive decomposition using lithium oleate or mixtures of sodium, potassium, or lithium soaps as emulsifier–catalysts. With lithium oleate the products were 90% tetraesters or higher; lower esters could be produced by blending the emulsifier system to include sodium and potassium oleate soaps.

James (209) described the preparation of sucrose monoesters from nonsolvent media at 125°C. The process involves the sucrosolysis of methyl esters, such as those obtained from tallow, with potassium carbonate highly emulsified through the use of an emulsifier system consisting of sucrose monoester and tallow fatty acids. Recently, a nonsolvent two-step process was developed by The Procter & Gamble Company (210) for the preparation of sucrose polyesters (SPE) that permits the use of less soap emulsifiers than heretofore based on the use of metallic sodium or potassium or their hydrides added to the mixture of sucrose, methyl esters, and soap at temperatures as low as 130°C. In the first step, methyl esters and sucrose at a 3:1 molar ratio are reacted with potassium soap emulsifier to form low-sucrose esters. In the second step, more methyl esters are added and reacted to produce SPE in yields up to 90% based on sucrose.

Komori et al. (211) discovered that the slightly water soluble sucrose diesters, previously used before about 1960 as oil-soluble dispersing agents, could be ethoxylated in the presence of an alkaline catalyst at 100–130°C to give water-dispersible products with good surface-active properties.

Fatty acid esters of amylose were prepared (and evaluated) by Gros and Feuge (212) by the acylation of starch suspended in a series of solvents with fatty acid chlorides.

4.3   ESTER INTERCHANGE (TRANSESTERIFICATION) (107, 108).

General Considerations   As little as twenty years ago the phrase "ester interchange in fats" was applied almost exclusively to interactions between separate triglyceride molecules; today, the scope of this class of reactions is broader. Ester interchange reactions, also called *transesterifications*, or more precisely, ester–ester interchanges, now include a great many intermolecular reactions. All combinations of interactions between monohydroxy alcohol esters, mono- and diesters of glycols, mono-, di- and triglycerides (Scheme 1, reactions 3d, 3d1, 3e, and 4), and the various esters of tetrahydroxy- and higher alcohols are possible. Although the ester–ester interchanges of materials intended for nonedible industrial uses are undoubtedly becoming increasingly important, interchanges between triglycerides today (1979) have food or food additive applications and command major interest. Significant developments in the improvement of physical properties of triglyceride products like margarine, shortening, and hard butters (synthetic cocoa butter), for example, have already been achieved. Much of the information on ester–ester interchange has resulted from research and development efforts on food products such as these; the principles and techniques are equally applicable to nonedible products.

Applied to fats and oils, ester–ester interchange may improve physical properties because it changes the arrangement of the acyl groups within components of the mixed triglycerides of naturally derived fats and oils. Similar but different changes in physical properties may be achieved through the use of partial or complete *hydrogenation*, where only the unsaturation of the triglyceride acyl groups is chemically altered, but not the position of the acyl groups on glycerol within the respective triglycerides. Hydrogenation and ester–ester interchange are *chemical* operations, as contrasted to fractionation in which changes in physical properties are the result of separation of components, a *physical* operation. Whatever the intended use of the products may be, reactions between two or more polyesters, including triglycerides, are the most important.

Ester–ester interchange may be effected without catalyst at high temperatues (250°C or more) or with an acid, alkaline, or metal catalyst. The most effective are the alkali metal alkoxides, which may be useful at temperatures as low as 0°C (213). Ester interchange may proceed at random, with the eventual attainment of an equilibrium composition corresponding to the laws of probability; or, by special techniques, it may be directed toward some degree of segregation of the fatty acid radicals according to their degree of unsaturation or their chain lengths, with or without concurrent separation of the esters into two fractions.

Because fats or fat mixtures often have some degree of organization—produced either naturally or artificially—with respect to the distribution of their component fatty acids, profound changes in composition will often result from a random molecular rearrangement. Some vegetable seed oils, for example, cottonseed oil, peanut oil, cocoa butter, tend (more or less) to have each fatty acid distributed evenly, insofar as that is possible, among the different glyceride molecules. Hence trisaturated glycerides generally do not occur in quantity in these oils unless the content of saturated fatty acids exceeds about two-thirds of the total. On the other hand, with random distribution of all acid radicals, a fat containing two-thirds saturated fatty acids will contain nearly 30% fully saturated glycerides, and even a fat containing as little as one-third saturated acids will have 3.7% fully saturated glycerides. Since the plastic range, and particularly the melting point, of a fat is highly dependent on the trisaturated glyceride content, it follows that random rearrangement of a seed oil will usually change the consistency considerably and raise the melting point. A mixture of two radically different fats is always very far from a random distribution of fatty acids and hence will respond to this type of interchange. In particular, a mixture of a highly saturated fat with a liquid oil or an oil of the coconut type will be greatly lowered in melting point by random rearrangement (see Figure 2.12) because the acids of the saturated fat become more widely distributed.

Typical results of random rearrangements, taken from various published articles and patents (214–220) are shown in Table 2.8. The greatest effects are apparent when the material treated is a seed oil or a mixture of fats differing widely in melting point. Palm oil is less affected than the

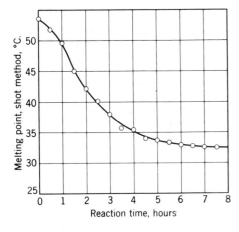

Figure 2.12 Change of melting point with reaction time in a mixture of 15% cottonseed oil stearine and 85% soybean oil subjected to rearrangement at 450°F with stannous hydroxide equivalent to 0.39% tin as a catalyst.

Table 2.8   Effect of random molecular rearrangement on melting point of various fats and fat mixtures

Fat	Melting Point (°F)		Ref.
	Before	After	
Soybean oil	19.4	41.9	214
Cottonseed oil	50.9	93.2	214
Oleo stock (beef fat)	121.1	120.2	214
Prime steam lard	109.4	109.4	214
Palm oil	103.7	116.6	215
Tallow (beef fat)	115.2	112.3	213
Coconut oil	78.8	82.8	215
Cocoa butter	94	126	215a
Kokum butter	109	146	215
Butterfat	68.4[a]	78.2[a]	216, 216a
10% Highly hydrog. cottonseed oil plus 60% coconut oil	136	106	217
25% Tristearin plus 75% soybean oil	140	90	218
50% Highly hydrog. lard plus 50% lard	135	123	219
15% Highly hydrog. lard plus 85% lard	124	107	219
25% Highly hydrog. palm oil plus 75% highly hydrog. palm kernel oil	122.3	104.5	220

[a] "Setting" point (°F).

seed oils, as far as melting point is concerned. This is true also of the animal fats, lard and tallow. More careful evaluation of these fats over a range of melting by means of micropenetration studies (Table 2.9) reveals, however, that profound changes in structure have undoubtedly taken place. From a practical standpoint, it seems that random molecular rearrangement might be useful chiefly in treating fats or fat mixtures to destroy trisaturated glycerides and bring about distribution of the saturated fatty acids in such a manner as to produce margarine oils or confectioners' fats (hard butters) of relatively low melting point and short plastic range (218, 220).

Compositional Changes.   The randomization nature of ester–ester interchange reactions shuffles the fatty acyl groups within a triglyceride to all the possible combinations. This may be visualized, as illustrated by the case of a typical triglyceride, SOL, as follows:

Interesterification Equilibrium Mixture

A simplified, abbreviated version of the preceding sequence is

$$SSS \rightleftharpoons (SUS \rightleftharpoons SSU) \rightleftharpoons (SUU \rightleftharpoons USU) \rightleftharpoons UUU$$

where S is a saturated acid and U is an unsaturated acid. It follows logically that the composition of such a randomly rearranged fat or oil can be calculated from probability considerations. If A, B, or C are the molar percentages of fatty acids A, B, or C, then the molar percentage of glycerides containing only one acid is

$$\% \, AAA = A^3 : 10{,}000$$

Table 2.9   Comparison of lard products before and after random rearrangement: consistency and performance of fats in standard pound cake test

Consistency and Performance	Leaf Lard		Hydrogenated Lard	
	Before	After	Before	After
Micropenetrations (mm/10) at				
70°F	62	65	40	53
80°F	80	130	60	125
90°F	140	225	110	285
Pound cake test (air incorporated, %)				
Period 1	62	225	170	265
Period 2	88	268	225	280
Period 3	80	315	212	325
Period 4	75	253	185	248
Leaf volume (cm³/lb)	850	1495	1265	1410

the molar percentage of glycerides containing two acids is

$$\% \; AAB \; = \; 3A^2B : 10,000$$

and the molar percentage of glycerides containing three acids is

$$\% \; ABC \; = \; 6ABC : 10,000$$

The use of these formulas for the detailed illustrative example where equimolar percentages of stearic, oleic, and linoleic acids are present results in the following triglyceride composition for the final randomized equilibrium mixture:

Stearic–stearic–stearic = $(33.3 \times 33.3 \times 33.3) : 10,000 = 3.7\%$
Oleic–oleic–oleic = $(33.3 \times 33.3 \times 33.3) : 10,000 = 3.7\%$
Linoleic–linoleic–linoleic = $(33.3 \times 33.3 \times 33.3) : 10,000 = 3.7\%$
Stearic–stearic–oleic = $(33.3 \times 33.3 \times 33.3)3 : 10,000 = 11.1\%$
Stearic–stearic–linoleic = $(33.3 \times 33.3 \times 33.3)3 : 10,000 = 11.1\%$
Stearic–oleic–oleic = $(33.3 \times 33.3 \times 33.3)3 : 10,000 = 11.1\%$
Stearic–linoleic–linoleic = $(33.3 \times 33.3 \times 33.3)3 : 10,000 = 11.1\%$
Oleic–oleic–linoleic = $(33.3 \times 33.3 \times 33.3)3 : 10,000 = 11.1\%$
Oleic–linoleic–linoleic = $(33.3 \times 33.3 \times 33.3)3 : 10,000 = 11.1\%$
Stearic–oleic–linoleic = $(33.3 \times 33.3 \times 33.3)6 : 10,000 = 22.2\%$

Norris and Mattil (219) confirmed that the statistical distribution prevails by examination of the randomization of an equimolar mixture of tristearin and triolein.

Mechanisms.   Sreenivasan (108) has outlined two distinct mechanisms for the ester–ester interchange occurring between fat triglycerides, as postulated separately by Weiss et al. (221) and Coenen (222).

The brown color generated in a triglyceride oil some time after the addition of an interchange catalyst like sodium methoxide (usually a milky white slurry suspended in dry oil) is usually associated with the formation of the active catalyst, a complex of the oil and the initiator (sodium methoxide). One reaction mechanism postulates the formation of an enolate ion as an intermediate by the action of the initiator on the glyceride (221); another postulates an addition complex of the initiator with the carbonyl group of the glyceride (222).

The induction of intramolecular and intermolecular interchange is a consequence of enolate ion formation as outlined in Fig. 2.13; the inter- and intramolecular interchanges are shown in Figs. 2.14 and 2.15, respectively. In both reactions a β ketoester is a necessary intermediate. These structures have characteristic infrared absorption maxima at 6.4 μ, which appears coincidentally during the reaction, but the usefulness and significance of this is somewhat negated by the absorption of carboxylate ion (as soaps) in that range (223). Figure 2.16 outlines the other mechanism. An alkylate ion adds onto a polarized ester carbonyl pro-

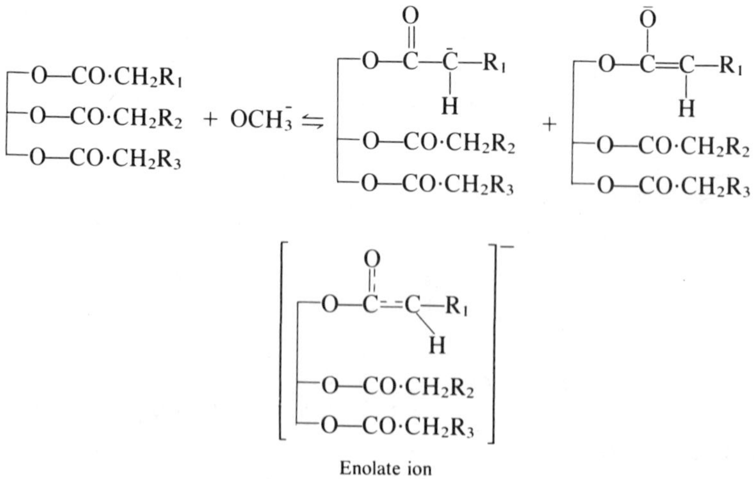

Enolate ion

Figure 2.13   Enolate ion formation (221).

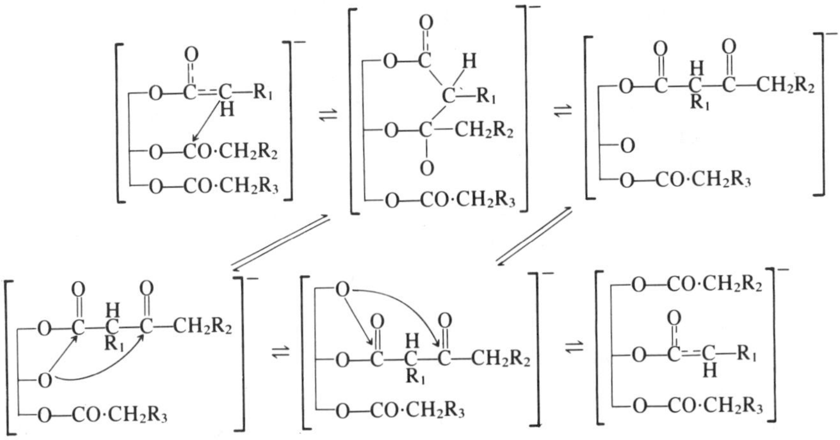

Figure 2.14   Mechanism for intramolecular ester–ester interchange (221).

ducing a diglycerinate intermediate that would react with another glyceride by abstracting a fatty acid forming a new triglyceride and regenerating the diglycerinate. A repetition of this process through a series of chain reactions would then continue until all available fatty acid groups have changed positions and randomization is complete. A simple mixture of S_3 and U_3 undergoing this transformation is outlined in Figure 2.17.

Both mechanisms lack confirmatory experimental support. Apparently, the formation of the brown-colored active intermediate is the rate-controlling and slow reaction; the fatty acid interchange catalyzed by it is the fast reaction. Reaction periods as fast as 3–5 min have been variously reported.

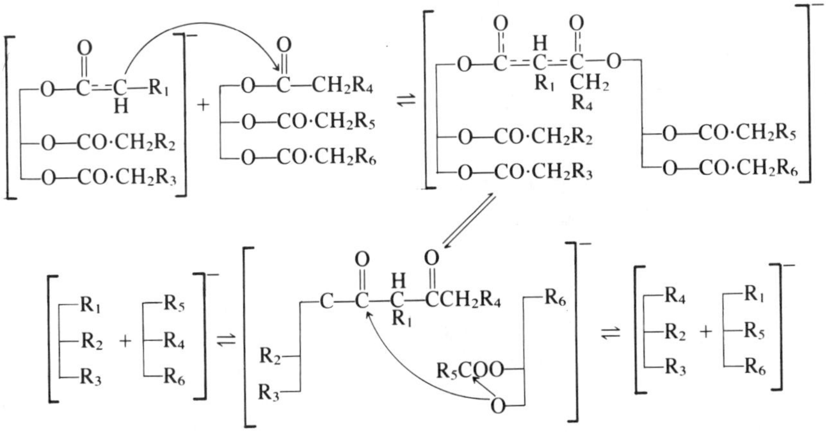

Figure 2.15   Mechanism for intermolecular ester–ester interchange (221).

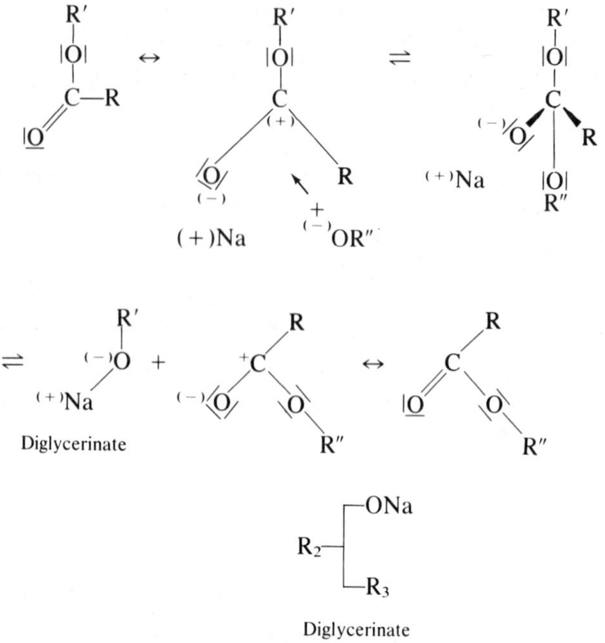

Diglycerinate

$$R_2 \begin{cases} -ONa \\ -R_3 \end{cases}$$

Diglycerinate

Figure 2.16 Mechanism of interesterification: carbonyl addition (222). R′ = diglyceride part of triglyceride, R = fatty acid, R′ = diglyceride part of triglyceride, R″ = CH₃,—C₂H₅,t-butylate.

$$S_3 + U_2ONa \underset{K}{\overset{3K}{\rightleftharpoons}} SU_2 + ONa$$

$$U_3 + S_2ONa \underset{K}{\overset{3K}{\rightleftharpoons}} S_2U + U_2ONa$$

$$SU_2 + U_2ONa \underset{3K}{\overset{2K}{\rightleftharpoons}} U_3 + SUONa$$

$$S_2U + S_2ONa \underset{3K}{\overset{2K}{\rightleftharpoons}} S_3 + SUONa$$

$$S_2U + U_2ONa \underset{K}{\overset{2K}{\rightleftharpoons}} SU_2 + SUONa$$

$$SU_2 + S_2ONa \underset{K}{\overset{2K}{\rightleftharpoons}} S_2U + SUONa$$

Figure 2.17 Chain reaction initiated by diglycerinate for fatty acid interchange (222).

154

Methods for Detecting Endpoints. The visual change in color that occurs during interesterification, namely, the intensification of a brownish coloration, is potentially a means of control but, to date, has not been developed for this purpose.

MELTING POINT.   The oldest and perhaps still the fastest method for controlling interchange reactions is by determination of melting point before during and after. A compilation of changes in melting point in interesterification is given in Table 2.8 indicating that, in general, vegetable oils show increases and co-randomization mixtures show decreases. It is, of course, desirable to know in advance whether the fat will melt lower or higher after randomization and the range expected. Unfortunately, determination of melting point is insufficiently accurate for several industrial applications.

DILATOMETRY.   Fulton et al. (224) have developed a method for the determination of solid fat content or solid content index (SCI) of fat mixtures related to temperature that has utility in monitoring the progress of ester–ester interchanges. The melting behavior of cocoa butter before and after interesterification as reflected by the SCI is shown in Figure 2.18. Figure 2.19 illustrates the opposite effect in co-randomization. A number of further examples demonstrating the application of SCIs to interesterification are given in Table 2.10.

In the case of cocoa butter, a fat with sharp melting characteristics and a steep SCI curve is converted into one with a wider melting range and a flat SCI curve. Solids present at 50°C are indicative of the formation

Table 2.10   Changes in SCI values due to interesterification (107, 222, 227–231)

	Before			After		
	10°C	20°C	30°C	10°C	20°C	30°C
Cocoa butter	84.9	80	0	52.0	46	35.5
Palm oil	54	32	7.5	52.5	39	21.5
Palm kernel oil	—	38.2	8.0	—	27.2	1.0
Hydrogenated palm kernel oil	74.2	67.0	15.4	65	49.7	1.4
Lard	26.7	19.8	2.5	24.8	11.8	4.8
Tallow	58.0	51.6	26.7	57.1	50.0	26.7
60% Palm oil–40% coconut oil	30.0	9.0	4.7	33.2	13.1	0.6
50% Palm oil–50% coconut oil	33.2	7.5	2.8	34.4	12.0	0
40% Palm oil–60% coconut oil	37.0	6.1	2.4	35.5	10.7	0
20% Palm stearine–80% lightly hydrogenated vegetable oil	24.4	20.8	12.3	21.2	12.2	1.5

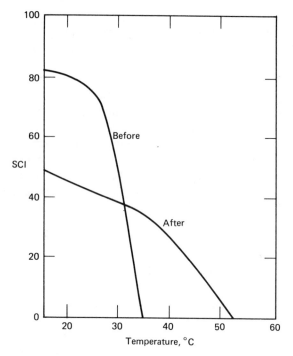

Figure 2.18   Solid content index (SCI) of cocoa butter before and after interesterification.

of higher melting glycerides by randomization. Thus although cocoa butter will melt completely in the mouth, randomized cocoa butter will not. In the case of the fat mixture shown in Figure 2.19, before randomization it consisted of a high proportion of higher-melting glycerides due to the palm stearine, but after the reaction these levels are reduced and mixed glycerides containing unsaturated fatty acids are formed. The result is a lowered SCI at higher temperatures and complete disappearance of solids at about 50°C. Effectively, a slow melting and pasty tasting mixture is converted into a clean tasting product with a sharper melting range.

Palm oil behaves like cocoa butter in showing an increase in melting point; it exhibits a higher content of solids at the higher temperatures. Lard and tallow, on the other hand, show miminum changes in solids content on interesterification; dilatometry is less useful with them. With co-randomizing mixtures, there is little change shown at 10°C and a moderate increase or decrease at 20°C. At 35°C, especially in the case of cocoa butter, there is a significant decline or disappearance of solids, and this is a logical choice of temperature for establishing a control method. The disadvantage of dilatometry is that it is timeconsuming, and any faster technique of measuring solid contents, either by NMR (225, 226) or otherwise, would be desirable. Nevertheless, SCIs represent a much-used method for following the course of ester–ester interchange reactions in the shortening and margarine industries.

GLYCERIDE COMPOSITIONAL ANALYSIS.   Since the fundamental change that occurs in interesterification involves specific triglyceride structures, any method of direct analysis for specific components can be employed for detecting endpoints of reactions. Such methods as thin layer chromatography (TLC) (232, 233), triglyceride GLC (231, 234), mass spectrometry (235), and pancreatic lipase analysis for fatty acid distribution of triglycerides (236, 237) have already been applied for this purpose.

Two excellent examples of the use of silver nitrate complexed–TLC, which separates triglycerides according to their unsaturation, have been reported by Freeman (232). The interesterification of pure triglycerides, such as POP and POS, was shown to reach equilibrium in a little over 45 min. The course of the co-randomization of 60% sunflower oil and 40% fully hydrogenated lard is summarized in Table 2.11. The reaction could be satisfactorily followed by measuring the disappearance of S_3 triglycerides.

Gas liquid chromatography of triglycerides is an acceptable method for

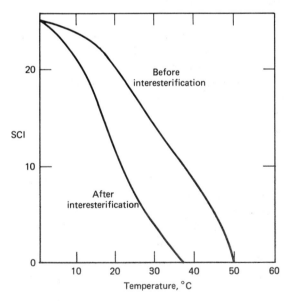

Figure 2.19   Effect of randomization of SCI of an 80:20 mixture of lightly hardened oil and palm stearine.

*Table 2.11   Selected glyceride changes during the co-randomization of 60%
sunflower oil and 40% fully hydrogenated lard with 0.2% NaOCH₃ at 70–90°C[a]*
(233)

Triglyceride Group	No. of Double Bonds	Minutes after Start				Difference[c] (%)
		0[b] (%)	20 (%)	40 (%)	60 (%)	
S_3	0	37.7	32.0	17.1	6.1	−31.6
S_2O	1	—	0.7	3.7	9.2	+9.2
S_2L	2	0.5	2.9	12.7	20.9	+20.4
SOL	3	4.7	5.2	11.2	14.0	+9.3
SL_2	4	11.4	11.8	16.2	20.2	+8.8
OL_2	5	20.0	20.2	16.5	9.5	−10.5
LLL	6	19.2	19.4	14.3	11.6	−7.6

[a] S, saturated; O, oleic; L, linoleic.

[b] Composition of starting mixture.

[c] Difference between randomized (60 min) and starting mixture.

following ester–ester interchanges. Rossel (231) used this technique in following palm kernel oil randomization. The method was fast (20–30 min), and in this case the composition of the oil closely approximated the calculated random distribution.

Mass spectroscopy provides a molecular-weight distribution of triglyceride mixtures. Cocoa butter, analyzed by this technique (235), showed the greatest change in POS and SOS triglyceride content, which were reduced significantly during the randomization. The method is applicable to corn oil, soybean oil, sunflower oil, and safflower oil.

Pancreatic lipase hydrolysis is predominantly a laboratory method not especially suited for industrial control. The method specifically hydrolyses the 1- and 3-position acyl groups of triglycerides and produces 2-monoglycerides. Thus if the fatty acid distribution of the 2-monoglycerides is the same as that of the total randomized fat, the reaction mixture can be assumed to be equilibrated. Results for lard and randomized lard are presented in Table 2.12 (236). Mixtures of soybean oil and soya stearine were also randomized and studied by this method (237).

Other methods are determination of cooling curves (221, 228), differential scanning calorimetry (231), and x-ray diffraction (238–240).

Applications

FAT MODIFICATION WITH ALKYL ESTERS.   The interesterification of palm oil triglycerides with alkyl esters has been described by Koslowsky (241).

By interesterifying palm oil with ethyl oleate (presumably derived from palm oil by alkanolysis and fractional distillation) in the presence of 0.1–0.5% sodium methoxide at temperatures not exceeding 60°C for about 30 min, a liquid triglyceride oil could be obtained, after removal by distillation of a saturated ethyl ester, that was suited for use as a salad oil (66.2% 18:1, 13.7% 18:2 acids; IV = 80). Subsequently, a hard fraction (91.3% 16:0, 2.3% 18:0, and only 4.4% 18:1 and 0.8% 18:2 acids) could be obtained by interchange of palm oil with a saturated ethyl ester recovered by distillation from the first step. A continuous industrial-scale plant with a capacity for the interesterification of 100 tons per 24 hours of palm oil is said to be contemplated (Shemen Oil Co., Haifa, Israel).

SHORTENINGS.   Because of its coarse structure, lard is the most investigated fat for this purpose (221, 228, 229, 238, 239). Lard contains a high proportion of 16:0 fatty acid in the 2-position of its disaturated (S_2U) triglycerides (236, 242). The proportion of 16:0 fatty acid in the 2-position is lowered from about 64 to 24% on randomization (Table 2.12). Concurrently, other changes also occur, and these produce a smooth-textured lard. Natural lard tends to crystallize in the β-phase (4.6 Å) (239), and randomized lard crystallizes in the β' (4.2 Å) phase (240). Randomization of lard improves its plastic range and thus makes it a better shortening than natural lard.

MARGARINES: HIGH-STABILITY MARGARINE BLENDS.   Short-chain acids (C_6–C_{14}) give better melting qualities, and long-chain acids (C_{20}–C_{22}) on the same triglycerides provide stiffening power in margarines. Both these properties can be combined using blends of randomized oils (243): 75 parts co-randomized 40% coconut oil–60% palm oil; 10 parts co-randomized 50% coconut oil–50% hydrogenated rapeseed oil IV 4; and 15 parts hydrogenated soybean oil IV 95. Solid content index values of the blend at

Table 2.12 Pancreatic lipase hydrolysis of lard and randomized lard (236)

Fatty Acid	Lard		Randomized Lard[a]	
	Total Fat	2-Position	Total Fat	2-Position
18:0	24.8	63.6	23.8	24.2
16:1	3.1	6.4	2.9	3.3
18:0	12.6	5.0	12.2	12.0
18:1	45.0	16.5	47.2	47.4
18:2	9.8	5.4	4.4	3.3

[a] 0.3% $NaOCH_3$, 90–100°C, 2 hours.

10, 20, and 33.3°C are 32, 19, and 1, respectively. Margarine made from this blend has good spreadability, high-temperature stability, and pleasant taste.

NUTRITIONAL MARGARINE BLENDS.   High polyunsaturated content and low to zero *trans*-acid-content margarines are desirable for nutritional reasons. List et al. (237) have reported on an interesterified blend of soybean oil and soybean stearine: for a randomized blend of 80 parts soybean oil and 20 parts of soy stearine, SCI values at 10, 21.1, and 33.3°C were, respectively, 8, 3.4, and 2.2; the polyunsaturated acid content, 18:2 + 18:3, was 51.5%, and *trans* acids were 1.6%. Similarly, a blend of 75 parts of soybean oil and 25 parts of fully hydrogenated cottonseed oil, interesterified with sodium methoxide (244), had the following characteristics: melting point, 39°C (102–104°F); essential fatty acids, 44.6%; and *trans* acids, 2%. Flavor and oxidative stability of the margarines made from interesterified oils were observed to be good.

CONFECTIONERY FATS.   The relatively high price and the occasional unavailability of cocoa butter have encouraged development of confectionery fat substitutes. Perhaps the most qualified substitutes are those resulting from blending interesterified lauric acid fats with certain common fats. Sreenivasan (108) points out that hydrogenated palm kernel oil is a hard butter melting at 46°C that produces a waxy feel in the mouth. On randomization its melting point is reduced to 35°C. By blending hydrogenated palm kernel oil with randomized product, a whole series of hard butters with highly desirable melting (rapid melt in mouth) qualities are obtained. Table 2.13 summarizes the thermal properties of a few of these blends. The big drop in the solids content between 20 and 38°C (mouth temperature) is responsible for the rapid melt in the mouth.

Table 2.13   *Confectionery fats from blends of hydrogenated and interesterified hydrogenated palm kernel oil* (108)

Fat	Melting Point (°C)	SCI			
		10°C	20°C	35°C	38°C
Hydrogenated palm kernel oil (HPKO)	46.8	74.2	67.0	15.4	11.7
Interesterified hydrogenated PKO (IHPKO)	35.0	65.0	49.9	1.4	1.1
50% HPKO–50%IHPKO	41.7	70.0	57.4	8.7	5.2
65% HPKO–35%IHPKO	44.2	71.0	59.7	10.2	6.7
80% HPKO–20%IHPKO	46.0	72.4	62.6	12.4	8.5

Other non-lauric-containing fats can also be blended with lauric fats to achieve a combination of desirable properties. Partially or fully hydrogenated lauric fats are co-randomized with fully hardened cottonseed oil (245, 246), or long-chain acid-containing oil, such as fully hydrogenated rapeseed and herring oils (247). Such co-randomized fats have better gloss-producing and quick set-up properties in surface-coating applications.

Directed Interesterification.   If one of the products of a reversible reaction is removed from the sphere of the reaction, the equilibrium shifts to regenerate more of the removed component. Thus selective crystallization to remove a trisaturated glyceride from the products of interesterification of a fat or mixed fats may be directed to the effective conversion of all the saturated fatty acids to trisaturated triglycerides (213). Such a process is termed *directed interesterification* (215, 248, 249). In the case of a triglyceride SOL, the composition of the randomized fat would be:

Solids	Stearic–stearic–stearic	33.33 mole %
Liquids	Oleic–oleic–oleic	8.33 mole %
	Linoleic–linoleic–linoleic	8.33 mole %
	Oleic–oleic–linoleic	24.99 mole %
	Oleic–linoleic–linoleic	24.99 mole %

The changes in solids content due to directed interesterification in the case of lard are illustrated in Figure 2.20. Crude lard has a low solids content at warm temperatures but needs addition of stearine for high-temperature stability. Directed interesterification produces a lard with increased solids content at high temperatures and, thus, an extended plastic range. Therefore, directed lard can be used as a shortening without addition of fully saturated triglycerides.

A flow diagram for the directed interesterification of lard using NAK (sodium potassium alloy) as a catalyst is shown in Figure 2.21 (108, 248). Freshly refined lard is dried to a moisture content of 0.01% and then cooled to 104–108°F (40–42°C). It is pumped into a mixer where the NAK catalyst is metered in, and the lard–catalyst mix is passed through a coil (hold time 15 mins) for randomization to occur. The randomized mixture is cooled to 68–71°F (20–22°C) by pumping it through a votator (ammonia cooled) with a residence time in the votator of 0.5 min. The crystallized mass is passed into a picker box in which the mixture is agitated for an average time of 2.5 mins. As a result of heat of crystallization the temperature rises to 81–83°F (27–28°C). It is again cooled to 70°F by passage through another ammonia-cooled votator. From here the mixture is

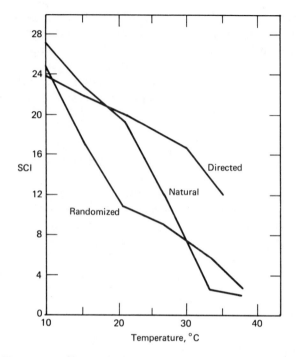

Figure 2.20 Changes in the SCI of lard due to rearrangement.

pumped through a series of crystallizers with gentle agitation and a holding time of 1.5 hours. The stock leaving the crystallizer is at a temperature of 86–90°F (30–32°C) and is treated with CO_2 and water in a high-speed mixer to eliminate the catalyst. The soap that forms is removed by centrifugation, and the lard is further washed to remove the soap. The washed lard is dried. The final SCI at 92°F should be about 14.

The directed interesterification and fractionation by crystallization of palm oil (250) has been applied to produce an improved salad oil. The solid contents of palm oil, randomized palm oil, and directed interesterified palm oil are shown in Figure 2.22. Randomization produces only small changes in the solids content of palm oil, but directed interesterification increases the solids at higher temperatures significantly. Fractional crystallization of palm oil at 20°C produces a liquid fraction with a cloud point of 5°C. Fractionation by crystallization after directed interesterification results in a liquid fraction with a cloud point of 2.7°C, and thus a better salad oil is obtained.

Dimethyl sulfoxide is a catalyst-activating solvent for interesterifications or directed interesterifications (251). Directed interesterification at

0°C for one day yielded cottonseed oil with an SCI of 11.4; this rose to 18.0 if the reaction was performed in the presence of 1% dimethyl sulfoxide. A practical application of this technique is the directed interesterification of sunflower and safflower oils at − 10°C in the presence of 1% dimethyl sulfoxide to produce stocks suited for margarine production (252). Safflower oil interesterified in this way had SCI of 4.2 at 0°C, 4.3 at 21.1°C, 2.8 at 33.3°C, and 0 at 40°C. Sunflower oil, similarly, had SCIs of 10.7, 6.0, 5.2, and 2.1 at 0, 21.1, 33.3, and 40°C, respectively. Margarines made from both of these oils had acceptable properties.

An innovation said to further improve interesterified oil performance is to cycle the temperature of the reaction mixture 1–15°C below and to at least 5°C above its cloud point (252). Sunflower oil, interesterified with 0.25% sodium, was cycled between − 5°C and 10°C for 24 hours to yield a product having higher solids content than a control sample chilled first to − 5°C and then held at 10°C for 47.5 hours.

Like directed crystallization, selective extraction may be used to upset equilibrium and thus induce directed interesterification. Such a partitioning is accomplished (253) with soybean oil dissolved in hexane and brought into contact with a countercurrent flow of dimethyl formamide, resulting

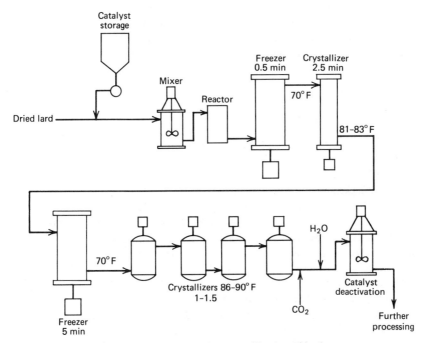

Figure 2.21   Directed interesterification of lard.

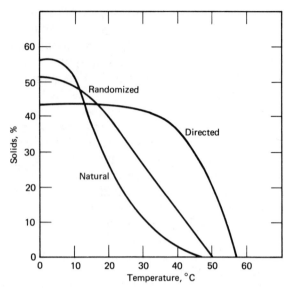

Figure 2.22   Solid contents in palm oil with random and directed interesterifications.

in fractionation according to glyceride unsaturation. Thus original oil containing 48.0% linoleic and 6.8% linolenic acid components could be converted to a fractionally extracted oil containing 47.9% linoleic and 3.0% linolenic acid components.

References

1. C. S. Castner (to Schuyler Development Corp.), U.S. Pat. 3,654,327 (April 4, 1972).
2. O. J. Ackelsberg, *J. Am. Oil Chemists' Soc.*, **35**, 635–640 (1958).
3. E. A. Lawrence, *J. Am. Oil Chemists' Soc.*, **31**, 542–544 (1954).
4. V. J. Muckerheide, *J. Am. Oil Chemists' Soc.*, **29**, 490–495 (1952).
5. M. D. Reinish, *J. Am. Oil Chemists' Soc.*, **33**, 516–520 (1956).
6. H. H. Mueller and E. K. Holt, *J. Am. Oil Chemists' Soc.*, **25**, 305–307 (1948).
7. V. Mills and H. K. McClain, *Ind. Eng. Chem.*, **41**, 1982–1985 (1949).
8. L. Lascaray, *Seifensieder-Zig.*, **64**, 122–126 (1937); *Fette, Anstrichmittel*, **46**, 628–632 (1939).
9. A Sturzenegger and H. Sturm, *Ind. Eng. Chem.*, **43**, 510–515 (1951).
10. L. Lascaray, *Ind. Eng. Chem.*, **41**, 786–790, (1949).
11. H. P. Kaufmann and M. C. Keller, *Fette, Seifen, Anstrichm.*, **44**, 42–47, 105–107 (1937).
12. E. Twitchell, U.S. Pat. 601,603 (1898); *J. Am. Chem. Soc.*, **22**, 22–26 (1900).
12a. S. D. Vaidya and J. G. Kane, *J. Oil Technol. Assoc. India*, **18**, 200–208 (1963).

12b. S. D. Vaidya, V. V. R. Subrahmanyam, and J. G. Kane, *Indian J. Technol.*, **4**, 301–304 (1966).

13. A. J. Stirton, E. M. Hammaker, S. F. Herb, and E. T. Roe, *Oil Soap*, **21**, 148–151 (1944).

14. C. B. Cox, *Trans. Inst. Chem. Eng. (London)*, **27**, 123–137 (1949).

15. L. Hartman, *J. Am. Oil Chemists' Soc.*, **30**, 349–350 (1953).

15a. M. R. Kallyanpur, V. V. R. Subrahmanyam, and J. G. Kane, *Indian J. Technol.*, **5**, 20–22 (1967).

16. K. E. Lunde (to Carad Corp.), U.S. Pat. 3,253,007 (May 24, 1966).

17. M. H. Ittner (to Colgate-Palmolive-Peet Co.), U.S. Pat. 2,139,589 (1938); Re. 22,006 (1942).

18. V. Mills (to Procter and Gamble Co.), U.S. Pat. 2,156,863 (1939).

19. H. L. Barneby and A. C. Brown, *J. Am. Oil Chemists' Soc.*, **25**, 95–99 (1948).

19a. M. R. Kallyanpur, V. V. R. Subrahmanyam, and J. G. Kane, *Indian J. Technol.* **5**, 63–65 (1967).

20. W. Connstein, E. Hoyer, and H. Wartenberg, *Ber.*, **35**, 3988–4006 (1902).

21. M. Loury and J. Max, *Bull. Mat. Grasses Inst. Colonial Marseille*, **29**, 25–34 (1945).

22. P. Desnuelle and P. Savary, *J. Lipid Res.*, **4**, 369–384 (1963); M. Kates, in K. Bloch, Ed., *Lipid Metabolism*, Chapter 5, Wiley, New York, 1960; M. H. Coleman, *Advances in Lipid Research*, Vol. 1, Academic, 1963, p. 2.

23. M. H. Coleman, *J. Am. Oil Chemists' Soc.*, **40**, 568–571 (1963).

24. F. E. Luddy, R. A. Barford, S. F. Herb, P. Magidman, and R. W. Riemenschneider, *J. Am. Oil Chemists' Soc.*, **41**, 693–696 (1964).

25. F. H. Mattson and E. S. Lutton, *J. Biol. Chem.*, **233**, 868–871 (1958).

26. R. J. VanderWal, *J. Am. Oil Chemists' Soc.*, **37**, 18–20 (1960); **40**, 242–247 (1963).

27. F. D. Gunstone, in *Progress in the Chemistry of Fats and Other Lipids*, Vol. IV, Chapter 1, Pergamon, New York, 1958; also in *Progress in Organic Chemistry*, Vol. IV, Chapter 1, Academic, New York, 1958. E. M. Meade, in *Progress in the Chemistry of Fats and Other Lipids*, Vol. IV, Chapter 2, Pergamon, New York, 1958. F. R. Earle, E. H. Melvin, L. H. Mason, C. H. Van Etten, I. A. Wolff, and Q. Jones, *J. Am. Oil Chemists' Soc.*, **36**, 304–307 (1959).

28. C. F. Krewson, J. S. Ard, and R. W. Riemenschneider, *J. Am. Oil Chemists' Soc.*, **39**, 334–340 (1962).

29. J. Sampugna, R. G. Jensen, R. M. Perry, Jr., and C. F. Krewson, *J. Am. Oil Chemists' Soc.*, **41**, 132–133 (1964).

30. J. L. Trauth, *Oil Soap*, **23**, 137–140 (1946).

31. J. W. Lawrie, *Glycerol and the Glycols*, 1928.

32. C. S. Miner and N. N. Dalton, *Glycerol*, ACS Monograph No. 117, 1953.

33. A. K. Tosh, "Glycerine Production and Refining," *J. Am. Oil Chemists' Soc.*, **35**, 615–623 (1958).

34. N. W. Ziels, "Recovery and Purification of Glycerol," *J. Am. Oil Chemists' Soc.*, **33**, 556–565 (1956).

35. D. M. Stromquist and A. C. Reents, "C.P. Glycerol by Ion-Exchange," *Ind. Eng. Chem.*, **43**, 1065–1070 (1951).

36. S. E. Zager and T. C. Doody, "Glycerol Removal from Aqueous Solutions by Ion-Exchange," *Ind. Eng. Chem.*, **43**, 1070–1073 (1951).

166   Fat Splitting, Esterification, and Interesterification

37. M. W. Formo, *J. Am. Oil Chemists' Soc.*, **31**, 548–559 (1954).
38. K. S. Markley, *Fatty Acids*, 2nd ed., Part 2, Chapter 9, pp. 757–984; Part 5, 1968, Chapter 9A, pp. 3459–3560.
39. L. Hartman, *Chem. Rev.*, **58**, 845–867 (1958).
40. J. Pelouze and A. Gelis, *Ann. Chim. Phys.*, **10**, 434–456 (1844).
41. M. Berthelot, *Compt. rend.* **36**, 27–29 (1853); **37**, 398–403, 403–406 (1853); **38**, 668–673 (1854); *Ann. Chim.*, **41**, 216–319 (1854).
42. R. O. Feuge, E. A. Kraemer, and A. E. Bailey, *Oil Soap*, **22**, 202–207 (1945).
43. T. P. Hilditch and J. G. Rigg, *J. Chem. Soc.*, 1774–1778 (1935); (to Imperial Chemical Industries, Ltd.), U.S. Pat. 2,073,997 (March 16, 1937).
44. R. O. Feuge, *J. Am. Oil Chemists' Soc.*, **39**, 521–527 (1962).
45. N. Ivanoff, *Chim. Ind.*, **53**, 41 (1945).
46. T. J. Patrick, Jr. and K. G. Johnson (to Monsanto Chemical Co.), U.S. Pat. 3,060,224 (October 23, 1962).
47. A. T. Gros and R. O. Feuge, *J. Am. Oil Chemists' Soc.*, **41**, 727–731 (1964).
48. L. Hartman, *J. Am. Oil Chemists' Soc.*, **43**, 536–538 (1966).
49. T. Bellucci and R. Manzetti, *Atti. Acad. Lincei*, **20**, I, 125–128 (1911).
50. R. Bhattacharya and T. P. Hilditch, *Proc. R. Soc. London, Ser. A*, **129**, 468–476 (1930).
51. Anonymous, *Synthetic Fats, Their Potential Contribution to World Food Requirements*, Publication No. 1, The Division of Nutrition of FAO, Nutritional Studies Series, 1949.
52. P. N. Williams, *Synthetic Fats, Chem. Ind. (London)*, **19**, 251 (1947).
53. E. Stossel, *Oil Gas J.*, p. 146 (August 18, 1945).
54. E. L. Baldeschwieler, *Production of Synthetic Fatty Acids and Edible Fats, Deutsche Fettsaurewerke, Witten*, OPB Report No. 225 (1945).
55. A. G. Mohan and W. R. Christian (to Nopco Chemical Co.), U.S. Pat. 3,071,604 (January 1, 1963).
56. E. W. Wilson and J. R. Hutchins (to Eastman Kodak Co.), U.S. Pat. 3,055,869 (September 25, 1962).
57. G. R. Bond, Jr., (to Houdry Process Corp.), U.S. Pat. 2,910,489 (October 29, 1959).
58. L. W. Lehman and W. J. Gauglitz, *J. Am. Oil Chemists' Soc.*, **41**, 533–535 (1964).
59. G. Barsky and V. Babayan (to E. F. Drew Co.), U.S. Pat. 2,988,483 (June 13, 1961).
60. A. L. Markman, M. Mirzabaeva, and A. I. Glushenkova, *Uzbeksk. Khim. Zh.*, **10** (1), 24–26 (1966).
61. I. Bellucci, *Gazz. Chem. Ital.*, **42**, I, 283–305 (1913).
62. I. Bellucci, *Atti. Accad. Lincei*, **20**, I, 235–238 (1911).
63. R. O. Feuge, R. K. Willich, and W. A. Guice, *J. Am. Oil Chemists' Soc.*, **40**, 260–264 (1963); R. O. Feuge and R. K. Willich (to U.S. Department of Agriculture), U.S. Pat. 3,119,849 (January 28, 1964).
64. M. F. Sorokin, Z. A. Kochnova, and I. S. Krivopalova, *Tr. Mosk. Khim.-Tekhnol. Inst.*, **1968**, No. 57, 61–65.
65. T. I. Koftun, *Neftyanik* **5**, No. 6, 14–15 (1960).
66. V. F. Belyaev, *Sbortsiya iz Rasivorov Vysokopolimerami i Uglyami, Belarussk. Gos. Univ.*, **1961**, 143–150.

67. P. J. Flory, *J. Am. Chem. Soc.*, **61**, 3334–3340 (1939).

68. B. K. Zeinalov, F. I. Garibov, A. B. Nasirov, and P. M. Kerimov, *Azerb. Khim. Zh.*, **1975** (5), 61–63.

69. D. A. Chernysheva, A. E. Ostaeva, and N. G. Polyanskii, *Tr. Tambovskogo Inst. Khim. Mashinostr.*, **1970**, No. 4, 172–175.

70. M. Decouzon and M. Naudet, *Bull. Soc. Chim. Fr.*, **1966**, 3541–3542.

71. A. A. Blagonravova and A. M. Lazarev, *J. Appl. Chem. U.S.S.R.*, **13**, 879–883 (1940).

72. H. Burrell, *Oil Soap*, **21**, 206–211 (1944); *Ind. Eng. Chem.*, **37**, 86–89 (1945); (to Heyden Chemical Corp.), U.S. Pat. 2,360,394 (1944).

73. J. C. Konen, E. T. Clocker, and R. P. Cox, *Oil Soap*, **22**, 57–60 (1945).

74. E. R. Mueller, P. L. Enness, and E. E. McSweeney, *Ind. Eng. Chem.*, **42**, 1532–1536 (1950).

75. M. Mirzabaeva, K. K. Kholmatov, A. I. Glushenkova, and A. L. Markman, *Maslo-Zhir. Promst.*, **1975**, (12) 16–17.

76. J. Perka and S. Ropuszynski, *Przem. Chem.*, **1972**, 51 (12) 796–800.

77. V. K. Babayan, H. Lehman, and M. Warman, *Polyglycerols and Polyglycerol Esters—Some Typical Derivatives and Applications*, at ACS Division of Industrial and Chemical Engineering, Philadelphia, April 6, 1964.

78. E. Mares and J. Pokorny, *Sb. Vysoke Skoly Chem-Technol. Praze Oddil. Pak. Potravinareske Technol.*, **4**, Part 1, 275–298 (1960).

79. M. R. Baichwal and J. K. Lalla, *Indian J. Technol.*, **1969**, 7 (8), 261–263.

80. J. D. Brandner, R. H. Hunter, M. D. Brewster, and R. E. Bonner, *Ind. Eng. Chem.*, **37**, 809–812 (1945).

81. E. W. Eckey and R. O. Alderson (to Procter & Gamble Co.), U.S. Pat. 3,093,481 (June 11, 1963).

82. Gevaert Photo-Producten N.V. (by J. T. Lemmerling), German Pat. 1,238,890 (April 20, 1967).

83. A. W. Ralston, *Fatty Acids and Their Derivatives*, Wiley, New York, 1948, pp. 492–498.

84. I. Roberts and H. C. Urey, *J. Am. Chem. Soc.*, **60**, 2391–2393 (1938); **61**, 2584–2587 (1939).

85. A. Skrabel, *Chemical Kinetics*, in *International Critical Tables*, Vol. 7, McGraw-Hill, New York, 1930, p. 138.

86. J. J. Sudborough and J. M. Gittins, *J. Chem. Soc.*, **93**, 210–217 (1908); **95**, 315–321 (1909).

87. Report of the AOCS Spectroscopy Committee, 1961–1962, *J. Am. Oil chemists' Soc.*, **40**, 14A, 32A (1963).

88. L. D. Metcalfe and A. A. Schmitz, *Anal. Chem.*, **33**, 363–364 (1961).

89. A. K. Lough, *Nature*, **202**, 795 (1964).

90. G. R. Jamieson and G. S. Reid, *J. Chromatogr.*, **17**, 230–237 (1965).

91. M. L. Vorbeck, L. R. Mattick, F. A. Lee, and C. S. Pederson, *Anal. Chem.*, **33**, 1512–1514 (1961).

92. M. Morgantini, *Boll. Lab. Chim. Provinciali* (*Bologna*) **15** (3), 254–261 (1964).

93. Ruhrchemie A.-G., British Pat. 957,679 (May 15, 1964).

94. F. V. Linchevskii and R. S. Simonova, *Tr. Nauchn.-Issled. Inst. Sintetich. Zhiro-zamenitelei i Moyushchikh Sredstv*, No. 3, 26–29 (1962).

95. V. I. Babaev, T. S. El'kina, and K. G. Besedina, *Maslob.-Zhir. Promst.*, **30** (2), 33–34 (1964).

96. V. V. Veselov, F. V. Linchevskii, and V. I. Morina, *Khim. Prom.*, **1962**, 558–560.

97. P. H. Eaves, J. J. Spadaro, and E. A. Gastrock, *J. Am. Oil Chemists' Soc.*, **36**, 230–234 (1959).

98. V. L. Hansley (to E. I. du Pont de Nemours & Co.), U.S. Pat. 2,177,407 (October 24, 1939).

99. Societe Anon. d'Innovations Chimiques "Sinnova" ou "Sadic," French Pat. 1,-402,133 (June 11, 1965).

100. D. McNicoll, *J. Soc. Chem. Ind.*, **40**, 124T (1921).

101. R. H. Anderson and D. H. Wheeler, *Oil Soap*, **22**, 137–141 (1945).

102. *Official and Tentative Methods of the American Oil Chemists' Society*, 3rd ed. (including additions and revisions—1976) Method Da-12-48.

103. A. Linder and V. Persson, *Svensk Papperstid.*, **52**, 331–337 (1949); *J. Am. Oil Chemists' Soc.*, **34**, 24–27 (1957).

104. W. Q. Braun, *J. Am. Oil Chemists' Soc.*, **37**, 598–601 (1960).

105. E. W. Eckey, *J. Am. Oil Chemists' Soc.*, **33**, 575–579 (1956).

106. J. C. Cowan, *J. Am. Oil Chemists' Soc.*, **27**, 492–499 (1950).

107. L. H. Going, *J. Am. Oil Chemists' Soc.*, **44**, 414A–422A, 455A, 456A (1967).

108. B. Sreenivasan, *J. Am. Oil Chemists' Soc.*, **55**, 796–805 (1978).

109. W. Normann, *Chem. Umschau*, **30**, 250–251 (1923).

110. G. L. Schwartz (to E. I. du Pont de Nemours and Co.), U.S. Pat. 1,588,299 (1925).

110a. E. M. Meade and D. M. Walder, *J. Am. Oil Chemists' Soc.*, **39**, 1–3 (1962).

110b. E. M. Meade and D. M. Walder, *J. Am. Oil Chemists' Soc.*, **39**, 4–6 (1962).

111. G. Barsky (to Wecoline Products, Inc.), U.S. Pat. 2,182,332 (1939).

112. A. Grün (to Georg. Schicht A.-G.), U.S. Pat. 1,505,560 (1924).

113. K. M. Gol'dberg, N. V. Prilutskaya, and M. M. Fal'kovich, *Lakrkrasoch. Mater. Ikh Primen.*, **1967** (3), 23–25.

114. E. W. Eckey (to Procter and Gamble Co.), U.S. Pat. 2,378,006 (1945).

115. M. M. Chakrabarty and K. Talapatra, *J. Am. Oil Chemists' Soc.*, **45**, 172–175 (1968).

116. M. M. Chakrabarty and M. K. Kundu, *Sci. Cult.*, **35** (7) 328–329 (1969).

117. R. O. Feuge and A. E. Bailey, *Oil Soap*, **23**, 259–264 (1946).

118. R. O. Feuge and A. Gros, *J. Am. Oil Chemists' Soc.*, **26**, 97–102 (1949).

119. R. G. Ackman, in *Progress in the Chemistry of Fats and Other Lipids*, Vol. 12, R. T. Holman, Ed., Pergamon, New York, 1972, pp. 165–284.

120. G. R. Jamieson, in *Topics in Lipid Chemistry*, Vol. 1, F. D. Gunstone, Ed., Wiley Interscience, Logos Press, New York, 1972, pp. 107–159.

121. N. Pelick and V. Mahadevan, in *Analysis of Lipids and Lipoproteins*, E. G. Perkins, Ed., American Oil Chemists' Society, Champaign, Illinois, 1975, pp. 36–62.

122. W. W. Christie, in *Topics in Lipid Chemistry*, Vol. 3, F. D. Gunstone, Ed., Wiley-Interscience, Logos Press, New York, 1972, pp. 171–197.

123. A. Kuksis, *Separation Purification Meth.*, **6**, 353 (1977).

124. Report of the Instrumental Techniques Committee, AOCS, 1962–1963, *J. Am. Oil Chemists' Soc.*, **41**, 158–160 (1964).

125. Report of the Instrumental Techniques Committee, AOCS, 1963–1964, *J. Am. Oil Chemists' Soc.*, **42**, 347–351 (1965).

126. W. R. Morrison and L. M. Smith, *J. Lipid Res.*, **5**, 600–608 (1964).

127. L. D. Metcalfe, A. A. Schmitz, and J. R. Pelka, *Rapid Esterification of Lipids for Gas Chromatography Analysis*, Armour Industrial Chemical Co., Chicago, 1965.

128. A. Kaderavek, *Olearia*, **16**, 5–10 (1962).

129. B. Doro and G. Gabucci, *Boll. Lab. Chim. Provinciali* (Bologna), **13**, 478–487 (1962).

130. J. I. Peterson, H. De Schmertzing, and K. Abel, *J. Gas Chromatogr.*, **3**, 126–130 (1965).

131. M. Loury, *Rev. Fr. Corps Gras*, **14** (6), 383–389 (1967).

132. E. J. Gauglitz and L. W. Lehman, *J. Am. Oil Chemists' Soc.*, **40**, 197–198 (1963).

133. C. R. Houle, in M. E. Stansby, Ed., *Fish Oils*, Avi, Westport, Conn., 1967, pp. 52–62.

134. T. Nakanishi and T. Nakae, *Nippon Nogei Kagaku Jaishi*, **36** (4), 361–364 (1962).

135. G. B. Bradshaw and W. C. Meuly (to E. I. du Pont de Nemours and Co.), U.S. Pats. 2,271,619 (1942) and 2,360,844 (1944). See also G. B. Bradshaw, *Soap Sanit. Chem.*, **18**, No. 5, 23–24, 69–70 (1942).

136. H. D. Allen and W. A. Kline, U.S. Pat. 2,383,579; C. J. Arrowsmith and J. Ross, U.S. Pats. 2,383,580–1; E. E. Dreger, U.S. Pat. 2,383,596; G. A. Glossop, U.S. Pat. 2,383,599; J. H. Percy, U.S. Pat. 2,383,614; W. R. Trent, U.S. Pat. 2,383,632–3; all issued in 1945 and assigned to Colgate-Palmolive-Peet Co.

137. H. J. Wright, J. B. Segur, H. V. Clark, S. K. Coburn, E. E. Langdon, and R. N. DuPuis, *Oil Soap*, **21**, 145–148 (1944).

138. V. N. Pathak and R. K. Bhatnagar, *J. Oil Technol. Assoc. India*, **18** (2), 194–199 (1963).

139. R. Celades and C. Paquot, *Chem. Phys. Appl. Surface Active Subst., Proc. 4th Int. Congr.* (1964; published 1967) **1**, 249–255.

140. E. H. Gruger, Jr., D. C. Malins, and E. J. Gauglitz, *J. Am. Oil Chemists' Soc.*, **37**, 214–217 (1960).

141. J. C. Wooten and E. S. Lutton, *J. Am. Chem. Soc.*, **81**, 1762–1764 (1959).

142. L. A. Goldblatt, D. A. Yeadon, and M. Brown, *J. Am. Chem. Soc.*, **77**, 2477–2479 (1955).

143. H. M. Fett, *J. Am. Oil Chemists' Soc.*, **40**, 81–83 (1963).

144. H. Birnbaum (to Star Kist Foods, Inc.), U.S. Pat. 3,443,965 (May 13, 1969).

145. N. Krog, *J. Am. Oil Chemists' Soc.*, **54**, 124–131 (1977).

146. E. H. Freund (to National Dairy Products Corp.), U.S. Pat. 3,293,272 (December 20, 1966).

147. D. Meisner, *Succinylated Monoglycerides*, The Bakers Digest, June 1969, pp. 38–41.

148. D. F. Meisner, K. Lorenz, and J. L. Jonas, "Succinylated Monoglycerides: Effects in Conventional and CM Bread," *Cereal Sci. Today*, 400–405 (September 1967).

149. G. D. Neu and W. J. Simcox, "Dough Conditioning and Crumb Softening in Yeast-Raised Bakery Products with Succinylated Monoglycerides," *Cereal Foods world*, 203–208 (April 1975).

150. J. J. Geminder, *J. Am. Oil Chemists' Soc.*, **41**, 92–94 (1964).

151. R. E. Egan and S. B. Lampson (to Ashland Oil and Refining Co.), U.S. Pat. 3,433,645 (March 18, 1969).

152. R. E. Egan, S. B. Lampson, and I. R. MacDonald (to Ashland Oil and Refining Co.), U.S. Pat. 3,490,918 (January 20, 1970).

153. J. Moncrieff and A. G. Oszanyi, "Development and Evaluation of a New Dough Conditioner," *Baker's Dig.*, 44–46 (August 1970).

154. I. A. MacDonald, "Functionality of Ethoxylated Mono- and Diglycerides in Yeast Raised Bakery Products," Paper No. 45, 54th Annual Meeting, American Association Cereal Chemists, (1969), abstracted in *Baker's Dig.*, **43**, 75 (1969).

155. J. Van Haften, at AOCS Short Course, *Industrial Fatty Acids*, Tamiment, Pa., June 13, 1979, Paper No. 33, *Fat-Based Food Emulsifiers*.

156. P. Savary, *Compt. Rend.*, **226**, 1284–1285 (1948).

157. C. M. Gooding and H. W. Vahlteich (to Best Foods, Inc.), U.S. Pats. 2,197,339–340 (1940).

158. A. S. Richardson (to Procter & Gamble Co.), U.S. Pat. 2,251,692 (1941). A. S. Richardson and E. W. Eckey (to Procter & Gamble Co.), U.S. Pat. 2,251,693 (1941).

159. C. Franzke, F. Kretzschman, D. Kubel, L. Zahn, and E. Hollstein, *Nahrung,* **11** (7–8), 639–643 (1967).

160. R. S. McKinney and L. A. Goldblatt, *J. Am. Oil Chemists' Soc.*, **34**, 585–587 (1957).

161. E. H. Eaves, J. J. Spadaro, V. O. Cirino, and E. L. Patton. *J. Am. Oil Chemists' Soc.*, **38**, 443–447 (1961).

162. R. B. R. Choudhury, *Sci. Cult. (India)* **23**, 476–477 (1958).

163. L. Hartman, *J. Am. Oil Chemists' Soc.*, **40**, 142 (1963).

164. J. B. Brandner and R. L. Birkmeir, *J. Am. Oil Chemists' Soc.*, **37**, 390–396 (1960).

165. A. Edeler and A. S. Richardson (to Procter & Gamble Co.), U.S. Pats. 2,206,167–8 (1940).

166. C. W. Christensen (to Armour and Co.), U.S. Pat. 2,022,493 (1935).

167. Bea Salvia, Miguel, Spanish Pat. 383,780 (September 18, 1970).

168. N. I. Gel'perin and I. N. Gula, U.S.S.R. Pat. 193,485 (September 14, 1965).

169. C. J. Arrowsmith and J. Ross (to Colgate-Palmolive-Co.), U.S. Pat. 2,383,581 (1945).

170. H. Birnbaum (to Hockmeister, Inc.), U.S. Pat. 2,875,221 (February 24, 1959).

171. H. Birnbaum and J. Lederer (to Hochmeister, Inc.), U.S. Pat. 3,102,129 (August 27, 1963).

172. R. W. Giddings and A. C. Davis (to Lever Brothers Co.), U.S. Pat. 3,095,431 (June 25, 1963).

173. W. G. Alsop and I. J. Krems (to Colgate-Palmolive Co.), U.S. Pat. 3,083,216 (March 26, 1963).

174. S. S. Chang and L. H. Wiedermann (to Swift & Co.), U.S. Pat. 3,079,412 (February 26, 1963).

175. R. R. Allen and R. L. Campbell, Jr. (to Anderson, Clayton & Co.), U.S. Pat. 3,313,834 (April 11, 1967).

176. W. D. Pohl, V. C. Mehlenbacher, and J. H. Cook, *Oil Soap*, **22**, 115–119 (1945). W. D. Pohle, and V. C. Mehlenbacher, *J. Am. Oil Chemists' Soc.*, **27**, 54–56 (1950). Official and Tentative Methods of the American Oil Chemists Society, 3rd ed. (including revisions and additions–1976), Method Cd-11-57.

177. G. Y. Brokaw, E. S. Perry, and W. C. Lyman, *J. Am. Oil Chemists' Soc.*, **32**, 194–197 (1955).

177a. L. Hartman, *J. Am. Oil Chemists Soc.*, **39**, 126–128 (1962).

178. E. W. Eckey and M. W. Formo, *J. Am. Oil Chemists' Soc.*, **26**, 207–211 (1949). E. W. Eckey (to Procter and Gamble Co.), U.S. Pat. 2,442,534 (1948).

179. J. A. Monick, *J. Am. Oil Chemists' Soc.*, **40**, 606–608 (1963).

180. I. R. K. Kochar and R. K. Bhatnagar (to Council of Scientific and Industrial Research), Indian Pat. 71,979 (April 7, 1962).

181. S. K. Dey, I. R. K. Kochar, and R. K. Bhatnagar, *Indian Oilseeds J.*, **6**, 215–225 (1962).

182. A. E. Rheineck, R. Bergseth, and R. Sreenivasan, *J. Am. Oil Chemists' Soc.*, **46**, 447–451 (1969).

183. J. Harwood (to Glidden Co.), U.S. Pat. 3,312,724 (April 4, 1967).

184. T. Fujita, I. Yanagisawa, and M. Mori, *Nippon Suisan Chuo Kenkyusho Hokoku*, **10**, 63–66 (1966).

185. J. Zajic and E. Mares, *Sb. Vys. Sk. Chem-Technol. Praze. Potraviny*, **18**, 67–74 (1967).

186. H. Szczepanska and T. Janowska, *Tluszcze, Srodki Piorace, Kosmet.*, **12** (4), 129–130 (1968).

187. R. Schoellner and L. Laebisch, *Fette, Seifen, Anstrichm.*, **69**, 582–588 (1967).

188. R. Schoellner and L. Laebisch, *Fette, Seifen, Anstrichm.*, **69**, 426–431 (1967).

189. G. Dalby (one half to B. T. Rauber), U.S. Pat. 3,251,870 (May 17, 1966).

190. R. Aldo Macchi, I. Gallardo deKuck, and F. Crespo, *Rev. Argent. Grasas Aceitas*, **9**, 34–39 (1967).

191. R. Reiser and A. Furman (to Research Corporation), U.S. Pat. 3,595,888 (July 27, 1971).

192. A. S. Richardson and E. W. Eckey (to Procter and Gamble Co.), U.S. Pat. 2,132,437 (1938).

193. J. Ross, A. C. Bell, C. J. Arrowsmith, and A. I. Gebhart, *Oil Soap*, **23**, 257–259 (1946).

194. R. O. Feuge, *J. Am. Oil Chemists' Soc.*, **24**, 49–52 (1947).

195. N. H. Kuhrt, E. A. Welch, and F. J. Kovarik, *J. Am. Oil Chemists' Soc.*, **27**, 310–313 (1950).

196. M. V. Lomonosov Moscow Institute of Fine Chemical Technology (by N. A. Preobrazhenski, T. K. Mitrofanova and L. G. Geine), U.S.S.R. Pat. 185,882 (September 12, 1966).

197. M. V. Lomonosov Institute of Fine Chemical Technology, Moscow (by T. K. Mitrofanova and N. A. Preobrazhenski), U.S.S.R. Pat. 187,768 (October 21, 1968).

198. S. Matsuyama, M. Takasago, and K. Hirokawa, *Kagaku to Kogyo (Osaka)*, **42**, 239–241 (1968).

199. C. Schmulinzon, A. Yaron, and A. Letan, *Riv. Ital. Sostanze Grasse*, **48**, 168–169 (1971).

199a. S. Matsuyama, M. Takasago, K. Horikawa, F. Fujiwara, and C. Sugimoto, *Kagaku to Kogyo (Osaka)*, **42**, 242–248 (1968).

200. H. J. Wright, J. B. Segur, H. V. Clark, S. K. Coburn, E. E. Langdon, and R. N. DuPois, *Oil Soap*, **21**, 145–148 (1944).

201. J. Mleziva and V. Hanzlik, *Fette, Seifen, Anstrichm.*, **60**, 197–201 (1958).

202. L. Osipow, F. D. Snell, and A. Finchler, *J. Am. Oil Chemists' Soc.*, **34**, 185–188 (1957).

203. W. C. York, A. Finchler, L. Osipow, and F. D. Snell, *J. Am. Oil Chemists' Soc.*, **33**, 424–426 (1956).

203a. E. J. Lorand, U.S. Pat. 1,959,590 (May 22, 1954).

204. L. Osipow, F. D. Snell, D. Marra, and W. C. York, *Ind. Eng. Chem.*, **48**, 1462–1463 (1956).

205. Thomas Hedley & Co., British Pat. 804,197 (November 12, 1958).

206. J. Terc, *Veda Vyzk. Prum. Potravin.*, **19**, 117–140 (1969).

206a. Ledoga S.p.A., Belgian Pat. 622,394 (December 28, 1962).

206b. R. U. Lemieux and A. G. McInnes, *Can. J. Chem.*, **40**, 2376–2393 (1962).

206c. E. G. Bobalek, A. P. deMendoza, A. G. Causa, W. J. Collings, and G. Kapo, *Ind. Eng. Chem. Prod. Res. Develop.*, **2**, 9–16 (1963).

206d. K. Kunugi, *Chem. Pharm. Bull.* (*Tokyo*), **11**, 478–482, 482–486, 486–489 (1963).

206e. Farbenfabriken Bayer A.-G. (by W. Hagge, G. Matthaeus, and M. Quaedvlieg), Belgian Pat. 612,041 (January 12, 1962).

206f. Howards of Ilford, Ltd., (by C. F. Cardy, F. Schild, and R. J. Wicker), British Pat. 915,578 (January 16, 1963).

206g. S. Sakawa and T. Kikuchi, Japanese Pat. 18710 (December 14, 1962).

206h. Farbenfabriken Bayer A.-G., British Pat. 925,718 (May 8, 1963).

207. L. I. Osipow and W. Rosenblatt, *J. Am. Oil Chemists' Soc.*, **44**, 307–309 (1967).

208. R. O. Feuge, H. J. Zeringue, Jr., T. J. Weiss, and M. Brown, *J. Am. Oil Chemists' Soc.*, **47**, 56–60 (1970).

209. K. James (Tate and Lyle, Ltd.), German Pat. 2,546,716 (April 29, 1976).

210. G. P. Rizzi and H. M. Taylor, *J. Am. Oil Chemists' Soc.*, **55**, 398–401 (1978); U.S. Pat. 3,963,699 (to Procter & Gamble Co.), (June 15, 1976).

211. S. Komori, M. Okahara, and K. Okamoto, *J. Am. Oil Chemists' Soc.*, **37**, 468–473 (1960).

212. A. T. Gros and R. O. Feuge, *J. Am. Oil Chemists' Soc.*, **39**, 19–24 (1962).

213. E. W. Eckey, *Ind. Eng. Chem.*, **40**, 1183–1190 (1948).

214. F. A. Norris and K. F. Mattil, *J. Am. Oil Chemists' Soc.*, **24**, 274–275 (1947).

215. E. W. Eckey (to Procter & Gamble Co.), U.S. Pat. 2,442,531 (1948).

215a. G. Jurriens and A. C. J. Kroesen, *J. Am. Oil Chemists' Soc.*, **42**, 9–14 (1965).

216. A. Kuksis, M. J. McCarthy, and J. M. R. Beveridge, *J. Am. Oil Chemists' Soc.*, **41**, 201–205 (1964).

216a. D. Weihe, *J. Dairy Sci.*, **44** (May), 944–947 (1961).

217. E. W. Eckey (to Procter & Gamble Co.), U.S. Pat. 2,378,005 (1945).

218. C. van Loon (to Naamlooze Vennootschap Anton Jurgens Margarinefabrieken), Dutch Pat. 16,703 (1927); U.S. Pat. 1,873,513 (1932).

219. F. A. Norris and K. F. Mattil, *Oil Soap*, **23**, 289–291 (1946).

220. C. M. Gooding (to Best Foods, Inc.), U.S. Pat. 2,309,949 (1943).

221. T. J. Weiss, G. A. Jacobson, and L. H. Wiedermann, *J. Am. Oil Chemists' Soc.*, **38**, 396–399 (1961).

222. J. W. E. Coenen, *Rev. Fr. Corps Gras*, **21**, 403–413 (1974).

223. L. J. Bellamy, *Infrared Spectra of Complex Molecules*, Wiley, New York, 1954, pp. 149–150.

224. N. D. Fulton, Jr., E. S. Lutton, and R. L. Wille, *J. Am. Oil Chemists' Soc.*, **31**, 98–103 (1954).

225. K. Van Putte, L. Vermaas, J. Van Den Enden, and C. Den Hollander, *J. Am. Oil Chemists' Soc.*, **52**, 179–181 (1975).

226. B. L. Madison and R. C. Hill, *J. Am. Oil Chemists' Soc.*, **55**, 328–331 (1978).

227. Thomas Hedley & Co., Ltd. (by J. J. Devlin and A. P. Walker), British Pat. 832,377 (April 6, 1960).

228. F. E. Luddy, S. G. Morris, F. Magidman, and R. W. Riemenschneider, *J. Am. Oil Chemists' Soc.*, **32**, 522–525 (1955).

229. B. F. Teasdale and G. A. Helmel (to Canada Packers, Ltd.), U.S. Pat. 3,174,868 (March 23, 1965).

230. H. H. Hustedt, *J. Am. Oil Chemists' Soc.*, **53**, 390–392 (1976).

231. J. B. Rossel, *J. Am. Oil Chemists' Soc.*, **52**, 505–511 (1975).

232. I. P. Freeman, *J. Am. Oil Chemists' Soc.*, **45**, 456–460 (1968).

233. D. Chobanov and R. Chobanova, *J. Am. Oil Chemists' Soc.*, **54**, 47–50 (1977).

234. A. Kuksis, L. Marai, and J. J. Myher, *J. Am. Oil Chemists' Soc.*, **50**, 193–201 (1973).

235. R. Hites, *Anal. Chem.*, **42**, 1736–1740 (1970).

236. L. J. Filer, Jr., F. H. Mattson, and S. J. Foman, *J. Nutr.*, **99**, 293–298 (1969).

237. G. R. List, E. A. Emken, W. F. Kowlek, T. D. Simpson, and H. J. Dutton, *J. Am. Oil Chemists' Soc.*, **54**, 408–413 (1977).

238. C. W. Hoerr and D. F. Waugh, *J. Am. Oil Chemists' Soc.*, **32**, 37–41 (1955).

239. E. S. Lutton, M. F. Mallery, and J. Burgers, *J. Am. Oil Chemists' Soc.*, **39**, 233–235 (1962).

240. L. H. Wiedermann, T. J. Weiss, G. A. Jacobson, and K. F. Mattil, *J. Am. Oil Chemists' Soc.*, **38**, 389–395 (1961).

241. L. Koslowsky, *Oléagineux*, **30**, 221–224 (1975).

242. H. K. Hawley and C. W. Holman, *J. Am. Oil Chemists' Soc.*, **33**, 29–35 (1956).

243. P. Seiden (to Procter & Gamble Co.), U.S. Pat. 3,353,964 (November 21, 1967).

244. D. Melnick and C. M. Gooding (to Corn Products Co.), U.S. Pat. 2,921,855 (January 19, 1960).

245. C. M. Gooding and C. A. Cravens (to Corn Products Co.), U.S. Pat. 3,085,882 (April 16, 1968).

246. R. J. Bell, R. L. Campbell, Jr., P. Gibson, and J. F. Sims (to Anderson, Clayton & Co.), U.S. Pat. 3,396,037 (August 6, 1968).

247. D. P. Kidger (to National Biscuit Co.), U.S. Pat. 3,361,568 (January 2, 1968).

248. G. W. Holman and L. H. Going (to Procter & Gamble Co.), U.S. Pats. 2,875,066–067 (February 24, 1959).

249. W. Stein, H. Rutzen, and E. Sussner (to Henkel & Cie), U.S. Pat. 3,232,971 (February 1, 1966).

250. A. Babin, *Oléagineux*, **29** (7), 375 (1974).

251. B. Sreenivasan (to Lever Brothers Co.), U.S. Pat. 3,748,348 (July 24, 1973).

252. B. Sreenivasan (to Lever Brothers Co.), U.S. Pat. 3,859,447 (January 7, 1975).

253. N. R. Artman (to Procter & Gamble Co.), U.S. Pat. 3,376,326 (April 2, 1968).

3

Extraction of Fats and Oils

The separation of oils and fats (1) from oil-bearing animal and vegetable materials constitutes a distinct and specialized branch of fat technology. The widely differing characteristics of fatty materials from diverse sources have given rise to extraction processes, such as rendering, pressing, and solvent extraction. All extraction processes, however, have the following common objectives: (*a*) to obtain the fat or oil uninjured and as free as possible from undesirable impurities; (*b*) to obtain the fat or oil in as high a yield as is consistent with the economy of the process; and (*c*) to produce an oil cake or residue of the greatest possible value.

Fatty animal tissues consist largely of fat and water that may be separated from the solid portions of the tissue and from each other with relative ease by one of the rendering processes. The extraction of vegetable oils is a more difficult matter. Vegetable materials, and in particular some of the oil seeds, contain a large proportion of solid material associated with the oil. Here, careful reduction of the material, followed by heat treatment and the application of heavy pressure, is required to obtain an efficient separation of the oil from the solids.

Even after the most efficient pressing, an oil cake will retain an appreciable amount of absorbed oil, usually 2.5–5% by weight. In the case of seed or other materials initially high in oil and low in solids content, the unextracted residue will contain only a small fraction of the total oil; however in seeds of low oil content, such as soybeans, it may contain as much as 15–20% of the total oil. For the processing of low-oil seeds solvent extraction is particularly valuable since it will reduce residual oil in the extracted seeds to less than 1%. The chief disadvantages of solvent extraction are the high initial cost of the equipment and the fact that some oil seeds disintegrate under the influence of the solvent and, consequently, are difficult to handle.

A number of more or less critical operations in oil milling are auxiliary to the actual expression or extraction. Wherever possible, it is desirable to decorticate oil seeds before the oil is removed so as to both increase the capacity of the extraction equipment and avoid loss of oil through absorption by the hulls. The seeds must then be rolled, ground, or otherwise reduced to fine particles. After they are reduced, they must be heated to make the walls of the oil cells permeable to the oil and to render the oil free flowing, except where solvent extraction is used; then heat treatment is not generally necessary. In processing cottonseed, special attention must be given to the inactivation of gossypol or other toxic constituents.

In extracting oil from oil seeds there are some major differences between common American and common European practice. They result from basic differences in the supply of raw materials. Most American mills operate on domestic oil seeds, and they are usually located close to production areas. Frequently, only one type of oil seed is processed. The quality of the seed is generally high, with relatively little variation in seed characteristics through the processing season or from one season to another. European mills, on the other hand, process imported raw materials almost exclusively, and each mill must be prepared to handle a variety of oil seeds differing widely in quality and processing characteristics. As a result, American milling practice has become highly specialized, with the object in each case being to perform a specific operation with the highest possible efficiency. In European mills it has been nec-

Table 3.1   Average yield of oil from commercial processing of common oil seeds (percent oil from seed of normal moisture content)[a]

Babassu (kernels)	63	Perilla seed	37
Castor beans	45	Poppyseed	40
Coconut (copra)	63	Rapeseed	35
Corn (germs)	45	Rice bran	14
Cottonseed	18	Safflower seed	28
Flaxseed	34	Sesame seed	47
Hempseed	24	Soybeans	18
Kapok seed	20	Sunflower seed	25
Oiticica (kernels)	60	Tea seed	48
Palm, African (kernels)	45	Tung	35
Peanuts	35		

[a] Yields are by mechanical expression for all except soybeans, cottonseed, and rice bran, which are by solvent extraction, and refer to whole or undecorticated seeds, unless otherwise stated.

essary to sacrifice some operating efficiency in favor of flexibility of operation.

The average yields of oil obtainable by commercial extraction methods from a number of common oil seeds are summarized in Table 3.1. Certain comparative data on whole seeds and kernels are found in Table 3.2. For information on yields from fruit pulps and animal sources, reference should be made to the specific fats and oils in other portions of this chapter and in other chapters.

It is probable that the first methods of fat extraction were rendering procedures practiced by primitive humans, following cooking techniques

Table 3.2 Approximate proportions of hulls and kernels in different oil seed, and oil contents of whole seed, kernels, and hulls

Oil Seeds	Percent Kernel	Percent Hull	Percent Oil in Whole Seed	Kernel	Hull
Usually decorticated					
Oil palm	25	75	—	48	—
Babassu	9	91	—	67	—
Cohune	10	90	—	67	—
Tucum	30	70	—	47	—
Murumuru	40	60	—	—	—
Tung	60	40	30	50	—
Oiticica	65	35	38	58	—
Cocoa beans	88	12	50	—	—
Castor beans	70–80	20–30	40–50	—	—
Cottonseed (delinted)	62	38	19	30	1–2
Peanuts	75	25	38	50	0.5–1
Sunflower seed	45–60	40–55	22–36	36–55	1–2
Kapok	60	40	20–25	40	—
Safflower	50	50	28–33	55–65	1.5–2
Soybeans[a]	93	7	18	19	0.6
Usually not decorticated					
Flaxseed	57	43	—	58	22
Perilla seed	68	32	34	—	—
Hemp seed	62	38	31	—	—
Rapeseed	82	18	42	—	—
Mustard seed	80	20	—	—	—

[a] Soybeans are now being dehulled at many mills so as to produce a 49% protein soybean meal especially suitable where low fiber content is important in a feed.

developed for the preparation of meats for food. The pressing of oil from olive pulp probably antedated the pressing of oil seeds, although seeds were processed by the Chinese and others at an early date using mechanical presses operated by wedges or levers. On the other hand, the more efficient hydraulic operation of mechanical presses was not adopted until early in the nineteenth century. The continuous screw press is a modern development, and the solvent extraction of oil seeds on a large scale was not a reality until after World War I. The entire subject was recently reviewed in the Proceedings of the World Conference on Oilseed and Vegetable Oil Processing Technology in March 1976 (2) and earlier in The Short Course on Unit Processes in the Fatty Oil, Soap, and Detergent Industries (3).

The residues from the processing of oil seeds or animal tissues for fat are generally high in protein content and are in good demand as animal feedstuffs. They have a limited use as a source of human food (soybean or cottonseed flour, and soybean protein isolates and concentrates), or as a source of industrial proteins (e.g., for making glues). The residues from castor beans and tung nuts are toxic unless specially treated; hence they are used only as fertilizer and the like.

Mechanical Pretreatment

1.1   PREPARATION OF ANIMAL MATERIALS

Fatty animal materials, as compared with oil seeds and other vegetable materials, require comparatively little preparation prior to the rendering operation. Fatty stock destined for the production of neutral, low-temperature-rendered fats, such as oleo stock or neutral lard, is carefully trimmed and washed before it is charged to the rendering units. Ordinary stock, such as that used in making prime steam lard, is not always washed and is less carefully trimmed.

In the larger establishments the stock to be rendered is sorted into different classes of material, partly to avoid mixing high-quality materials with those of low quality, and partly because some stocks, such as those containing large bones, require more severe rendering than others.

In either dry rendering or steam rendering, separation of the fat is more rapid if the fatty stock is first cut into small pieces, although this operation is ordinarily omitted in steam rendering. Prolonged wet rendering under pressure will disintegrate even large bones or whole carcasses; thus the preparation of stock for this process is not critical.

Rotary hashers, similar in principle to ordinary household food choppers, are used for the reduction of stock that is free of bones. The degree of reduction is usually much coarser than that employed in the processing of oil seeds; the dimensions of the hashed pieces may be measured in large fractions of inches or even in inches. Most animal materials disintegrate quite readily. Whale blubber is particularly tough and requires more drastic treatment. Blubber presses, consisting of heavy corrugated rolls, are now in use (4, 5). Passage of large chunks of blubber through these rolls reduces them to semifluid condition and decreases the rendering time.

1.2   Preparation of Oil Seeds (5a–d)

Cleaning.   The first step in the processing of oil seeds is cleaning to separate foreign material. Sticks, stems, leaves, and similar trash are usually removed by means of revolving screens or reels. Sand or dirt is also removed by screening. Permanent or electromagnets installed over a conveyor belt are used for the removal of tramp iron. Special "stoners" are employed for taking out heavy stones and mud balls from shelled peanuts. A pneumatic system will include an aspirating area where the light material is pulled through and the heavier material (usually the oil-bearing seed) gravitates out. As stated previously, the cleaning of oil seeds is preferably carried out before the seeds are placed in storage; often, however, it is not, since adequate cleaning capacity is costly.

Dehulling and Separation of Hulls.   Wherever practicable, oil seeds are preferably decorticated before they are extracted. The hulls of oil-bearing seeds are low in oil content, usually containing not more than about 1%, although contamination with kernels will, of course, increase the oil content with resultant loss of available oil. If the hulls are not removed from the seeds before the latter are extracted, they reduce the total yield of oil by absorbing and retaining oil in the press cake and, in addition, reduce the capacity of the extraction equipment.

The hulling machines used for the decortication of medium-sized oil seeds with a flexible seed coat, such as cottonseed, peanuts, and sunflower seed, are of two principal types: bar hullers and disc hullers.

The rotating member of a bar huller is a cylinder equipped on its outer surface with a number of slightly projecting, longitudinally placed, sharply ground, square-edged knives or "bars." Opposed to the cylinder over an area corresponding to about one-third of its surface is a concave member provided with similar projecting bars. The seeds are fed between the rotating cylinder and the concave member, and the hulls are split as the seeds are caught between the opposed cutting edges. The clearance between the cutting edges may be adjusted for seeds of different sizes.

The disc huller is more or less similar in principle to the bar huller, except that the cutting edges consist of grooves cut radially in the surfaces of two opposed and vertically mounted discs, one of which is stationary and the other rotating. The seeds are fed to the center of the discs and are discharged at their periphery by centrifugal force. With either type of huller the condition of the seed is somewhat critical. Wet seeds are difficult to split cleanly and may clog the huller, particularly if it is of the disc type. On the other hand, if the seeds are very dry, the kernels may disintegrate excessively.

Different seeds vary considerably in the readiness with which they fall out of the split hulls. Peanuts, for example, are loose in the shell and separate readily. Cottonseed kernels or "meats" are more adherent to the hull; consequently, the hulls are customarily passed through a hull beater to detach small meat particles after the first separation of hulls and meats by screening. The separation systems used for cottonseed, peanuts, and so on consist of various combinations of vibrating screens and pneumatic lifts. It is necessary not only to separate the hulls from the meats, but also to separate and recycle a certain proportion of uncut seeds that escape the action of the huller. In the case of cottonseed the following separations are commonly carried out: (a) separation of large meat particles from hulls and uncut seed by screening; (b) separation of hulls from uncut seed by an air lift; (c) separation of small meat particles from hulls by beating and screening; and (d) separation of hull particles from meats by air.

In practical mill operation, especially on cottonseeds, the greatest yield of oil is obtained by nicely balancing the degree of separation attained. If an attempt is made to separate hulls from the meats too cleanly, there will be a loss of oil as a result of meats being carried over into the hulls. If an excessive proportion of hulls is left in the kernels, there will likewise be an undue loss of oil from absorption by the hulls. Under certain conditions, there may be an appreciable loss of oil due to absorption by the hulls as the latter come into contact with the oily meat particles during the separation operation. It is generally advisable to effect the separation of kernels and hulls as quickly as possible after the seeds are hulled to avoid excessive contact between hulls and kernels or kernel particles (6).

Cottonseed are invariably delivered to the mills from the gins without removal of their coating of short fibers or linters and must be delinted before they are hulled. Delintering machines (known as *linters*) are similar in principle and appearance to cotton gins, consisting essentially of a revolving assembly of closely spaced circular saws that pick the lint from the seed. The fibers are removed from the saw teeth by a revolving cylindrical brush or by an air blast that suspends them in an air stream in which they are conveyed through pipes to collection equipment. The lint is not ordinarily removed from the seed in a single operation but is taken off in two or three cuts. Each successive cut is of lower grade than the cut preceding it since increasing quantities of hull material are removed by the saws as delinting proceeds and fiber length is decreased.

Previously soybeans were seldom decorticated before processing for oil (except where the meal was destined for human consumption) because of mechanical difficulties and because the hull constitutes only a small part of the seed and is relatively nonabsorbent. Today, dehulling is common. It is usually accomplished by first cracking the beans on cracking rolls and then separating the hulls from the kernels in two stages:

1. Hulls are screened from kernels and uncracked beans and aspirated at the end of the top deck of a double-deck shaking screen. Uncracked beans are returned to the cracking rolls while the kernels are put on a second deck of finer mesh screen, at the end of which hulls are again aspirated. Fines are joined to the whole kernel flow.
2. Hulls that have been aspirated contain some kernel particles; therefore, these are subjected to an air separation using a gravity table to separate the light hulls from the heavier kernels. Depending on the degree of separation required, a "middling" fraction may be taken, and this is broken down on another gravity table.

A somewhat different system makes use of simultaneous grinding and aspiration to dehull the beans and separate the hulls. In general, the choice of system depends on the processor, who may use many modifications of general techniques for his or her purposes.

Small oil seeds, such as flaxseed, perilla, rapeseed, and sesame, are often handled without decortication in view of processing difficulties; however, considerable attention has been focused on sunflower, safflower, and rapeseed. The separation of sunflower seed into two fractions according to size, and the subsequent dehulling of each fraction reportedly leads to improved yields (7). Successful decortication of sunflower and safflower has also been reported by Galloway (8).

The various palm kernels, such as oil palm or African palm kernels, babassu kernels, and cohune kernels, constitute a special class of oil seeds, since they are of relatively large size and are surrounded by a particularly hard, thick shell. Because of the low cost of labor in the producing regions, the large size of the nuts, and the refractory nature of the shells, these nuts are often cracked and the kernels separated by hand. The entire production of Brazilian babassu kernels, amounting in some seasons to over 80,000 tons, has in the past been separated in this manner.

In Africa, nuts of the oil palm, which are less thick-shelled than most of the American palm nuts, are apparently hand cracked to some extent; but on the plantations of Indonesia and Malaysia they are usually machine cracked. In one type of machine the nuts are fed to the center of a rotor provided with curved baffles, along which the nuts are flung out against a heavy steel housing and broken by impact. Another type of machine is simply a special type of hammer mill. The rotor consists of a frame supporting four heavy steel paddles; the nuts are dropped into the path of the paddles and cracked by impact.

After the nuts are cracked, they are dropped to rotary screens where some separation of kernels and shells is obtained. A considerable proportion of shell fragments, however, cannot be separated by screening. Because of the high density of the shells, air separation like that used on cottonseed and peanuts is likewise ineffective in producing a further separation. There are two methods in vogue for separating palm kernels from shell fragments of a size comparable to that of the kernels. The dry method takes advantage of the fact that the kernels are rounded and roll easily, whereas the pieces of shell are flat and sharp edged and hence, do not roll as readily on an inclined surface.

Dry separators consist of inclined belts provided with sharp projections that move continuously upward. When a mixture of kernels and seeds is fed onto the surface of the belt, the kernels roll down the belt and are collected at the lower end, whereas the fragments of shell are caught on the projections and carried over the top of the belt into a separate bin. Means must be provided for recycling material after both the cracking and separating operation since neither cracking nor separation is complete after one passage of the material through the machines.

The alternative method of separation consists of floating the kernels from the more dense shells in salt or mud baths. A recent modification utilizes a hydrocyclone to separate the light kernels from shells and some nuts that are returned for further processing. The final kernel stream is strained on a vibrating strainer and then sent to a dryer for about 10 hours. The dried nuts will average 6–7% moisture and around 50% palm kernel oil.

The American palm nuts of the *Attalea* family, including the babassu and cohune, are excessively thick shelled and extremely difficult to decorticate by machinery. The babassu is particularly troublesome because it contains several kernels, each of which is enclosed in a separate cavity within the shell. Whereas the splitting of an oil palm nut or most cohune nuts along a single plane of cleavage will usually free the kernel, similar splitting of a babassu nut may not release a single one of its four to eight kernels.

A number of different machines have been devised for cracking American palm nuts. The machines designed for round nuts of the coyol type have either been of the centrifugal or hammer-mill design or have utilized the positive action of mechanically or hydraulically operated hammers striking against the nut as it is confined against a stationary anvil member. Some of the machines designed for cohune or babassu nuts employ chisel-like cutting edges to open the shell. One type of babassu-opening machine has opposed cutting edges that split the nut into a number of segments like those of an orange. Other machines for cohune and babassu nuts employ the hammer-mill principle. These machines break up the kernel rather badly, and thorough drying of the kernels is relied on to inhibit excessive enzyme action in the broken kernels during shipment.

In the case of any variety of palm nut, adequate drying of the nuts prior to cracking is mandatory to ensure that the kernel will not adhere to the shell. Green or undried kernels fill the shell cavity tightly and adhere very strongly. Nuts are commonly placed in silos, after being separated from the fibers, where they remain for about 10 hours to dry further in warm air (8a). Aside from the fact that it is necessary for efficient decortication, thorough drying of the nuts will, of course, minimize the danger of deterioration in the kernels from enzyme action.

Reduction of Oil Seeds. The extraction of oil from oil seeds, either by mechanical expression or by means of solvents, is facilitated by reduction of the seed to small particles.

Opinion is divided as to whether the grinding or rolling of oil seeds actually disrupts a large proportion of the oil-bearing cells. The assumption of extensive cell breakage has in the past been based chiefly on the fact that seed flakes yield a large fraction of "easily extractable" oil on treatment with solvents and a smaller fraction (usually 10–30%) of oil that is extracted with much greater difficulty (9, 10). The former fraction was presumed to come from broken cells. It has been shown (11), however, that seeds (soybeans) that are cracked rather than rolled, with a minimum of crushing, likewise yield a large fraction of oil that is easily washed out with solvents. Furthermore, Woolrich and Carpenter (12) could observe little disruption of cells in rolled cottonseed flakes examined under the microscope. As an argument against extensive cell destruction, they pointed out that the cells of cottonseed are only 0.001–0.0015 in. in diameter, whereas the thickness of rolled cottonseed particles is not less than 0.005 in. On the other hand, Shchepkina's (13) rather high estimates of the proportion of broken cells were made from a count of free aleurone grains in flake samples.

In any event, it appears that many oil cells remain intact after even the most careful reduction, and the walls of these cells are made permeable to the oil only by the action of heat and moisture in the cooking operation. The cell walls, however, will be more readily acted on by heat and moisture if the seed particles are small.

Obviously, rolling seed or seed particles into thin flakes will facilitate solvent extraction both from the disruptive effect of rolling and by reducing the distances that solvent and oil must diffuse in and out of the seed during the extraction process. Early work indicated that the rate-controlling factor in the solvent extraction of seed flakes was probably the internal resistance of the flakes to the molecular diffusion of solvent and oil (14). If this is the case, the extraction rate should theoretically be in indirect proportion to the square of the flake thickness; doubling the thickness, for example, should quadruple the time required for reduction of the residual oil to a given level. The data of Karnofsky (11) on the extraction of soybean flakes by percolation with hexane (Figure 3.1) roughly bear out this expectation.

Later work by Othmer and Agarwal (15), using countercurrent batch extraction, has permitted a calculation of the oil in meal from measurements of the increase in miscella concentration; namely,

$$\frac{-dC}{dt} = kF^{3.97}C^{3.5}$$

The extraction rate is proportional to the 3.97 power of the flake thickness F and to the 3.5 power of the residual oil C. In practice, this would mean that if flake thickness were reduced to one-third of its former value, the extraction rate would be increased by $(1/3)^{-3.97}$ or approximately eightyfold. Similarly, with constant flake thickness, if the oil content dropped from 20% to one-tenth of that, or 2%, the extraction rate would decrease to less than three-thousandths of its initial value. Other factors should be considered, however, such as the mechanical strength of the flakes, the resistance offered by the flake mass to the flow of solvent, and the ease with which miscella may be washed from flakes. Consequently, for solvent extraction, seeds are not usually rolled to the least possible thickness.

Hammer mills, attrition mills, and other devices are sometimes used for the preliminary reduction of large oil seeds, such as copra and palm or babassu kernels; but for the final reduction it is the almost invariable practices in the United States to use milling rolls. These are generally

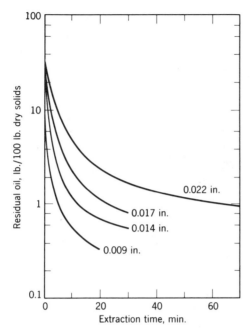

Figure 3.1   Relation of flake thickness to extraction rate in the solvent extraction of soybean flakes by percolation with hexane (11).

considered to be more economical to operate than other types of mill. Also, thin flakes to which oil seeds are reduced by smooth rolls are more satisfactory for hydraulic pressing then the irregularly shaped particles obtained by grinding. Flaking rolls are essential for preparing oil seeds for continuous solvent extraction since no other form of mill is capable of forming particles that are thin enough to extract readily yet large enough and coherent enough to form a mass through which the solvent will freely flow.

A roll assembly commonly used for the reduction of cottonseed, flaxseed, and peanuts in the mills of the southern United States (Figure 3.2) consists of a series of five rolls placed one above the other. The seed is introduced by a feeding mechanism between the two top rolls. It passes back and forth between adjoining pairs of rolls as it travels from the top to the bottom of the assembly; hence it is rolled four times. Each roll supports the weight of all the rolls above it, so that the seed particles are subjected to progressively increasing pressure as they pass from one pair of rolls to another. Although the lower rolls are smooth, the top roll is commonly corrugated to ensure that the seed will be "nipped" as fast as

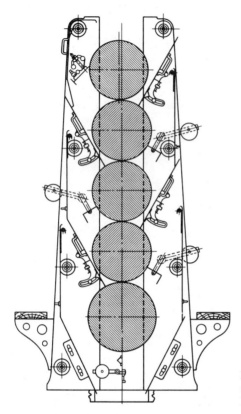

Figure 3.2   Five-high crushing rolls, cross section. (Courtesy of The French Oil Mill Machinery Company.)

they are fed to it. A popular five-high roll assembly consists of four upper rolls each 14 in. in diameter by 48 in. in width, and a bottom roll 16 in. by 48 in. in size, operating at a peripheral speed of about 630 ft/min. This unit has a rated capacity of 80 short tons of cottonseed or 300 bushels of flaxseed in 24 hours. However, the actual capacity in any case depends on the flake thickness that is obtained. Detailed data on the capacity and efficiency of cottonseed flaking rolls have been published by Wamble (16).

Cottonseeds are usually rolled to a thickness of 0.005–0.010 in. where mechanical pressing is to be used. With solvent, flake thickness is seldom less than 0.008–0.010 in. The repeated passage of the material through the rolls results in considerable breaking up of the individual flakes, but this is not particularly disadvantageous in the case of seeds that are to be mechanically expressed. Small oil seeds, such as flaxseed and sesame, are usually rolled in preparation for expression.

In the preparation of oil seeds for expression in expellers* or screw presses, the production of thin particles is not so essential as for hydraulic pressing since heat is generated and seed particles are broken up by the intense shearing stresses developed in the barrel of the expeller. Soybeans to be processed in expellers are usually cracked by corrugated cracking rolls into particles averaging 10–16 mesh in size and are then expressed without rolling or further reduction. Palm kernels, copra, peanuts, and so on are handled in expeller plants both with and without rolling. Cottonseed are usually rolled before expeller processing.

The rolls used for flaking soybeans or other oil seeds for solvent extraction are normally somewhat different in design from those just described. Since large, coherent flakes are desired, the flaking operation is commonly carried out by a single passage of the whole or cracked seeds through the rolls; therefore, only one pair of rolls is provided. The rolls are mounted side by side rather than being superimposed and are equipped with heavy springs to maintain the pressure of one roll against the other (17). Since the clearance between rolls of this type is adjustable, flakes of quite uniform thickness are produced. In modern deep bed extractors of large capacity, flakes of 0.016-in. thickness pack well into the extractor permitting a longer extraction time that compensates for the thicker flakes (17).

A reasonably high moisture content is required in oil seeds that are to be formed into thin, coherent flakes. Very dry seeds do not flake well. For solvent extraction, cracked soybeans are adjusted to a moisture content of 10–11% and flaked while still hot and slightly plastic; that is, while at a temperature of 160–170°F (17). In some cases the cracked beans are steamed for a short time prior to flaking.

Heat Treatment of Oil-Bearing Materials

The heat treatments given oil-bearing materials may be divided into two categories according to whether they are alone productive of oil or merely serve to facilitate the subsequent expression of oil by mechanical means. The term "rendering" is generally applied to treatment designed to remove all or most of the fat from fatty animal tissues or other materials with a high ratio of fat to solid matter. The heat treatment applied to oil seeds and similar materials prior to pressing is more commonly termed

---

* From this point on the term "expellers" is used to include screw presses and expellers since these machines are very similar.

"cooking." Some processing methods are a combination of rendering and cooking.

In the case of either rendering or cooking, the principal object of the heat treatment is the same, that is, to coagulate the proteins in the walls of the fat-containing cells and make the walls permeable to the flow of oil. The flow of oil from the oil-bearing material is also assisted by the lowered viscosity of the oil at elevated temperatures. Since oil-containing materials are never completely dry, heat treatment is inevitably associated with various effects due to the presence of moisture, even when water is not added in the processing operation. Water must be present for the above-mentioned protein coagulation to take place. Anhydrous proteins do not readily coagulate or exhibit other evidence of heat denaturation. In some cases water also assists in the displacement of oil from the surfaces of solid materials through a superior physiochemical affinity for the latter.

Before going into the specifics of obtaining oil from oil-bearing material, it should be noted that many advances have been made in processing in the last 20–25 years. Rendering of animal fats has probably changed the least, but oil seed processing has changed markedly. In the United States, where many small cottonseed mills existed and hydraulic pressing was not uncommon, consolidation has taken place with fewer but larger mills processing almost entirely by continuous pressing or some form of solvent extraction. Likewise, hydraulic pressing of soybeans is rare, if it exists at all; and with the possible exception of one or two specialty mills, solvent extraction has completely replaced continuous pressing. Prepressing, once unheard of in the United States, is now common to cottonseed. Also, as the soybean crop has largely replaced cottonseed for oil, solvent extraction has become increasingly important, especially with the advent of undenatured or only partially denatured soybean flakes for edible (human) uses.

In this chapter considerable attention has been given to mechanical expression for any help it may give to small mill installations where solvent extraction is uneconomical or impractical because of available labor or safety considerations. Even one serious mistake in a solvent mill may result in loss of life as well as property.

2.1   RENDERING OF ANIMAL FATS (18)

Fatty animal tissues free from muscle or bone are usually 70–90% fat; the remainder consists of water plus a small amount of connective tissue. The latter is made up largely of proteins; hence the residue from rendering ("tankage," "cracklings," "stick," etc.), like the residue from the processing of oil seeds, is essentially a protein concentrate that is used principally as an animal feed.

The product of highest fat content (92–95%) obtained in meat packing establishments is leaf fat from hogs. The internal fat from cattle used for the manufacture of oleo stock contains 60–80% fat. Back fat and other so-called cutting fat from pork runs 80–85% fat. A considerable amount of lard and tallow is obtained, however, from bone stock and other low-fat material, which may not contain more than 10–15% fat. Under certain circumstances, whole carcasses of large animals may be rendered for inedible fat recovery and conversion of the residue to tankage.

Most of the fish oil produced comes from the rendering of whole small oily fishes, such as sardines and herring, which contain 10–20% oil. Whales, however, which give an average oil yield in the neighborhood of 30,000 lb per animal, are trimmed of their fatty tissues or blubber, which contains about 70% fat and is rendered separately from bones or flesh.

Methods of rendering are dictated by the nature of the fatty stock, as well as the characteristics desired in the rendered fat and the rendering equipment available.

Dry Rendering. "Dry" rendering is one of the simpler methods of fat extraction. It is distinguished from "wet" rendering in that the expulsion of fat is accompanied by dehydration of the fat and fatty tissues, so that the latter are essentially dry at the end of the operation. The frying of bacon, to cite a familiar example, is essentially a dry-rendering process.

Dry rendering is normally carried out in horizontal steam-jacketed tanks with a large charging opening in the center of the top and an agitator. The agitator has paddles attached by arms to a horizontal shaft. After the charge (5000–10,000 lb) is dried to the desired moisture level, the contents are discharged into a steel box equipped with a perforated liner, and all possible free liquid is drained off. The residue is pressed, and the fat obtained is combined with the drained fat. After settling, centrifuging, or filtering, it is ready for market. The residue is ground as a protein supplement for animal and poultry feed.

The cooking or drying operation may be carried out at atmospheric, superatmospheric, or reduced pressures. Best yields are obtained under vacuum, but most plants operate at atmospheric pressure. Recent development in cooking includes increased agitation (34–40 rpm versus 15–20 formerly), permitting a decrease in cooking time of about 25%.

In the Anderson C-G continuous inedible rendering process, coarsely ground solids are mixed with liquid fat to form a pumpable slurry that is dried under vacuum and the fat separated centrifugally. The solids removed are conveyed to expeller presses where the fat content is reduced from 25–35% to as low as 7%. A high quality tallow and meal is claimed, using only about one-half the steam required by other conventional systems (19).

Dry rendering is preferred for inedible products where flavor and odor are secondary and the production of large quantities of high-quality residue is important.

Wet Rendering.   Wet rendering is used for edible products where color, flavor, and keeping qualities are of prime importance and the relative percentage of residue is small. It is carried out in the presence of a large amount of water. Separated fat was formerly removed by skimming, but centrifugal methods are widespread today. There are two varieties of wet rendering: low temperature, which is conducted at temperatures up to the boiling point of water, and high temperature or steam rendering, which is carried out under pressure in closed vessels.

Most of the animal fat produced in the United States is rendered by the steam process. The lard produced by this method of rendering is known as "prime steam lard." In addition to lard, tallow and whale oil are also usually steam rendered.

The apparatus used in United States packing houses is a vertical cylindrical steel autoclave or digester with a cone bottom, designed for a steam pressure of 40–60 psi and a correspondingly high temperature. The vessel is filled with the fatty material plus a small amount of water, and steam is admitted to boil the water and displace the air. The vessel is then closed except for a small vent, and the injection of steam is continued until the operating temperature and pressure are attained; the time required for digestion varies with the temperature and also the nature of the charge. The usual digestion time is 4–6 hours. Under the influence of the high temperatures employed, the fatty materials in the digester disintegrate to some extent; there is a very efficient separation of the fat, which rises to the top of the vessel, leaving a layer of solids (tankage) and "stick water" in the bottom. Pressure is then slowly relieved and the fat–water interface is adjusted to the level of a draw-off cock on the side of the vessel. The fat is drawn off and purified from traces of water and solid material by settling or occasionally by centrifuging. Eventually it may be filtered.

In the steam rendering of high-fat stock, 99.5% or more of the fat in the raw material is ordinarily recovered. The fat that is not recovered consists of a small residue in the tankage plus a very small amount that remains in the "stick water." The usual packing house "killing" and "cutting" fats yield about 80 and 70% lard, respectively, plus 2–3% each of dry tankage and dry "stick" or solid residue from the evaporation of "stick water." The dry tankage and stick will ordinarily contain about 10–12% and 1.5–2% fat, respectively. Both are high in protein content; tankage from good stock may analyze as high as 70–72% protein, and stick may be as high as 90% or more.

The advantages of steam rendering are that an efficient recovery of fat is obtained in relatively simple equipment and that it is adaptable to a wide variety of materials. There is little tendency for proteins and other substances to dissolve or disperse in the fat in the presence of water; hence the fatty stock may contain a large proportion of nonfatty tissue. Bony stock can be handled by this process since it is effectively disintegrated by prolonged treatment with steam under elevated pressure.

Steam rendering is less rapid and less efficient than dry rendering from the standpoint of heat consumption, however, and a large amount of water must be evaporated to recover the nonfatty residue in a concentrated form. Some hydrolysis of fat occurs during steam rendering; the free fatty acids content of prime steam lard is seldom less than about 0.35%. At 47 lbs pressure (20), development of free fatty acids is at the rate of about 0.06%/hour. The acidity in any case depends on the rendering time and temperature and the storage temperature and duration of storage of the fatty stock before it is processed (18, 21). By careful scheduling of operations, killing fats may be rendered reasonably soon after the animals are slaughtered, but carcasses are chilled to 32–36°F before cutting fats are available. The stability of lard toward oxidation bears no relation to the acidity and appears to depend principally on processing and handling subsequent to rendering (21).

One of the major recent developments in rendering has been the discovery that antioxidants added before rendering greatly enhance the stability of the fat produced (22, 23). Sims and Hilfman (24) studied the stabilization of lard and edible beef fats during pressure steam rendering. Antioxidants tested included butylated hydroxyanisole (BHA), butylated hydroxytoluene (BHT), propyl gallate and citric acid combinations, and a mixture of BHA and BHT. Best results for a given stabilizer level were obtained with the individual phenolics BHA and BHT. Poorer results were obtained with the mixtures in propylene glycol.

There are several modifications of continuous wet rendering operations in which attempts are made to obtain a better product and a better protein residue than the usual pressure tank produces. Some of these are described in the paragraphs that follow.

KINGAN PROCESS (25).   This process is based on the release of oil from tissue through comminution to subcellular dimensions. Raw material is finely ground, pumped through an appropriate heat exchanger, and re-ground in a hammer mill, and the fat is then separated from the protein and waste by a special type of centrifuge.

TITAN EXPULSION SYSTEM (26).   The fat stock is quickly minced and rendered in a combined mixer–boiler apparatus ("The expulsor"). It is then strained (to remove tissue, which is subsequently pressed) and the strained emulsion pumped to continuous three-phase separators where a low-moisture clarified oil is drawn off and separated sludge is intermittently discharged. Results in Russian and Hungarian installations have been described (27).

DELAVAL CENTRIFLOW PROCESS (28, 29).   Cell rupture is accomplished by mechanical disintegration (first minimizing temperature as required) in a specially designed disintegrator. Then cracklings are removed from the fat mass by a "desludger" centrifuge, after which the liquid phase is heated, deodorized, and centrifuged to produce purified oil and glue water.

SHARPLES PROCESS (30).   This is based on the mechanical rupturing of the fat tissue, followed by two-stage centrifugal separation. A "Super-D-Canter" separates the protein tissue from the liquid fat and discharges it as a dry meaty solid. A second centrifuge called an "Autojector Clarifier" removes protein and water from the fat, intermittently discharging sludge. By suitably low temperatures (115–120°F) a noncoagulated protein material is produced as one of the products, it is claimed, and this product appears to have possibilities as an edible meat product (31).

IMPULSE RENDERING (32–34).   This process is used mainly for preparing and defatting bones for glue. Fat stock, especially crushed bones, is continuously disintegrated in a high-speed hammer mill under an excess of flowing cold water. The intense impact sets the fat free. The discharge from the mill is allowed to settle in a cold-water tank from which the fat is continuously skimmed off. The ground bone is continuously removed from the bottom and transferred to another tank containing hot water (70–95°F), where more fat is separated. It is claimed that better quality fat and higher protein residue are obtained by this process.

The production of marine oils is rather similar to animal fat rendering (35). It varies with the type of fish processed and whether vitamin A and D oils or high-quality fish meal is the prime objective. With the synthetic production of vitamins D_2 and D_3 as well as vitamin A, the current trend is to fish meal and growth factor production. Details of commercial processing methods are given by Bailey (5).

Digestive Rendering Processes. Considerable attention has been given to wet rendering of animal fats with the assistance of added chemicals or enzymes that promote the separation of fat by hydrolyzing and dissolving the connective tissue.

Deatherage (36) has described in detail laboratory and pilot plant experiments on the alkali rendering of lard and beef fats. The best results are obtained when the fat is digested at 85–95°C for 45 min to 1 hour with a 1.75% sodium hydroxide solution. After digestion is complete, the fat is separated from the aqueous liquid, which contains a small amount of undigested solids, by centrifuging, and washed, first with 2–5% salt solution, and then with water. Fat recovery is equivalent to or better than that obtained by steam rendering without significant hydrolysis or darkening of the fat or production of the typical cooked flavor of steam lard. The process is best adapted to reasonably fresh fat; stocks in which any considerable amount of hydrolysis has occurred are difficult to process because of the excessive formation of soap in the aqueous phase. Soap is derived only from free fatty acids in the fat; under the mild conditions of the digestion, there appears to be no appreciable saponification of neutral fat. The fat is, of course, alkali refined as it is rendered; hence it is produced substantially free of acidity. A typical lard has a free fatty acids content of 0.01% and Lovibond color of 2 yellow and 0.3 red. Later reports indicate that 1.2 and 2.0% sodium hydroxide (30% of fatty material) gives improved yields over conventional wet rendering (37, 38).

The rendering of slaughterhouse waste by ammonia plus ammonium diacid phosphate under pressure for peptonization and separation into aqueous and fat phases has been patented (39).

A recent publication (40) reports reduced "fruitiness" and free fatty acids in olive oil from adding alkaline materials to olive pulp during grinding or working. Also, the manufacture of a good-quality olive oil has been claimed (41) by drying olive pomace to 5% moisture, mixing with soda ash, and extracting with carbon disulfide.

Alkali rendering has been found better than steam, water, or acid digestion for recovering vitamin A from fish livers with an oil content of 30% and upward. Partial removal of antioxidants does not impair the stability of the vitamin (42).

The use of proteolytic enzymes in rendering is described in a number of patents; it does not, however, appear to have been used commerically, except perhaps in the recovery of fish liver oils (4). The patent of Parfentjev (43) covers the digestion of fish livers with pepsin at a low pH and a low temperature. The process of Keil (44) for the recovery of lard or other animal fats involves digestion of the fatty stock with a proteolytic enzyme of vegetable origin, for example, 0.005–0.020% papain at a pH of 6.0–7.5, followed by heating to 140–185°F to separate the fat. Halmbacher (45) has patented the use of an enzyme, such as papain or ficin, with a cysteine activator, to decrease digestion time while increasing yield. Other publications deal with treatment of eggs with papain (46), fish rendering (47), and rendering of coconut meats (48).

2.2   COOKING OF OIL SEEDS

General Considerations. It is universally recognized that oil seeds yield their oil more readily to mechanical expression after cooking, but a complete explanation of why this is so is lacking. It is certain that the changes brought about by cooking are complex and that they are both chemical and physicochemical in nature.

The oil droplets in a cottonseed or similar oil seed are almost ultramicroscopic in size and are distributed throughout the seed. One effect of cooking is to cause these very small droplets to coalesce into drops large enough to flow from the seed. An important factor in this phase of the process is the heat denaturation of proteins and similar substances. Before the proteins become coagulated through denaturation, the oil droplets are virtually in the form of an emulsion. Coagulation causes the emulsion to break, after which there remains only the problem of separating gross droplets of oil from the solid material in the seed. Since the surface of the seed particles is highly extended, surface activity figures prominently in the displacement of the oil. Cooking, in turn, has a profound influence upon the surface activity of the material. The primary objects of the cooking process may, therefore, be summarized as follows: (*a*) to coagulate the proteins in the seed causing coalescence of oil droplets and making the seed permeable to the flow of oil and (*b*) to decrease the affinity of the oil for the solid surfaces of the seed so that the best possible yield of oil may be obtained when the seed are subsequently pressed.

Important secondary effects of cooking are drying of the seeds to give the seed mass the proper plasticity for efficient pressing, insolubilization of phosphatides and possibly other undesirable impurities, destruction of molds and bacteria, increase of the fluidity of the oil through increase in temperature, and, in the case of cottonseed, detoxification of gossypol or related substances (49).

One factor that obviously affects the affinity between the seed and the oil and is amenable to control in the cooking operation is the moisture content of the seed. Very dry seeds cannot be efficiently freed of their oil; however, it is impossible to say just how moisture inhibits wetting between the seed and the oil. It may be that the cooking process produces a film of adsorbed liquid water on the seed surfaces that displaces the oil. On the other hand, the water may be in a more nearly "bound" state, and its presence in the seed in this condition may serve to make the seed surface relatively lipophobic. The optimum moisture of cooked seed varies widely according to the variety of the seed and the method to be used for expression. On cottonseed, for example, 5–6% moisture is best for hydraulic pressing, whereas about 3% is optimum for expellers or screw presses; this level needs to be closely controlled for best results. At moistures of 4% and higher, excessive amounts of oil are left in the cake. Soybeans are ordinarily dried to 2.5–3% moisture before pressing in expellers; copra and sesame seed require moisture levels of about 2%.

Many substances in oil seeds are surface active, such as phosphatides and free fatty acids. The degree to which they are present or become active during cooking doubtless influences the tendency of the seed to adsorb and retain oil. It is generally observed that damaged oil seeds give lower yields of oil than undamaged seeds of equivalent oil content. The tendency of damaged seed to retain oil tenaciously is probably due to their high content of free fatty acids or other surface-active agents.

Effect on Quality of Oil and Oil Cake.   In addition to its effect upon the yield of oil, the method of cooking also markedly determines the quality of both the oil and the oil cake. Cooking is particularly important in its relation to the refining loss of the oil. A large part of the oil lost in caustic refining consists of neutral oil, which is emulsified in the foots. Certain surface-active agents naturally present in the oil favor this emulsification; others appear to inhibit it (50). The relative proportions of the two classes of substances in the oil depend to a great extent on the operation of the cooker.

There is little published information on the identity of the surface-active agents in crude oils, but it appears that the substances responsible for high refining losses are generally phosphatides or related substances. The presence of gossypol in cottonseed oil is generally assumed to contribute to the production of hard foots and a low refining loss (50); however, in mill-scale experiments by Wamble and Harris (51) conducted at five different screw-press mills, it was concluded that there was no apparent relation between the gossypol content of the crude oil and the refining loss or refined color.

Normal cooking variations have little effect on oil color or refining loss, although with widely varying cooking conditions considerable differences are noted. Thus Eaves et al. (52) showed that oils prepared by solvent extraction from raw, tempered, or cooked cottonseed flakes varied in yield of crude oil but the yield of neutral oil was virtually unaffected. Crude oil from raw flakes was highest in impurities and lowest in neutral oil, crude from tempered flakes was lower in impurities and higher in neutral oil, and crude from cooked flakes was outstandingly low in impurities and high in neutral oil. As a practical matter and based on our present trading rules, it is not necessarily economical to produce a low-refining-loss oil; that is, unless oil penalties are sufficient to counteract any change in crude oil yield, it may be to the processor's advantage to avoid partially refining the oil during preparation for extraction.

King et al. (53) have studied the effect of pH during cooking of cottonseed on the properties of the meals and oils. They concluded that oils made from meats cooked at low pH were high in gossypol and were subject to color reversion during storage, whereas oil from meats cooked at high pH levels had a lower refining loss, were low in gossypol, and were not subject to color reversion on storage.

In good cooking practice, flaked cottonseed meats are brought to approximately 12–15% moisture by the time they are in the top kettle of the cooker, where the temperature is increased rapidly to 190°F or higher, to inactivate the enzyme systems and prevent free fatty acid rise during cooking (54). Heating should be continued in the presence of not less than 12% moisture until the temperature reaches about 220°F; after this, the object is to reduce the moisture content to a value suitable for efficient pressing. This normally requires temperatures from 240 to 270°F, depending on the amount of venting used. An average final temperature is probably 260°F.

Overcooking of oil seeds has been recognized as undesirable for some time since it may produce abnormally dark oil and cake. There is also evidence that prolonged or drastic cooking tends to be injurious to the nutritive properties of the cake. With cottonseed, for example, it has been shown (55) that increasing maximum cooker temperature or cooking time decreases the feed efficiency for chicks and the relative protein efficiency for rats. Likewise, soybean meal has been shown to lose nutritive value for chicks as heating time is increased.

On the other hand, the nutritive value of soybean meal is definitely improved by moderate cooking. This is due to the coincident inactivation of specific heat-labile factors (trypsin inhibitor, hemagglutinin, saponin, goitrogenic factor, anticoagulant factor, diuretic principle, and lipoxidase). This subject has been summarized by Liener (55) and more recently by Cowan (56).

To improve their palatability and nutritive value, solvent-extracted soybean flakes intended for animal feeding are invariably toasted before they are shipped from the extraction plant. Until fairly recently this was done by adding moisture and cooking in a conventional "stack" cooker after the solvent was removed. With improved desolventizing techniques, desolventizing and toasting are largely accomplished simultaneously by injecting live steam into the solvent-laden flakes as they leave the extractor (57); the steam condenses on the cooler flakes, thus furnishing heat to boil off solvent while at the same time adding moisture. Thus by the time all of the solvent is removed, there has been sufficient moist heat treatment largely to inactivate the heat-labile antinutritional factors mentioned above. If the oil seed residue is destined for industrial protein use, this type of desolventizing is avoided and a quick heat treatment just sufficient to remove the solvent is carried out, often using superheated solvent vapor or reduced pressure.

One of the prime purposes of cooking cottonseed is to bring about destruction or deactivation of a principle toxic to certain animals (particularly swine and poultry); this principle has been generally identified as the complex polyphenolic compound, gossypol. Boatner and co-workers (58) have shown that gossypol is associated in the seed with several related compounds and that one or more of these may be actually responsible for the bulk of the observed toxicity, inasmuch as separated whole pigment glands are much more toxic than purified gossypol (59). At the present time, most people attribute all the toxicity of cottonseed to free gossypol, possibly because no one has seen a low free-gossypol meal that was toxic. On the other hand, there have been numerous cases of nontoxic meals containing levels of free gossypol above that normally considered toxic (60). A partial explanation for this may be the method of analysis for gossypol. Besides the number of other materials closely resembling gossypol chemically, it has now been established that gossypol that is chemically combined may be partially liberated in the analytical prodcedure, giving abnormally high free-gossypol figures. Thus a special method is now official for aniline-treated cottonseed meal.

The toxic principle, whatever it may be, is contained in the cottonseed pigment glands, from which it may be extracted by ether, acetone, butanol, and other polar solvents. Hexane and similar nonpolar solvents will not extract it from the intact pigment glands, but if these glands are ruptured by moisture, wet heat, or polar solvents, the liberated "gossypol" is readily extracted. The toxic principle is quite stable to dry heat. Solvent extraction using a mixture of acetone–hexane–water (53:44:3) has been proposed as one method for obtaining a low gossypol meal without unwanted protein denaturation by heat (61).

Lyman et al. (62), who have made a special study of the cooking of cottonseed in relation to detoxification, recommend for hydraulic pressing that meats be brought to a moisture content of at least 14.5% before cooking, that the cooking period be at least 90 minutes, and that the final temperature be not less than 115°C (239°F). A review of the work of many other investigators, however, indicates that for expeller processing the initial moisture content may be lowered somewhat (to about 12%) and the cooking time greatly reduced. In this connection it is important to bear in mind that the term "cooking" usually is used to cover wet cooking plus a drying to moisture levels around 3%. Actually, these are two distinct processes, as Dunning has pointed out (63), and after the wet-cooking step is completed, subsequent drying may be done instantaneously (by flashing) or very slowly. In normal practice, however, both processes are accomplished in a stack cooker where there is a rather gradual reduction in moisture content and, thus, a gradual transition from cooking to drying. Final cooking temperature thus varies considerably with the amount of venting or aeration of the cooked flakes. For example, a final cooking temperature of 240 or 260°F will yield the same final moisture content if adequate venting is used with lower temperature.

Cooking for Hydraulic and Continuous Pressing.   The cooking of oil seeds is usually carried out in "stack cookers" (Figure 3.3). These consist of a series of four to eight closed, superimposed, cylindrical steel kettles each usually 72–132 in. in diameter and 1.5–2.5 ft high. Each kettle is normally jacketed for steam heating on the bottom (and sometimes on the sides) and is equipped with a sweeptype stirrer mounted close to the bottom and operated by a common shaft extending through the entire series of kettles. There is an automatically operated gate in the bottom of all but the last kettle for discharging the contents to the kettle below; the bottom kettle feeds into a cake former or a continuous press. The top kettle may be provided with spray jets for the addition of moisture to the seed, and each of the lower kettles is provided with an exhaust pipe with natural or forced draft for the removal of moisture; thus it is possible to control the moisture of the cooking seed, not only with respect to final moisture content, but also at each stage of the operation.

In practice, the rolled meats are delivered at a constant rate to the top kettle by means of a conveyor. After a predetermined period of cooking in that kettle, the charge of meats is automatically dropped to the kettle below so that there is a continuous progression of meats downward through the cooker. The gates that govern the flow of meats from one kettle to another are normally opened and closed automatically by a mechanism that engages the meats at a specific level in each kettle. Thus the time that the meats charge remains in each kettle is determined by the

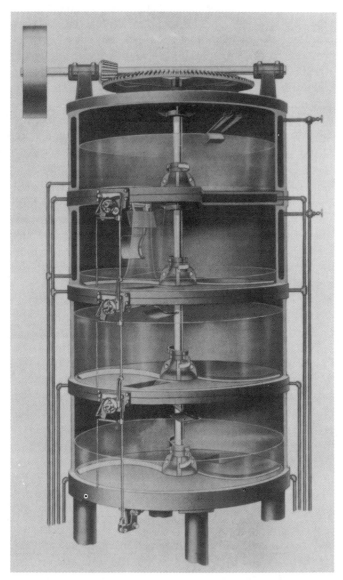

Figure 3.3 Phantom view of four-high stack cooker. (Courtesy of The French Oil Mill Machinery Company.)

meats levels for which the kettles are set. An 85-in, five-high cooker, a common size, has a rated capacity of about 90 tons of cottonseed (calculated upon the basis of the whole seed) per 24 hours.

Steam pressure on the upper stacks of a stack cooker is usually maintained at a relatively high value, for example, 70–90 psi, in order to provide quick heating. On the lower stacks it is usually reduced somewhat, since there it is necessary to only maintain the heated meats at cooking temperature. Cottonseed meats are usually kept in the cooker for 30–120 min and leave at a temperature of 230–270°F. Seed of good quality are normally cooked longer than poor seed, which tend to darken on prolonged cooking. Peanuts are often cooked for a shorter period.

In continuous operation of a stack cooker, material first in is not always first out. This has been noted by Alderks (49) and can be easily demonstrated by the use of added corn kernels, salted flakes, dyed flakes, and so on. So-called "cooking time" represents an average cooking time, with some material remaining in the cooker much longer and some material only a fraction of the average time; however, this does not appear to affect efficient mechanical pressing adversely.

Oil seeds are usually moistened before cooking, or during the early stages of cooking, unless they are initially fairly high in moisture, and their moisture content is then reduced in the cooker. An initial moisture content of 9–14% is common in the top kettle of the cooker. This stays relatively constant in the top two to four kettles where the actual "cooking" takes place. In the bottom kettles drying is the objective with increased temperatures and venting commonly employed. The final moisture content depends on the material processed and on whether cooking is to be followed by hydraulic pressing or expeller or screw pressing. For the former, 5–6% is used for cottonseed; for the latter, a drier product, around 3% moisture, is preferred.

Pressure cooking (64) appeared promising at one time and equipment was installed in several mills. Today, however, this type of cooking is not believed to be in use in any commercial installation in this country.

Another specialized type of cooking, the Skipin process (65), was developed in Russia about 45 years ago but has had no acceptance in this country, where quality and efficiency standards are apparently much higher.

It should be noted that although cooking in a stack cooker has been stressed here, it is also possible to accomplish the same objective using horizontal-jacketed tubes ("conditioners") through which the material is conveyed by suitable means. In general, these are more commonly used in conjunction with some stack cooking rather than as a substitute for the latter.

Mechanical Expression of Oil

3.1   BATCH PRESSING

In recent years increased mechanization and higher labor costs have made hydraulic pressing of oil seeds uneconomical in practically all cases. Today, there is no appreciable volume of soybeans or cottonseeds hydraulically pressed. Pressing of other oil seeds appears doomed to the same fate.

The oldest method of oil extraction comprises the application of pressure to batches of the oil-bearing material confined in bags, cloths, cages, or other suitable devices.

Levers, wedges, screws, and so on have been used as a means of applying pressure in the more primitive styles of presses, but modern presses are almost invariably actuated by a hydraulic system; thus the term "hydraulic pressing" is often used in reference to batch pressing in general. There is a limited use of mechanically operated presses for special purposes where only a relatively light pressure is required, as in the pressing of partially solidified oleo stock or lard to yield oleo oil or lard oil.

Batch presses may be divided into two main classes: the "open" type, which requires the oily material to be confined in press cloths; and the "closed" type, which dispenses with press cloths and confines the material in some species of cage. Open-type presses may be subdivided into plate presses and box presses; closed types may be classified as pot presses or cage presses.

The completeness with which the oil is recovered by mechanical expression is influenced by a number of factors related to the affinity of the oil for solid material in the seed. These include the moisture content, the method of cooking, and the chemical composition of the seed; damaged seed generally retain oil more tenaciously than seed of good quality. With a given lot of seed, cooked and ready for pressing, the oil yield will depend on the rate at which pressure is applied, the maximum pressure attained, the time allowed for oil drainage at full pressure, and the temperature or the viscosity of the oil.

Attempts have been made to establish a correlation between oil recovery from different seeds and such factors as pressure, pressing time, and temperature or viscosity. Koo (66) and Baskerville et al. (67) proposed empirical equations designed to permit a calculation of the fraction of oil extracted from seed from data on the pressing time, pressure on the cake, viscosity of the oil, and so on, with all other factors assumed constant. Later work (68), however, indicated that other factors are involved, all of which are not mutually independent, so that it may not be possible to develop a single equation that will correlate all the processing variables.

Hickox (69) has summarized four years of work in this connection at the Engineering Experiment Station of the University of Tennessee, including data from some mill-scale tests. He concludes that for hydraulic pressing of cottonseed:

1. The hull content of meats to be pressed should be kept as low as possible since increased hulls lower extraction efficiency and press capacity.
2. Pressure should be applied slowly at first, more slowly than is customary.
3. The total pressure on the cake need not be increased over 2000 psi unless the final cake thickness is over 1 in. For thin cakes, increasing the pressure has no effect on the residual oil.
4. The cake should be kept as thin as economical considerations and throughput of the mill will permit.
5. The moisture content of the cake should be controlled carefully (i.e., within a few tenths of 1%) in order to obtain minimum residual oil.
6. Since the top and bottom cakes in the press are cooler than the middle cakes, it is desirable to raise their temperature by appropriate means to obtain maximum extraction efficiency.
7. Preferably, pressing should be carried out at a temperature of 205°F, about 30° higher than typical mill operation.

Open-Type Presses.   The frame of an open or Anglo-American Press (Figure 3.4) consists of four heavy, vertical steel columns fastened at the top and bottom to heavy end blocks. Within the open cage formed by the columns, and suspended from the top of the press, are a series of horizontal steel plates. These plates closely fill the space enclosed by the columns. They are equally spaced at intervals of about 3–5 in and are suspended, one from the other, by linkages permitting the entire assembly to become compressed in the pressing operation. Below the plate assembly and attached to a ram operated from below is a heavier bottom plate. The material to be pressed is formed into rectangular cakes that are placed between the various suspended plates. Raising the ram compresses the series of cakes and causes the oil to fall into a drip pan resting on the bottom block. The stress created by the application of pressure is directed against the top block and is translated into longitudinal stress on the four columns.

In ordinary plate presses the oil-seed flakes are completely wrapped in press cloths and placed between the plates without the use of accessory devices to restrain the cake mass as it is pressed. The surface of the plates,

Figure 3.4   Plate press. (Courtesy of The French Oil Mill Machinery Company.)

however, are usually either corrugated or covered with hair mats to assist in the drainage of the oil and to overcome cake creepage. Box presses are provided with a special boxlike arrangement (Figure 3.5) that encloses the cake on its two long sides and simplifies the wrapping of the cake. The complete press box includes a corrugated drainage rack, a perforated and corrugated steel drainage mat that rests on the drainage rack and underneath the cake, and steel "angles" that project from the underside of each plate to form the sides of the box enclosing the cake below. With

Figure 3.5   Press box assembly for use in box presses. (Courtesy of The French Oil Mill Machinery Company.)

this arrangement, it is necessary for the press cloth only to enclose the cake on the top, bottom, and ends; thus folding of the press cloth in two directions is avoided, and very heavy, durable cloths may be used. Standard-size press boxes are about 2 in deep, 25 in long, 14 in wide at the back, and 14⅜ in wide at the front, being slightly widened from back to front to facilitate insertion and removal of the cake. Presses are usually constructed with either 15 or 16 boxes. Plate presses of an equivalent size have 24 plates and, hence, have a greater capacity than box presses.

Presses similar to those just described are generally provided with a 16-in ram operating at a pressure of 4000–4500 psi; hence the pressure on the cake is 1650–1850 psi. It is important to build up pressure on the cakes gradually. In order to conduct the initial stage of compression more rapidly than the later stages, the hydraulic system operating the presses is provided with automatic valves which deliver oil at 500 lb pressure to the ram until an equivalent pressure is built up on the cake and, thereafter, delivers the maximum pressure of about 4000 lb. The time allowed for drainage of the oil after the maximum pressure is reached is somewhat variable among different mill operators. However, a typical press cycle is as follows: for charging the press, 2 min; for attaining maximum pressure, 6 min; for draining, 26 min; for discharging the press, 2 min; total, 36 min. The capacity of a 15-box press operated under these conditions is approximately 11 short tons of whole cottonseed or whole peanuts per 24 hours.

According to Baskerville and Wamble (70), the average press cycle in mills processing cottonseed in the United States was probably 30 minutes or less; their calculations indicate that the economically optimum cycle is approximately 50% longer.

An essential accessory to the operation of either plate or box presses is a cake former for automatically delivering a proper quantity of flakes from the cooker and forming the flakes into a cake of the proper size and shape within the press cloth. Cake formers are designed to press the flakes into a coherent mass without the application of sufficient pressure to start the oil from them. They are hydraulically operated. Mechanically operated cake strippers are also provided for removing the somewhat adherent press cloths from the spent press cake. Charging and discharging the presses is carried out entirely by hand, however. An operator is also required for both the cake former and the cake stripper, as neither is fully automatic.

The edges of the cake coming from an open-type press are soft and higher in oil content than the remainder of the cake. Consequently, it is the usual practice to slice or beat off these edges in a mechanical cake trimmer and rework the trimmings through the presses.

Plate presses are usually preferred for flaxseed; box presses are standard equipment in cottonseed or peanut mills. The press cloths used with box presses are woven from human hair, camels' hair, or nylon. A wide variety of materials are used for the cloths used in plate presses, including cotton, wool, and hair.

Closed-Type Presses.   Cage presses confine the oil-containing materials within a strong perforated steel cage during the pressing operation and, thus, largely dispense with the use of press cloths. They may be operated at higher pressures than are practicable with open presses. They are particularly suitable for the expression of copra, palm kernels, and other oil seeds that are high in oil content and low in fiber and, hence, are inclined to flow and burst the press cloths of open presses. Castor beans or other seed that are to be processed without heat treatment can be pressed satisfactorily only in presses of this type as very high pressures are required to extract the oil efficiently from cold seeds. They are desirable for mills that process many varieties of oil seeds because they can be used on practically any oil seed or other oily material.

Cages for this type of press are built in both round and square forms. They are usually made up from a number of closely spaced steel bars or slotted steel plates, supported inside a heavy frame or ringed with heavy steel bands. The channels through which the oil escapes increase in size from the interior of the cage outward to minimize any tendency for them to become clogged with solid particles. The cages are operated in a vertical position in a frame similar to that of the Anglo-American press. Oil is expressed from the charge by forcing a closely fitting head up into the cage from below by means of a hydraulically operated ram. The upper end of the cage may be closed solidly; then pressure is applied only to one end of the charge. Alternately, the cage may float between the lower ram and an opposed head entering from above. In the latter case, pressure is applied to both ends of the seed mass. Cage presses are designed to attain pressures of 6000 psi or more.

Since there is a marked tendency for the oil flow in the compressed cake to be longitudinal rather than radial, the cage cannot be packed solidly with the oil seed but must be charged with layers of seed separated by drainage plates and press cloths. Auxiliary equipment is required for filling the cages and discharging the cake. This requirement and the rather elaborate and heavy design of the cages make the initial cost of this type of pressing equipment relatively high. In large installations the cages are usually made removable from the presses, and filling and discharging presses are provided, in addition to a number of finishing presses. A cage carriage is provided for transferring the heavy cages from one press to another.

The pot press is a special form of cage press used for the extraction of cocoa butter or other fats that are solid at ordinary room temperature. In this press the cage is replaced by a series of short, superimposed, steam-heated cylinder sections or "pots." The walls of the pots are solid, and drainage takes place through perforated plates and filter mats in the bottom of each section. Pot presses are usually designed for pressures intermediate between those employed in open presses and cage presses, although they can be built for virtually any desired pressure. The advantages of pot presses are that they can be heated and that they can handle very soft, nonfibrous material, such as fruit pulp, at high pressures without forcing large quantities of solid material into the oil. Their capacity is small, however, in relation to their size and cost, and they require more hand labor to operate than other types of press.

Some oil seeds of high oil content, such as copra, are difficult to express satisfactorily in batch equipment by a single pressing. In some places it is customary to break up the oil cake derived from the first pressing and subject it to a second pressing with or without intervening moisture or heat treatment for the recovery of residual oil. Such practice, of course, requires a double reduction of the seed and also yields an oil of inferior quality from the second pressing. In American practice, the double pressing of oil seed is generally considered obsolete. Oil seeds that cannot be reduced to a low oil content by a single pressing in hydraulic presses are preferably processed in continuous screw presses or expellers.

3.2 CONTINUOUS PRESSING

Continuous expellers or screw presses are now used to the almost complete exclusion of hydraulic presses for the mechanical extraction of soybeans, cottonseed, flaxseed, and peanuts in this country. They are also used extensively throughout the world for the expression of copra, palm kernels, peanuts, cottonseed, flaxseed, and almost every other variety of oil seed.

A screw press is essentially a continuous device for gradually increasing the pressure on material fed to it as the latter progresses inside a closed barrel, with provisions for the oil to drain out as it is squeezed from the feedstock. A column or plug of compressed meal is formed at the discharge end of the barrel, acting like a hydraulic presshead with new cake being formed at the end as cake is expelled past a choke device. Fresh feed is forced in by feed worms against the frictional resistance of the plug at the choke, thus creating a hydraulic pressing equivalent to that of a hydraulic press ram. Labor required is much less than that necessary for hydraulic pressing, but this is at the expense of higher power requirements and maintenance costs; however, the greater oil yield (3–4% oil in cake versus 6–10% in hydraulic cake) and reduced labor more than make up for the increased power and maintenance. Today, hydraulic pressing is rare in the United States and screw press plants are increasingly being used as prepress for extraction systems or are being replaced by direct solvent plants. Expellar plants, because of their simplicity and freedom from flammable solvent hazards, are the next step up from hydraulic pressing in underdeveloped countries or where plant size is necessarily small.

The first successful mechanical screw press, called an "expeller" (model No. 1), was made by V. D. Anderson (now Anderson IBEC, a division of International Basic Economy Corporation) in 1900. It was soon used to express the oil from flaxseed and whole cottonseed. About 1910 the Krupp Works was licensed to manufacture these machines in Germany, where they were used primarily as a forepress unit ahead of hydraulic presses. In the United States interest was primarily in expressing as much oil as possible from seed in one operation so improvements were made resulting in an "RB" (roller bearing) expeller in 1926 and later the "Duo" and "Super Duo" (Figure 3.6) types. In 1933 a "screw press" was introduced by the French Oil Mill Machinery Company, a modern

Figure 3.6 Twin-motor Super Duo Expeller 55 with 14-in. conditioner. (Courtesy of Anderson IBEC.)

version of which is shown in Figure 3.7. Today these two companies are the leading manufacturers of continuous screw presses in this country.

The Anderson machines (expellers) utilize a vertical cage to express the most easily removable oil, followed by a horizontal cage (Figure 3.8) for attainment of the high pressure necessary for removal of most of the remaining oil. The French "screw presses" use only a horizontal cage where pressure is gradually built up to a maximum. Another point of difference in machines of the two manufacturers is in the method of cooling. Expellers are cooled by product oil, after removal of "foots" in a screening tank and cooling in heat exchanges to reduce the temperature to approximately 120°F. Screw presses, on the other hand, are equipped with water-cooled shafts and water-cooled ribs in the bar cages.

Originally, expellers operated on flaked raw materials that were cooked in a horizontal cooker while screw presses employed the stack cooker used for hydraulic pressing. Currently, stack cookers are commonly used with either type of machine, sometimes with preliminary cooking in a horizontal cooker. The trend in the newer installations is to use one large cooker to feed two or more presses.

In both types of machine the pressure necessary to force the oil out of the cooked flakes is obtained by means of continuously rotating worm shafts and worms, with a choke mechanism by means of which cake thickness is controlled. The main worm shaft and worms are designed to exert a pressure of 5–8 tons/in^2 on the seed being processed and at the same time to convey the seed through and out of the pressure chamber. Several different worm shafts may be employed, depending on the material to be processed and whether expression is to be complete or merely prepressing preliminary to later solvent extraction. Some worm arrangements are shown in Figure 3.9.

The drainage barrel is made up of rectangular bars which fit into a heavy barrel bar frame (Figure 3.10). The individual bars in the drainage barrel

Figure 3.7 French screw press with cooker. (Courtesy of the French Oil Mill Machinery Company.)

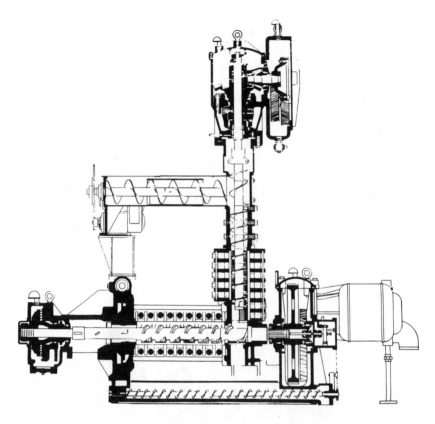

Figure 3.8 Sectional view of Anderson Super Duo Expeller showing conditioner, vertical barrel, and horizontal barrel. (Courtesy of Anderson IBEC.)

are separated by bar spacing clips; the specific spacings depending on the type and preparation of the material being processed. For example, in an expeller processing cottonseed, the spacing of bars in the main barrel may be 0.010 in. in the feed section, 0.0075 in. in the center section, and 0.010 in. in the discharge section. The same sections processing copra may have bar spacings of 0.030, 0.020, and 0.010 in. The spacing of the bars not only permits the drainage of oil from the material being processed, but also acts as a coarse filter medium for the solids.

In the last 30 years, extraction efficiency with expellers and screw presses has greatly improved as the result of better seed preparation and machine changes. Initially, improved cooking lowered oil refining loss and bleach color so that "expeller" oil could now be handled like hydraulic oil. This change in preparation first permitted the doubling of the capacity of an expeller press (Table 3.3) (71).

Anderson expellers were modified in the vertical section and the horizontal section was increased in length to as much as 84 in (72). Capacity for direct (single-stage) pressing reached 50 tons/day.

The French mechanical screw presses (72) evolved through 9-, 11-, and 22-in water-cooled extensions and now cover the largest capacity range of any manufacturer, up to 110 tons/day single pressing and 460 tons/day prepress. Water cage cooling is used with provision for steam heating on startup. Claimed improvements include a two-speed water-cooled main shaft and provision for easy change of shaft speed. Cottonseed cake containing 3–3.5 oil is not uncommon now as a result of these new developments.

Tonnage may also be increased without loss in efficiency by having a minimum amount of hulls in the expeller feed. Since the expeller appears to handle a certain *volume* of feed, removal of hulls makes it possible to increase capacity by removing essentially nonextractable material from the feed. This also minimizes wear from the highly abrasive hulls. This increased tonnage with high efficiency naturally increases the power requirements with motor horsepowers up to 125 now being used as com-

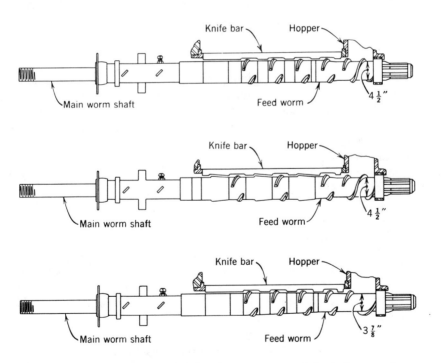

Figure 3.9 Examples of three different worm arrangements of the Anderson main worm shaft. (Courtesy of Anderson IBEC.)

Figure 3.10   Assembly of barrel bars with spacers in a barrel frame. (Courtesy of Anderson IBEC.)

pared to 40–60 previously. Suitably strengthened gear boxes and so on are also required. With the high capacity and efficiency thus possible, coupled with lower installed equipment cost as compared to solvent extraction, expeller operations compare favorably in many cases with the more efficient solvent extraction.

In addition to many older references on continuous pressing of oilseeds, there are two recent papers presented at the World Conference on Oilseed and Vegetable Oil Processing Technology (72, 73), as well as one by Bredeson (73a).

3.3   LOW-PRESSURE PRESSING

For the prepressing of oil seeds prior to extraction, ordinary high-pressure screw presses may be operated at low pressure and at increased capacity. Specially designed machines are much more satisfactory, however; in new installations these are normally used.

In this country most of the plants solvent-extracting cottonseed do so by the prepress route. At first this was probably the result of problems in handling "fines" and difficulties in detoxifying the extracted flakes when direct extraction was used. Other advantages of prepressing include the need for only a minimum-sized solvent plant, since most of the oil is removed in the prepressing step, and the production of meal of high protein quality. Disadvantages are higher initial equipment costs if, for example, soybeans must also be processed in equivalent tonnage in the same plant and, normally, higher power requirements and repairs. Prepress solvent extraction of cottonseed has been discussed by Pons et al. (74) and by Rea and Wamble (75). The nutritional value of the meal has been discussed by Chang et al. (76). Prepress extraction of soybeans is described by Dunning and Terstage (77).

The largest Anderson prepress expeller, the "11-66" (Figure 3.11), uses a single horizontal main worm shaft that runs through a barrel 66-in long by 12-in wide. The capacity is 200 tons/day.

French manufactures three prepresses with capacities of 100, 170, and 460 tons/day. The latter is driven by a 600-hp motor and has a 16-in-diameter feed section and two 4-section main cages of 14 in diameter (Figure 3.12). Their B-2100 machine (Figure 3.13) is being used for cottonseed, copra, sunflower seed, safflower, sesame, and corn germ. It will prepress 125 short tons or the meats from 180 metric tons of whole cottonseed. French machines possess several features that are claimed to minimize down time, provide for rapid changes in revolutions per minute,

Figure 3.11   Anderson 11-66 prepress expeller. (Courtesy of Anderson IBEC.)

Figure 3.12 French H$_2$-6600 screw prepress. (Courtesy of the French Oil Mill Machinery Company.)

and give improved extraction efficiency. On most of French's new range of screw presses, wedge bars have been discarded, providing an increase in available drainage area (72).

Expellers and screw presses of the same design as those used for oil seeds are sometimes used for pressing whale or seal flesh or fish, and for processing meat scraps; but these materials are more commonly handled in screw presses specially designed for the purpose. These are generally of lighter construction than the machines built for oil seed extraction and are operated under lower pressure.

3.4 CENTRIFUGAL EXPRESSION

The removal of oil from an oil-bearing material by centrifugation has been a standard method only in the case of palm fruit; however, recent developments in the rendering of animal fats make full use of centrifugal separation of oil. The centrifugal recovery of palm oil is discussed in a later part of this chapter.

Figure 3.13   French B-2100 screw prepress. (Courtesy of the French Oil Mill Machinery Company.)

Solvent Extraction

4.1   APPLICATION

Although extraction with solvents constitutes the most efficient method for the recovery of oil from any oil-bearing material, it is relatively the most advantageous in the processing of seeds or other material low in oil. The minimum oil content to which oil cake can be reduced by mechanical expression is approximately the same for all oil seeds, that is, about 2–3%. Consequently, the oil unrecoverable by mechanical expression, in terms of percentage of the total oil, increases progressively as the oil content of the seed decreases. The latest available comparative yields of oil from representative seeds of low, medium, and high oil content by the two methods of processing are shown in Table 3.3. Substitution of solvent

Table 3.3   Comparison of solvent extraction with mechanical pressing in the United States (78)

	Soybeans	Cottonseed[a]	Flaxseed (Estimated)
Oil yield per ton			
Hydraulic	—	308	None
Continuous screw press	300[b]	327	674
Solvent extraction	362	376	710
Total, United States	358	339	699
Percent increased oil yield			
Solvent over screw press	12.1	11.5	5.3
Solvent over hydraulic	—	22.1	—

[a] Solvent extraction includes prepress solvent extraction.

[b] Hydraulic included with screw press.

extraction for pressing methods increases the yield of oil from soybeans by 12.1%; in the processing of cottonseed the increase is 11.5% and in the case of flaxseed, only 5.3%. These figures are industry-wide averages. The increase of oil yield for soybeans and cottonseed by solvent extraction over the most efficient mechanical processing today is appreciably less.

Of all the common oil seeds, soybeans are solvent extracted most easily. In the crop year 1957–1958 it was estimated by the Production and Marketing Administration of the U.S. Department of Agriculture (78) that 93.2% of the soybeans processed were solvent extracted, with an average yield of 362 lb of oil per ton, as compared with 6.8% screw pressed, with an oil yield of 300 lb/ton. Official figures are no longer available, but it is believed that essentially all soybeans are solvent extracted at the present time.

Cottonseed flakes disintegrate more readily and occasion more trouble from the production of fines; peanuts and flaxseed disintegrate very badly. By "prepressing" or "forepressing" the seed in low-pressure screw presses to remove a portion of the oil it is possible, subsequently, to solvent extract high-oil-content seeds that are difficult or impossible to handle in their original form in conventional equipment. In Europe and where European practices have prevailed, it is the general practice to extract whole soybeans but to prepress other oil seeds. In the United States prepressing is used commercially on cottonseed, flaxseed, peanuts, safflower, sunflower, and corn germ.

Solvent extraction finds some use in the recovery of animal fats. The tankage or cracklings from dry rendering are often solvent extracted, usually in batch extractors. The recovery of fat from garbage is frequently carried out by means of solvent extraction since the low fat content of this material makes other methods of recovery difficult. Garbage is extracted in batch equipment of special design.

Materials containing scarce or expensive oil are often solvent extracted even when the operation is relatively difficult. Examples are castor oil, olive oil, and wheat germ oil residues from mechanical pressing. Solvent extraction may be used to obtain a fat-free residue or a residue in which proteins are not heat denatured, rather than for the primary purpose of improving the yield of oil. Thus, for example, cocoa is solvent extracted to produce a residue that may serve as a source of theobromine. Solvent-extracted meal is preferred for the manufacture of protein adhesives, fibers, or plastics, since there is much less denaturation of the protein in this meal than in that obtained by cooking and mechanical pressing.

Since minimum heat treatment is involved, oil produced by solvent extraction is of maximum quality, and the meal contains protein subjected to a minimum of damage due to the effects of heat. On the other hand, there are several disadvantages:

1. Solvent extraction equipment is relatively expensive compared to other extraction systems.
2. Except where nonflammable solvents can be used, there is the ever-present danger of fire and explosion.
3. Low-oil-content meal tends to be dusty, with attendant problems.
4. As in the case of cottonseed, unheated flakes from the direct extraction of raw flakes may contain material that is toxic to nonruminants and is not removed or inactivated by the relatively mild processing, thus requiring further treatment.

4.2 PRINCIPLES AND THEORY

Methods of Achieving Contact with Solvent.   The laboratory extraction of an oil-bearing material in an ordinary Butt extraction tube is an example of solvent extraction in its simplest form. In this extraction procedure the pure solvent is delivered continuously to the top of the mass of material to be extracted and is percolated through the mass by gravity until the removal of oil is substantially complete. Although this method is effective in the laboratory, it is highly inefficient. Complete extraction can be accomplished only by the use of a large volume of solvent relative to the volume of oil extracted, and this solvent must eventually be recovered from the oil. Even in the most efficient extraction plants, charges for steam and water for solvent recovery constitute a substantial part of the operating costs (79); if the solvent/oil ratio is high, such charges may easily become prohibitive. A prime object in modern solvent extraction practice is, therefore, to reduce the solvent content of the final miscella or oil–solvent mixture to the lowest possible figure. In the best continuous soybean extraction plants the solvent/oil ratio may be as low as 1:1 by weight, whereas by simple percolation an equivalent extraction could hardly be accomplished with several times as much solvent.

Efficiency is somewhat improved if the continuous percolation of fresh solvent is replaced by prolonged treatment of the oil seeds or other material with successive portions of solvent. Each portion is recirculated through the material being extracted until equilibrium or near equilibrium is established between the oil content of the solid material and that of the solvent; that is, until free miscella is as rich in oil as the miscella absorbed within the solid particles. When this condition is attained, the free miscella is drained off, a fresh batch of solvent is brought into the system, and the operation is repeated. Extraction is thus continued in successive cycles of recirculation and drainage until the oil content of the material is reduced to the value desired.

Although batch extraction by means of percolation is satisfactory for some materials, it is not generally adaptable to the large-scale processing of oil seeds. It is virtually impossible to charge large extraction chambers with oilseed flakes without uneven compacting of the material and consequent channeling and incomplete extraction; hence batch extractors for oil seeds are generally provided with some means of mechanically mixing the solvent and the seed particles. From the standpoint of efficiency in maintaining a low solvent/oil ratio, however, it is immaterial whether the solvent and the oil seeds are brought into equilibrium with respect to oil content by circulating the solvent through the seeds while the latter are contained in a tower or by simply intermixing the solvent and seeds in a chamber of suitable design. The system of extraction by means of successive batches of pure solvent is generally referred to as *multiple extraction*.

The last portions of miscella recovered in the multiple extraction process will naturally be very lean in oil; hence these portions may well be substituted for fresh solvent in the initial treatment of fresh seed. In this way each portion of solvent is made to perform a double duty, and the amount of solvent to be recovered eventually from the oil is decreased accordingly. A batch extraction system set up in such a manner as to utilize the principle of solvent reuse to the best possible advantage is designated as a *batch countercurrent system*, where a battery of extractors is provided and the solvent is used to treat the contents of each extractor in succession. Each time a batch of miscella is drained from an extractor, it is used to treat a batch of seeds that have previously been extracted with a richer miscella. On the other hand, the drained seeds are each time extracted with a leaner miscella. Thus the seeds are treated with batches of solvent of progressively decreasing oil content, until they are finally extracted with a fresh solvent and discharged, while the solvent is brought into contact with batches of seed of progressively increasing oil content until it finally encounters fresh seed and is then discharged as the finished miscella. In this way the miscella is brought out of the system at a uniformly high oil content. If there are a large number of extractors in the battery, the effect approximates that of mixing the solvent and oil in continuously moving countercurrent streams.

Although batch countercurrent extraction may theoretically be brought to an efficiency approaching that of continuous countercurrent extraction by sufficiently increasing the number of extractors, the system thereby becomes unnecessarily cumbersome; in practice, therefore, solvent extraction is carried out on the largest scale only in continuous systems that are entirely automatic in operation. Such systems achieve the highest economy of steam, power, labor, and materials. Their adaptability is limited only by the mechanical difficulties involved in moving the seed mass and the miscella in opposite directions with free intermixing and in effecting a final separation of the miscella and the seed particles.

If it is assumed in batch extraction that a constant volume of miscella is retained by the seeds after each drainage period, and if this volume is known, one may calculate the number of extractions required to reduce the oil content of the seeds to any given level (80, 81), in the case of either multiple or batch countercurrent extraction. Actually, however, the retention of miscella is not usually constant but is variable for different solvent/oil ratios, presumably because of the effect on drainage of such factors as viscosity and surface tension of the miscella. This circumstance renders calculations highly involved, but Ravenscroft (82) has introduced a graphical method for estimating the number of extractions required for a given recovery of oil that is applicable in the case of variable oil retention. Ravenscroft assumes that miscella is retained only on the surface of the solid particles and refers to this retention as *entrainment*. However, his treatment is equally valid for miscella absorbed within the particles so long as there is equilibrium between the miscella within the particles and without. In the case of continuous countercurrent extraction, differences in oil concentration in the miscella within and without the seed cannot be ignored, since here equilibrium conditions do not exist. Equations for extraction under conditions of nonequilibrium have been developed by Ruth (83) and Grosberg (84).

Extraction Rates. Although the following pages are concerned with fundamental extraction theory, it must be remembered that there are many important factors in extraction, some of which are difficult to evaluate quantitatively. Thus although there are still differences of opinion about the theory, large commercial units are nevertheless in operation working efficiently under conditions somewhat at variance with theory. For example, although thin flakes are theoretically attractive, deep-bed extractors like the Rotocel and the French are used in high-capacity operations with flakes of only about 0.016 in thickness, whereas in lower-capacity plants flakes as thin as 0.010 in may be used. Supposedly, the relatively thick and dense flakes pack better into the extractor, permitting a longer extraction time and thus compensating for the thicker flakes (6).

In practice, the design of large-scale solvent extraction apparatus must be determined by the rate at which equilibrium is attained between a lean miscella outside the seed particles and oil and solvent within the particles. The attainment of equilibrium may be quite slow, particularly as the oil content of the seed (on a dry, solvent-free basis) falls toward the low level (usually below 1.0%) demanded by efficient commercial operation. Modern investigations indicate that the rate at which equilibrium is approached (and hence, in effect, the extraction rate) is influenced by a number of factors, including: the intrinsic capacity for diffusion of solvent and oil, which is determined primarily by the viscosities of the two; the size, the shape, and the internal structure of the seed particles; and, at low seed oil levels, the rate at which the solvent dissolves nonglyceride substances that are oil soluble but dissolve less readily than the glyceride portion of the oil.

In a homogeneous oil-impregnated material consisting of thin platelets of uniform thickness whose total surface area is substantially that of the two faces, the theoretical extraction rate, based on simple diffusion, has been given by Boucher et al. (14) as follows:

$$E = \frac{8}{\pi^2} \sum_{n=0}^{n=\infty} \frac{1}{(2n+1)^2} \exp\left[-(2n+1)^2 \left(\frac{\pi}{2}\right)^2 \left(\frac{D\theta}{R^2}\right)\right]$$

where E is the fraction of the total oil unextracted at the end of time θ (in hours), R is one-half the plate thickness (in feet), and D is the diffusion coefficient (in square feet per hour).

Except at low values of θ, the preceding equation takes the approximate form (85):

$$E = \frac{8}{\pi^2} \exp\left[-\frac{\pi^2 D\theta}{4R^2}\right]$$

or

$$\log_{10} E = -0.091 - 1.07\frac{D\theta}{R^2}$$

Hence at the lower values of E, a plot of log E against θ gives a straight line with a slope dependent on the diffusion coefficient and the plate thickness. It is to be emphasized that the equation is valid only when all platelets have the same thickness, and average plate thickness cannot be used for a material of nonuniform thickness.

Working with porous clay plates impregnated with phosphatide-free soybean oil and with tetrachloroethylene as a solvent, Boucher et al. (14) found that experimentally determined extraction rates checked closely with theory; a typical theoretical extraction curve is shown (curve A) in Figure 3.14. A lack of correspondence between extraction rates and Reynolds number of the flowing solvent, over a wide range of the latter, indicated that liquid-film resistance to the transfer of oil to the solvent was inconsequential as compared to resistance to diffusion within the plates. The diffusion coefficient was found to be simply a function of the product of the viscosities of solvent and oil; under the particular conditions of their tests it could be represented by the formula

$$D = 12.96 \times 10^{-6}(\mu_o\mu_s)^{-0.46}$$

where μ_o and μ_s refer to the viscosities, in centipoises, of oil and solvent, respectively. The numerical values in the formula are, undoubtedly, re-

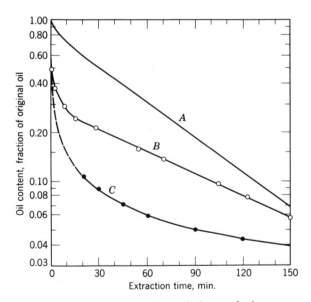

Figure 3.14 Solvent extraction curves: (A) theoretical curve for homogeneous oil-impregnated platelets of uniform thickness (14); (B) peanut slices, 0.026 in thick 14% moisture content, extracted at 25°C with commercial hexane (85); (C) cottonseed flakes, 0.017 in thick, 10–25 mesh, 11.6% moisture content, extracted at 150°F with commercial hexane (86).

lated to the structure of the plates and, hence, can be considered specific only for the lot of plates used in the tests. Tests involving extraction with solvent–oil mixtures as well as pure solvents show that the diffusion coefficient is independent of the composition of the solvent, in terms of relative proportions of solvent and oil. It can, of course, be expected to increase with an increase in temperature or with the use of a less viscous solvent than tetrachloroethylene.

The experiments of Fan et al. (85) with peanuts carefully sliced with a microtome show that the relationships developed by Boucher et al. (14) are also applicable to at least one oil seed, provided that structural considerations are not complicated by crushing of the seed to form flakes. A typical extraction curve is shown (curve *B*) in Figure 3.14. As required by diffusion theory, there is a linear relationship between the logarithm of the residual oil content and the extraction time after a short period has elapsed. During this period, however, a large proportion of the oil is extracted very rapidly.

Fan et al. (85) carried out a mathematical analysis that indicated that this deviation from theory with respect to rapidly extractable oil was caused by the opening of a certain number of oil-bearing cells in slicing the oil seeds, plus the occurrence of void spaces in the seeds after drying. Thus they agreed with Osburn and Katz (87) that the major obstacle to extraction is probably diffusion through the cell walls and that the initial rapid extraction is to be attributed to cell destruction. The proportion of easily extractable oil was found to decrease rapidly with increase in the slice thickness. In the case of curve *B* of Figure 3.14, which represents the extraction of peanut slices 0.026 in thick, the extraction curve became linear after about 76% of the oil was extracted; in other experiments with flakes of similar moisture content, there was linearity with 0.032-in flakes after about 51% of the oil was extracted, and linearity with 0.040-in flakes after about 30% was extracted. With flakes of constant thickness, there was a progressive decrease in the content of rapidly extractable oil with increase in the moisture content.

Fan et al. (85) found that the diffusion coefficient decreased considerably with increase in the moisture content (in the range of 10–22% moisture, about 0.4 cm^2/sec per 1% moisture). With commercial hexane (Skellysolve B) at 24–26°C and peanut slices with 13% moisture, the calculated diffusion coefficients averaged about 7×10^{-9} cm^2/sec.

In the extraction of oil seed flakes formed commercially by rolling, there appear to be factors that still further complicate the extraction rate. Extraction curves not only reveal a very large fraction of easily extractable oil but tend to be continuously concave upward from the time axis; curve *C* of Figure 3.14, constructed from the laboratory data of Wingard and Shand (86), is typical (cf. Figure 3.1). In practice, extraction in the range of about 0.5–5.0% residual oil (on the basis of the dry, solvent-free meal) is so slow that it actually controls the overall extraction rate and extractor design (11). On a semilog plot the concavity of the extraction curve is so great in this region that, actually, a nearer approximation to a straight line is obtained with a log–log plot as in Figure 3.15 (86).

A variety of explanations can be offered for the large deviation of curve C (Figure 3.14) from the form of curve A or curve B. It has been pointed out by King et al. (88), as well as by Osburn and Katz (87), that structural heterogeneity leading to the simultaneous operation of two different diffusion processes with different diffusion coefficients could account for the shape of the curves. The analysis of soybean flake extraction curves by Osburn and Katz suggested that at 80°F, 70–90% of the oil was extracted with trichloroethylene with the relatively high diffusion coefficient of about 4×10^{-6} ft^2/hour, whereas the remaining 10–30% was extracted with the lower diffusion coefficient of about 5×10^{-7} ft^2/hour. It was further suggested that the larger portion of readily available oil was derived from cells ruptured in rolling, whereas the smaller portion of difficultly extractable oil was in cells that remained intact. It seems unlikely, however, that such extensive cell destruction could occur. If lack of structural homogeneity is the proper explanation, it appears that it is

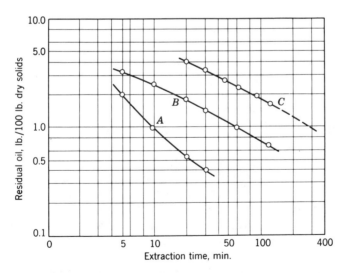

Figure 3.15   Representative curves for solvent extraction of (A) soybean flakes, (B) flaxseed flakes, and (C) cottonseed flakes (86). (*Note*: flakes are not necessarily all of the same thickness; hence extraction rates are not comparable.)

probably of a different kind. Other kinds of heterogeneity can be conceived, including nonuniformity of flake thickness.

As a result of extensive experimental work on the extraction of oil seed flakes, Karnofsky and coworkers (11, 89) have advanced the hypothesis that the slow final extraction rate is at least in part the result of decreased solubility of the last portions of oil. It is well known that oil seeds subjected to repeated extraction with a solvent yield fractions of oil toward the end of the process that are much higher in phosphatides and other nonglyceride materials than the first fractions (11, 90); hence these materials are obviously less soluble. That the difficulty of extracting the last portions of oil from oil seeds may be related to the chemical composition of the "oil" was suggested previously by Goss (91). The hypothesis of Karnofsky and coworkers is supported by the observation (89) that the last portions of oil are removed much more readily if the oil seeds are first given a "soaking" period, even in relatively strong miscella, and that no difficulty is encountered in recovering the last portions of oil from oil seeds reconstituted from extracted oil and oil-free residue.

Acceptance of the "difficultly soluble oil" theory does not vitiate many of the basic conclusions to be derived from the preceding theories based on simple diffusion with free miscibility of solvent and oil. If free miscibility does not exist in the latter stages of extraction, this means simply that the effective concentration of solute is not the concentration of "oil" in the solid seed material but a lower concentration that is limited by the solubility of the "oil" in the solvent. The rate of diffusion will be less than observed in the earlier stages, not because the diffusion coefficient has decreased, but because the "oil" content of the solid material is no longer a proper measure of its instantaneous content of diffusible material. The diffusion or extraction rate will, for example, still be inversely proportional to the square of the flake thickness.

It may be noted that with simple diffusion an increase in the extraction temperature can be expected to increase the extraction rate by lowering the viscosities of solvent and oil, but that with incomplete solubility of the oil an additional effect can be anticipated through an increase in the solubility. This may account for the great effect of temperature observed by Karnofsky (11); in one case increasing the temperature from 100 to 192°F reduced by 80% the time required to lower the oil content of cottonseed flakes to 3% with heptane. This is a rather greater effect than would be predicted from a simple decrease in viscosity, according to the data of Boucher et al. (14).

According to Wingard and Phillips (92), the time required to reduce oil seeds to 1% residual oil content varies with a power of the temperature that, with cottonseed, soybean, and flaxseed flakes and hexane as a solvent, ranges from -1.9 to -2.4; hence a log time/log temperature plot yields a straight line.

It is evident in commercial practice and has been confirmed by closely controlled laboratory experiments that different oil seeds differ markedly in the rate at which flakes of a given thickness can be extracted to a low residual oil content. The relation of particle size of the oil seed to extraction rate has been clarified by a laboratory investigation reported by Coats and Wingard (93), who found that the hexane extraction of soybeans, cottonseed, flaxseed, and peanuts, as either flakes or cracked particles (grits), conformed to the mathematical formula $T = KD^n$, where T is the time required to reduce the material to a residual oil content of 1.0% (on a dry, solvent-free basis), D is the flake thickness or grit diameter, and K and n are constants. Consequently, a T/D plot on a log–log scale yields a straight line, with a slope equal to n. Approximate values found for n were, for four samples of soybean flakes, 2.3–2.5; for two samples of cottonseed flakes, 1.5; for one sample of flaxseed flakes, 7; for one sample of peanut flakes, 3.2; for two samples of soybean grits, 5.5; for one sample of cottonseed grits, 4; and for one sample of corn germ grits, 3.4. With T expressed in minutes and D expressed in units of 0.010 in each, approximate values for K were: for soybean flakes 6–20; for cottonseed flakes, 140 and 270; for flaxseed flakes, 3600; for peanut flakes, 1.4; for soybean grits, 2.5 and 10; for cottonseed grits, 40; and for corn germ grits, 1.6.

Note that K in the equation $T = KD^n$ is a measure of the ease of extraction (of flakes 0.010 in thick), whereas n is a measure of the influence of flake thickness on the extraction rate. Thus soybean flakes extract more readily than cottonseed flakes of equivalent thickness and cottonseed flakes, in turn, extract more readily than flaxseed flakes (Figure 3.16). The extraction rate of flaxseed is highly sensitive to flake or particle thickness, whereas that of soybean is less so, and cottonseed are less sensitive to flake thickness than either. Soybean grits of a given diameter extract more easily than flakes of equivalent thickness, but their extraction rate is more dependent on their thickness. The value of n for soybean flakes was found to be substantially the same for different lots of beans of varying moisture content flaked by different methods and that the data of King et al. (88) obtained by trichloroethylene extraction indicated a value for n exactly in line with the results with hexane, as well as a comparable value for K.

There is evidence (88) that large seed particles rolled to form flakes of a definite thickness can be extracted more rapidly than small particles, presumably because they undergo greater internal disruption in the rolling process.

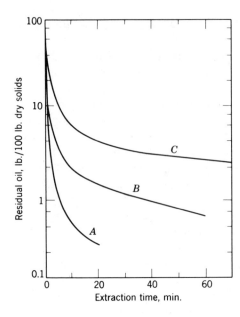

Figure 3.16 Relative extraction rates of different oil-seed flakes: (*A*) soybeans, 0.0075 in thick; (*B*) cottonseed, 0.0095 in thick; (*C*) flaxseed, 0.0075 in thick (11).

Laboratory and pilot plant work on soybeans by Othmer and Agarwal (94) has led them to conclude that for hexane extraction of soybeans:

1. The oil extracted, the residual oil, and the rate of extraction are all independent of the concentration of oil in the solvent; that is, there can be no advantage in countercurrent extraction.
2. The rate of extraction is proportional to:
   a. Residual oil$^{3.5}$.
   b. Flake thickness$^{-3.97}$; that is, increasing flake thickness by 3 times decreases rate by 80 times.

Commerical application of point 1 is seen in the "filtration extraction" process for cottonseed (95) in which a slurry of cooked cottonseed flakes and miscella is held for a time before the rich miscella is washed off on a continuous horizontal filter.

A later paper by Othmer and Jaatinen (96) extends their work to other solvents and shows that solvents other than hexane and acetone do not extract at a rate independent of miscella concentration.

Some practical and theoretical aspects of the solvent extraction of soybeans are discussed by Myers (96a).

Any review of the literature on extraction uncovers several inconsistencies and contradictions that are due largely to the experimental techniques used. Results obtained under one method may not be obtained under another; consequently, the conclusions will differ, although each may be consistent with the data on which it was based. Some investigators mix solvent or miscella with flaked oil seeds under conditions of good agitation, whereas others believe that a percolation-type extraction more closely parallels commercial plant conditions. Some use pure solvent and others use a combination of miscellas, such as would be used in the plant. Some prefer to calculate the residual oil content of the extracted material from the enrichment of the miscella rather than separate and analyze the extracted solids. Some consider solvent extraction as consisting of two parts—extraction proper and washing—where washing is considered rather inconsequential, with primary emphasis on the attainment of equilibrium between oil seed flakes and miscella. Efficient washing, however, is a requirement in a commercial plant; and if washing time is unduly long, it increases the total time the flakes are in the extractor just the same as if more time were needed for extraction proper. Future researchers might well consider solvent extraction as including everything that happens between the time oil seed flakes are contacted by miscella or solvent and the time they enter the solvent removal equipment. It would also seem that actual analysis of spent flakes is to be preferred to any calculations of what the oil content would be if the flakes were removed from the system and washed instantaneously.

4.3 EXTRACTION STANDARDS

Chiefly from experience with soybeans, it is accepted that commercial solvent extraction must reduce the oil content of the dry solid residue to less than 1.0%, and preferably to about 0.5%, to be efficient.

In making guarantees on soybean extraction equipment, manufacturers usually specify that the analysis be made on extracted flakes before toasting, since it is generally acknowledged that there is an increase in the apparent "oil" content (petroleum ether-soluble material) during toasting. Moreover, since appreciable toasting takes place during the removal of solvent from extracted flakes in desolventizer–toasters, the analysis should preferably be made on spent flakes that have been desolventized without the use of steam or heat. Unfortunately, this is seldom practical from a safety standpoint; thus, meal analysis is usually relied on.

In any given installation the objective is usually to reduce the oil in meal to the lowest possible level; in actual practice this is probably desirable. However, published evidence shows that the last portions of "oil" (or petroleum ether-soluble material) in the meal are largely not oil at all but phosphatides and other nonglyceride impurities. Bull and Hopper (90) have reported a phosphatide content of 18.62% in the last 1.1% of material extracted from a sample of soybean flakes with commercial hexane at 40°C. Similarly, Karnofsky (11) has reported that a similar fraction had a refining loss of 81.5%. More research is needed on the quality of the total oil extracted versus the degree of extraction.

4.4 SOLVENTS FOR OIL EXTRACTION

The most common solvents used in the United States for oil and fat extraction are light paraffinic petroleum fractions. The more popular products are cuts of fairly narrow boiling range, which are distinguished according to the chain length of their principal components. One manufacturer lists the following general ASTM boiling ranges for four types of naphtha: pentane type, 88–97°F; hexane type, 146–156°F; heptane type, 194–210°F; and octane type, 215–264°F (97). The hexane-type naphtha is the most widely used and the one generally preferred for oil-seed extraction, although the heptane-type product is also suitable for use in most modern plants. The pentane type finds limited use in the extraction of heat-labile products, such as pharmaceuticals; the higher-boiling products are required for the extraction of castor oil, which is not freely miscible with hydrocarbons except at elevated temperatures.

A preliminary laboratory study of a number of pure hydrocarbons with respect to their overall desirability as solvents for the extraction of cottonseed has been published by Ayers and Dooley (98), who consider paraffinic hydrocarbons preferable to cyclic hydrocarbons and find methylpentanes the best of the former group. A commercially available methylpentane-type naphtha is listed as having a typical boiling range of 139–145°F.

Although solvent losses in American plants employing hexanes for extraction are higher than in European plants using a higher-boiling product, they are not excessive and in well-operated soybean-extraction plants do not exceed about 2 gallons for each ton of seed processed. At this rate, charges for solvent loss are less than for labor, power, or steam (78).

The American-produced extraction naphthas are substantially free of nitrogen- or sulfur-containing compounds and unsaturated hydrocarbons and leave a residue on evaporation of less than 0.0016%. They are sufficiently stable to be reused indefinitely, cheap, and available in practically unlimited quantities. The only serious disadvantage to their use is their extreme flammability. Rather elaborate precautions are required to avoid fire or explosion hazard in the plants in which they are used. The proper safety measures have been discussed by Bonotto (99), MacGee (97, 100), and most recently by Heilman (101) and Critchfield (102).

The history, composition, and characteristics of American extraction naphthas have been reviewed by MacGee (97, 100, 103). Detailed analyses of a popular hexane product have been published by Griswold et al. (104) and MacGee (103).

Because of the potential fire and explosion hazard involved when hydrocarbon solvents are used for extraction, there has always been a great deal of interest in nonflammable solvents. Trichloroethylene, boiling at 188°F, is such a solvent (105, 106). During World War II it was used in one large extraction plant in England because of its safety. In the United States, however, its use was confined to a few small soybean extraction plants. Originally, it was thought that the use of a nonflammable solvent such as trichloroethylene would result in less danger of fire, less need for skilled labor, and lower initial investment cost. Experience has shown, however, that:

1. Although safe from a fire and explosion standpoint, the toxicity of the solvent requires that it be handled carefully.
2. Its relatively high cost is not counterbalanced by proportionately lower solvent losses.
3. Corrosion is a serious problem despite attempts at stabilization.
4. Oil produced must be exhaustively stripped of solvent, since small amounts of residual solvent will interfere with subsequent hydrogenation of the oil unless special precautions are taken (107), thus skilled labor is required.
5. The solvent reacts with cysteine in the soybeans to give S-dichlorovinyl cysteine, which causes aplastic anemia and disease symptoms in cattle that are identical to the effect of trichloroethylene-extracted meal when fed to cattle (108).

At one time carbon disulfide was widely used in Europe for the extraction of olive press cake to recover the inedible product termed olive oil "foots" or "sulfur olive oil." However, carbon disulfide is not a desirable solvent; it has never been used in the United States, and in Europe its use has declined in favor of petroleum naphthas that yield an edible extract. Acetone has been used to some extent for the recovery of oil from wet materials, such as fish livers, as has ethyl ether (82).

Interesting laboratory developments include oil seed extraction processes employing isopropyl alcohol (109–111) and ethyl alcohol (112–114) as solvents. The former, unlike petroleum naphthas, effectively extracts gossypol from cottonseed and, thus, offers a possible means of detoxifying the residue from this seed without severe heat treatment; however, the miscella must be purified of phosphatides, carbohydrates, and other non-glyceride extractives if it is to yield a crude oil of low refining loss and good refined color. When cooled moderately, alcohol miscellas separate into two layers consisting principally of oil and of solvent; hence, with such a solvent, equipment and steam for the evaporation and recovery of solvent may be greatly reduced.

Eaves et al. (115) have investigated the extraction of cottonseed by five commercial solvents (hexane, benzene, ethyl ether, acetone, and butan-one) and concluded that none compared favorably with hexane as an extractant for cottonseed. A combination of solvent extraction with ace-tone and miscella refining has been described by Vaccarino (116).

When cottonseed was first extracted directly, gossypol remained in the meal, causing a serious complication. Several wet toasting methods were evolved to correct this, but most successful was extraction with an ace-tone–hexane–water mixture (53:44:3) developed at the U.S. Department of Agriculture, Southern Regional Laboratory in New Orleans (117). This removed gossypol from the meal and transferred it to the oil, from which it could be easily removed by conventional alkali refining. Meals made by this method produced superior growth in chicks.

4.5   TYPES OF EXTRACTOR

Batch Extractors.   Batteries of batch extractors are still in use in Europe for the recovery of oil from oil seeds or mechanical press residues. In modern plants, however, batch equipment is used principally in the form of small units for the recovery of pharmaceutical oils or other expensive oils; for the extraction of spent bleaching earth; for the processing of meat scraps, cracklings, and garbage; or for other purposes where the tonnage of material handled does not justify the expense of installing continuous extractors.

Batch extractors vary greatly in design. An extractor that was popular in the castor oil industry consists of a large horizontal drum (18 by 8.5 ft) mounted on rollers by means of which the drum can be rotated on its longitudinal axis. Inside the drum is a horizontal, perforated, metal strainer covered with a filter mat of burlap, which extends the length of the drum and divides it into two compartments, one much smaller than the other. The large compartment receives a charge of 10–12 tons of solid material through which solvent is percolated to drain into the smaller compartment by gravity, from which it is continuously pumped during the drainage period. Four to six successive extractions suffice to reduce the oil content of castor pomace from about 15 to 1.5%. A common

European extractor, somewhat similar but of a stationary vertical design with internal mixing arms, has been described by Goss (91).

The extractor commonly used for the extraction of garbage consists of a vertical cylindrical kettle, with a large ratio of diameter to depth, equipped with a vaportight cover, a steam jacket, and a vertical low-speed agitator. The most popular unit is about 4.5 ft high and 10 ft in diameter and takes a charge of 3–5 tons of material. This extractor is suitable also for the extraction of other relatively wet materials, as the material may be dried and extracted in the same vessel.

Solvent systems are used to some extent for the extraction of fish liver oils, as well as for fish oil (3, 4). Various other types of batch extractor have been described (118). The extraction of miscellaneous oil-containing materials, as well as oil seeds, has been developed to a much higher degree in Europe than in the United States.

Continuous Extractors.   As a result of the shortage of fats and oils after World War I, the Germans sought better ways to get the most out of their Manchurian soybeans, and two continuous extractors were developed. The Bollman or basket extractor was patented in Germany in 1919 and 1920 (119), and the Hildebrand or U tube was patented in 1931–1934 (120). The former represents a percolation type and the latter, an immersion type.

Percolation-Type Extractor.   In this type, liquid solvent or miscella is distributed over a bed of flakes or cake, where it percolates down through the bed and exits the bed at the bottom through some type of supported filtering device, such as a perforated plate, a mesh screen, or a wedge wire screen bar system (Figure 3.17). In the earliest models, large baskets were supported on endless chains within a gastight housing. The flaked seeds were conveyed by a screw into a closed charging hopper at the top of the housing, and the completely filled conveyor tube served as an effective vapor seal against the solvent vapors inside the extractor. The baskets were continuously and very slowly raised and lowered at the rate of about one revolution per hour. As each basket started down the descending side of the apparatus, a charge of seed was automatically dropped into it from a charging hopper (Figure 3.18). Extraction was effected by the percolation of solvent through the seed during their passage from the top to the bottom and again to the top of the apparatus. As the baskets containing the spent and drained flakes ascended to the top of the housing on the opposite side from the charging hopper, they were automatically inverted and the contents dumped into a discharge hopper,

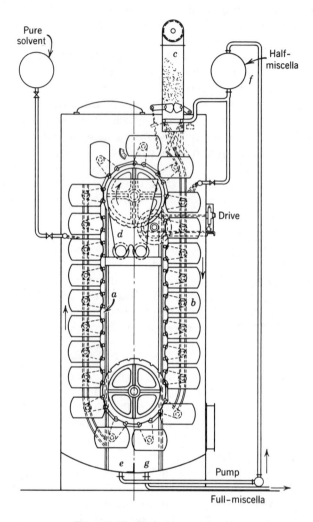

Figure 3.17   Basket-type extractor.

from which they were conveyed by means of screw conveyors to the meal driers.

Fresh solvent at the rate of approximately one pound of solvent per pound of seed was sprayed into a basket near the top of the ascending line of baskets, from which it percolated by gravity through the lower baskets in countercurrent flow. The miscella from this side, termed *half miscella*, was collected in a sump in the lower part of the housing. A pump continuously withdrew it from the sump and sprayed it into the topmost basket of the descending line. From this basket it percolated downward through lower baskets like the fresh solvent introduced on the other side of the system and was collected in a separate sump as *full miscella*. The full miscella was freed from fine seed particles and solvent to yield the finished oil, by means that are described later.

These early extractors were tall and bulky, susceptible to chain breakage and were difficult to service quickly and safely. Later, they were modified into square and rectangular (horizontal) types where, in addition to solvent or miscella draining vertically from one basket to another, it could be pumped to individual baskets in the horizontal sections. In this way recirculation of miscella was used with generally improved efficiency. The horizontal extractor also permits one-floor operation and can be housed at minimum cost. Horizontal chain and basket extractors are

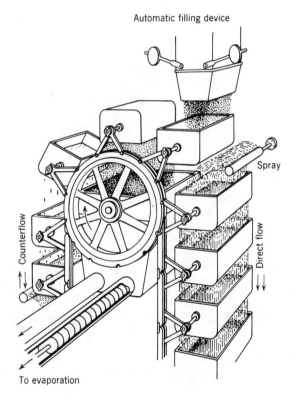

Figure 3.18 Interior of basket-type extractor, showing schematically filling and dumping of baskets and flow of solvent.

available from Gianazza in Italy, Lurgi in Germany, H. L. S. Ltd. in Israel, and Bernardini (used in combination with their immersion extractor) in Italy.

A rotary-type percolation extractor was developed by Blaw-Knox (now Dravo) and called the "Rotocel" (Figure 3.19) (121, 122). Similar in principle to the basket extractors described previously, it carries baskets in a rotary motion in a single horizontal plane. Miscella percolates through the baskets and falls into compartments in the bottom of the extractor housing, where it is picked up by a series of pumps and recirculated countercurrently to the flakes. The first commercial unit was placed in operation on soybeans in 1950 and had a capacity of 250 tons/day. Current models have capacities to 3000 tons/day and are generally regarded as compact and flexible. This type of extractor is licensed by Simon-Rosedown and Krupp. The EMI rotary extractor, in use since 1963, is quite similar.

Of opposite design is the French Stationary Basket Extractor (123), also licensed by Speichem, in which the liquid manifolds and solids hopper

Figure 3.19 Rotocel extractor. (Courtesy of Blaw-Knox Company, Chemical Plants Division).

rotate and the cells and perforated doors are fixed. These also reach capacities approximating 3000 tons/day of soybeans.

The De Smet extractor (124) is a third type of percolation employing a horizontal endless perforated belt. Cells are formed by creating ridges of flakes periodically to act as barriers to the backflow of miscella, which otherwise could result in loss of extraction efficiency. These have not found popularity in the United States.

The Crown Iron Works (125) extractor is another type of percolation that combines immersion. It consists of a chain conveyor unit in which a double drag chain and flight move inside a stationary casing, conveying the extracting solids over two sections of wedge wire screen. Originally designed as a small extraction plant for a nonflammable solvent (trichlorethylene), these have expanded until they now handle 2000 tons/day in some units.

A fifth type of percolation extractor is the filter type of Wurster and Sanger (Figure 3.20) (126). In this unit the final extraction and washing is done on a horizontal rotary filter where vacuum is employed to reduce

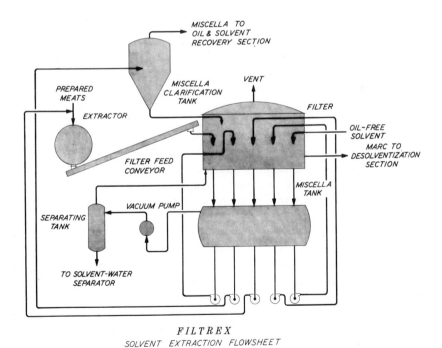

FILTREX
SOLVENT EXTRACTION FLOWSHEET

Figure 3.20 Flow sheet of FILTREX solvent extraction system. (Courtesy of Wurster & Sanger, a Division of Jacobs Engineering Company, Chicago, Illinois.)

the solvent content of the extracted meal. This process also involves a special mild cooking, followed by *crisping* (evaporative cooling by flashing to the atmosphere 1–3% steam from the cooked material) to obtain an easily filterable porous particle with nonoil components bound to the meal fraction. This material is then soaked in weak miscella that extracts the bulk of the oil. Next, the material is transferred to the horizontal rotary filter, where a relatively thin meal layer is contacted (countercurrently) with weak miscella and finally with oil-free hexane. The process was first developed at the U.S. Department of Agriculture, Southern Regional Laboratory on cottonseed (95).

The advantages claimed for this extractor include: low fines in miscella, superior-quality crude oil, less solvent to evaporate from the meal, low gossypol meal, and applicability to other oil seeds, such as soybeans, peanuts, safflower, sunflower, and castor beans (127). The first commercial plant was built for Mississippi Cottonseed Products Company at Greenwood, Mississippi in 1953. Other plants installed or being installed operate on cottonseed, soybeans, rapeseed, flaxseed, rice bran, sesame seed, sunflower seed, safflower, and press cake.

Immersion-Type Extractors.   In an immersion extractor, solids to be extracted are transported, usually by chain or screw conveyor, through a pool of liquid (solvent). The Hildebrand extractor was probably the earliest example of this type (120). It consisted of two vertical tubes interconnected at the bottom by a third horizontal tube, with motor-driven screws to propel the flakes down one tube, across, and up the other tube countercurrent to the flow of solvent. Because of the working given the flakes by the screws, flake disintegration and fines production is relatively extensive; for this reason it is unsuitable for direct extraction of seeds, such as cottonseed, although it has been reasonably successful in processing soybeans. The original equipment is no longer manufactured and the trend has been away from this type of extractor, at least in the United States.

Bonotto patented an extractor (128) that had a column divided into a number of sections by a revolving assembly of horizontal plates attached to a central shaft. The plates were provided with a series of staggered slots through which the flakes, introduced at the top of the column, proceeded downward by gravity, countercurrent to a rising flow of solvent. Stationary scraper arms placed just above each plate provided gentle agitation of the flake mass to prevent packing or bridging and assist in moving the flakes through the slots. Originally, this extractor employed a screw discharge with choke mechanism to compress the spent flakes and seal the bottom of the column against the escape of solvent. This has the advantage of squeezing most of the entrained solvent from the spent flakes. On seeds other than soybeans, however, this discharge mechanism was unsatisfactory and was generally replaced with an inclined side tube, up which flakes are carried by a Redler or drag-link conveyor through a set of squeezing rolls.

The Allis-Chalmers and V. D. Anderson extractors were modifications of the Bonotto apparatus. Because of problems with "fines" in the miscella, they were operated more successfully on prepressed cake. Likewise, the Kennedy extractor (129), carrying perforated blades that moved solids through a series of horizontal extraction sections, countercurrent to solvent, has not found acceptance in this country. Today, the Bernardini extractor is offered in a vertical design for use in combination with their percolation extractor (130).

4.6   RECOVERY OF SOLVENT

Recovery from Miscella.   Normally, miscella from the solvent extractor is freed of finely divided solid material before it is processed for oil recovery. Although it is possible to clarify the oil after the solvent is removed (80), the presence of fines complicates the operation of packed distillation columns, and this method also leads to excessive losses of oil entrained in the fines unless the latter are well washed. It is generally not considered advantageous to recycle a large amount of separated fines through the extraction equipment. Recently, however, increasingly widespread use of "disk and doughnut"-type stripping columns has made the handling of miscellas containing fines much easier and has decreased emphasis on complete removal of fines from miscella before distillation.

The amount of fines to be handled varies greatly with different oil seeds and different types of extractor. Basket-type or percolation extractors processing soybeans, now increasingly common in the United States, produce very little fines and the miscella is simply filtered, usually through leaf-type filters, which must be cleaned only occasionally. In processing nonprepressed seed, such as cottonseed, through extractors of the immersion type a considerable proportion of fines is obtained. It must be handled with special equipment and washed to reduce the oil content if a high overall extraction efficiency is to be maintained. Here the tendency was to rely on continuous centrifuges of the Bird type to remove most of the solid material and filtration to provide the final clarification. None of the clarification systems thus far introduced appears to be altogether satisfactory where large amounts of fines are to be handled; current problems in the extraction of high-oil-content seeds are largely a matter of fines prevention or efficient fines handling rather than a matter of difficulty in reducing the oil content of the seed.

The original Hansa–Mühle plant effected recovery of solvent from the oil in three stages. The miscella containing 20–25% oil was passed in series through two steam-heated pot stills to reduce the solvent content to about 50%, then through a falling film evaporator, where it was reduced to 5–10%, and finally through one or two steam-stripping columns of the packed type, where the last portions were removed. In the American-built plants the pot stills have been replaced by miscella preheaters and rising film evaporators with entrainment separators. In some cases the falling film evaporator has been retained or replaced with a small rising film evaporator operated with sparging steam; in others, miscella from the first large rising film evaporator goes directly to the stripping column. Usually the stripping column is maintained under reduced pressure with a steam ejector. The American equipment is more compact, and with it the miscella is kept hot for a shorter period. In the extraction of cottonseed oil, in particular, prolonged heating of the oil or miscella will "set" the color and produce a permanently dark oil (131); however, no difficulty in this respect has been encountered in plants using the rapid evaporation equipment described earlier.

Many soybean processors operate degumming and lecithin recovery plants in connection with solvent extraction plants. In the past, some plants have purposely allowed some condensation of stripping steam to occur in the stripping column to hydrate the phosphatides and deliver an oil ready for passage through the degumming centrifuges.

Systems comprising of a fatty oil and a hexane or other hydrocarbon solvent exhibit a considerable negative deviation from the ideal; that is, the vapor pressure of the solvent is lower than that calculated from its molar concentration in the miscella and the vapor pressure of the pure solvent on the basis of Raoult's law. Below a solvent concentration of about 10% by weight, the boiling point becomes so high that steam stripping is essential in the final stages of solvent recovery.

Boiling point and vapor pressure data on mixtures of commercial hexane with cottonseed and peanut oils have been published by Pollard et al. (132). Values for the boiling points of cottonseed–hexane mixtures at different pressures, as derived from Pollard's smoothed data, are shown in Table 3.4.

Figure 3.21 plots vapor pressure curves for pure hexane, for commercial hexane experimentally determined, and for a 10% commercial hexane miscella, both as calculated for an ideal solution and as actually determined. The plots, which are on the conventional basis of log vapor pres-

Table 3.4   Boiling points (in degrees Fahrenheit) of mixtures of cottonseed oil and commercial hexane (132)

Oil in Mixture (wt. %)	Pressure (mm)				
	760	610	460	310	160
0	152	140	124	105	80
50	158	145	130	111	82
60	162	150	133	114	85
70	171	157	140	120	91
80	186	172	154	132	102
85	201	186	167	144	113
90	231	210	189	163	129
92	248	226	203	177	142
94	273	250	224	192	156
95	289	268	238	203	165
96	—	—	254	215	177
97	—	—	271	230	190
98	—	—	—	248	207
99	—	—	—	272	229

sure versus reciprocal of the absolute temperature to give straight lines, show that below a temperature of about 200°F (corresponding to a $1/T$ value of 15.16), the actual vapor pressure curve of the miscella is a straight line, with a slope equal to that of the vapor pressure curve of the pure solvent or of the ideal curve but considerably below the latter. Assuming

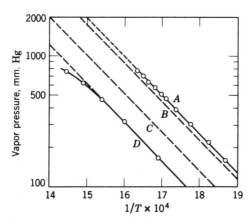

Figure 3.21 Vapor pressure curves: (*A*) commercial hexane: (*B*) pure hexane; (*C*) 10% commercial hexane in mixture with cottonseed oil, ideal curve; (*D*) same mixture, actual curve, according to data of Pollard et al. (132).

an average molecular weight for the oil of 865 and for the solvent of 86, the mole % of solvent in a mixture containing 10% solvent by weight is 52.8; hence the ideal vapor pressure is at any temperature 52.8% of that of the solvent alone. The *activity coefficient* or the ratio of actual vapor pressure to ideal vapor pressure over the linear portion of the actual vapor pressure curve may be determined from Figure 3.21; at a value of $1/T$ of 16, for example, the actual vapor pressure is 310 mm, whereas the ideal vapor pressure is 520 mm; hence the activity coefficient is 310/520 = 0.60.

Similar plots and calculations may be made for miscellas of other concentrations where experimental values have been determined at temperatures below 200°F; the graphical data are shown in Figure 3.22. Figure 3.23 shows activity coefficients in terms of the composition of the miscella. Data subsequently published by Smith and Wechter (133) on the vapor pressures of mixtures of soybean oil with a practical grade of hexane in the range of 75–120°C and 2–36 mole % or 0.25–5.25 wt. % (Figure 3.24) indicate activity coefficients generally between 0.50 and 0.60. For leaner miscellas, where boiling occurs at reasonable temperatures and steam stripping is not required, the data in Table 3.4 are directly applicable. The values from which Table 3.4 was derived were determined in

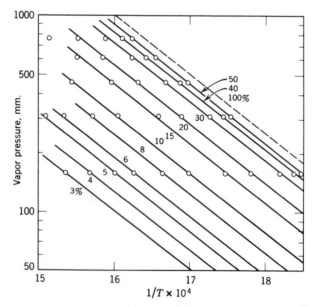

Figure 3.22 Vapor presure curves of commercial hexane in different concentrations by weight in mixtures with cottonseed oil, according to data of Pollard et al. (132).

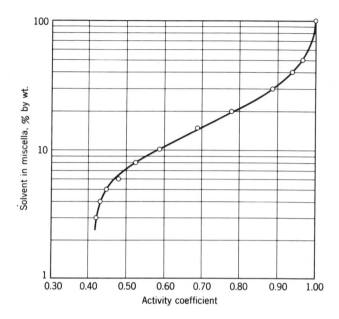

Figure 3.23 Activity coefficients of commercial hexane in in admixture with cottonseed oil, according to data of Pollard et al. (132).

a laboratory apparatus with vigorous stirring to prevent superheating of the solvent. In ordinary apparatus, considerable superheating may be expected.

The effect of differences in boiling points of various hexanes on vapor pressures of the solvent or solvent fraction in a miscella has been studied by Smith (134), who has also determined the vapor pressure of hexane–soybean oil solutions at high solvent concentrations (135).

In making practical stripping calculations, it should be recognized that the last portions of solvent consist in part of heavy ends of lower volatility than the original solvent.

Recent improvements in recovery of solvent from miscella have included double-effect and dual-evaporation systems. Generally speaking, these two systems make use of hot vapors from other parts of the oil and meal recovery equipment so as to obtain maximum efficiency from the steam used and to reduce steam consumption to a fraction of that previously required (136).

Unfortunately, most existing solvent extraction plants were designed for low capital investment at some sacrifice of operating costs. Recently, energy costs have skyrocketed with adverse effects. In a 1976 paper (137)

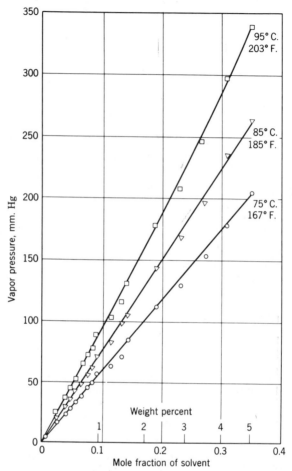

Figure 3.24   Vapor pressure curves for low concentrations of hexane in soybean oil (133).

it was estimated that the energy costs for that year would amount to about one-third of total production costs of soybean solvent extraction, with steam accounting for about two-thirds of the total energy extraction costs. Desolventizing–toasting and meal drying–cooling are the major steam users, so attention should be concentrated here (137). Opportunities are naturally greater in new than in existing plants; but, in any case, a thorough engineering review should be worth while.

Recovery from Extracted Flakes.   Formerly the standard equipment, originally of German design, for desolventizing of the extracted flakes consisted of a series of horizontal steam-jacketed tubes ("Schneckens") through which the flakes were propelled by screws. For the final removal of the last traces of solvent or "deodorization" of the flakes, a similar but larger tube was provided through which the flakes passed counter-current to a current of stripping steam. In recent years this type of meal desolventizing has been largely replaced by "desolventizer–toasters," similar to stack cookers, in which both live and indirect steam are used (138). In this equipment live steam is injected into the top and often also into lower kettles, where it evaporates most of the hexane as it, in turn, is condensed. This adds moisture to the meal, minimizing dust carryover to the condenser, and results in a combination of solvent removal and toasting. From an operating standpoint this system is generally preferred to the "Schneckens" originally employed.

In some cases desolventizing is accomplished by superheated solvent vapors. This and "flash" desolventizing (139) are of interest mainly where it is desired to avoid toasting and to minimize denaturation of the protein as, for example, when undenatured or partially denatured flakes are required because of protein specifications.

When hydrocarbon solvents are used, the separation of condensed solvent and condensed stripping steam from the "deodorizer" and from the miscella stripping column is facilitated by the comparative immiscibility of the solvent and water; a continuous decanter or settling tank is used to separate the solvent, which is reused without further treatment.

The extractor proper, all solvent and miscella tanks, and the various solvent condensers are all vented to a vent condenser or condensers that are protected by special means from loss of solvents. In some plants the vent condensers are refrigerated; in others, they communicate with the atmosphere through charcoal-filled adsorbers that are periodically steamed for the recovery of solvent or through a packed column down which a small side stream of oil is diverted as an adsorbing agent. As indicated previously, solvent losses in the more efficient plants operating with a hexane-type solvent do not exceed 2 gal of solvent (about 11.5 lb) for each short ton of flakes extracted.

Vapors from the "deodorizer" and air from a final meal cooler may carry considerable dust, lint, and so on, particularly if the extracted material is inclined to powder. Scrubbers, cyclone separators, and similar devices of various designs are used to collect dust and avoid fouling of the condensers and other portions of the solvent recovery system, as well as the formation of emulsions in the solvent–steam condensate separator.

4.7 AUXILIARY EQUIPMENT

Although auxiliary equipment naturally varies with any particular installation and with the type of raw material processed, a diagram of typical direct extraction and prepress systems is shown in Figure 3.25.

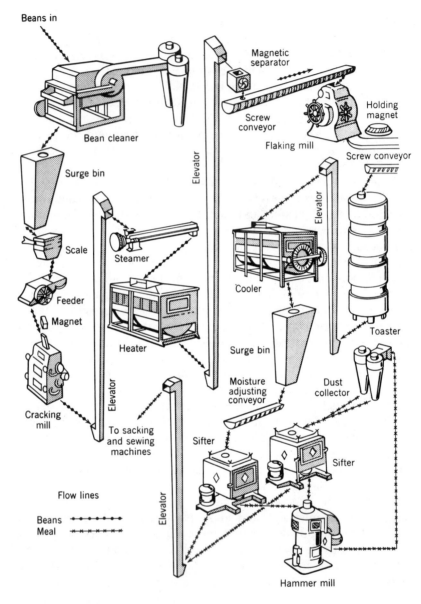

Figure 3.25 Diagram of equipment for preparation, prepressing, solvent extraction, and meal handling of oil seeds. (Courtesy of The French Oil Mill Machinery Company.)

Recovery of Oil from Fruit Pulps

The only fruit pulp oils of commercial importance are olive oil and palm oil. These oils must be recovered by techniques somewhat different from those employed for the treatment of either fatty animal tissues or oil seeds. The extraction of these oils are, therefore, considered apart from the different processes that have been discussed previously.

5.1   EXTRACTION OF OLIVE OIL

The extraction of olive oil is in general carried out by less efficient and less modern methods than most other vegetable oils, since the industry is highly decentralized and is distributed throughout the growing countries in many small establishments.

Notwithstanding the fact that olives are seldom transported long distances from the producing groves to the extraction plants, they are often subjected to considerable abuse prior to being processed. The rather general practice of bruising the fruit during harvesting and later storing it for protracted periods accounts for the fact that olive oil usually has a rather high content of free fatty acids.

The equipment used for processing olives for oil varies greatly from one country to another and even from mill to mill within the same country. Commonly, the olives are washed to remove impurities (140) before breaking down the vegetable structure by milling. Next, the dispersed drops of oil in the milled paste are united into a continuous oily paste (malaxation). Pressure extraction follows, employing disk presses (most common), screw presses, or centrifugation. The crude oil from the presses must be separated from about 70% water and a small quantity of solids. The cake is extracted by hexane.

5.2   EXTRACTION OF PALM OIL

At one time, most of the palm oil on the market was extracted by African natives by exceedingly primitive methods. In recent years, however, the production of oil from wild-growing trees has been much exceeded by that from the plantations of Indonesia, Malaysia, and the Republics of the Congo. The plantation oil has generally been extracted by modern methods and is much superior in quality to the older African oil. Good plantation oil consistently runs lower than 5% in free fatty acids; whereas, native-produced oil was not uncommonly as high as 15, 25, or even 50% in free fatty acid content.

Commonly, the palm clusters are sterilized to stop acid formation. A picker then separates the fruits from the stems and the fruits are washed in boiling water. Mashing then detaches the pulp from the nut. Presses dry the fibrous and oily mixture without breaking the nuts. Continuous presses are in general use. Raw juices contain about 35% palm oil plus a mixture of water and deposits formed from sand and vegetable residue. The oil cakes from the presses contain the dehydrated fibers of fruits from which the palm nuts, still whole, are recovered. Centrifugation finally effects separation of the oil which is strained and may be reclarified in a second centrifuge (8a).

The oil cake is broken up to separate the nuts from the fibers (which will dry by the heat in the conveyor jacket). The heavy nuts are then separated from fiber by air, after which the nuts are dried and crushed, freeing the kernel. Shells are separated and the kernels are dried to 6–7% moisture (5% oil) before being extracted in an oil mill.

References

1. General references: Proceedings of the World Conference on Oilseed and Vegetable Oil Processing Technology, *J. Am. Oil Chemists' Soc.*, **53**, 248–301 (1976); W. J. Wolf and J. C. Cowan, *Food Technol.* (critical reviews in), **2**, 81–158 (April 1971); O. H. Alderks, in A. E. Bailey, Ed., *Cottonseed and Cottonseed Products*, Chapter 15; Interscience, New York, 1948; L. F. Langhurst, in K. S. Markley, Ed., *Soybeans and Soybean Products*, Chapters 14 and 15; Interscience, New York, 1950; A. C. Wamble, in A. E. Bailey, Ed., *Cottonseed and Cottonseed Products*, Chapter 14; Interscience, New York, 1948, W. R. Woolrich and E. L. Carpenter, *Mechanical Processing of Cottonseed,* Engineering Experimental Station, University of Tennessee, Knoxville, 1935; R. P. Hutchins, *J. Am. Chem. Soc.*, **33**, 457–462 (1956); American Oil Chemists' Society, "Short Course on Engineering Aspects of Processing Edible Oils," *J. Am. Oil Chemists' Soc.*, **30**, 473–582 (1953); American Oil Chemists' Society, "Short Course on Inedible Fats and Fatty Acids," *J. Am. Oil Chemists' Soc.*, **31**, 485–600 (1954); R. T. Holman, W. O. Lundberg, and T. Malkin, Eds., *Progress in the Chemistry of Fats and Other Lipids*, Vol. 5, Pergamon, New York, 1958.

2. "Proceedings of the World Conference on Oilseed and Vegetable Oil Processing Technology," *J. Am. Oil Chemists' Soc.*, **53**, 221–461 (1976).

3. R. Stokes, L. M. Reid, R. P. Hutchins, and J. W. Dunning, *J. Am. Oil Chemists' Soc.*, **33**, 453–470 (1956).

4. H. N. Brocklesby, *The Chemistry and Technology of Marine Animal Oils*, Fisheries Res. Board Canada, Ottawa, 1941.

5. B. E. Bailey, N. M. Carter, and L. A. Swain, Marine Oils, *Bull. No. 89,* Fisheries Research Board of Canada, Ottawa, 1952.

5a. J. G. Fawbush, *J. Am. Oil Chemists' Soc.*, **54**, 474A (1976).

5b. J. M. Ridlehuber, *J. Am. Oil Chemists' Soc.*, **54**, 477A (1976).

5c. A. Garcia-Serrato, *J. Am. Oil Chemists' Soc.*, **54**, 483A (1977).

5d. P. Molabrigo, *J. Am. Oil Chemists' Soc.*, **54**, 485A (1977).

6. R. P. Hutchins, *J. Am. Oil Chemists' Soc.*, **53**, 279–282 (1976).

7. N. P. Kovalenko and L. K. Bogancova, *Maslob.-Zhir. Promst.*, **34,** (10) 3 (1968).

8. J. P. Galloway, *J. Am. Oil Chemists' Soc.*, **53,** 271 (1976).

8a. Bernard de Ramecourt, *J. Am. Oil Chemists' Soc.*, **53,** 256–258 (1976).

9. A. M. Goldovskii and M. Podol'skaia, *Maslob.-Zhir. Delo*, **10,** (4) (1934).

10. J. O. Osburn and D. L. Katz, *Trans. Am. Inst. Chem. Eng.*, **40,** 511–531 (1944).

11. G. Karnofsky, *Proceedings of a Six Day Short Course in Vegetable Oils*, American Oil Chemists' Society, University of Illinois, 1948, pp. 61–69.

12. W. R. Woolrich and E. L. Carpenter, *Mechanical Processing of Cottonseed*, Engineering Experimental Station, University of Tennessee, Knoxville, 1935.

13. O. Shchepkina, *Maslob.-Zhir. Delo*, **10** (4) (1934).

14. D. F. Boucher, J. C. Brier, and J. O. Osburn, *Trans. Am. Inst. Chem. Eng.*, **38,** 967–993 (1942).

15. D. F. Othmer and J. C. Agarwal, *Chem. Eng. Prog.*, **51,** 372–378 (1955).

16. A. C. Wamble, in A. E. Bailey, Ed., *Cottonseed and Cottonseed Products*, Chapter 14, Interscience, New York, 1948.

17. R. P. Hutchins, *J. Am. Oil Chemists' Soc.*, **33,** 457–462 (1956).

18. For a review of rendering as carried out in meat packing establishments in the United States, see H. C. Dormitzer, *J. Am. Oil Chemists' Soc.*, **33,** 471–473 (1956); C. B. Rose, *J. Am. Oil Chemists' Soc.*, **31,** 498–503 (1954); and W. C. Ault, *Progress in the Chemistry of Fats and Other Lipids*, Vol. 5, Chapter 9, Pergamon, New York, 1958.

19. Anderson IBEC, Strongsville, Ohio.

20. F. C. Vibrans, *J. Am. Oil Chemists' Soc.*, **26,** 575–580 (1949).

21. C. E. Morris, *Oil Soap*, **13,** 60–62 (1936).

22. L. R. Dugan, L. Marx, P. Ostby, and O. H. M. Wilder, *J. Am. Oil Chemists' Soc.*, **31,** 46–49 (1954).

23. W. M. Gearhart and B. N. Stuckey, *J. Am. Oil Chem. Soc.*, **32,** 287–290 (1955).

24. R. J. Sims and L. Hilfman, *J. Am. Oil Chem. Soc.*, **33,** 381–383 (1956).

25. W. R. Dayen, K. M. Christensen, and R. E. Morse, *Food Technol.*, **7,** 421–423 (1953).

26. J. E. Thompson, *Food Eng.*, **26,** 72–73, 102–105 (1954).

27. *Food Sci. Technol. Abstr.*, **2,** 11N417; *Chem. Abstr.*, **52,** 17553 (1958).

28. DeLaval Technical Bulletin No. 5504E.

29. F. E. Sullivan, *J. Am. Oil Chemists' Soc.*, **36,** 70 (1959).

30. Sharples Corporation, Philadelphia, Pa.

31. F. P. Downing, *J. Am. Oil Chemists' Soc.*, **36,** 319–321 (1959).

32. I. H. Chayen, U.S. Pat. 2,635,104 (1953).

33. Chem. Eng. News, **30,** 5266–5268 (1952); ibid., **37,** 42–43 (1959).

34. R. Chayen, *Chem. Proc. Eng.*, **41,** 14 (1960).

35. H. N. Brocklesby and J. R. Patrick, *Progress in the Chemistry of Fats and Other Lipids*, Vol. 5, Pergamon, New York, 1958, pp. 140–148.

36. F. E. Deatherage, *J. Am. Oil Chemists' Soc.*, **23,** 327–331 (1946); U.S. Pat. 2,456,684 (1948).

37. K. E. Youssef, A. Y. Latfi, and A. A. Esmail, *Acta Agronomica Academiae Scientarum Hungaricae*, **23,** (1/2), 178 (1974); *Food Sci. Technol. Abstr.*, **8,** 6N218 (1976).

38. E. Nosova, *Myasnaza Ind. S.S.S.R.*, **30,** (1), 14 (1959); *Chem. Abstr.*, **54,** 6992i.

39. H. Haneschka and R. Helinger, Austrian Pat. 159,739 (1940).

40. P. M. Rousseau, *Oléagineux,* **10,** 183 (1955).

41. G. Proto, Italian Pat. 486,850 (1953).

42. L. Hartman, *J. Am. Oil Chemists' Soc.,* **27,** 409–411 (1950).

43. I. A. Parfentjev, U.S. Pat. 2,395,790 (1946).

44. H. L. Keil, U.S. Pat. 2,423,102 (1947).

45. P. Halmbacker, U.S. Pat. 2,527,305 (1950).

46. I. Nakamoto, Japanese Pat. 154,087 (1942).

47. C. E. Lane, U.S. Pat. 2,550,570 (1951).

48. *Food Eng.,* **24,** (10), 143 (1954).

49. O. H. Alderks, in A. E. Bailey, Ed., *Cottonseed and Cottonseed Products,* Chapter 15, Interscience, New York, 1948.

50. H. D. Royce and F. A. Lindsey, *Ind. Eng. Chem.,* **25,** 1047–1050 (1933).

51. A. C. Wamble and W. B. Harris, *Tex. Eng. Expt. Sta. Bull. No. 136,* September 1954.

52. P. H. Eaves, L. J. Molaison, E. F. Pollard, and J. J. Spadaro, *J. Am. Oil Chemists' Soc.,* **34,** 156–159 (1957).

53. W. H. King, L. T. Wolford, F. H. Thurber, A. M. Altschul, A. B. Watts, C. W. Pope, and Jean Conly, *J. Am. Oil Chemists' Soc.,* **33,** 71–74 (1956).

54. A. C. Wamble and W. B. Harris, *Tex. Eng. Expt. Sta. Res. Rept. No. 47,* April 1954.

55. I. Liener, in A. M. Altschul, Ed., *Processed Plant Protein Foodstuffs,* Chapter 5, Academic, New York, 1958.

56. J. C. Cowan, in B. E. Caldwell, Ed., *Soybeans,* Chapter 20, American Society of Agronomy, Inc., Madison, WI, 1973.

57. R. P. Hutchins, U.S. Pat. 2,695,459 (1954).

58. C. H. Boatner, in A. E. Bailey, Ed., *Cottonseed and Cottonseed Products,* Chapter ·6, Interscience, New York, 1948.

59. E. Eagle, L. E. Castillon, C. M. Hall, and C. H. Boatner, *Arch. Biochem.,* **18,** 271–277 (1948).

60. E. Eagle, H. F. Bialek, D. L. Davis, and J. W. Bremer, *J. Am. Oil Chemists' Soc.,* **31,** 121–124 (1954); ibid., **33,** 15–21 (1956).

61. G. E. Mann, F. L. Carter, V. L. Frampton, A. B. Watts, and Charles Johnson, *J. Am. Oil Chemists' Soc.,* **39,** 86–90 (1962).

62. C. M. Lyman, B. R. Holland, and F. Hale, *Ind. Eng. Chem.,* **36,** 188–190 (1944).

63. J. W. Dunning, *Oil Mill Gaz.,* **54,** 20–22 (July 1950).

64. J. Leahy, *South. Power Ind.,* **57** (10), 37–44 (1939); R. W. Morton, *Mech. Eng.,* **62,** 731–735 (1940).

65. C. H. Boatner, in A. E. Bailey, Ed., *Cottonseed and Cottonseed Products,* Interscience, New York, 1948, pp. 340–342.

66. E. C. Koo, *Ind. Res.* (*China*), **6,** 9–14 (1937).

67. W. H. Baskerville, A. J. Glass, and A. H. Morgan, *Oil Mill Gaz.,* **51,** 56–63 (May 1947).

68. C. L. Carter, *Oil Mill Gaz.,* **56,** 40–44 (March 1952).

69. G. H. Hickox, *J. Am. Oil Chemists' Soc.,* **30,** 481–486 (1953).

70. W. H. Baskerville and A. C. Wamble, *Tenn. Univ., Agr. Expt. Sta. Bull. No. 13,* 1945.

71. J. W. Dunning, *J. Am. Oil Chemists' Soc.*, **30**, 486–492 (1953).

72. L. H. Tindale and S. R. Hill-Hass, *J. Am. Oil Chemists' Soc.*, **53**, 265–270 (1976).

73. J. A. Ward, *J. Am. Oil Chemists' Soc.*, **53**, 261–264 (1976).

73a. D. K. Bredeson, *J. Am. Oil Chemists' Soc.*, **54**, 489A (1977).

74. W. A. Pons, Jr., F. H. Thurber, and C. L. Hoffpauir, *J. Am. Oil Chemists' Soc.*, **32**, 98–103 (1955).

75. H. E. Rea, Jr., and A. C. Wamble, *Tex. Eng. Expt. Sta., Res. Rept. No. 11*, February 1950.

76. W.-Y. Chang, J. R. Couch, C. M. Lyman, W. L. Hunter, V. P. Entwistle, W. C. Green, A. B. Watts, C. W. Pope, C. A. Cabell, and I. P. Earle, *J. Am. Oil Chemists' Soc.*, **32**, 103–109 (1955).

77. J. W. Dunning and R. J. Terstage, *J. Am. Oil Chemists' Soc.*, **31**, 28–29 (1954).

78. *Market. Res. Rept. No. 360* (September 1959), AMS, U.S. Department of Agriculture.

79. H. P. J. Jongeneelen, *J. Am. Oil Chemists' Soc.*, **53**, 291 (1976).

80. L. F. Hawley, *Ind. Eng. Chem.*, **9**, 866–871 (1917).

81. L. Silberstein, *Ind. Eng. Chem.*, **20**, 899–901 (1928); S. D. Turner, *Ind. Eng. Chem.*, **21**, 190 (1929).

82. E. A. Ravenscroft, *Ind. Eng. Chem.*, **28**, 851–855 (1936).

83. B. F. Ruth, *Chem. Eng. Progr.*, **44**, 71–80 (1948).

84. J. A. Grosberg, *Ind. Eng. Chem.*, **42**, 154–161 (1950).

85. H. P. Fan, J. C. Morris, and H. Wakeham, *Ind. Eng. Chem.*, **40**, 195–199 (1948).

86. M. R. Wingard and W. C. Shand, *J. Am. Oil Chemists' Soc.*, **26**, 422–426 (1949).

87. J. O. Osburn and D. L. Katz, *Trans. Am. Inst. Chem. Eng.*, **40**, 511–531 (1944).

88. C. O. King, D. L. Katz, and J. C. Brier, *Trans. Am. Inst. Chem. Eng.*, **40**, 533–556 (1944).

89. H. B. Coats and G. Karnofsky, *J. Am. Oil Chemists' Soc.*, **27**, 51–53 (1950).

90. W. C. Bull and T. H. Hopper, *Oil Soap*, **18**, 219–222 (1941).

91. W. H. Goss, *Oil Soap*, **23**, 348–354 (1946).

92. M. R. Wingard and R. C. Phillips, Abstracts of Papers, 41st Annual Meeting of American Oil Chemists' Society, May 1–3, 1950.

93. H. B. Coats and M. R. Wingard, *J. Am. Oil Chemists' Soc.*, **27**, 93–96 (1950).

94. D. F. Othmer and J. C. Agarwal, *Chem. Eng. Progr.*, **51**, 372–378 (1955).

95. K. M. Decossas, H. G. Many, O. J. McMillan, Jr., E. A. Gastrock, and E. F. Pollard, *Ind. Eng. Chem.*, **49**, 930–935 (1957); H. W. Haines, Jr., G. C. Perry, and E. A. Gastrock, *Ind. Eng. Chem.*, **49**, 920–929 (1957).

96. D. F. Othmer and W. A. Jaatinen, *Ind. Eng. Chem.*, **51**, 543–546 (1959).

96a. N. W. Myers, *J. Am. Oil Chemists' Soc.*, **54**, 491A (1977).

97. A. E. MacGee, *Oil Mill Gaz.*, **51** (August 1947); *Cotton Gin Oil Mill Press* (August 9, 1947).

98. A. L. Ayers and J. J. Dooley, *J. Am. Oil Chemists' Soc.*, **25**, 372–379 (1948).

99. M. Bonotto, *Oil Soap*, **14**, 30–33 (1937).

100. A. E. MacGee, L. J. Weber, and C. H. Senter, *J. Am. Oil Chemists' Soc.*, **25**, 279–295 (1948).

101. J. E. Heilman, *J. Am. Oil Chemists' Soc.*, **53**, 293–294 (1976).

102. C. E. M. Critchfield, *J. Am. Oil Chemists' Soc.*, **53**, 295–298 (1976).

103. A. E. MacGee, *J. Am. Chemists' Soc.*, **26**, 176–179 (1949).

104. J. Griswold, C. F. Van Berg, and J. E. Kasch, *Ind. Eng. Chem.*, **35**, 854–857 (1943).

105. I. J. Duncan, *J. Am. Oil Chemists' Soc.*, **25**, 277–278 (1948).

106. O. R. Sweeney and L. K. Arnold, *J. Am. Oil Chemists' Soc.*, **26**, 697–700 (1949); O. R. Sweeney and L. K. Arnold, U.S. Pat. 2,497,700 (1950).

107. F. A. Norris, K. F. Mattil, and W. J. Lehmann, *J. Am. Oil Chemists' Soc.*, **29**, 28–32 (1952).

108. L. L. McKinney, F. B. Weakley, A. C. Eldridge, R. E. Campbell, J. C. Cowan, J. C. Picken, Jr., and H. E. Biester, *J. Am. Chem. Soc.*, **79**, 3932–3933 (1957); ibid., **81**, 909–915 (1959).

109. W. D. Harris and J. W. Hayward, *Tex. Agr. Expt. Sta., Bull. No. 121*, September 1950, p. 72; W. D. Harris and J. W. Hayward, *J. Am. Oil Chemists' Soc.*, **27**, 273–275 (1950).

110. A. C. Beckel and J. C. Cowan, U.S. Pat. 2,584,108 (February 5, 1952).

111. F. C. Magne and E. L. Skau, *J. Am. Oil Chemists' Soc.*, **30**, 288–291 (1953).

112. R. K. Rao, M. G. Krishna, S. H. Zsheer, and L. K. Arnold, *J. Am. Oil Chemists' Soc.*, **32**, 420–423 (1955).

113. R. K. Rao and L. K. Arnold, *J. Am. Oil Chemists' Soc.*, **33**, 82–84, 389–391 (1956).

114. R. K. Rao and L. K. Arnold, *J. Am. Oil Chemists' Soc.*, **35**, 277–281 (1958).

115. P. H. Eaves, L. J. Molaison, C. C. Blak, A. J. Crovetto, and E. L. D'Aquin, *J. Am. Oil Chemists' Soc.*, **29**, 88–94 (1952).

116. C. Vaccarino, *Oléagineux*, **13**, 233–236 (1958).

117. G. E. Mann, F. L. Carter, V. L. Frampton, A. B. Watts, and C. Johnson, *J. Am. Oil Chemists' Soc.*, **39**, 86–90 (1962).

118. A. van der Werth, in H. Schönfeld, Ed., *Chemie und Technologie der Fette und Fettprodukte*, Vol. 1, Springer, Vienna, 1936, pp. 677–748.

119. H. Bollman, German Pats. 303,846 (1919) and 322,446 (1920); British Pat 156,905 (1921); Hansa, Mühlenbau und Industrie, A.-G., German Pat. 670,283 (1939); British Pat. 507,465 (1939).

120. K. Hildebrand, German Pats. 528,287 (1931) and 547,040 (1932); U.S. Pat. 1,961,420 (1934).

121. G. Karnofsky, *J. Am. Oil Chemists' Soc.*, **26**, 570–574 (1949).

122. K. McCubbin and G. J. Ritz, *Cotton Gin Oil Mill Press*, **52** (6), 40–42 (1950); *Chem. Ind.*, **66**, 354–356 (1950); G. Karnofsky, *Chem. Eng.*, **57** (8) 108–110 (1950).

123. E. D. Milligan, *J. Am. Oil Chemists' Soc.*, **53**, 286–290 (1976).

124. Extraction DeSmet S.A., Edegem—Antwerp, Belgium.

125. Crown Iron Works Company, Minneapolis, Minn.

126. Wurster and Sanger, Division of Jacobs Engineering Co., Chicago, Ill.

127. W. W. Haines, Jr., G. C. Perry, and E. A. Gastrock, *Ind. Eng. Chem.*, **49**, 920–929 (1957).

128. M. Bonotto, U.S. Pats. 2,086,181 (1937), 2,112,805 (1938), 2,156,236 (1939), and 2,184,248 (1939).

129. A. B. Kennedy, U.S. Pat. 1,628,787 (1927); F. Lerman, A. B. Kennedy, and J. Loshin, *Ind. Eng. Chem.*, **40**, 1753–1758 (1948).

130. E. D. Milligan, *J. Am. Oil Chemists' Soc.*, **53**, 286–290 (1976).

131. H. L. E. Vix, E. F. Pollard, J. J. Spadaro, and E. A. Gastrock, *Ind. Eng. Chem.*, **38**, 635–642 (1946).

132. E. F. Pollard, H. L. E. Vix, and E. A. Gastrock, *Ind. Eng. Chem.*, **37**, 1022–1026 (1945).

133. A. S. Smith and F. J. Wechter, *J. Am. Oil Chemists' Soc.*, **27**, 381–383 (1950).

134. A. S. Smith, *J. Am. Oil Chemists' Soc.*, **28**, 356–361 (1951).

135. A. S. Smith and B. Florence, *J. Am. Oil Chemists' Soc.*, **28**, 360–361 (1951).

136. R. P. Hutchins, *J. Am. Oil Chemists' Soc.*, **33**, 457–462 (1956).

137. H. P. J. Jongeneelen, *J. Am. Oil Chemists' Soc.*, **53**, 291 (1976).

138. N. F. Kruse, U.S. Pat. 2,585,793 (1952).

139. O. L. Brekke, G. C. Mustakas, M. C. Raether, and E. L. Griffen, *J. Am. Oil Chemists' Soc.*, **36**, 256–260 (1959).

140. *Olive Oil Technology,* Food and Agriculture Organization of the United Nations, Rome, 1975.

4
Refining and Bleaching

Crude fats and oils produced by rendering, expression, or solvent extraction contain variable amounts of nonglyceride impurities. The impurities in high-grade animal fats and certain vegetable oils, such as coconut and palm kernel oils, consist principally of free fatty acids. In most vegetable oils, as well as in animal fats rendered from low-grade materials, however, there are significant amounts of other substances. Thus Bailey (1) reported the presence of fatty acids, glycerides, phosphatides, sterols, tocopherols, hydrocarbons, pigments (gossypol and chlorophyll), sterol glucosides, and protein fragments, as well as resinous and mucilaginous material in crude cottonseed oil.

Not all impurities in crude oils are undesirable. The sterols are colorless and heat stable and, for all practical purposes, inert; hence they pass unnoticed unless present in unusually large amounts. Tocopherols perform the important function of protecting the oil from oxidation. For this reason, they may be classed as highly desirable constituents of most oil and fat products. Most of the other impurities are objectionable, however, since they render the oil dark-colored, cause it to foam or smoke, or are precipitated when the oil is heated in subsequent processing operations. The object of refining and bleaching is to remove the objectionable impurities in the oil with the least possible damage to either the glycerides or the tocopherols or other desirable impurities, and with the least possible loss of oil.

As used here, the term "refining" refers to any purifying treatment designed to remove free fatty acids, phosphatides, or mucilaginous material, or other gross impurities in the oil; it excludes "bleaching" and also "deodorization." The term "bleaching" is reserved for treatment designed solely to reduce the color of the oil. Very little material is removed from the oil by bleaching, and bleaching treatment is commonly applied to oils after purification has been largely accomplished by refining. "Deodorization" is the term used for treatment intended primarily for the removal of traces of constituents that give rise to flavors and odors. Deodorization usually follows refining and bleaching. Refining and bleaching are closely related processes and are treated together in this chapter; deodorization is accomplished by the use of entirely different techniques that are covered in another chapter.

In Europe the most common method of refining—which involves the use of alkalies to react with free fatty acids in the oil— is ordinarily called *neutralization*. The term "refining" covers processing through deodorized oil.

General Considerations

1.1 REFINING AND BLEACHING METHODS

By far the most important and generally practiced method of refining is treatment of the oil with an alkali. Alkali refining effects an almost complete removal of free fatty acids, which are converted into oil-insoluble soaps. Other acidic substances likewise combine with the alkali, and there is some removal of impurities from the oil by adsorption on the soap formed in the operation. Moreover, all substances that become insoluble on hydration are removed.

The alkali most commonly employed for refining oils is caustic soda, which is much more effective in its decolorizing action than weaker alkalies. Caustic soda has the disadvantage, however, of saponifying a small proportion of neutral oil in addition to reaction with free fatty acids; for this reason, other alkalies, such as sodium carbonate, have been tried but have not been accepted to any extent. Organic bases, such as the ethanolamines, have also been proposed as refining agents because of their selective action toward the free acids in the oil.

Certain oil impurities, such as phosphatides, proteins, or protein fragments, and gummy or mucilaginous substances, are soluble in the oil only in an anhydrous form and can be precipitated and removed if they are simply hydrated. Hydration is accomplished by steaming the oil or mixing it with water or a weak aqueous solution. It may also occur when the oil is stored with access to the atmosphere; hence some vegetable oils tend to become purified naturally in storage, with precipitation and settling of so-called "foots".

Since free fatty acids are much more volatile than glycerides, it is possible to remove them from an oil by steam distillation at a high temperature under reduced pressure. So-called "steam refining" is similar to ordinary steam deodorization; with some fats it is possible to combine the two in one operation.

Liquid–liquid extraction, employing furfural or propane, has been used to some extent but is not generally economical or satisfactory from a quality standpoint (2).

Oils, particularly those intended for use in paints, are sometimes "acid refined" by treatment with strong sulfuric acid. Acid refining does not reduce the free fatty acid content, but it chars and precipitates phosphatides and similar impurities.

The standard method of bleaching is by adsorption or treatment of the oil with bleaching earth or carbon. The various chemical bleaching methods applied to inedible fats all depend on oxidation of the pigments to colorless or lightly colored materials. Most of the refining methods enumerated accomplish considerable reduction of the color of the oil. The carotenoid pigments are not altogether stable to heat and are converted to colorless compounds by hydrogenation; hence some bleaching effects are incidental to the operations of hydrogenation and deodorization.

1.2   Effect of Refining and Other Processing Treatment on Specific Impurities

Alkali refining of an oil with caustic soda readily reduces the free fatty acid content to 0.01–0.03%. With a weaker alkali, such as sodium carbonate, however, it is difficult to get the free fatty acids below about 0.10%. Steam refining with suitable equipment will also reduce the free acids to 0.01–0.03%. Degumming of a crude oil (with water) will often cause a substantial reduction in its titratable acidity, but not in its actual content of free fatty acids. Bleaching with earths or carbon has little effect on the acidity of oils except in the case of certain acid-activated earths; these may increase the acidity appreciably, for example, 0.05–0.10%, particularly if the oil is soapy or if the contact time is prolonged and the oil is not well dried. Free fatty acids can be completely removed from oils by chromatographic adsorption on alumina or silica gels, but a high adsorbent/acids ratio is required (10–30:1), and the process is uneconomic (3).

Alkali refining effects essentially complete removal of phosphatides from oils. In fact, it has been stated (4) that the phosphorus content of a refined oil is an acceptable criterion of the overall efficiency of refining and that it should not exceed about 0.5 ppm (0.00005%). This corresponds to about 0.0015% phosphatides. The phosphorus content of degummed soybean oils is of the order of 0.0003–0.0009%; that of nondegummed extracted oils may be as high as 0.10%.

Alkali refining, and to a lesser extent degumming, also brings about the removal of a certain amount of nonphosphatide oil-soluble material, including carbohydrates and protein fragments, although information as to specific compounds is not available. McGuire et al. (5) found in laboratory experiments that a crude soybean oil containing 0.019% nitrogen was reduced to 0.00044% nitrogen by one water washing and to 0.00024 and 0.00019% by a second and a third water washing. Alkali refining and bleaching of the original crude and the once, twice, and thrice water-washed oil yielded products with 0.0009, 0.00016, 0.00012, and 0.00009% nitrogen, respectively. Deodorization further reduced the nitrogen to about 0.00005% in all cases. Samples of commercial soybean, cottonseed, and corn salad oils were found to contain 0.00002, 0.00003, and 0.00014% nitrogen, respectively. Much of the nitrogen was in the form of phosphatides. The ratio by weight of phosphorus to nitrogen in soybean oil phosphatides is about 2.7:1 (6); ratios of 2.4–3.1:1 have been reported for purified cottonseed oil phosphatides (7, 8).

Acid refining likewise removes phosphatides and related substances, or "break material," quite completely. At the high temperatures employed in steam refining the oil will break, and insoluble material will be precipitated if phosphatides are present in any considerable amount. If a moderate amount of phosphatides is present, no visible separation of solid material will occur, but the oil will become dark colored, presumably from the decomposition of associated carbohydrates. If the phosphatide content is less than about 0.02%, no such effect is observed (9).

Treatment of an oil with bleaching earth is quite effective in removing phosphatides and the various mucilaginous materials referred to as "gums," "slimes," and so on. A crude fish oil, for example, which can be hydrogenated only with great difficulty because of the poisoning effect of such materials on the catalyst, can be readily hydrogenated after a heavy treatment with an acid-activated earth. Nickel hydrogenation catalysts are even better adsorbents than are bleaching earths for many impurities, and they are often effective after they have become largely inactive for hydrogenation; hence treatment of an oil with a "spent" catalyst is often a useful purification method (10).

Gossypol and related pigments of cottonseed oil readily combine with caustic soda and thus are removed more or less completely by alkali refining. Oils that are presumably colored only by carotenoid pigments are also lightened by alkali refining, although it is probable that the pigments are physically adsorbed on the soap formed by the alkali rather than chemically combined. The adsorptive capacity of the soap is limited, however; palm oil, which is very strongly colored by carotene, is apparently little affected in color by alkalies. Caustic soda is reported to have no effect on the color of tung oil (11).

Acid refining ordinarily reduces the red and yellow color of vegetable oils to a greater extent than does alkali refining (11), particularly if the oil is of relatively poor quality. Most vegetable oils lose considerable red and yellow color from heat treatment alone, although, as is seen later, such oils may exhibit a somewhat complicated behavior if heat treatment is accompanied by oxidation. Cottonseed oil darkens when heated to a high temperature and the color becomes "set," that is, impossible to remove by treatment with alkalies or bleaching earth.

Palm oil exhibits certain characteristics of behavior ascribed to its high content of β-carotene. When the oil is hydrogenated even slightly, its deep orange-red color is replaced with a light yellow or yellow–red color similar to that of other refined vegetable oils. When it is heated to a high temperature, the original deep color disappears and solid "break" material is precipitated. After reduction, the carotenoid pigments of palm oil do not regain their color as the oil becomes oxidized (12).

It is well known that carotenoid pigments are readily adsorbed. The yellow–red color of most vegetable oils is reduced without difficulty by treatment of the oil with bleaching earths or earth and carbon. Oils from badly damaged seeds or animal tissues, however, which contain brown pigments evidently derived from decomposed proteins and carbohydrates, may be very resistant to bleaching by adsorption.

The fact that refined vegetable oils usually have a number of pigments and exhibit only general absorption in the red–yellow region, rather than definite absorption maxima, has thus far largely precluded any investigation of refining or bleaching in terms of the removal of any particular red or yellow pigment. On the other hand, green color in oils is apparently caused only by chlorophyll or related compounds that have well-defined maxima at about 6600 Å (chlorophyll A) or 6400 Å (chlorophyll B). Consequently, the effect of various processing treatments on the chlorophyll content is known quantitatively.

Pritchett et al. (13) found that in a normal crude soybean oil with a chlorophyll content of 1000–1500 μg/l (ca. 0.0001–0.00015%), this chlorophyll content was reduced about 25% by alkali refining, even though chlorophyll is ordinarily considered stable to alkalies and unstable to acids. By subsequent bleaching with acid-activated earth (prior to hydrogenation), it was possible to reduce the chlorophyll concentration to a very low value, for example, 15 μg/liter; however, many processors prefer to give conspicuously green oils a light alkali treatment at a minimum temperature, followed by a heavy bleaching treatment.

Although acid adsorbents are generally more efficient bleaching agents than neutral materials, the effect of acidity is particularly marked in bleaching to remove green color. Hinners et al. (14) have shown in a study of soybean oil bleached with five different earths that the bleaching efficiency is not determined by the "free" hydrogen or hydroxyl ion concentration of the earth, as calculated from the pH of an aqueous suspension, but is a function of the ion adsorption capacity, as measured by the difference between the "free" ion concentration and the concentration of titrable ion.

O'Connor et al. (15) have published spectral absorption curves through both the visible and ultraviolet regions of crude cottonseed, soybean, peanut, sesame, rice bran, and okra seed oils and of the same oils following the successive steps of refining, adsorption bleaching, and steam deodorizing.

Limited hydrogenation gives the appearance of intensifying the green color of an oil through its bleaching effect on the red and yellow pigments. Pritchett et al. (13), however, have shown that partial hydrogenation of a soybean oil (to an iodine number of 75) can be expected to reduce the chlorophyll content about 67% and that substantially complete hydrogenation will reduce it to about 5% of its original value. Hydrogenation also appears to affect the chlorophyll chemically; in the case of soybean oil it causes the absorption maximum at 6600 Å to shift to about 6400 Å, corresponding to a conversion of chlorophyll A to chlorophyll B.

Oxidation has an important effect on the color of fats and oils. While oxidation bleaches the carotenoid pigments, it develops the color of other types of coloring material and in some cases apparently even produces colored compounds of a quinoid nature. The partial oxidation of vegetable oils causes an increase in their red–yellow color, most of which is apparently due to formation of the chroman-5,6-quinones described by Golumbic (16). Cottonseed oil is particularly prone to darken on oxidation (Figure 4.1). Some lots of cottonseed and soybean oil darken so readily that the darkening tendency is noticeable in bleaching. The poor bleach colors are obtained because new pigments develop as the old ones are adsorbed.

In a study of the bleaching of cottonseed and soybean oils, King and Wharton (17) have produced evidence that oxidation not only develops new pigments but also stabilizes existing pigments against adsorption, with the absorbent itself strongly catalyzing the oxidation reactions.

Inedible animal fats of high free fatty acid content may become very dark colored through reaction of their free acids to form colored compounds. Traces of iron and some other metallic contaminants greatly favor color development in such fats.

Certain pigments, such as those in low-grade tallows and greases and in field-damaged soybean oil, are very refractory to ordinary refining and bleaching treatment but may be removed effectively by liquid–liquid extraction.

Alkali refining with caustic soda removes a substantial, although minor,

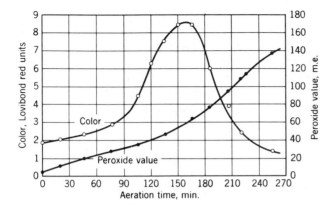

Figure 4.1 Effect of oxidation (aeration at 120°C) on the color of refined and bleached cottonseed oil.

proportion of the sterols of vegtable oils; hence caustic soapstock is a good source of sterols (18). Caustic refining has been reported (19) to remove as much as 10–20% of the tocopherols of soybean oil. Removal is presumably by adsorption on the soap that is formed. Since free fatty acids and gums are largely eliminated by prior treatment with soda ash, the caustic soda soapstock from the second stage of the continuous soda ash–caustic soda refining process is reported to be particularly rich in both sterols and tocopherols (4, 20). A minor proportion of both the sterols and tocopherols in oils is removed by steam refining or deodorization. The amount in any case depends on the severity of the process with respect to time, temperature, and stripping steam flow; these parameters vary greatly from one processor to another. Very drastic deodorization is reported to be capable of reducing the unsaponifiable content of soybean oil by 30–60% (21).

Alkali refining may significantly reduce the vitamin A content of fish liver oils through adsorption of the vitamin on the soaps formed *in situ* in the oil (22); however, this is reported to be obviated by refining in a solvent (23). Chromatographic adsorption is a standard technique for the separation of tocopherols from oils, although ordinary bleaching with earths or carbon appears to abstract no significant amount of tocopherols. According to Buxton (24), the treatment of fish liver oils with carbon removes antioxidants and renders the vitamin A in the oil unstable. Hydrogenation has no adverse effect on tocopherols or other antioxidants but saturates vitamin A and carotene and destroys their biological activity unless special precautions are taken.

Metallic contaminants in fats and oils are presumably in the form of metal soaps. Refining with caustic soda is effective in removing certain heavy metals; however, treatment of the oil with adsorbents appears to be a more certain method of removal. O'Connor et al. (25) found refining and bleaching more effective than refining alone for reducing the iron content of a hydrogenated cottonseed oil product; refining and bleaching accomplished practically complete removal of nickel, whereas refining alone reduced the nickel content only from 10.1 to 6.7 ppm. "Postbleaching" of oils after hydrogenation to remove traces of nickel is a common practice. In dealing with reesterified oils containing zinc or tin catalysts, Feuge et al. (26) found that refining with caustic soda in 0.40% excess reduced the zinc content from 0.036 to 0.0001% and that further treatment with bleaching earth reduced it to less than 0.00001%, whereas either refining alone or refining and bleaching reduced the tin content from 0.116 to less than 0.00032%.

A common method of removing metallic contaminants, which is particularly useful as an adjunct to the deodorization process, is by means of so-called metal scavengers or compounds that are capable of forming inactive complexes with iron or other heavy metals. The most popular compounds that have been used in this country and abroad for many years are certain acids, such as phosphoric acid and organic acids (e.g., citric and tartaric). Commerical lecithin is also effective at deodorizing temperatures because of its phosphoric acid group. More recently, it has been demonstrated that certain polyhydroxy compounds, such as sorbitol and sugar and sugar derivatives, are also metal scavengers (27). The effect of metal scavengers and techniques for their use have been reviewed in articles by Dutton et al. (28). In the series of experiments referred to previously, Feuge et al. (26) found that treatment with strong phosphoric acid was as effective as refining and bleaching for tin removal but that it was less effective with zinc, reducing the zinc content of the oil to only 0.0094%. Metal scavengers are particularly useful for eliminating contamination with iron.

Alkali refining itself introduces sodium soaps as a contaminant in oils. In general, European practice relies on very thorough washing of the refined oil for soap removal, whereas in the United States it is felt that a low soap content is best insured by treatment with bleaching adsorbents and thorough drying of the oil during the bleaching operation to reduce soap solubility. The analysis of oil for low concentrations of residual soap is somewhat uncertain, but it appears that in good practice the soap is reduced in bleaching to about 5–10 ppm. Soap is strongly adsorbed by hydrogenation catalysts; hence hydrogenated oils are virtually free of this impurity. Spent catalysts are good adsorbents for soap; in processing soapy oils it may be advantageous to start the hydrogenation with a spent catalyst, adding an active catalyst later, with or without removal of the spent catalyst.

1.3 REFINING LOSSES

Much of the technology of fat and oil refining is concerned with the minimization of oil losses rather than the thoroughness of purification. In ordinary alkali refining with caustic soda there is always a considerable amount of neutral oil saponified by the alkali or entrained in the soapstock. This oil is recoverable only as a low-grade material and, therefore, represents a direct monetary loss to the refiner.

The amount of neutral oil lost in alkali refining depends primarily on the amount and kind of impurities in the oil. It is relatively low in the refining of such oils as coconut and palm kernel oil and animal and marine fats that are low in phosphatides and similar impurities. In kettle refining coconut oils, for example, the total loss usually does not exceed about 1.4 times the amount of free fatty acids removed. On the other hand, vegetable oils, such as cottonseed and soybean oil, can seldom be refined in batches with a loss less than 3.0 times the free fatty acids content, even when the oil is relatively free of impurities; more commonly, the refining loss of low-acid oils is 5–10 times as great as the free fatty acids content.

High refining losses in vegetable oils free from solid material are generally attributed to the presence of phosphatides. Possibly other gummy and surface-active materials assist in emulsifying neutral fat and contributing to high losses; however, some impurities produce a hard soapstock and low refining loss (29). Gossypol, which is normally present in crude cottonseed oil, has been claimed to have a beneficial effect in refining (30, 31). Wamble and Harris (32), however, were unable to demonstrate a good correlation between gossypol content and refining loss. Likewise, Norris (33) found added pure gossypol to have an unpredictable effect on cottonseed oil refining losses (official cup method). Other phenolic substances not found in crude oils also have been claimed to reduce the refining loss (34). The composition of a crude oil with respect to surface-active constituents is determined, of course, in the extraction process and particularly during the cooking of the seed.

Newer, continuous, alkali-refining processes have been designed to reduce the loss of neutral oil by saponifiction as well as by entrainment in the soapstock. Neutral fats and oils are actually quite resistant toward reaction even with quite strong lyes, and a small amount of caustic soda solution mixed into a neutral oil will undergo substantially complete reaction only after a somewhat prolonged period at an elevated temperature. This is shown clearly by the data published by Thurman (35) and shown in Figure 4.2, where the three curves show the effect of lye strength, temperature, and contact time on the saponification of neutral oil in terms of the percentage of the total excess caustic reacting, that is, the excess of caustic above that required to react with 0.075% of free fatty acids. The lye used in each experiment contained sufficient sodium hydroxide to neutralize the free acids, plus 0.153 gram per 100 grams of oil, or a sufficient amount, if totally reacted, to saponify 1% of the neutral oil.

A method was developed by Wesson (36) and modified by Jamieson (37) for estimating the total amount of fatty acids and other substances that are normally removed from an oil by alkali refining. The content of these substances in the oil, often referred to as the *absolute loss* or the

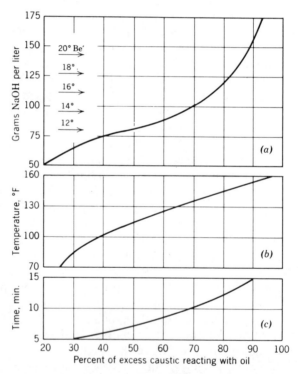

Figure 4.2 Effect of (*a*) strength of lye, (*b*) temperature, and (*c*) contact time on the completeness of reaction of caustic soda solution with neutral oil. Values are in terms of percentage of total available caustic reacting under the following conditions: (*a*) temperature, 130°F, contact time, 10 mins; (*b*) contact time, 10 mins, 14.1° Bé lye; (*c*) temperature, 130°F, 14.1° Bé lye; 0.153% excess NaOH used in all cases.

Wesson loss, is of some significance in refining theory and practice, inasmuch as it represents a minimum value beyond which the refining loss of the oil cannot be reduced. The Wesson loss in crude cottonseed, soybean, peanut, and corn oils is ordinarily 1–3% greater than the free fatty acid content of the oil. In a study of the Wesson loss on 39 hydraulic cottonseed oil samples and 29 expeller soybean oil samples, King and Wharton (38) found the Wesson loss of the cottonseed oil to average 1.70% higher than the free fatty acid content and the Wesson loss of the soybean oil to average 3.3% higher. Extremes were 0.7–2.7% and 1.4–6.4%, respectively. The ranges for cottonseed oil samples were 0.9–2.8% in free fatty acid content and 3.4–9.4% in refining loss by the official American Oil Chemists' Society laboratory cup method, and those for the soybean oil samples were 0.5–1.4% in free fatty acids and 4.3–12.3% in refining loss.

Later, Linteris and Handschumaker (39) described a simple chromatographic method for crude vegetable oils that appears to give a more accurate measure of neutral oil than the Wesson loss. It is more rapid and convenient than the Wesson method and yields an oil containing the normal refined oil pigments; consequently, the oil may subsequently be examined for color and subjected to a microbleaching test. Unfortunately, neither the chromatographic loss nor the Wesson loss gives comparable results in all cases, and neither method shows a consistent correlation with the cup refining loss used for trading purposes. Linteris and Handschumaker (39) concluded that the Wesson method gives high results for neutral oil in some instances as a result of hydrolysis of phospholipids by the alkali. Hartman and White (40) found that in the case of refined oils some neutral glycerides are adsorbed on the column in the absence of free fatty acids and nonglyceride material, but this condition does not occur in crude vegetable oils. This chromatographic method has been modified and is now an official method (Ca-9f-57) of the American Oil Chemists' Society (41).

1.4 APPLICATIONS

Some oils and fats are seldom given any kind of purifying treatment. Butterfat, oleo oil, and olive oil are neither refined nor bleached in the ordinary course of manufacture, although the refining process may occasionally be applied for the reclamation of off-grade or badly deteriorated materials.

Most of the lard on the market is not refined in the proper sense of the term. The product known to the trade as "refined lard" has merely been clarified and solidified, with the application of a light bleaching treatment, in some cases, by means of a bleaching clay or carbon. A few special lard products which are hydrogenated or deodorized are also alkali refined. Small amounts of tallow or oleo stock are consumed without being refined, and limited quantities of unrefined oleostearine are used in the manufacture of blended-type shortenings and margarines of the "puff–paste" type. Inedible animal fats used in the manufacture of soaps are sometimes refined. In the treatment of soap fats refining is chiefly valuable because of its beneficial effect on the color of the fats. Soap fats are almost invariably bleached.

In some parts of the world, particularly in Africa, Southern Europe, and the Orient, considerable quantities of oil seeds yielding a relatively pure oil, such as peanut, sesame, rapeseed, and soybean, are cold pressed; the resulting oil, like olive oil, is used for edible purposes without further processing treatment. Oils that are to be deodorized and marketed as neutral salad or cooking oils, as well as vegetable oils to be converted into shortenings or margarines, are almost invariably alklai refined. Free fatty acids are objectionable in all these products because they tend to smoke when the fat is heated. The higher fatty acids are so insoluble as to be practically tasteless; hence oils containing large proportions of these may be rendered at least temporarily neutral by deodorization alone. The fatty acids of coconut oil and other lauric acid oils are sufficiently soluble, however, to impart a taste; therefore, these oils must be rendered substantially free of free acids before they are used in food products.

Vegetable paint oils are not always alkali refined since a moderate content of free acids is unobjectionable in these oils. They are frequently acid or water washed to remove material that would precipitate under heat treatment. The removal of this so-called "break" material is essential in varnish oils, which are usually refined with alkali.

Whale, fish, and other marine oils, with the exception of fish liver oils, are refined before they are manufactured into edible products since such products are invariably hydrogenated and deodorized. Fish liver oils are preferably not alkali refined because of vitamin A adsorption of soaps formed *in situ* in the oil.

Desliming or Degumming

As used here, the terms "desliming" and "degumming" refer to refining treatment designed to remove phosphatides and certain other ill-defined "slimes" or "mucilaginous materials" from the oil but does not significantly reduce the acidity of the oil.

2.1 DEGUMMING BY HYDRATION

The degumming of oils intended for use in edible products, if it is done at all, is almost always accomplished by hydrating the phosphatides and similar materials to make them insoluble in the oil.

Crude soybean oil, particularly that obtained by solvent extraction, is very often degummed before refining. As mentioned previously, gums, if left in the oil, tend to produce high refining losses and occasion trouble from settling out in storage tanks and tank cars. In addition, the gums, after reasonably simple and inexpensive refining treatment, are marketable as commercial soya lecithin. Other vegetable oils, such as corn oil, sunflower seed, and peanut oil, also yield gums suitable for conversion to commercial lecithin. However, the market is generally well supplied by the standard soybean oil product; hence these oils are less commonly degummed than soybean oil in the United States. The degumming of crude cottonseed serves to reduce the refining loss, but there is at present no market for the dark-colored resinous gums; thus the process is generally considered uneconomic. However, one oil mill has degummed cottonseed oil in miscella (42) and added these nutritious phosphatides back to the meal, where they function to increase oil content, minimize dustiness, and facilitate pelleting.

Degumming should be completed before the oil is shipped or placed in storage; in addition, a better lecithin is produced from fresh oil. Degumming is, therefore, usually carried out at the oil mill or extraction plant rather than at the refinery. In all cases separation of the hydrated gums from the oil is accomplished with continuous centrifuges similar to those used for the separation of soapstock in continuous refining. In some extraction plants the gums are allowed to hydrate from steam condensed in the last stage of steam stripping for removal of solvent from the miscella; in others, a small proportion of water (about 1%) is mixed continuously into the substantially dry oil. The minimum quantity of water consistent with a free-flowing product is desired, as any excess of water tends to make the separated gums ferment more readily. In any case, water must be removed from the final lecithin product. It is possible to obtain gums containing as little as 20–25% water. A small quantity of oil is also occluded with the gums, and hence there is a slight loss over and above the amount of impurities separated.

A typical soybean oil will yield about 3.5% of gummy material that is 25% water and 75% oil soluble; the latter will contain about 33% oil and about 67% acetone insoluble (phosphatides, etc.). If lecithin is to be reclaimed, the oil should, of course, be free of meal or other solid material before it is hydrated. As hydrating agents, mineral acids, acid salts, organic acids, and many other materials have been suggested (2) but are seldom used because they interfere with the quality of the separated gums, damage the oil, or pose problems for handling in the plant.

According to Kantor (43), the content of foots or gums in linseed oil varies directly with the specific gravity of the oil (from 2.0% at a specific gravity of 0.9340 to less than 0.2% at a specific gravity of 0.9315). This permits the optimum amount of hydrating water (equal to the foots content) to be determined without difficulty. The dosage of water is rather critical, inasmuch as the gums bind water within very specific limits.

Degumming is not ordinarily applied to animal fats because these are very low in phosphatides. Dry-rendered fats, however, that have been produced with excessive dehydration of the parent tissues and are consequently dark colored from the presence of proteinaceous matter can sometimes be lightened by a hydrating treatment.

Temperatures for degumming are not highly critical, although separation of the hydrated gums occurs more readily if the viscosity of the oil is reduced by operating at 90–120°F.

A novel process that reportedly degums soybean oil completely has been described (44, 45). A small amount of acetic anhydride is mixed with the crude oil and sufficient water added for gum hydration. After a short tempering and heating time, the hydrated gums are centrifugally separated and the degummed oil is water washed to remove traces of acetic acid and then vacuum dried continuously. The degummed oil so obtained is claimed to be completely break free, requiring no caustic refining prior to bleaching and deodorization to produce a stable product. This process has not gained acceptance in the United States, perhaps partly because prooxidant metals are not removed as thoroughly as in alkal refining.

The use of ammonia degumming has been patented by Clayton (46). Ammonium hydroxide is a volatile nonsaponifying alkali. Consequently, it yields a neutralized oil and also a lecithin product. Refining with ammonia eliminates acidulation of soapstock and avoids decomposition of the phosphatides, a source of choline, inositol, and so on, and the phosphatides may be further purified or added back into the meal or into feeds. However, unless a caustic soda wash is included, the refined oil color will be darker than that produced by the soda ash or caustic soda process (2).

2.2   PREPARATION OF COMMERCIAL LECITHIN

The conversion of soybean oil gums to the various grades of commerical lecithin has been described by Langhurst (47), Stanley (48), and Van Nieuwenhuyzen (49). Material from the degumming centrifuges is dried to a low moisture content, preferably under 1%, to improve preservability and fluidity. Horizontal film dryers or vertical thin film dryers are commonly used. Excessive temperatures and/or contact times must be avoided if a high-quality light-colored product is to be obtained. The dried product is cooled below 121°F (50°C) to prevent subsequent darkening. Lecithin may be stored for months at 65–85°F (20–30°C) without significant change in quality.

According to tentative specifications of the National Soybean Processors' Association (50), the six common commerical grades consist of plastic-consistency and fluid-consistency products, each of which may be unbleached, single-bleached, or double-bleached. Bleaching is usually accomplished with hydrogen peroxide, which may be added either during the degumming step with water used for hydration or in the bleaching kettle. The fluid-consistency product is made by adding 2–5% of mixed soybean oil fatty acids, methyl esters, or other free fatty acids to the plastic-consistency material. Specifications call for a minimum content of acetone-insoluble material of 65 and 62% for the plastic and fluid grades, respectively; and for maximum moisture and benzene-insoluble contents of 1.0% and 0.3%, respectively, in all cases. Maximum acid values are 30 for the plastic grade and 32 for the fluid grade. The Gardner color of a 5% solution in colorless mineral oil must not exceed 10 for unbleached lecithin, 7 for single-bleached lecithin, or 4 for the double-bleached product.

2.3   ACID REFINING

Strong mineral acids are rarely employed to degum edible oils but have been used on rapeseed for burning oil, for pretreatment of fats before splitting, and for fish and whale oil difficult to hydrogenate (2). Sulfuric acid requires special attention to avoid charring and sulfonation and is usually diluted for use. Hydrocholoric acid has been used, particularly on peanut oil (51, 52). Phosphoric acid is effective and much easier to handle, and is often used preliminary to alkali refining. Sullivan (53) claims improvement by treating crude oils with 0.1–0.4% phosphoric acid until addition compounds have been formed. Phosphates and organic acids have also been used to a limited extent. In general, only phosphoric acid is used commercially in the United States, and mainly as a pretreatment before alkali refining.

2.4   REMOVAL OF BREAK MATERIAL BY HEAT TREATMENT

Heat treatment is seldom, if ever, employed alone for the removal of phosphatides of "break matcrial" from oils. However, the heat treatment incidental to the steam refining process will cause the precipitation of such material.

Alkali Refining

The technology of alkali refining is concerned with the proper choice of alkalies, amounts of alkalies, and refining techniques, to produce the desired purification without excessive saponification of neutral oil, and with methods for the efficient separation of refined oil and soapstock.

In general, the art of refining was much more elaborately developed in Europe and in areas influenced by European practice than it was in the United States because of the greater variety of oils handled by European processors and their need for dealing with much poor-quality oil that is difficult to refine properly. On the other hand, American practice in the rapid and continuous refining of cottonseed and soybean oils of good quality has reached a high degree of perfection and efficiency that is unmatched by batch methods. Present-day emphasis here and abroad is on continuous refining wherever processing volume on a single feedstock justifies the cost. Once refining conditions are set for an oil, a very large quantity can be refined with high efficiency and minimum attention.

3.1   REFINING WITH CAUSTIC SODA

Selection of Lye.   The selection of the proper amount and strength of lye for refining is highly important in the case of any oil or fat that is to be refined with caustic soda. It might be supposed that the lye to be used could be quite simply determined on the basis of the free fatty acid content of the oil, but such is not the case. Although a relatively pure fat like lard can be expected to exhibit a more or less predictable behavior in refining, according to its free acid content, the presence of pigments and surface-active substances in most vegetable oils makes their reaction to alkali treatment extremely variable. In addition to differences in the oil produced by different mills, as a result of variations in methods of processing the oil seeds, there are well-defined variations in the characteristics of oils according to their geographical origin and other variations according to climatic conditions that occur from season to season. Variability in the reaction of the oil to refining and bleaching is particularly marked in the case of cottonseed oil (54).

In the choice of both lyes and refining methods the refiner is invariably guided by preliminary refining tests conducted in the laboratory (54). The official laboratory refining methods developed under the auspices of the

American Oil Chemists' Society (41) and adopted by various trade organizations serve not only to evaluate shipment of crude cottonseed oil
and other oils on the basis of their worth to the refiner, but also provide
an indication of the results to be expected from comparable treatment of
the oil in the plant. Thus by refining samples of an oil with different
amounts and strengths of lye, and with different times of stirring and so
on, and noting the refining loss and color of the oil obtained in each case,
it is possible to determine the optimum conditions for refining the oil in
the plant.

It is customary to measure the strength of lye solutions for refining in
terms of their specific gravity expressed in degrees Baumé. The refiner
ordinarily employs one of a series of lyes ranging from about 10 to 30°
Bé, in 2° steps, with oils of good quality usually being refined with 12,
14, or 16° Bé lye. The actual sodium hydroxide content of lyes of different
strengths is shown in Table 4.1.

Ordinarily, the refiner uses lye containing a sufficient amount of sodium
hydroxide to produce the color desired in the oil and uses the strength
of lye that will produce the lowest refining loss with the desired color.
In general, the best results are obtained with relatively weak lyes on low
free fatty acid oil and with stronger lyes on high-acid oils; but the exact
lye can be determined only by trial. American refineries customarily
calculate the amount of lye required to neutralize the free fatty acids in
the oil and to this amount add an excess that varies with the characteristics
of the oil. The excess is expressed in terms of dry sodium hydroxide, on
a percentage basis, calculated on the weight of the oil. Part A of Table

Table 4.1   Sodium hydroxide content of lyes of different
degrees Baumé (41)

Degrees Baumé at 15°C	Sodium Hydroxide Content (%)
10	6.57
12	8.00
14	9.50
16	11.06
18	12.68
20	14.36
22	16.09
24	17.87
26	19.70
28	21.58
30	23.50

4.2 gives the percentages of lyes of various strengths required to neutralize the different percentages of free acids in the oil. Part B gives the additional amounts of lye required to provide sodium hydroxide in different degrees of excess.

The use of Table 4.2 may be clarified by an example. Thus it may be desired to know the percentage of 16° Bé lye required to refine an oil containing 2.0% free fatty acids, with 0.45% excess sodium hydroxide. Referring to Table 4.2, it is seen that the neutralization of 2.0% free acids, calculated as oleic, requires 2.57% of 16° Bé lye. An excess of 0.45% corresponds to 4.07% of the same lye. The total amount of lye required is the sum of the two, or 6.64%.

Animal and marine fats of good quality, and coconut oil or other vegetable oils very low in gums and pigments, may usually be refined satisfactorily with 0.10–0.20% excess sodium hydroxide. Other vegetable oils and low-grade animal fats require sodium hydroxide in increasing excess as their content of gums and mucilaginous materials increases and as the necessity for decolorizing becomes greater. Cottonseed oil, for example, is seldom refined with less than about 0.25% excess, and the use of 0.50–0.60% excess is not uncommon, even in dealing with oils of relatively good quality. The amounts of lye specified for official laboratory refining tests by the National Cottonseed Products Association (55) for cottonseed, peanut, corn, and coconut oils and by the National Soybean Processors Association (50) for soybean oil may be considered more or less representative of good refining practice, although in many cases substantially less lye may be used, particularly if the color of the refined oil is not a prime consideration. For good hydraulic cottonseed oil containing 0.5–1.0% free fatty acids, the prescribed excess of sodium hydroxide (in the form of 12 or 14° Bé lye) is about 0.45%. For oils of increasingly poorer quality greater excesses and stronger lyes are specified, with the minimum excess rising to 0.74% (14 and 18° Bé) at 4.0% free fatty acids, and to 1.28% excess (20 and 26° Bé) at 15.0% free fatty acids.

Good expeller soybean oils are tested by refining with two different amounts of 12° Bé lye, with the sodium hydroxide excess amounting to about 0.15 and 0.20% excess in the two cases. Extracted soybean oil is refined with 14° Bé lye using 87.5 and 66.6% of the maximum used for hydraulic and expeller soybean oil, respectively. The preceding applies only to soybean oil that has not been degummed. Degummed soybean oil is given two test refinings, with 12° Bé lye and with 0.10 and 0.20% excess sodium hydroxide, respectively, regardless of the acidity of the oil.

The official refining methods for peanut oil specify 12 and 16° Bé lyes for peanut oils not over 3.0% in free fatty acids, and 16 and 20° Bé lyes

Table 4.2   Calculation of lyes for refining.
A.   Percentages of lyes of different strengths required to neutralize the free fatty acids in oils of varying acidity (free fatty acids calculated as oleic).

Free Fatty Acids (%)	Percent Lye of Baumé Strength				
	12° Bé	14° Bé	16° Bé	18° Bé	20° Bé
0.6	1.07	0.90	0.77	0.67	0.59
0.7	1.24	1.05	0.90	0.78	0.69
0.8	1.42	1.20	1.03	0.89	0.79
0.9	1.60	1.35	1.16	1.00	0.89
1.0	1.78	1.50	1.29	1.11	0.99
1.1	1.95	1.65	1.41	1.23	1.09
1.2	2.13	1.80	1.54	1.34	1.19
1.3	2.31	1.95	1.67	1.45	1.29
1.4	2.48	2.10	1.80	1.56	1.39
1.5	2.66	2.25	1.93	1.67	1.49
1.6	2.84	2.40	2.06	1.79	1.58
1.7	3.02	2.54	2.18	1.90	1.68
1.8	3.20	2.69	2.31	2.01	1.78
1.9	3.37	2.84	2.44	2.12	1.88
2.0	3.55	2.99	2.57	2.23	1.98
2.1	3.73	3.14	2.70	2.35	2.08
2.2	3.91	3.29	2.83	2.46	2.18
2.3	4.08	3.44	2.96	2.57	2.28
2.4	4.26	3.59	3.08	2.68	2.37
2.5	4.44	3.74	3.21	2.80	2.47
2.6	4.61	3.89	3.34	2.91	2.57
2.7	4.80	4.04	3.47	3.02	2.67
2.8	4.97	4.19	3.60	3.13	2.77
2.9	5.15	4.34	3.72	3.24	2.87
3.0	5.32	4.49	3.85	3.36	2.97
3.2	5.68	4.78	4.10	3.58	3.16
3.4	6.04	5.18	4.35	3.80	3.36
3.6	6.39	5.48	4.61	4.03	3.56
3.8	6.75	5.78	4.87	4.25	3.76
4.0	7.10	6.08	5.12	4.47	3.95
4.2	7.45	6.38	5.38	4.70	4.15
4.4	7.80	6.68	5.64	4.92	4.35
4.6	8.16	6.98	5.89	5.15	4.55
4.8	8.52	7.28	6.15	5.37	4.74
5.0	8.88	7.47	6.42	5.60	4.94

Table 4.2   *(Continued)*

B.   Percentages of lyes of different strengths required to provide different excesses of lye (excess calculated as dry sodium hydroxide).

Excess (%)	Percent Lye of Baumé Strength				
	12° Bé	14° Bé	16° Bé	18° Bé	20° Bé
0.05	0.62	0.53	0.45	0.39	0.35
0.10	1.25	1.05	0.90	0.79	0.70
0.15	1.87	1.58	1.35	1.18	1.05
0.16	2.00	1.69	1.44	1.26	1.12
0.17	2.12	1.79	1.53	1.34	1.19
0.18	2.25	1.90	1.62	1.42	1.26
0.19	2.28	2.00	1.71	1.50	1.33
0.20	2.50	2.10	1.81	1.58	1.39
0.21	2.63	2.21	1.90	1.66	1.46
0.22	2.75	2.31	1.99	1.74	1.53
0.23	2.88	2.42	2.08	1.81	1.60
0.24	3.00	2.52	2.17	1.89	1.67
0.25	3.13	2.63	2.26	1.97	1.74
0.26	3.25	2.73	2.35	2.05	1.81
0.27	3.38	2.84	2.44	2.13	1.88
0.28	3.50	2.94	2.53	2.21	1.95
0.29	3.63	3.05	2.62	2.29	2.02
0.30	3.75	3.15	2.71	2.37	2.09
0.31	3.88	3.26	2.80	2.44	2.16
0.32	4.00	3.36	2.89	2.52	2.23
0.33	4.13	3.47	2.98	2.60	2.30
0.34	4.25	3.57	3.07	2.68	2.37
0.35	4.37	3.68	3.16	2.76	2.44
0.36	4.50	3.78	3.25	2.84	2.51
0.37	4.62	3.89	3.34	2.92	2.58
0.38	4.75	3.99	3.43	3.00	2.65
0.39	4.88	4.10	3.58	3.07	2.72
0.40	5.00	4.21	3.61	3.15	2.79
0.41	5.13	4.31	3.70	3.23	2.86
0.42	5.25	4.42	3.80	3.31	2.93
0.43	5.38	4.52	3.89	3.39	3.00
0.44	5.50	4.63	3.98	3.47	3.06
0.45	5.63	4.73	4.07	3.55	3.13
0.46	5.75	4.84	4.16	3.63	3.20
0.47	5.88	4.85	4.25	3.70	3.27
0.48	6.00	4.95	4.34	3.78	3.34
0.49	6.13	5.16	4.43	3.86	3.41
0.50	6.25	5.26	4.52	3.94	3.48

for poorer oils; excesses of sodium hydroxide vary from 0.25 to 0.47% for the better oils, and do not exceed 0.55% for oil with 10.0% free fatty acids. For corn oils, 16° Bé lye is specified in all cases, with excesses of about 0.25 and 0.36% for the better oils. For coconut oil the official refining test method specifies the use of 20° Bé lye only, and the amount used being in all cases approximately one-tenth greater than that theoretically required to neutralize the acidity of the oil, calculated in terms of oleic acid.

Drying oils are generally refined with somewhat less excess lye than edible oils.

As is explained later, in batch refining it is sometimes necessary to use salt, sodium carbonate, or other electrolyte in conjunction with caustic soda to ensure adequate salting or graining out of soapstock. Sodium silicate has often been used with caustic soda in the refining of oils to "weight" the foots and promote their settling, yielding sloppy foots.

Batch Refining by the Dry Method. The dry method of refining is the one generally practiced in the United States where any edible fat or oil is to be batch refined by means of caustic soda. This method is termed "dry" because the oil is treated with a relatively strong lye and the soapstock or "foots" are recovered in a solid or semisolid form from the cooled oil. It is distinguished from the "wet" method of refining in which the soapstock is washed to the bottom of the refining kettle with considerable quantities of water and recovered in the form of a fluid solution. Dry refining has the advantage of being rapid and convenient and of producing a concentrated soapstock and a refined oil relatively free of soap or moisture. It is particularly adapted to the refining of cottonseed oil, which in most cases produces a firm soapstock, free of any considerable amount of occluded oil. It is less suitable for treatment of oils, such as linseed or degummed soybean oil, which produce soft, sloppy foots that do not readily settle to a firm mass.

The equipment required for batch refining is simple, consisting of an open tank or kettle equipped with an agitator, steam coils for heating, and a conical bottom. It is common for refining kettles to hold a full tank car of oil, or 60,000 lb. The agitator consists almost invariably of a central vertical shaft to which are attached a series of horizontal paddle arms; the latter are placed in staggered positions down the shaft so that they will reach all portions of the kettle charge and are inclined at a 45° angle, to give a lifting action when they are in motion. Common agitator speeds are about 40 rpm for rapid agitation and 8 rpm for slow agitation; the former must be sufficiently vigorous to bring about intimate mixing and emulsification of the oil and lye, whereas the latter is designed only to keep the contents of the kettle moving and to maintain particles of soap in suspension in the oil while they undergo melting and coalescence (56). Many refiners prefer continuously variable speed drives, although two-speed drives are common and appear to be satisfactory for most purposes. The heating coil must be designed to raise the temperature of the batch rapidly; usually, but not always, the cone of the kettle is steam jacketed to assist in melting and discharging of the foots after refining is completed.

The first stage of refining is carried out with the oil at atmospheric temperature, or in the case of a fat, at a temperature just high enough to keep the material molten and liquid. Higher temperatures are avoided, partly because they tend to increase the amount of neutral oil saponified, but more because lighter refined oil colors are obtained at relatively low temperatures. For the refining of cottonseed oil, where color removal is particularly important, an initial temperature of 68–75°F is preferred and specified for the official laboratory refining tests although no marked disadvantage appears to result from a temperature up to about 90°F. If the oil contains occluded air after it is pumped to the refining kettle, it must be settled long enough for the air to rise to the surface and escape; otherwise, the foots will entrain sufficient air to float partially and thus will not settle properly to the bottom of the kettle.

After the charge of oil is at the proper temperature and free from air, the agitator is started at high speed and the proper amount of lye is rapidly run in. The lye is usually distributed fairly evenly over the surface of the oil, although an elaborate distribution or spraying system is not necessary. Agitation is then continued until the alkali and oil are thoroughly emulsified. With some oils the best results are obtained if the mixing period is relatively short, for example, 10–15 min. At the end of the mixing period the agitator is reduced to a low speed, sufficient only to keep the contents of the kettle stirred, and heat is applied to bring the temperature of the charge up to 135–145°F as rapidly as possible. Under the influence of heat, the emulsion breaks and the soapstock separates from the clear oil in the form of small flocculent particles which tend to coalesce as stirring is continued. After the desired degree of "break" is obtained, agitation is stopped, heat is turned off the kettle, and the soapstock, or foots, is allowed to settle to the bottom of the kettle by gravity.

Thorough settling of the soapstock before the oil is drawn off is essential for low refining loss. With cottonseed oil or other oil that has a high refining loss or a tendency for much neutral oil to be occluded in the soapstock, a settling time of about 10–12 hours is minimum, and the batch is usually settled overnight. On the other hand, oils, such as lard or tallow, can be refined quite satisfactorily with a settling time of 1–4 hours.

When the contents of the kettle are well settled, the refined oil is drawn off the top through a swinging suction pipe, leaving the soapstock in the form of a more or less coherent mass at the bottom. Usually two suction lines of different sizes are provided, of which the larger is used to remove the bulk of the oil, and the smaller is used for a final skimming operation. Removal of the last portions of oil by judicious use of the suction pipe and by manipulating the foots with a pole or paddle requires considerable skill on the part of the operator. If the amount of soapstock is large, a portion is usually dropped through a large, quick-opening valve in the bottom of the cone into an open soapstock-receiving tank under the kettle before skimming is completed. In many plants the skimmings are not mixed with the bulk of the refined oil but are diverted to a separate tank or kettle where they are again settled. The soapstock, too, is often heated and skimmed for oil recovery in the soapstock receiver. A dark-colored oil is obtained, however, and it must be rerefined.

Although dry refining produces a relatively clean and clear oil, the refined oil contains traces of moisture and soap that should be removed before the oil is put into storage. As a cleaning-up procedure, a common practice is to filter the refined oil through spent bleaching earth. The use of spent earth, already saturated with oil, avoids loss of oil through retention on the earth. If the oil is bleached before it is stored, it is, of course, dehydrated and freed of soap in the bleaching operation.

These procedures are not essentially altered if the fat to be refined is tallow or other animal fat, although the entire operation must, of course, be carried out at a temperature above the solidifying point of the fat.

Much care must be exercised in the refining of oils containing excessive proportions of free fatty acids, that is, above 15–20%. Unskillful treatment of such an oil may result in the saponification of so large a proportion of the charge that it will be impossible to effect a separation of the foots and the oil. In some cases it may be advisable to refine high-acid oils in two stages, with only partial neutralization accomplished in the first stage. Generally, poor oils of very high acidity can be handled more satisfactorily by some variation of the wet refining method.

Batch Refining by the Wet Method. In European refineries, where many of the oils refined may be relatively high in free fatty acids and produce soft soapstocks, the "wet" refining method is commonly practiced. In general, it involves heating the oil charge to a relatively high temperature, for example, 150°F, mixing in the lye, and washing down the precipitated soapstock with a spray of hot water directed onto the surface of the oil. In some cases salt, sodium carbonate, or another electrolyte is added to assist in breaking the emulsion of soapstock and oil and to aid in graining out and settling the soapstock. Several successive water washes are required to complete the substantial removal of soap from the oil. After each wash the oil must, of course, be thoroughly settled. Innumerable variations of the method are practiced according to the idiosyncrasies of the particular oil being processed; little published information is available concerning the details of the method followed in different cases.

The wet method is particularly advantageous for refining oils of a very high degree of acidity. For neutralizing extracted olive oil, which may average about 20% free fatty acids, Lewis (57) recommends one of the following alternative methods:

1. The oil is heated to 140–150°F and emulsified with 10% of its own weight of water. Strong (45° Bé) lye is then added in the calculated amount to neutralize the free fatty acids exactly, with agitation discontinued and the soapstock allowed to settle immediately after the lye is added.
2. The oil is heated to 125°F with agitation, and sufficient (20° Bé) lye is added to combine exactly with the free acids. It is then heated to 135°F, 10–15% of 10% soda ash solution is added, heating is continued to 160°F, and agitation is stopped. It is reported that oils containing as high as 35% free fatty acids can be refined by these methods, with refining losses somewhat in excess of twice the free fatty acid content.

The wet refining method is little used in the United States except for the refining of coconut or other lauric acid oils. Usually, the oil at about 150°F is treated with 20° Bé lye in sufficient amount to give about 0.10% excess sodium hydroxide, plus salt in the dry form or in the form of a strong brine, equivalent to about 0.10% sodium chloride per 1.0% free fatty acids in the oil. The salt is used because of the small excess of sodium hydroxide and the fact that coconut oil soaps require a relatively large concentration of electrolyte to cause them to grain out. After a short period of agitation the foots are washed down with hot water. Refining losses amount uniformly to about 1.4 times the quantity of free fatty acids neutralized.

In the batch refining of linseed oil, according to Kantor (43), the oil is commonly treated according to a typical dry refining method; but after the soapstock has settled for 0.5–2 hours, the batch is washed down with two successive 5% portions of water. Then, after a final settling period of 12–24 hours under heat, the soapstock is drawn off in a fluid form.

Refining equipment for the wet method is not essentially different from that employed in refining by the dry method, except that it is quite common practice to use closed kettles, which may also serve as vacuum bleachers for the refined and washed oil.

Continuous Caustic Refining.   The batch method of caustic-refining vegetable oils, such as cottonseed and soybean, has been largely superseded in the United States by continuous methods in which the separation of oil and foots is carried out in centrifuges. The continuous method has the double advantage of greatly reducing the time of contact between oil and alkali and of effecting a very efficient separation of foots and oil. Consequently, it reduces to a minimum the loss of neutral oil through saponification or occlusion in the soapstock and at the same time produces a refined oil of as good grade as batch methods do.

The first continuous plants installed in the United States for refining cottonseed oil were sold under a guarantee that they would effect a 30% saving in refining loss over kettle refining. A report by Tyler (58) (Table 4.3) indicated a slightly smaller average saving in the refining of oils produced during the test period.

Although centrifugal refining was proposed as early as 1923 by Hapgood and Mayno (59), successful continuous systems were not developed until about 10 years later as a result of the work of James (60) and of Clayton and Thurman and associates (61). The first system installed in the United States was described by James (62).

EQUIPMENT.   In continuous refining gravity settling is replaced by a much stronger force caused by the rapid rotation of a centrifuge. The force is equal to mass M times acceleration A, and acceleration is equal to the square of the angular velocity w times the radius r. The theory is discussed by Cowan (63) and Foust (64).

In general, centrifuges are of two types, tubular bowl and disc bowl. In the tubular bowl type (Figure 4.3), a high rotation speed develops centrifugal forces of the order of 13,000 times gravity, but capacities are relatively low, and without any automatic removal system only small concentrations of solids can be handled. Also, in these open centrifuges, the heavy phase outlet diameter is varied by means of discharge ring dams located at the top of the bowl. A temporary shutdown and partial dismantling is required for a change of discharge rings. Although these machines were extensively used at one time and are still desirable in locations where production is small or where conditions are not suitable for installation, operation, and maintenance of large machines, their general use is rapidly diminishing. The disc bowl is a larger machine, rotating at a lower speed and developing a centrifugal force up to 7000 times gravity.

Probably the most recent development has been the SRPX-317 machine (Figure 4.4) made by DeLaval. This unit has a capacity of 25,000–50,000 lb/hour depending on the feed, and is equipped with a 30-in-diameter bowl

Table 4.3 *Comparison between batch and continuous centrifugal refining of crude cottonseed oil (58)*

Test No.	Free Fatty Acid of Crude Oil (%)	Color Lovibond Red		Bleach Color Lovibond Red		FFA Content (%)		Refining Loss (%)		Saving in Loss by Refining Continuous (%)
		Batch	Cont.	Batch	Cont.	Batch	Cont.	Batch	Cont.	
1	1.4	7.3	6.9	1.6	1.6	0.011	0.015	6.7	4.8	28.3
2	3.5	10.7	9.5	2.3	2.3	0.009	0.025	11.7	9.0	23.1
3	0.9	6.2	6.9	1.7	1.7	0.005	0.015	5.9	4.2	28.8
4	1.8	7.4	7.4	1.7	1.7	0.013	0.018	8.2	5.7	30.5
5	3.7	11.0	9.5	2.3	2.3	0.015	0.024	11.7	8.9	23.9
6	2.5	8.2	8.1	1.9	1.9	0.017	0.020	9.2	6.7	27.2
7	1.7	7.5	7.7	1.9	2.0	0.015	0.020	7.5	5.3	29.3
8	1.9	7.3	8.1	1.6	1.9	0.015	0.018	8.0	5.9	26.2
9	2.7	9.5	9.3	2.4	2.2	0.018	0.025	9.4	6.6	29.8
10	3.1	10.3	9.0	2.3	2.2	0.018	0.027	10.8	8.6	20.3
Average	2.32	8.5	8.2	2.0	2.0	0.014	0.021	8.91	6.57	26.3

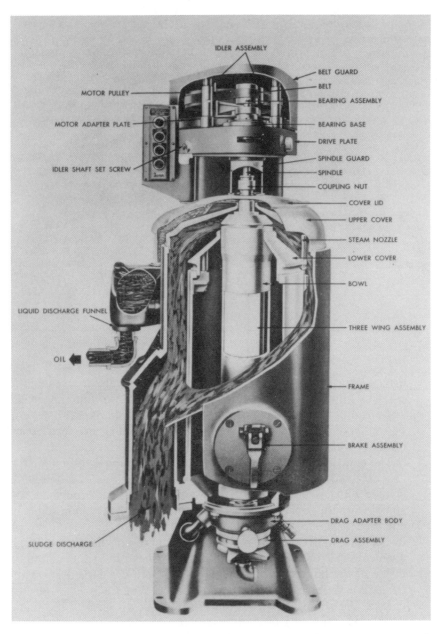

Figure 4.3 Sharples vegetable oil refining supercentrifuge. (Courtesy of the Sharples Division of Pennwalt Corporation, Oak Brook, Illinois.)

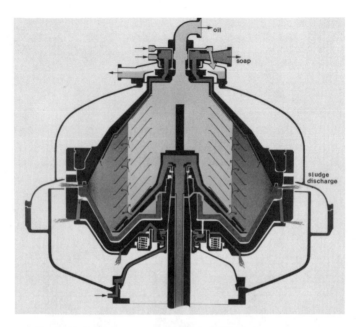

Figure 4.4 DeLaval SRPX 317 hermetic refining centrifuge. (Courtesy of Sullivan Systems, Inc., a subsidiary of DeLaval Separator Company, Tiburon, California.)

driven by a 50-hp motor, the bowl sealed to operate at pressures to 125 psi. The unit is compact, occupying only 4 × 5 ft of floor space. It is of self-cleaning design, and any accumulated solids in the bowl are capable of discharge while the unit is operating, thus avoiding any need for manual cleaning and its consequent additional labor costs.

The centrifuge can be used for either degumming or refining. In either case, the feed product enters the bowl at the bottom of a hollow driven spindle with a sealing device. The flow moves upward through the rotating spindle into the separator bowl. The light oil, the heavy immiscible liquid, and any solids are separated, with both the light and the heavy phases discharging under operating pressure at the top of the bowl through mechanical seals. The solids, such as meal fines, are accumulated on the inside of the bowl in the sludge space and are discharged intermittently through a series of slots in the bowl wall. When it is necessary to discharge, the sliding bowl bottom is forced downward by the liquid hydraulic pressure, the discharge slots are exposed, and the accumulated solids or sludge are discharged quickly.

The operating water, which controls the opening and closing of the sliding bowl bottom, is supplied from an outside separate water source to a pressure reducing valve. By regulating the water supply pressure, the size of the shot or discharge and time interval are controlled at the desired range. As a result of the tremendous hydraulic force exerted downward by the liquid in the bowl, this discharges rapidly when the bowl is open. This operation of opening, cleaning, and closing takes only 3–4 sec, according to reports from the manufacturer. In the meantime, the centrifuges are still operating at normal speeds.

The neutral zone diameter, that is, the zone between the two phases, depends primarily on the gravity difference between the two phases and the diameter of the heavy phase outlet. The change of this zone in a hermetic centrifuge, such as the SRPX-317, is easily accomplished by varying the pressure on a control valve in the oil outlet. An increase in the back pressure on the outgoing oil moves the zone upward toward the periphery of the bowl and will give a cleaner oil phase. It will contain a lower content of soap or gums. This adjustment in pressure can be made while the separator is operating. By tracking the differential pressure between the inlet and oil discharge pressure, the bowl can be self-monitoring, discharging the accumulated sludge as required to maintain operating efficiency.

In addition to the common hollow-bowl and disc-bowl separators, a new type of centrifugal separator (Figure 4.5) has been successfully used in degumming, refining, and water washing (65a). It consists essentially of a horizontally-positioned rotor, mounted on a shaft with heavy-duty ball bearings, force-feed lubricated. The machine is constructed of stainless steel with heavy welded steel base and rotor cover. Inside the rotor are many contacting elements that provide for intimate mixing, coupled with low liquid velocity and controlled settling rates. The main contacting elements are claimed to provide several times as much coalescing surface per unit volume processed as do conventional machines. The heavy and light liquids are kept separate by pressure-balanced mechanical seals, two on each side of the rotor. Design factors reportedly allow retention of the advantageous characteristics of centrifugal contactors: a high ratio of holding time with low flow rate, a high ratio of travel time to settling time, and effective operation with relatively low gravities by machines of any practicable size.

The operation of the machine when used as a countercurrent liquid contactor is as follows (Figure 4.5): The light liquid enters through the shaft and is directed to the outer periphery of the rotor. The heavy liquid entering through the shaft at the opposite end is directed to the center of the rotor. Centrifugal force moves the heavy liquid outward and displaces

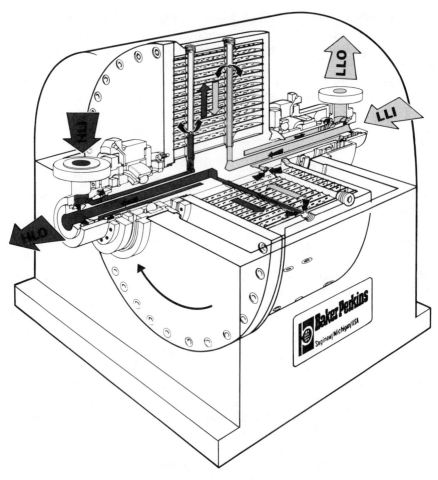

Figure 4.5   Cutaway view of Podbielniak contactor. (Courtesy of Baker-Perkins Inc., Saginaw, Michigan.)

the light liquid towards the center of the rotor. The contacting elements are so designed that passage through the orifices provides multistage mixing and separation.

When a semisolid, such as soapstock, is processed, it coalesces on the inner surfaces of the contacting elements; as it increases in amount it moves to the outer periphery of the rotor where it is deposited on the V-shaped annulus, which directs it to the spill-over sides of the rotor. From there it moves to the shaft and is discharged through a seal-shaft annulus.

The light liquid moves through the contacting elements toward the shaft and is discharged. A back-pressure regulator automatically maintains a pressure on the light liquid attempting to leave. The effect of this pressure is to control the position of the interface (comparable to the ringdam in an ordinary centrifuge), separating the light and heavy liquid, that is, oil and soapstock in the rotor. The position of the interface can be varied at will while the machine is operating. It is claimed that one unit can handle degumming separations, simultaneous rerefining and water washing, or simultaneous refining and water washing of degummed oil. It also finds use in the acidulation of soapstock (65b).

THE REFINING PROCESS.   A schematic view of a continuous refining system is shown in Figure 4.6. In general, the process consists of continuously mixing crude oil with a dilute caustic soda solution and then heating to obtain a "break" in the emulsion. Continuous centrifuges then separate the soapstock from the refined oil, which is now mixed with hot water ("water washing") and again centrifuged. In this way the oil is separated from the water phase, which contains small amounts of soap and other undesirable impurities. The water-washed refined oil, containing traces of moisture, is sent to a vacuum dryer and finally to storage.

In more detail the process may be described as follows (Figure 4.6) (66): Crude oil is commonly stored in a large feed tank holding about a day's run to obtain smooth continuous operation. When operating conditions are determined and perhaps modified after a short time, the remainder of the feed is processed essentially without change. Oil and caustic are accurately proportioned as, for example, using ratio controllers for a smooth nonpulsating flow of product to the mixers. Commonly, 3–4 horizontal compartmented mixers or vertical disc and doughnut mixers are connected in series, with provisions for bypassing one or more depending on the oil being processed, thus assuring flexibility, since oils vary in their mixing requirements.

The oil–caustic mixture must now be heated to assist in getting a good "break." Close temperature control is essential. The hot mix goes to a primary centrifuge that separates the refined oil from the soapstock. At this point the oil will carry a small amount of impurities that may be observed through a sight glass or by using a turbidity instrument. A high-speed test tube centrifuge may also be used as a control (67). Sullivan suggests that solids in the refined oil be maintained at about 0.1 plus or minus 0.05%.

The refined oil is next heated to 165–180°F and mixed with about 10–20% of water at 190–200°F. The mixer used will vary according to

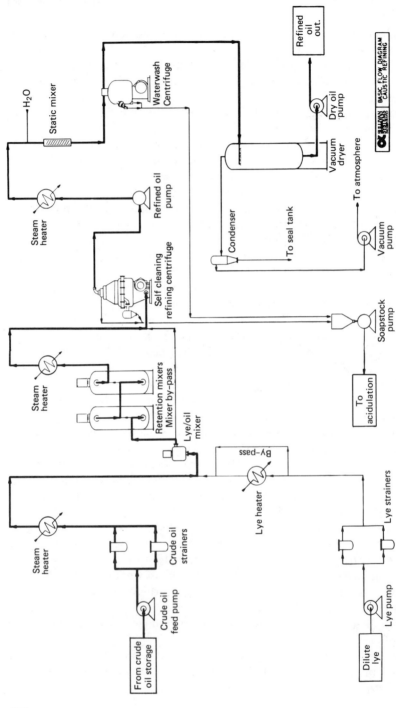

Figure 4.6 Basic flow diagram of caustic refining. (Courtesy of Sullivan Systems, Inc., a subsidiary of DeLaval Separator Company, Tiburon, California.)

284

the refiner's choice, but intimate mixing is essential. Water-washing centrifuges separate the oil as a light phase and the water–soap solution along with insolubles as the heavy phase. Single or double water washing may be used. One wash will ordinarily remove 90% of the soap.

Water for water washing should be softened and free of copper and iron. The washed oil is dried in a continuous vacuum dryer (27–28 in Hg). A common type comprises a vertical chamber with spray nozzles mounted in the upper section for discharging oil into the evacuated unit. Oil leaving the dryer will normally contain less than 0.1% moisture, frequently 0.05%. Soap, as sodium oleate, should not exceed 50 ppm (68).

MISCELLA REFINING.   A unique development (69, 70) is the refining of crude oil at solvent extraction plants before the solvent is removed, that is, while the oil is still in the miscella form. In this procedure miscella, instead of crude oil, is intimately mixed with caustic soda solution to neutralize the free fatty acids, coagulate the phosphatides, and remove the bulk of the coloring matter. A high-speed centrifuge separates the refined miscella from the soapstock. Some of the advantages of miscella refining are (a) reduced refining loss, (b) lighter-colored refined oil, and (c) elimination of the water-washing step.

Miscella refining is not without its disadvantages, however. These include:

1. All equipment must be totally enclosed and explosionproof. This increases investment considerably.
2. Equipment must be carefully maintained and operated to avoid excessive solvent loss.
3. Refining must be carried out at the solvent mills to be effective and economical, and these mills must have a ready market for the refined oil that they produce.
4. There are difficulties in obtaining efficient contact between the caustic soda solution and the miscella—full coagulation of the phosphatides and satisfactory decolorization do not occur in the course of ordinary mixing. Remedies include the use of a homogenizer similar to that used for homogenization of milk (71) and the addition of small amounts of chemical additives (72–78).
5. Neutralization and decolorization appear to be most effective when the concentration of the miscella is about 50% oil. Thus removal of solvent from miscella must take place in two stages, a preliminary concentration to about 50% oil, followed by refining, and a final removal of solvent. An integrated refining process for cottonseed oil

that involves going from crude oil to salad oil while still in the miscella form has been described by Cavanagh (79).

OTHER ALKALI REFINING METHODS

1. In the *Laval short-mix process* (80), introduced in Europe, where the relatively high acidity of many oils made it necessary to avoid the long contact time and the large excess of caustic used in the straight caustic process, oil is heated to 75–90°C prior to the addition of caustic soda, causing an immediate "break" between the oil and the soapstock, thus reducing emulsification losses. Likewise, saponification losses are reduced by the short contact time (about 30 sec) between the oil and the caustic. In normal European practice solvent extracted oil is degummed before refining, and a pretreatment with phosphoric acid is used to remove so-called "nonhydratable phosphatides" that are resistant to water degumming. This permits refining with smaller excesses of caustic than normally used in the straight caustic process. With cottonseed and some other oils of dark color, a second caustic treat may be used, but losses are relatively small since the bulk of the soapstock has been removed. By the phosphoric acid treatment, the natural calcium and magnesium content of oils tends to be precipitated as insoluble phosphates of high density. Here a self-cleaning centrifuge that can discharge solids intermittently is advantageous.

2. In the *ultra-short-mix process* (80), useful for oils of high acidity (e.g., palm oil), caustic soda is introduced directly into the hollow spindle of the centrifuge where a special mixing device is located. The very short contact time permits use of stronger caustic without excessive saponification.

3. In the *Zenith process* (81), a European method that involves a pretreatment with phosphoric acid, followed by sludge removal if necessary, the oil is neutralized as it rises as droplets through a column of caustic. Weak alkali is used, thus keeping the soap in solution and greatly reducing emulsion formation and saponification of neutral oil. Water washing of the neutralized oil is unnecessary. Low losses are claimed.

4. In the *sodium carbonate process* (82–86), since carbonate does not attack neutral glycerides, refining losses should be improved if free fatty acids are first neutralized by carbonate and then a light caustic treat used for color, thus reducing saponification. Several variations have evolved, designed to correct rather serious problems such as

the liberation of carbon dioxide, causing frothing, and making soapstock separation difficult. Although this method was rather widely
used in the United States at one time, the disadvantages of this refining have so outweighed the advantages that it is not used to any
extent at present.

REREFINING.   The dark-colored crude oils obtained from damaged vegetable oil seeds can be converted to a reasonably light color more satisfactorily by means of a double refining than by employing an excessive
amount of lye in a single refining. Off-grade cottonseed oils are usually
refined first with a normal lye containing about 0.60–0.80% excess sodium
hydroxide and, before bleaching, are rerefined with a smaller percentage
of lye that is usually stronger than that employed originally. For some
reason, a better reduction in color is obtained in many cases if the oil is
stored for some time before rerefining than if it is rerefined immediately.

Refiners differ in their preferred method for rerefining, using a variety
of alkalies. In batch refining, rerefining is practiced before the water-
washing stage. Likewise, in continuous refining an additional rerefining
stage is normally introduced between the neutralizing and water-washing
stage.

DETERMINATION OF REFINING LOSS.   Because of the economical importance of refining loss, processors have always tried to measure their loss
as frequently as feasible, with varying success. Weighing crude oil in and
refined oil out is an obvious choice. It is best when large volumes of the
same oil are handled on a daily basis with no change in processing. In
the total fat method the soapstock is analyzed for free fatty acid and
neutral oil. If the nonglyceride components of the original crude are
known, one can calculate whether any additional loss of oil in soapstock,
over theory, is occurring, as could result from saponification of neutral
oil.

A later development was the sodium balance method (87). This is based
on the premise that all sodium in the feed appears as sodium in the
soapstock and in the refined oil. By analyzing these three components,
the percent refining loss can be calculated from the formula:

$$\% \, RL = \frac{\% \, Treat \, (\% \, Na_r - \% \, Na_s) - 100(\% \, Na_o)}{\% \, Na_s - \% \, Na_o}$$

where RL is the refining loss, $\%$ Treat is the percent reagent used per
weight of crude oil, $\%$ Na_r is the percent sodium in reagent, $\%$ Na_s is the

percent sodium in soapstock, and % Na_o is the percent sodium in refined oil. This method requires only 1–4 oz of sample of each product. No flow data are needed. The method is rapid and appears to be as reliable as the methods described earlier.

Electronic efficiency systems represent the latest method for determining refining loss (88, 89). These systems are based on a simultaneous volumetric comparison of an inflow–outflow meter by means of positive displacement meters. Compensation is made for temperature and moisture. Refining loss is displayed continuously as percent loss or gain. It can also be recorded permanently. A major advantage of these continuous reading systems is that the attention of the plant personnel is focused on refining loss and the effect of processing changes becomes immediately apparent.

4.1   STEAM REFINING

Conventional oil refining usually involves the basic steps of free fatty acid removal, bleaching for reduction in pigment control, and final steam deodorization to produce a bland, stable, finished, edible oil. Although caustic soda is very satisfactory for deacidification, with certain high free fatty acid crude oils there is excessive loss of neutral oil by this treatment. Steam distillation, on the other hand, removes essentially only the free fatty acid, thus appreciably lowering overall refining loss. This technique was suggested by Hefter over 70 years ago (90) and has been used in Europe for certain high free fatty acid oils for many years. Normally, however, distillation was used to reduce the free fatty acid to about 0.2 to 0.5% with conventional alkali refining completing the acid removal.

To be successful, steam refining must be practiced on oils naturally very low in phosphatides or from which phosphatides have been removed. Moreover, the oils cannot be heat sensitive as, for example, cottonseed, which becomes dark on heating and resistant to subsequent color reduction.

With the increased production of lauric acid and palm oils, steam refining has gained new prominence. A plant in the western United States is reportedly now in production of a high-quality deodorized palm oil using the steam refining–deodorization technique (91). An acid pretreatment is used to prepare the oil for final distillation and deodorization.

Various treatments have been proposed to remove phosphatides and other undesirable materials from oil prior to steam distillation (92–95). At least in the case of soybean oil (96) there are socalled "nonhydratable phosphatides" that are not removed in the normal degumming process and that appear to be quite resistant to any conventional treatment except alkali refining. A plant in Israel has reportedly processed soybean oil by steam refining, but this has not been done in the United States as far as the author is aware.

Recently, a process has been patented (97) in which oil is treated with phosphoric acid and then, without removing the acid, bleaching earth is added and the mixture heated to a relatively high temperature (325–500°F). The mixture is then cooled and filtered prior to deodorization. Results vary with the amount and the type of bleaching earth used. Failure to completely remove the phosphoric acid can result in oils with a "melon" flavor. At least one plant designed along these general lines has been sold, and other proposals are being offered.

4.2   LIQUID–LIQUID EXTRACTION

Extraction of free fatty acids from oils with alcohols or other solvents having a greater affinity for free acids than for glycerides has often been proposed (98). Two of the better-known methods are the furfural and propane processes. The furfural process has been used commercially by the Pittsburgh Plate Glass Company (99) on linseed, soybean, tall, and other oils. In this method a furfural solution of oil is allowed to separate into two layers—a bottom solvent-predominant layer containing a concentration of unsaturated glycerides (extract phase) and an upper oil-predominant fraction containing a concentration of saturated glycerides (raffinate). An iodine number spread of 30–60 units may be obtained, and in this way soybean oil, for example, is given more "drying" properties. Thus the method is designed more for glyceride fractionation than for removing impurities from crude oil.

The propane extraction (Solexol) process (100–102) makes use of the fact that pigments containing many active groups are less soluble in propane than are glycerides, which have relatively fewer active groups. Since the solubility of the oil fractions changes with the temperature, the amount of dissolved oil and the corresponding amount of insoluble phase can be varied as desired by suitable adjustment of the temperature. The principal disadvantage of this method is probably the very high initial investment cost connected with high-pressure solvent equipment. As a refining method, this process is particularly advantageous with inedible tallows and greases, field-damaged vegetable oils, or other very dark oils that are difficult or impossible to bleach effectively by older methods. At the present time, however, there does not appear to have been the quality and yield advantages to liquid–liquid extraction sufficient to justify the high investment and operating costs when compared with the more conventional alkali refining methods.

4.3 REDUCTION OF ACIDITY BY REESTERIFICATION

The patent literature contains a number of proposals for reducing the free fatty acid content of oils by esterification, that is, by reacting the oil with glycerol or monoglycerides to convert the free acids back to the triglycerides (103–106). This method has apparently found limited commercial application in Europe. It appears useful only for highly acid oils, since glycerol does not react with free fatty acids quantitatively. Moreover, esterification must be followed by other treatment to decompose mono- and diglycerides and to remove excess combined glycerol, if this reagent is used in excess, or to remove a residue of free fatty acids, if insufficient glycerol is used for complete esterification. As elevated temperatures are required for the reaction, the oil to be treated must presumably be free of gums. In addition, glyceride structure may be affected since the glyceride structure produced is purely random when glycerides are formed by esterification.

Treatment and Disposal of Soapstock

The soapstock from alkali refining is a source of valuable fatty acids, but it also poses a disposal problem. With the greatly increased use of detergents over soaps, there is little use for soapstock as is. Normally, it is acidulated to produce free fatty acids. These are used as a high-energy ingredient in feeds or for chemical uses. For the latter, competition from tall oil is of major importance. In 1952, for example, no crude tall oil was used as a raw material for fatty acids (107), but in recent years over a billion pounds per year have been produced. Production of tall oil, a byproduct in paper manufacture, now appears to be leveling off as a result, at least in part, of the increased use of hardwood chips and recycled material in the manufacture of paper.

The value of soapstock is determined by its total (combined and free) fatty acids content, and this varies considerably according to the method of refining. Soapstock from batch refining seldom falls below 40% in total fatty acids and often runs as high as 50%. The product from continuous caustic soda refining is usually between 35 and 40% in total fatty acids. Soapstock is almost always shipped from the refinery either in a raw form as it comes from the refining equipment or in the concentrated form known as "acidulated" soapstock. Raw soapstock contains sufficient water to ferment readily. The official rules of the National Cottonseed Products Association (55) and the National Soybean Processors Association (50) provide for trading in the two commodities on the basis of a total fatty acids content of 50 and 95% for raw and acidulated soapstock, respectively. Shipments of acidulated soapstock may be rejected if the total fatty acids content falls below 85%; the corresponding minimum for raw soapstock is 35% under the National Cottonseed Products Association rules and 30% under the National Soybean Processors Association rules.

The composition of acidulated soapstock depends on the proportion of unsaponified oil and free caustic in the raw material. Usually, there is sufficient free alkali to saponify most of the neutral oil present so that the fatty acids are largely converted to the free form on acidification. The free fatty acids content of an acidulated soapstock containing 95% total fatty acids is seldom below 80% and is often 85–90% or more. If care is taken to have an excess of alkali and the charge is well boiled before the acid is added, virtually all the fatty acids may be liberated. To ensure a high total fatty acids content in acidulated soapstock, many processors pretreat the raw soapstock with more alkali, so as to saponify any remaining neutral oil, before going to the acidulation step.

In a simple batch system, soapstock is pumped from the refining area into large wooden, plastic, or lead-lined vats, where it is diluted with water and then contacted with sulfuric acid until the mixture contains free acid. It is then boiled with steam until the emulsion is broken and the soap split. On allowing to settle, the oily material rises to the top and the acid water (containing free sulfuric acid, sodium sulfate, and water-soluble impurities) forms a lower layer. This aqueous layer is drawn off and the oily layer boiled with fresh water to wash out residual sulfuric acid and other water-soluble matter. Further treatment of the aqueous washings is usually necessary to neutralize the mineral acid and to comply with existing environmental regulations.

Continuous acidulation is an obvious improvement if volume considerations make it economical. Keith et al. (108) described a continuous acidulation system using spent mineral acid at 175–195°F in the presence of a dispersing agent. A preliminary purification of the soapstock before acidulation was accomplished by treating with strong caustic, graining, and washing. Later, Crauer (109) described the continuous acidulation of diluted soapstock featuring control of the sulfuric acid by pH cells in the reaction mixture and a heat exchanger after the reaction mixture to cool the mix before centrifugation. Complete systems are now offered by the DeLaval Separator Company in Poughkeepsie, New York, and by the Sharples Division of the Pennwalt Corporation in Oak Brook, Illinois.

Crauer has also published further on a continuous treatment of refinery waste waters (110).

Bleaching

5.1 COLOR STANDARDS

American refiners usually determine the colors of the lighter refined and bleached oils, and also of shortenings and other oil and fat products, by matching in a suitable tintometer a 5.25-in column of the melted fat against red and yellow Lovibond color glasses. The red glasses are standardized by the National Bureau of Standards in terms of the Priest-Gibson N″ color scale (111). The N″ scale approximates, but does not exactly follow, the scale adopted by the manufacturers of the glasses (112); the latter varies slightly from one set of glasses to the other. Yellow glasses are not standardized since relatively large variations in yellow are imperceptible to the eye; in matching the color of a sample of oil, it is necessary to only approximate the yellow color to obtain a satisfactory match with the red glasses.

For most purposes, and for most oils, the depth of color of the fat or oil is satisfactorily expressed in terms of red units, according to the preceding scheme. The Lovibond system of color measurement is unsuitable, however, for oils that are excessively dark or that contain colored substances other than red and yellow in considerable concentration. Among the vegetable oils that may contain colored substances other than red and yellow are olive oil, which generally has a greenish cast due to the presence of chlorophyll, and soybean oil, which may also contain considerable chlorophyll if expressed from green beans. Oils obtained from damaged oil seeds are often brown after refining and hence are difficult or impossible to match with Lovibond glasses.

Because of these inadequacies of the Lovibond system and because of difficulties in obtaining suitable Lovibond glasses, American refiners have in recent years considered using the spectrophotometric evaluation of oil colors for both research and plant control work. Actually, the spectral transmission of an oil at 525–550 nm is not only more reproducible than Lovibond red readings but can also, in most oils, be closely correlated with the latter. In 1950 a spectrophotometric method for the determination of oil color was adopted as a tentative method for refined and bleached vegetable oils by the American Oil Chemists' Society (113). It provided for a determination of the optical density of the oil at 460, 550, 620, and 670 nm in a 21.8-mm cell, with the following equation being used for calculation of a photometric color that approximates color expressed in Lovibond red units:

$$\text{Photometric color} = 1.29A_{460} + 69.7A_{550} + 41.2A_{620} - 56.4A_{670}$$

where A is the absorbance.

This method has not replaced the Lovibond method for general use, however, partly because the older system is firmly entrenched and because color is intimately related to the economic value of the oil. Also, in a number of cases, the photometric method does not agree with Lovibond readings (114); however, the method is now official (41).

Inedible tallows and greases are often too dark or too strong in green or other off colors to be graded in terms of the Lovibond system, even when a short column (e.g., 1.25 in) of oil is used. For the approximate evaluation of the colors of such fats, the Fat Analysis Committee of the American Oil Chemists' Society has provided an arbitrary system of color standards (115) consisting of sealed vials of solutions of various inorganic salts. The color of any fat in question is determined by comparing a melted sample with the various standards until the nearest match is obtained. The complete set of standards consists of 26 vials, numbered from 1 to 45, in odd numbers, and divided into five series. Numbers 1 to 9, inclusive, are prepared for the grading of light-colored fats; numbers 11, 11A, 11B, and 11C are for very yellow fats; numbers 13 to 19 are for dark, reddish fats; numbers 21 to 29 are for greenish fats; and numbers 31 to 45 are for very dark fats. Spectral data on the various standards have been published by Urbain and Roschen (116).

The so-called FAC color system is somewhat confusing, inasmuch as the different series of standards are to some degree independent; thus there is no orderly increase in color from the lowest- to the highest-numbered members of the set. Some of the standards numbered from 21 to 29, for example, may actually be lighter in color than standards numbered from 13 to 19. In addition, fats are often encountered that do not match any of the standards. The FAC colors are, of course, not additive.

Drying oils are commonly graded for color according to the Lovibond system or by matching with the 1933 Gardner color standards (117), which likewise consist of solutions of inorganic salts and are numbered from 1 (the lightest) up to 18. The best pale commercial oils or bodied oils ordinarily have Gardner colors in the range of about 3–6.

Roughly, an FAC color of 1 is equivalent to a Gardner color of 3, which is equivalent to a Lovibond color (5.25-in column) of 27 yellow, 2.7 red. Other approximate equivalents among the lighter colors are as follows: FAC No. 3 equals Gardner No. 5 equals Lovibond 70 yellow, 7.0 red;

FAC No. 5 equals Gardner No. 7 equals Lovibond 100 yellow, 10.5 red; and FAC No. 7 equals Gardner No. 9.

Certain grades of salad or cooking oils are purposely made rather dark, but the oil used in almost all other edible products is invariably reduced to a Lovibond red color of about 2.5 or below. Certain high-grade shortening products run consistently below 1.0 unit in red color; however, somewhat higher bleach colors are permissible as considerable color reduction takes place in hydrogenation and deodorization. In general, the edible oil processor is concerned with bleaching to reduce the Lovibond red color of refined oils from about 4.0–9.0 to about 1.5–2.5 units, and at the same time remove green pigments almost completely if they are present. Soapmakers' fats are generally quite dark in color and require more drastic bleaching than do edible oils. The standards for bleaching vary greatly according to the product and its method of manufacture. In the manufacture of the better white soaps, color standards approach those of the edible industry; that is, Lovibond red colors in the range of 5–6 units are required.

5.2   BLEACHING BY ADSORPTION

Adsorbents.   The most important adsorbent used in bleaching fats and oils is bleaching earth or clay. Natural bleaching earth, otherwise known as *fuller's earth* from its ancient use in the "fulling" or the scouring of wool, comprises various earths or clays consisting basically of a hydrated aluminum silicate. The mineralogical characteristics of these earths have been discussed by Kerr (118) and Nutting (119). Attempts have been made to correlate the chemical composition of earths with their bleaching ability, but without success. An earth almost devoid of adsorptive capacity may be almost identical in composition with very active earth; hence these materials can be evaluated only by actual tests. The earths used for bleaching fatty oils are the same as those used in a slightly different form and in much greater volume for the bleaching of petroleum products.

Within recent years natural bleaching earths for use with fats and oils have been supplanted to a considerable degree by acidactivated clays. The raw materials used for the manufacture of this type of bleaching clay consist for the most part of bentonites or montmorillonite, which have little or no decolorizing power in the raw state (120). By treatment with sulfuric or hydrochloric acid, however, the surface of the clay is so altered that its bleaching power will in most cases considerably exceed that of natural clays. The acid treatment undoubtedly extends the surface of the clay and probably also causes important changes in its chemical or physicochemical nature. Acid-activated clays retain more oil per unit weight of clay than do natural earths, but their use generally leads to a lower overall loss of oil because they are more active. The apparent densities of natural earth, activated earth, and activated carbon are approximately 50, 30, and 25 lb/ft^3, respectively.

Activated clays are sold in neutral grades for bleaching edible oils and in slightly acid, but more active grades, for difficultly bleachable edible or inedible oils. The former, like the natural earths, do not materially increase the free fatty acid content of the oils on which they are used, whereas the latter hydrolyze the oil slightly, increasing its free acid content by a few hundredths of a percent.

The greatest usefulness of activated clays is in the treatment of off-grade oils. Certain types of color are extremely difficult to remove except by activated clays. For example, as mentioned previously, the green color due to chlorophyll in some soybean oils is much more responsive to a slightly acid earth than one of the ordinary type because the pigment is unstable under acid conditions.

Besides bleaching clay, the only adsorbent used to any extent on fatty oils is activated carbon. Because of its relatively high cost and its very high oil retention, carbon is rarely used alone on most vegetable oils, but oil refiners frequently employ it in admixture with bleaching clay, in a ratio of about 10–20 parts by weight of clay to 1 part of carbon. Such a mixture is often considerably more effective than bleaching clay alone.

Carbon is also a superior adsorbent for traces of soap in refined oils. It is particularly effective in removing the red, blue, and green pigments of coconut and palm kernel oils and the better grades of animal fats, and it is popular for use in connection with diatomaceous earth in clarifying and mildly bleaching lard (121, 122). Unlike bleaching earths, carbon imparts no foreign flavor or odor to the oil treated.

In bleaching most oils, the cost of the adsorbent is exceeded by that of the oil lost by retention in the spent adsorbent. This oil is difficult to recover and, after recovery, is usually badly oxidized and of poor quality; hence many refineries discard their spent earth without treatment. The retentiveness of an adsorbent is to some degree proportional to its activity since both properties are related to the nature and the extent of the adsorbing surface. The less active fuller's earths may retain up to 30% of their own weight of oil, but acid-activated earths usually have a retention of up to 70% (123). With drying oils, which oxidize and polymerize more readily, the retention is normally higher.

Because of its very porous nature, carbon retains much greater amounts of oil than do any of the clays, and the addition of even 5 or 10% of carbon to a bleaching clay will materially increase the oil retention of the latter. The choice of an adsorbent depends in most cases on striking a balance among the three factors of cost, activity, and oil retention (124). Although laboratory bleach tests and retention tests will give some indication of the value of a bleaching earth or carbon, it is necessary to resort to tests on a large scale to evaluate these materials accurately. The uncertainty in applying laboratory bleaching data to plant operations is in marked contrast to the utility of laboratory refining tests that furnish a very reliable indication of the behavior of an oil in the plant.

The amount of adsorbent required for any given bleaching operation will vary greatly with the activity and the nature of the adsorbent, the variety of the oil, the color of the unbleached oil, and the color desired in the bleached oil. In general, however, the range of clay used in the United States is 0.15–3% (125). Only in extreme cases are higher amounts employed; amounts well under 1% are common. Small amounts of activated carbon may be added to the bleaching clay, where chlorophyll pigments impart a greenish color to the oil.

Theory of Adsorption Bleaching: General Considerations.   Bleaching of oils by adsorption involves the removal of pigments that are either dissolved in the oil or present in the form of colloidally dispersed particles. From the standpoint of adsorption theory, it is immaterial whether the pigments are dissolved or merely dispersed.

The mechanics of adsorption are somewhat controversial; opinions differ as to the extent to which adsorption is a physical and/or a chemical phenomenon. The bond of attraction between adsorbent and color body is relatively weak as evidenced by the fact that the coloring matter can be readily removed from earth used in laboratory bleaching by extraction with acetone, isopropyl alcohol, or benzene at room temperature (126). Further, the extracted clay can be used again to bleach more oil with virtually the same adsorptive capacity it had originally. These observations would indicate that the adsorption mechanism is probably physical.

In any event, it is sufficient to recall that adsorption is a surface phenomenon, depending on a specific affinity between the solute and the adsorbent. The mathematical expression relating adsorption to residual solute concentration at a single temperature was developed by Freundlich (127):

$$\frac{x}{m} = Kc^n$$

where x is the amount of substance adsorbed, m is the amount of adsorbent, c is the amount of residual substance, and K and n are constants.

The equation may also be written in the form

$$\log \frac{x}{m} = \log K + n \log c$$

Thus a plot of x/m versus c on a log–log scale will produce an adsorption isotherm that is a straight line with a slope equal to n, and x/m will be equal to K when c equals 1.

The Freundlich equation is valid for any method of color measurement as long as the units of measurement are additive and proportional to the actual concentration of coloring materials in the oil. With fatty oils, either the Lovibond red color or the optical density at a specific wavelength conforms to this requirement. Ordinarily (and throughout the following discussion) the unit weight of adsorbent (the quantity m) is taken as 1 part per 100 parts of oil. With the concentration of adsorbent so expressed, the adsorption isotherm and the values of K and N are independent of the units used for measuring color or pigment concentration.

Results of a typical test to determine the bleach colors produced by different percentages of an adsorbent are shown in Figure 4.7. In this test the oil was a refined cottonseed oil with an initial red Lovibond color of 8.0 units, and the adsorbent was an acid-activated bleaching clay. The adsorption isotherm calculated from the results of the same test is shown in Figure 4.8. The value of K for this particular test was 1.14, and the value of n was 0.84.

From a practical standpoint, K is a general measure of the activity or decolorizing power of the adsorbent, whereas n is an indication of its

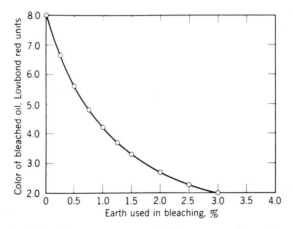

Figure 4.7   Typical bleaching test on cottonseed oil (percent earth used versus bleach color of the oil).

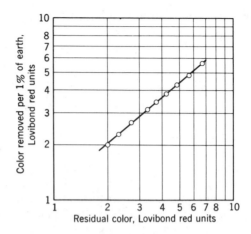

Figure 4.8 Typical adsorption isotherm for the bleaching of cottonseed oil (same test as shown in Figure 4.7).

characteristic manner of adsorption. If two adsorbents have different capacities for adsorbing color but adsorb in the same characteristic way, that is, if they exhibit different values for K but the same value for n, the relative amounts required to effect a given degree of decolorization will be in inverse proportion to the values of K. For example, if the following equation is found to apply to earth A:

$$\frac{x}{m} = 0.5c^{0.5}$$

and the following equation to earth B:

$$\frac{x}{m} = 1.0c^{0.5}$$

bleaching with earth A will require twice as much earth as bleaching with earth B, and this relationship will hold at any level to which decolorization may be carried.

The value of n determines the range of decolorization within which the adsorbent exhibits its greatest relative effect. If n is high, the adsorbent will be relatively effective in removing the first portions of color from the oil but relatively inefficient as an agent for effecting a very high degree of decolorization. If n is low, the reverse is true.

The hypothetical bleaching curves for adsorbents yielding different combinations of K and n (Figure 4.9) illustrate the principles mentioned previously. Under all circumstances it is desirable to use an adsorbent that gives a high value of K, corresponding to a low position of the bleaching curve in Figure 4.9. In general, a high value of n is also desirable, although not at the expense of a high value of K. It is to be particularly noted that an adequate comparison of two adsorbents cannot always be made without specifying the level to which decolorization is to be carried. In Figure 4.9, for example, adsorbent C is superior to B in bleaching down to a color of 2.1 units, but below this color level B is superior to C.

In the bleaching of fats and oils, values of both K and n vary widely according to the adsorbent, the particular oil bleached, and the method of bleaching. Values recorded by different observers, principally in laboratory tests, are shown in Table 4.4. Ordinarily, bleaching of oil in the plant requires considerably less earth than does bleaching of the same oil in the laboratory. The difference is in the so-called "press bleach", or additional decoloration as the oil passes through the bed of adsorbent retained in the filter press used for earth removal. Although no proof of the nature of press bleaching has ever been given, it may be presumed to be a concentration effect leading to a new equilibrium between adsorbent and pigments in the zone where the effective concentration of adsorbent is very high in relation to the oil.

It is evident from the Freundlich equation that bleaching clay or carbon that has reached equilibrium with respect to the coloring matter in a light oil will still have adsorptive capacity for the color in a darker oil. In other

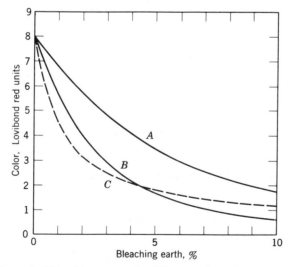

Figure 4.9 Theoretical bleaching curves for different values of K and n: (a) $K = 0.5$, $n = 0.5$; (b) $K = 1.0$, $n = 0.5$; (c) $K = 0.6$, $n = 1.2$.

Table 4.4 Bleaching of fatty oils by means of adsorbents; values for K and n in the Freundlich equation as reported by different observers[a]

Observer	Oil	Bleaching Agent	Bleaching Conditions	Method of Color Measurement[b]	K	n
Hassler and	Cottonseed	Natural earth	Lab. atm.	A	0.6	0.45
Hagberg	Cottonseed	Carbon	Lab. atm.	A	0.20	2.2
(128)	Coconut	Natural earth	Lab. atm.	A	0.5	1.3
	Coconut	Carbon	Lab. atm.	A	7.2	1.8
Bailey[c]	Cottonseed	Natural earth[d]	Lab. atm.	A	0.6	0.4
	Cottonseed	Activated earth	Lab. atm.	A	1.14	0.84
	Cottonseed	Activated earth	Lab. atm.	A	0.90	0.34
	Cottonseed	Activated earth	Plant atm.	A	1.6	0.73
King and	Cottonseed	Natural earth	Lab. atm.	A	2.0	0.42
Wharton	Cottonseed	Natural earth	Lab. vac.	A	3.3	0.39
(130)	Cottonseed	Activated earth	Lab. vac.	A	4.0	0.39
	Soybean	Natural earth	Lab. vac.	A	2.3	0.36
Hiners et al.[e]	Soybean	Activated earth	Lab. vac.	B[f]	0.25	0.33
	Soybean	Activated earth	Lab. vac.	B[f]	0.58	0.33
	Soybean	Activated earth	Lab. vac.	B[f]	1.10	0.33
Sierra talc[g]	Tallow	Activated earth	Lab. atm.	C[h]	0.66	0.77
	Tallow	Activated earth	Lab. atm.	D[i]	0.85	0.80
Stout et al.	Cottonseed	Natural earth	Lab. atm.	E[h]	0.29	2.21
(134)	Cottonseed	Activated earth	Lab. atm.	E[h]	0.45	2.16
	Cottonseed	Mg silicate	Lab. atm.	E[h]	0.10	4.00
	Soybean	Natural earth	Lab. atm.	E[h]	1.00	1.21
	Soybean	Activated earth	Lab. atm.	E[h]	3.12	1.48
	Soybean	Mg silicate	Lab. atm.	E[h]	1.26	1.70

[a] Concentration of earth expressed as parts by weight per 100 parts of oil.
[b] A = Lovibond red units; B = spectral, 660 mμ; C = spectral, 470 mμ; D = spectral, 520 mμ; and E = spectral, 475 mμ.
[c] A. E. Bailey, unpublished data.
[d] Official bleaching earth of American Oil Chemists' Society.
[e] H. F. Hinners, J. J. McCarthy, and R. E. Bass, Oil Soap, **23,** 22–25 (1946).
[f] Measure of green color or chlorophyll content.
[g] Sierra Talc & Clay Company, Los Angeles, CA, Leaflet AF-1, 1948.
[h] Measure of yellow color.
[i] Measure of red color.

words, the greatest efficiency of bleaching will theoretically be attained in countercurrent operation in a system wherein fresh oil is treated only with used adsorbent and fresh adsorbent is brought into contact only with oil that has been partially bleached. True continuous countercurrent bleaching can scarcely be obtained with the adsorbent in the form of a powder, but it is possible to conduct the operation in a multiplicity of stages, with the adsorbent being moved from one stage to the other countercurrent to the movement of oil. With an infinite number of stages, the effect is, of course, equivalent to that of continuous countercurrent flow. Actually, the theoretical benefits are considerable even when the number of stages does not exceed two or three.

Results obtained in the batch bleaching of certain vegetable oils in two countercurrent stages have been reported by Hassler and Hagberg (128). In spite of its potential advantages, batch countercurrent bleaching is seldom practiced principally because of the readiness with which oxidation of retained oil occurs when bleaching earth is exposed to the air in transferring the adsorbent from one stage to another. Countercurrent operation is more feasible in the continuous bleaching of oils within a closed system, and two-stage systems are actually in commercial use. The advantage of countercurrent bleaching depends on the curvature of the bleaching curve (Figure 4.9) or, in other words, on the value of n, being greater the greater the curvature or the higher the value of n.

Unless bleaching is conducted with the rigid exclusion of oxygen, the normal color reduction brought about by the adsorption of pigments may have superimposed on it effects due to oxidation that will considerably modify the results. Reference has been made previously to the darkening of oils from heating accompanied by oxidation. This effect appears to be accelerated in the presence of bleaching earth. On the other hand, during the course of atmospheric bleaching tests of soybean oil in the laboratory (129), the addition of earth before the oil is heated serves to inhibit heat darkening. This apparent anomaly is explained by King and Wharton (130) as a result of oxidative stabilization of pigments against adsorption; in other words, it may be assumed that addition of the earth early in the process enables it to adsorb pigments before oxidation occurs to reduce their affinity for the earth. Their theory of oxidative stabilization of pigments is supported by experiments that demonstrated that oils bleached much better under vacuum than when exposed to the atmosphere, even when they were characteristically "nonreverting" in color, that is, when mild oxidation of the oil alone brought about a reduction rather than an increase in the color.

Obviously, in the presence of oxygen, color changes are complex. Simultaneously, there may be darkening of existing pigments, development of pigments from colorless precursors, destruction of other pigments, modification of pigments to reduce their adsorbability, and adsorption of the different pigments and pigment products to different degrees.

Ordinarily bleaching has little effect on the acidity of oils. Neutral or slightly acid clays often reduce the acidity very slightly through preferential adsorption of the free fatty acids. On the other hand, acid clays may increase the acidity measurably (several hundredths of 1%), especially if moisture or soap is present or if the contact time with the oil is prolonged. Odeen and Slosson (120) suggest that the clay decomposes soap, adsorbing the sodium ion and leaving the fatty acid free. In treating soapy oils they recommend heating the batch to 220°F, allowing it to stand for an hour or two, and then skimming off the surface layer of separated soap before proceeding with the bleaching operation.

Mitchell and Kraybill (131) have demonstrated by means of ultraviolet adsorption spectra that commercial bleaching commonly produces 0.1–0.2% of conjugated fatty acids in the glycerides of cottonseed, corn, soybean, and linseed oils, through the isomerization of nonconjugated fatty acids. Their observation is of some interest in its relation to the stability of bleached oils in view of the readiness with which oxidation occurs in conjugated acids and the autocatalytic nature of the oxidation reaction in fats. Since some degree of prior oxidation was found to be prerequisite to this isomerization, it would seem that it might be maintained at a minimum by deaerating the oil before bleaching and carrying out the bleaching under a vacuum.

Continuous vacuum bleaching should be particularly effective in inhibiting isomerization since the latter presumably requires appreciable time, and time of contact between earth and oil is much shorter in continuous bleaching than in bleaching by the batch system. In comparing plant results with continuous vacuum bleaching versus open-kettle batch bleaching, King and Wharton (132) have reported significantly better flavor stability in soybean oil processed by the former method.

Besides decolorizing, treatment of an alkali-refined oil with bleaching earth serves the important function of largely removing traces of soap. The efficiency of soap removal during bleaching appears to depend on the thoroughness with which the oil and the earth are dehydrated during the operation, as the oil retains soap tenaciously only in the presence of dissolved moisture. For this reason, a low soap content is favored by vacuum bleaching and, particularly, by continuous vacuum bleaching where moisture removal is facilitated by spraying the oil and clay slurry into an evacuated chamber. In a series of comparative plant tests, King and Wharton (132) found that batch atmospheric bleaching reduced the soap in refined oil from an average of 103 to 32 ppm, whereas continuous vacuum bleaching effected an average reduction from 114 to 15 ppm.

The presence of some moisture seems to be essential for effective bleaching action. All bleaching earths contain a substantial amount of bound moisture that is released only at somewhat elevated temperatures. Bleaching earths that have been completely dehydrated by heating to a high temperature are inactive. The presence of moisture in the oil at the beginning of the bleaching operation has no adverse effect. Work by Rich (133) indicates that a substantial improvement in batch decolorization of tallow is obtained by using an activated (but not a natural) adsorbent when water is added to the system.

In general, there is no highly critical temperature for optimum bleaching results, and in most plants bleaching is carried out uniformly at a temperature in the neighborhood of 195–230°F. Some activated earths, however, yield slightly better results at a lower temperature; hence if the operation is carried out under vacuum, so that dehydration of the oil and earth constitutes no problem, temperatures as low as 170–180°F are recommended. In a study of atmospheric bleaching of cottonseed and soybean oil in the laboratory, Stout et al. (134) found that optimum temperatures for the removal of yellow color (475 nm) varied from 212 to 223°F for different activated earths and activated carbon and from 221 to 270°F for different natural earths. A temperature not in excess of 170°F is recommended for the bleaching of red oil (commercial oleic acid) with activated earth. On the other hand, temperatures in the range of 300–350°F are usually used in bleaching palm oil because of the instability of carotene to heat. Magnesium silicate adsorbents that are occasionally used for bleaching fatty oils require much higher temperatures than do ordinary bleaching earths (about 400°F) for optimum effect.

Bleaching adsorbents equilibrate with the pigments in oils quite rapidly with reasonably efficient stirring of the slurry; for all practical purposes, a contact time of 10–15 min is ample. Pigments are adsorbed irreversibly; they are not removed to any large extent even when the spent adsorbent is extracted with a nonpolar solvent, such as petroleum naphtha, although they may be removed with a polar solvent, such as acetone. In commercial operation the spent adsorbent in the form of a cake in the filter press is usually blown for a prolonged period with air and steam to recover as much as possible of the entrained oil. The recovered oil or "press steamings" is a dark-colored, partially oxidized product that cannot be incorporated in the bleached oil but must be rerefined.

Feuge and Janssen (135) have studied solvent (hexane) bleaching. They obtained lighter colors, required less earth, and found efficiency to increase as more dilute solutions were used. Countercurrent column bleaching was done successfully with elution of the pigments by means of acetone so as to reuse the earth. Loss of oil in the earth was minimized since "holdup" was a mixture of hexane plus oil and not oil alone.

Before going on to bleaching procedures some other factors should be mentioned. Bleaching probably removes other material besides pigments and thorough bleaching could improve oil quality in the opinion of some processors. In the past "heat bleach" has often been taken into account so that oil was not bleached to the level required in the deodorized product but to a level that after any heat bleach during deodorization would meet specifications. This saving in bleaching costs may be illusory without controlled tests since very little is known about the nature of many of the minor constituents of oil. Also, when "heat bleach" occurs, particularly in palm oil, there is some question as to the nutritive value of the heated carotenoid pigments (136). Many processors prefer to reduce the initial carotene content by bleaching before relying on any heat treatment for the final decolorization (137).

Finally, when bleaching reduces metal content, this helps flavor stability at the same time by reducing the load on any metal chelating agents used.

Batch Bleaching. The oldest method of bleaching, which is still followed in many plants, involves the use of open cylindrical cone-bottom kettles with mechanical agitators and steam heating coils. Such kettles are preferably not larger than about 60,000 lb in capacity, as it is desirable to complete the separation of earth from the oil reasonably soon after the earth is added. The agitator should be designed to maintain the earth in suspension and provide efficient stirring without splashing or aeration at the surface. Heating should be as rapid as possible, and the total heating period should never exceed about 1 hour.

Most operators add the bleaching earth or mixture of earth and carbon to the kettle in the desired amount somewhat before the top bleaching temperature of about 195–230°F is reached, for example, at 160–180°F. Often the earth is mixed in a concentrated slurry with a portion of the oil in a separate small tank that is placed in a dustproof room or provided with dust-collecting equipment. After heating is completed, agitation is continued for 15–20 min and pumping of the oil through the filter press is started. The first oil through the press is returned to the kettle for clarification and to build up a press cake and attain a maximum "press bleaching effect." After a minimum color is achieved in the recirculating oil, the latter is diverted to bleached oil holding or storage tanks.

The cake of spent earth in the filter press is blown with air and steam to recover as much as possible of the entrained oil. Blowing practices vary in different plants. A common procedure is to blow lightly with air for a few minutes until most of the free oil in the press chambers is displaced and then blow with dry steam for 30–45 min at about 15–45 lb of pressure. It is preferable to use presses that have a discharge into a closed line to avoid blowing a fog of oil particles into the pressroom. The blow line goes into a small closed tank vented to the outside atmosphere; from this tank condensed water is drawn off and the recovered oil is pumped back to the refining plant for reprocessing. When an acid-activated earth is used, the press should be cleaned immediately after blowing to avoid acid injury to cotton press cloths if they are used.

Because of the greater protection afforded the oil against oxidation, batch bleaching is usually conducted under vacuum in the more modern plants. A common vacuum–bleaching vessel has a capacity of about 30,000–40,000 lb. It is cylindrical in form, with dished bottom and cover, equipped with a motor-driven agitator and heating and cooling coils. The agitator, unlike that for open kettles, should be designed to roll the charge and constantly bring fresh material to the surface to assist in deaeration. To provide a larger surface and more splashing at the surface, European processors frequently use horizontal, cylindrical vessels, although they are uncommon in this country. The oil inlet should be designed to splash the oil into the evacuated vessel as the latter is charged. A two-stage steam ejector capable of maintaining a vacuum of 27–28.5 in is used.

Operation of the batch vacuum bleacher does not differ greatly from that of the open bleaching kettle. The bleaching earth should be as free from occluded air as possible. Thus moving earth by compressed air or vacuum and suction followed by air should be avoided. Oil should also be deaerated before the earth is added. Some operators add the adsorbent at the beginning of the heating period; others prefer to have the oil at bleaching temperature (usually 195–230°F) before it is added as dehydration of the charge is thereby facilitated. The earth may be pulled from a hopper into a vessel by vacuum through a 3- to 4-in line, as it will flow almost like a liquid. After the usual 15–20-min period of agitation, the batch is cooled to 160–180°F and filtered as described earlier. Alternately, the oil to be filtered is pumped to an enclosed filter, such as the "Funda" type. This is a tall cylindrical vessel containing a vertical shaft carrying a number of hollow horizontal plates. The oil goes through the wire cloth on the plate surface and flows through the interior of the plate and shaft to a receiving tank. When bleaching earth has accumulated on the surface of the cloth until the gap between the plates has been filled, the central shaft is then rotated mechanically, spinning off the filter cake, which can be removed by a small screw conveyor in the bottom of the vessel. Considerable labor savings are thus possible.

Continuous Bleaching.   Continuous vacuum bleaching protects the oil from the harmful effects of oxidation even more effectively than does batch vacuum bleaching since better deaeration is effected by spraying the oil into a vacuum than can ordinarily be obtained by agitating a large batch under vacuum. Also, the oil and the earth are more completely deaerated, and the contact time between the two is reduced, thus reducing the soap content of the bleached oil, minimizing free fatty acid development when acid earths are used, and producing oil of improved flavor stability (132). Economy in earth usage and oil retention is achieved by avoiding oxidation and, in one process, by filtering the feed oil through partially spent earth to achieve two-stage countercurrent operation. By effecting heat exchange between the feed and the bleached oil, some saving of heat is possible.

The continuous vacuum-bleaching process of King et al. (132, 138) is shown in Figure 4.10. Feed oil from storage is mixed continuously with adsorbent in metered amounts, and the resulting slurry is sprayed into the top section of an evacuated tower to flash off dissolved air and free moisture. It is then withdrawn from the tower, heated to bleaching temperature, and resprayed into a second bottom section of the tower to remove bound moisture that is released from the earth only after it is heated. A small amount of stripping steam in each section provides agitation and assists in the removal of moisture and air. From the second section the oil–clay mixture is pumped through closed filter presses to remove the clay and thence through a cooler to storage.

A somewhat similar continuous bleaching process that omits the second spraying effect has been patented by Robinson (139).

A continuous countercurrent bleaching system (140) was described by the Votator Division. This would appear to be a major improvement over other processes on a theoretical basis. Unfortunately, the system's complexity outweighed the savings in bleaching earth and the process is no longer sold. A concurrent system offered by the EMI Corporation and typical of a modern installation is shown schematically in Figure 4.11.

Activated clay and filter aid are continuously fed from the clay bin into the slurry tank at a controlled rate adjusted to suit the characteristics of the oil. The slurry of oil, clay, and filter aid is pumped continuously to the vacuum bleacher and sprayed into the head space to obtain complete deaeration prior to heating.

The deaerated slurry is heated to the desired bleaching temperature under automatic temperature control, by internal steam coils; and the bleacher is maintained under vacuum by a steam jet ejector system. The vacuum bleacher is compartmentalized to provide two stages of mixing for intimate contact and adequate time. The slurry is pumped from the

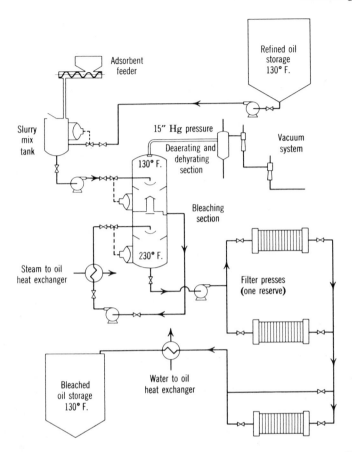

Figure 4.10   Flow diagram: continuous vacuum bleaching process according to King et al. (132).

vacuum bleacher through a cooler and into one of the two filters, which are provided for alternate use. The filtered pretreated oil is discharged continuously to the steam refining deodorizing process or to intermediate storage.

When a filter reaches its cake capacity, the slurry flow is diverted to the other filter. The oil is discharged from the out-of-service filter and the cake is blown with steam, emptied into the cake chute, and conveyed to the discharge point. The filter is then precoated, using the precoat tank and pump, in preparation for the next filter cycle. Steamings from the filter are collected in the steamings tank and reworked into the slurry tank.

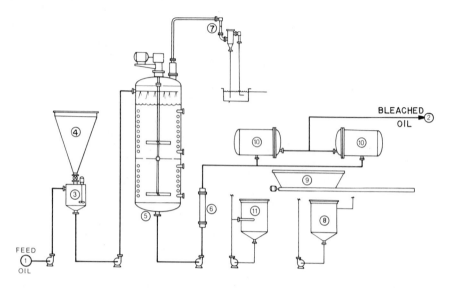

Figure 4.11 Votator continuous-vacuum bleaching plant. 1, feed oil; 2, bleached oil; 3, slurry tank; 4, clay bin; 5, vacuum bleacher; 6, cooler; 7, steam jet ejector system; 8, steamings tank; 9, cake chute; 10, filters; 11, precoat tank. (Courtesy EMI Corporation, Des Plaines, Illinois.)

Recovery of Oil from Spent Bleaching Earth.   As a result of present-day restrictions in the United States imposed by environmental control laws, disposal of wet clay has discouraged use of the water-phase method, although at least one such oil reclaiming plant is now located on the West Coast (142). The normal disposal now used in the United States for spent bleaching clay is to dump it in a landfill, although this method is not without problems and concerns (142). Also, after recovery of most of the oil in the water separation process, the spent clay still needs to be taken to a landfill.

Solvent extraction is an obvious method for recovering oil efficiently from spent earth. With enclosed filters, hexane solvent can extract oil from the earth in several steps before recovery of the oil by evaporation of the miscella. The spent cake should be extracted promptly to avoid oxidation and darkening of the recovered oil (143). If this is done, the recovered oil is almost equal in quality to the bulk of the bleached stock. On the other hand, considerable capital investment is required and safety requirements are increased. The process is economical only where large volumes of spent earth need to be processed. It is not used in the United States (142).

Clays containing unsaturated oil will rapidly oxidize to the point of causing spontaneous combustion, creating an odor problem as well as a fire hazard if not properly blanketed for air exclusion. Landfill area may also be a problem. In addition, the waste clay needs to be covered with soil as soon as possible so as to exclude air and minimize combustion hazards.

5.3   CHEMICAL BLEACHING

In earth bleaching, pigments are removed from oil. In chemical bleaching the pigments are allowed to remain but are oxidized to a colorless or less-colored form.

Along with the effect on pigments, it must be remembered that when an oxidizing agent is added to an oil, anything that can oxidize under the conditions used will oxidize regardless of what one wishes to happen. Moreover, a slight oxidation in terms of iodine number could be serious in terms of flavor stability since concentrations of materials in parts per million can markedly affect the flavor of oils and fats (144). The effect on flavor stability of an edible oil naturally varies with the oil and the conditions used but is generally regarded as undesirable. Consequently, edible oils and fats are never bleached chemically.

Before the widespread use of detergents, chemical bleaching was used rather extensively for bleaching palm oil or dark-colored animal fats for soapmaking. Thomssen and Kemp (145) have described the chrome bleaching of palm oil as follows: If the oil contains an appreciable amount of settlings or solid material, these are first removed by boiling the charge with a 10% salt solution and wet steam and allowing it to settle. Bleaching is conducted in a lead-lined tank, equipped with perforated coils for the injection of both steam and air. The charge consists of 1 ton of oil. The oil is brought to a temperature of 110°F, and 40 lb of fine dry salt are sprinkled into the tank. This is followed by the addition of 40 lb of concentrated commercial hydrochloric acid and 17 lb of sodium dichromate dissolved in 45 lb of the same acid. The latter solution is added slowly over a period of about 3 hours; the charge is agitated with air during the addition of the dichromate solution and for 1 hour thereafter. At the end of this time agitation is stopped and the aqueous phase is allowed to settle to the bottom of the tank, from which it is drawn off. About 40 gal of water is then added, and the charge is agitated and heated with open steam to 150–160°F, after which the operation is completed by allowing the contents of the tank to settle overnight.

The bleaching of inedible tallows and greases with chlorine dioxide generated *in situ* by the action of sulfuric acid on sodium chlorite has been described by Woodward and coworkers (146). The fat, at 210°F, is agitated for 0.5–1.0 hour with 0.1% sodium chlorite and sufficient sulfuric acid to lower the pH to 4 or below, after which the pH is raised to 8 by the addition of 8° Bé caustic soda solution, the watery layer is drawn off, and the fat is dried. The corrosive effects of acid on the equipment can be avoided by using gaseous chlorine to liberate the chlorine dioxide or by using dry chlorine dioxide as such.

References

1. A. E. Bailey, Ed., *Cottonseed and Cottonseed Products,* Interscience, New York, 1948.

2. A. J. C. Andersen, Ed., *Refining of Oils and Fats for Edible Purposes,* 2nd ed., Pergamon, New York, 1962.

3. M. Loury, *Bull. Mater. Grasses Inst. Colonial Marseille,* **27,** 151–160 (1943).

4. M. Mattikow, *J. Am. Oil Chemists' Soc.,* **25,** 200–203 (1948).

5. T. A. McGuire, F. R. Earle, and H. J. Dutton, *J. Am. Oil Chemists' Soc.,* **24,** 359–361 (1947).

6. C. R. Scholfield, H. J. Dutton, F. W. Tanner, Jr., and J. C. Cowan, *J. Am. Oil Chemists' Soc.,* **25,** 368–372 (1948).

7. B. H. Thurman, U.S. Pat. 2,201,061 (1940).

8. H. S. Olcott, *Science,* **100,** 226–227 (1944).

9. A. E. Bailey and R. O. Feuge, *Oil Soap,* **21,** 286–288 (1944).

10. I. Taussky, U.S. Pat. 2,413,009 (1946).

11. B. H. Thurman, *Ind. Eng. Chem.,* **24,** 1187–1190 (1932).

12. P. Dubouloz and S. Lalemont, *Bull. Mater. Grasses Inst. Colonial Marseille,* **29,** 35–38 (1945).

13. W. C. Pritchett, W. G. Taylor, and D. M. Carroll, *J. Am. Oil Chemists' Soc.,* **24,** 225–227 (1947).

14. H. F. Hinners, J. J. McCarthy, and R. E. Bass, *Oil Soap,* **23,** 22–25 (1946).

15. R. T. O'Connor, E. T. Field, M. E. Jefferson, and F. G. Dollear, *J. Am. Oil Chemists' Soc.,* **26,** 710–718 (1949).

16. C. Golumbic, *J. Am. Chem. Soc.,* **64,** 2337–2340 (1942).

17. R. R. King and F. W. Wharton, *J. Am. Oil Chemists' Soc.,* **26,** 201–207 (1949).

18. N. F. Kruse, E. B. Oberg, W. E. Mann, H. R. Kraybill, and K. E. Eldridge, U.S. Pat. 2,296,794 (1942).

19. H. W. Rawlings, N. H. Kuhrt, and J. G. Baxter, *J. Am. Oil Chemists' Soc.,* **25,** 24–26 (1948).

20. M. Mattikow, U.S. Pat. 2,415,301 (1947); B. H. Thurman, U.S. Pat. 2,415,313 (1947).

21. R. H. Neal, U.S. Pat. 2,351,832 (1944).

22. H. N. Brocklesby, *The Chemistry and Technology of Marine Animal Oils,* Fisheries Research Board of Canada, Ottawa, 1941; B. E. Bailey, N. M. Carter, and L. A. Swain, *Bull. Fisheries Res. Board Can., No. 89,* 1952.

23. H. J. Passino, *Ind. Eng. Chem.,* **41,** 280–287 (1949).

24. L. O. Buxton, *Ind. Eng. Chem.*, **34**, 1486–1489 (1942).

25. R. T. O'Connor, D. C. Heinzelman, and M. E. Jefferson, *J. Am. Oil Chemists' Soc.*, **25**, 408–414 (1948).

26. R. O. Feuge, E. A. Kraemer, and A. E. Bailey, *Oil Soap*, **22**, 202–207 (1945).

27. A. K. Epstein, H. L. Reynolds, and M. L. Hartley, U.S. Pats. 2,140,793–794 (1938).

28. H. J. Dutton, A. W. Schwab, H. A. Moser, and J. C. Cowan, *J. Am. Oil Chemists' Soc.*, **25**, 385–388 (1948); ibid., **26**, 441–444 (1949).

29. J. J. Ganucheau and E. L. D'Aquin, *Oil Soap*, **10**, 49–50 (1933).

30. H. D. Royce and F. A. Lindsey, *Ind. Eng. Chem.*, **25**, 1047–1050 (1933).

31. A. I. Skipin and M. Sokolova, *Maslob.-Zhir. Delo.* **10** (8), 4–11 (1934).

32. A. C. Wamble and W. B. Harris, *Tex. Eng. Expt. Sta., Bull. No. 136*, September 1954.

33. F. A. Norris, unpublished results.

34. M. S. El'gort and N. V. Ionova, *Maslob.-Zhir. Delo*, **12**, 548–549 (1936).

35. B. H. Thurman, *J. Am. Oil Chemists' Soc.*, **26**, 580–583 (1949).

36. D. Wesson, *J. Oil Fat Ind.*, **3**, 297–305 (1926).

37. G. S. Jamieson, *Vegetable Fats and Oils*, 2nd ed., Reinhold, New York, 1943.

38. R. R. King and F. W. Wharton, *J. Am. Oil Chemists' Soc.*, **25**, 66–68 (1948).

39. L. L. Linteris and E. Handschumaker, *J. Am. Oil Chemists' Soc.*, **27**, 260–263 (1950).

40. L. Hartman and M. D. L. White, *J. Am. Oil Chemists' Soc.*, **25**, 177–180 (1952).

41. American Oil Chemists' Society, Official and Tentative Methods (1977).

42. J. K. Sikes, *J. Am. Oil Chemists' Soc.*, **34**, 72–75 (1957).

43. M. Kantor, *J. Am. Oil Chemists' Soc.*, **27**, 455–462 (1950).

44. L. P. Hayes and H. Wolff, *J. Am. Oil Chemists' Soc.*, **33**, 440–442 (1956).

45. N. W. Myers, *J. Am. Oil Chemists' Soc.*, **34**, 93–96 (1957).

46. B. Clayton, U.S. Pats. 2,769,827 (1956) and 2,686,794 (1954).

47. L. F. Langhurst, in K. S. Markley, Ed., *Soybeans and Soybean Products*, Chapter 15, Interscience, New York, 1950.

48. J. Stanley, in K. S. Markley, Ed., *Soybeans and Soybean Products*, Chapter 16, Interscience, New York, 1950.

49. W. Van Nieuwenhuyzen, *J. Am. Oil Chemists' Soc.*, **53**, 425–427 (1976).

50. National Soybean Processors' Association, *Yearbook and Trading Rules, 1977–1978*, Chicago, Ill., 1977.

51. British Pat., 326,539; *Chem. Abstr.*, **24**, 4948.

52. British Pat., 366,996; *Chem. Abstr.*, **27**, 2056.

53. F. E. Sullivan, U.S. Pat. 2,702,813 (1955).

54. A. E. Bailey, Ed., *Cottonseed and Cottonseed Products*, Interscience, New York, 1948, pp. 369–370.

55. National Cottonseed Products Association, *Rules Governing Transactions Between Members, 1975–1976*, Memphis, Tenn.

56. C. B. Cluff, *Oil Fat Ind.*, **4**, 168–171 (1927).

57. R. R. Lewis, *J. Am. Oil Chemists' Soc.*, **24**, 315–316 (1948).

58. L. D. Tyler, *Food Ind.*, **20**, 1456–1459 (1948).

59. C. H. Hapgood and G. F. Mayno, U.S. Pat. 1,457,072 (1923).
60. E. M. James, British Pat. 407,995 (1934); French Pat. 743,449 (1933); Canadian Pat. 355,720 (1936); U.S. Pat. 2,050,844 (1936).
61. B. Clayton, U.S. Pats. 2,100,276–77 (1937); B. Clayton, W. B. Kerrick, and H. M. Stadt, U.S. Pats. 2,100,274–75 (1937) and 2,137,214 (1938); B. H. Thurman, U.S. Pat. 2,150,733 (1939).
62. E. M. James, *Oil Soap*, **11,** 137–138 (1934).
63. J. C. Cowan, *J. Am. Oil Chemists' Soc.,* **53,** 344–346 (1976).
64. A. S. Foust, L. A. Wenzel, C. W. Clump, L. Maus, and L. B. Andersen, *Principles of Unit Operations,* Wiley, New York, 1960.
65a. W. J. Podbielniak, A. M. Gavin, and H. R. Kaiser, *J. Am. Oil Chemists' Soc.,* **33,** 24–26 (1956); ibid., **36,** 238–241 (1959); C. M. Doyle, H. R. Kaiser, and A. M. Gavin, ibid., **36,** 277–280 (1959). H. R. Kaiser and C. M. Doyle, **37,** 4–7 (1960); N. W. Myers, ibid., **34,** 93–96 (1957); W. J. Podbielniak, G. J. Ziegenhorn, and H. R. Kaiser, ibid., **34,** 103–106 (1957).
65b. D. R. Todd and J. E. Morren, *J. Am. Oil Chemists' Soc.,* **42,** 172A, 178A (1965).
66. F. E. Sullivan, *J. Am. Oil Chemists' Soc.,* **45,** 564A (1968).
67. W. D. Pohle, R. L. Gregory, and S. E. Tierney, *J. Am. Oil Chemists' Soc.,* **40,** 703–707 (1963).
68. H. Goff, Jr. and F. E. Blachly, *J. Am. Oil Chemists' Soc.,* **34,** 320–323 (1957).
69. L. S. Crauer and H. Pennington, *J. Am. Oil Chemists' Soc.,* **41,** 656–659 (1964).
70. G. C. Cavanagh, *J. Am. Oil Chemists' Soc.,* **53,** 361–363 (1976).
71. G. C. Cavanagh, U.S. Pat. 2,789,120 (1957).
72. B. H. Thurman, U.S. Pat. 2,260,731 (1941).
73. L. P. Hayes and H. Wolff, *J. Am. Oil Chemists' Soc.,* **33,** 440–442 (1956).
74. M. Mattikow, U.S. Pat. 2,576,957 (1951).
75. R. G. Folzenlogen, U.S. Pat. 2,563,327 (1951).
76. F. E. Sullivan, U.S. Pat. 2,702,813 (1955).
77. N. W. Myers, *J. Am. Oil Chemists' Soc.,* **34,** 93–96 (1957).
78. B. Clayton, U.S. Pat. 2,190,588 (1940).
79. G. C. Cavanagh, *J. Am. Oil Chemists' Soc.,* **33,** 528–531 (1956).
80. B. Braae, *J. Am. Oil Chemists' Soc.,* **53,** 353–357 (1976).
81. Y. Hoffman, *J. Am. Oil Chemists' Soc.,* **50,** 260A (1973).
82. M. Mattikow, *Oil Soap*, **19,** 83–87 (1942).
83. M. Mattikow, *J. Am. Oil Chemists' Soc.,* **25,** 200–203 (1948).
84. F. E. Sullivan, *J. Am. Oil Chemists' Soc.,* **32,** 121–123 (1955).
85. F. H. Smith and A. V. Ayres, *J. Am. Oil Chemists' Soc.,* **33,** 93–95 (1956).
86. B. Clayton, U.S. Pat. 2,641,603 (1953).
87. L. S. Crauer and F. E. Sullivan, *J. Am. Oil Chemists' Soc.,* **38,** 172–174 (1961).
88. Sullivan Systems Inc., Tiburon, Calif.
89. Elliott Automation Company, Inc., Cincinnati, Ohio.
90. G. Hefter, *Technologie der Oele und Fette I.,* 654, Berlin, 1905.
91. F. E. Sullivan, *J. Am. Oil Chemists' Soc.,* **53,** 358–360 (1976).

92. H. Bock, U.S. Pat. 3,354,188 (1967).

93. T. K. Mag, *J. Am. Oil Chemists' Soc.*, **50**, 251–254 (1973).

94. F. E. Sullivan, *Chem. Process.*, 76, (April 15, 1974).

95. F. Taylor, U.S. Pat. 3,895,042 (1975).

96. A. Hvolby, *J. Am. Oil Chemists' Soc.*, **48**, 503–509 (1971).

97. F. Taylor, U.S. Pat. 3,895,042 (1975).

98. H. Bollman, U.S. Pat. 1,371,342 (1921).

99. R. L. Kenyon, collaboration with S. W. Gloyer, and C. C. Georgian, (staff industry collaborative report), *Ind. Eng. Chem.*, **40**, 228–236, 1162–1170 (1948); S. W. Gloyer, *J. Am. Oil Chemists' Soc.*, **26**, 162–166 (1949).

100. A. W. Hixson and R. Miller, U.S. Pats. 2,219,652 (1940); 2,226,129 (1940); 2,247,496 (1941); 2,344,089 (1944); and 2,388,412 (1945).

101. J. T. Dickinson, O. Morfit, and Van Orden, U.S. Pat. 2,454,638 (1948).

102. H. J. Passino and J. M. Meyers, U.S. Pat. 2,467,906 (1949).

103. H. Schlink et Cie, German Pats. 315,222 and 334,659 (1921).

104. E. R. Bolton and E. J. Lush, British Pat. 163,352 (1921); U.S. Pat. 1,419,109 (1922).

105. F. Gruber, French Pat. 677,711 (1930).

106. I. G. Farbenindustrie A.-G., German Pat. 563,203 (1932).

107. E. Scott Pattison, *Industrial Fatty Acids and Their Applications*, Reinhold, New York, 1968; *Fats and Oils Situation*, U.S. Department of Agriculture, January 1964.

108. F. W. Keith, V. G. Bell, and F. H. Smith, *J. Am. Oil Chemists' Soc.*, **32**, 517–519 (1955).

109. L. S. Crauer, *J. Am. Oil Chemists' Soc.*, **42**, 661–663 (1965).

110. L. S. Crauer, *J. Am. Oil Chemists' Soc.*, **47**, 210A (1970).

111. K. S. Gibson and F. K. Harris, *Nat. Bur. Std. Sci. Papers No. 547 (1927)*; K. S. Gibson and G. W. Haupt, *J. Res. Nat. Bur. Std.*, **13**, 433–451 (1934); *Oil Soap*, **11**, 246–250, 257–260 (1943); H. J. McNicholas, *Oil Soap*, **12**, 167–178 (1935).

112. I. G. Priest, *Oil Fat Ind.*, **6** (9), 27–29 (1929).

113. American Oil Chemists' Society Color Committee, G. W. Agee, Chairman, *J. Am. Oil Chemists' Soc.*, **27**,. 233–234 (1950).

114. V. C. Mehlenbacher, *The Analysis of Fats and Oils*, The Garrard Press, Champaign, Ill., 1960.

115. American Oil Chemists' Society, Official and Tentative Methods (1962).

116. W. M. Urbain and H. L. Roschen, *Oil Soap*, **16**, 124–126 (1939).

117. H. A. Gardner, *Physical and Chemical Examination of Paints, Varnishes, Lacquers and Colors*, 10th ed., Institute of Paint and Varnish Research, Bethesda, Md., 1946.

118. P. F. Kerr, *Am. Mineralogist*, **17**, 192–198 (1932).

119. P. G. Nutting, *U.S. Geol. Surv., Circ.*, **3**, 11, 17, 20 (1933).

120. H. Odeen and H. D. Slosson, *Oil Soap*, **12**, 211–215 (1935).

121. J. P. Harris, Ed., *Active Carbon in the Decolorizing, Deodorizing, and Purifying of Oils, Fats, and Related Products*, Industrial Chemical Division, West Virginia Pulp and Paper Co., New York, 1944.

122. A. B. Cummins, L. E. Weymouth, and L. L. Johnson, *Oil Soap*, **21**, 215–223 (1944).

123. H. B. W. Patterson, *J. Am. Oil Chemists' Soc.*, **53**, 339–341 (1976).

124. For a discussion of the calculation of relative bleaching cost, with a nomograph, see R. B. Langston and A. D. Rich, *Oil Soap*, **23**, 182–184 (1946).

125. E. A. Goebel, *J. Am. Oil Chemists' Soc.*, **53**, 342–343 (1976).

126. A. D. Rich, *J. Am. Oil Chemists' Soc.*, **41**, 315–321 (1964).

127. H. Freundlich, *Colloid and Capillary Chemistry*, translated from the 3rd German Ed. by H. S. Hatfield, Methuen, London, 1926.

128. J. W. Hassler and R. A. Hagberg, *Oil Soap*, **16**, 188–191 (1939).

129. C. J. Robertson, R. T. Munsberg, and A. R. Gudheim, *Oil Soap*, **16**, 153–157 (1939).

130. R. R. King and F. W. Wharton, *J. Am. Oil Chemists' Soc.*, **26**, 201–207 (1949).

131. J. H. Mitchell and H. R. Kraybill, *J. Am. Chem. Soc.*, **64**, 988–994 (1942).

132. R. R. King and F. W. Wharton, *J. Am. Oil Chemists' Soc.*, **26**, 389–392 (1949).

133. A. D. Rich, *J. Am. Oil Chemists' Soc.*, **31**, 374–376 (1954).

134. L. E. Stout, D. F. Chamberlain, and J. M. McKelvey, *J. Am. Oil Chemists' Soc.*, **26**, 120–126 (1949).

135. R. O. Feuge and H. J. Janssen, *J. Am. Oil Chemists' Soc.*, **28**, 429–432 (1951).

136. M. Loncin, *Oléagineux*, **30** (2), 77–80 (1975).

137. A. Athanassiadis, paper presented at the Malaysian International Symposium, June 1976.

138. R. R. King, S. E. Pack, and F. W. Wharton, U.S. Pat. 2,428,082 (1947).

139. A. A. Robinson, U.S. Pat. 2,483,710 (1949).

140. W. A. Singleton and C. E. McMichael, *J. Am. Oil Chemists' Soc.*, **32**, 1–6 (1955).

141. Chemtron Process, Division of Chemtron Corporation, 1978.

142. K. S. Watson and C. H. Meierhoefer, *J. Am. Oil Chemists' Soc.*, **53**, 437–442 (1976).

143. E. M. James, *J. Am. Oil Chemists' Soc.*, **35**, 76–83 (1958).

144. T. H. Smouse and S. S. Chang, *J. Am. Oil Chemists' Soc.*, **44**, 509–514 (1967).

145. E. G. Thomssen and C. R. Kemp, Modern Soap Making, MacNair-Dorland Co., New York, pp. 30–32, 1937.

146. J. B. Tuttle and E. R. Woodward, *Chem. Met. Eng.*, **53** (5), 114–115 (1946); E. R. Woodward and G. P. Vincent, *Soap Sanit. Chem.*, **22**, (9), 40–43, 137, 139, 141 (1946).

5

Cooking Oils, Salad Oils, and Salad Dressings

Introduction

Lipids used in the preparation of food products can be divided into two classes based on their consistency at about 25°C: (*a*) liquid oils, such as soybean, cottonseed, sunflower, safflower, peanut, olive, and rapeseed oils and (*b*) solids and semisolids, such as margarine, shortening, lard, palm oil, coconut oil, and palm kernel oil. In the preparation of some foods it is of no particular consequence whether the lipid material is liquid or solid, but in others the consistency is important. For example, in the preparation of green salad the object is to provide an oily coating for the various ingredients; hence a liquid oil must be used. On the other hand, baked food products generally require plastic fats for the incorporation of air needed for leavening. However, with the modern development of emulsifiers, in many of these areas it has been possible to use liquid oils with a suitable emulsifier (1, 2) to achieve this goal.

In the past, for reasons related to both history and climate, there were distinct geographical divisions of people into fat or oil consuming groups. Inhabitants of central and northern Europe derived their lipids wholly from domestic animals. Consequently, the food habits and the cuisine of these people developed around the use of plastic fats. On the other hand, the main form of lipids used by people from warmer climates of southern Europe, Africa and Asia were liquids. In the Western Hemisphere, plastic fats are the most widely used edible materials because of the predominantly northern European extraction of the early settlers and because it is eminently suited to large-scale raising of domestic animals. Even today, this distinction is largely true. A certain amount of shift has taken place in both groups for either health (implication of fats in atherosclerosis) or culinary (use of baked goods, especially in Africa and Asia) reasons. In

315

general, however, there is somewhat greater utilization of liquid oils in the countries that largely consume plastic fats than plastic fats in liquid oil-consuming countries.

Liquid oils in general are suitable for all classes of cooking except those that require the production of a highly developed dough structure. However, they are unsuitable for the production of plastic products, such as cream icings or fillings.

In addition to their household uses, cooking oils, are in considerable demand for commercial deep fat frying of many food products that are consumed immediately after frying. Although in the past fats were the preferred frying media for packaged foods, such as potato and corn chips which remain in storage for considerable periods, in recent times the use of liquid oils in such operations has significantly increased. Products like doughnuts still require plastic fats for frying because liquid oils tend to give them a greasy appearance.

All cooking oils are of vegetable origin. The only liquid oils from animal sources are of marine origin; but marine oils in their natural form are not very desirable because of their high degree of unsaturation. In various parts of the world, considerable quantities of marine oils are used in food products after hydrogenation.

Natural and Processed Cooking and Salad Oils

Cooking oils are used either in their natural state or after processing depending on local taste, custom, and nutritional belief. In the old world oils were used in their natural state. In occidental countries, however, the oil is processed to a bland state. This geographical difference in the use of oils is probably due to the different methods of obtaining the oil from its source. Where the oil is used in its natural state, it is obtained by rather inefficient expression methods that yield oil containing little nontriglyceride substances. On the other hand, high-pressure expression or solvent extraction used to obtain higher yields of oil also produce an oil with considerable amounts of nontriglyceride materials, such as coloring bodies and gums. These oils, because of the presence of these nontriglyceride substances and their strong flavor, must be suitably processed to make them acceptable and edible. Because of their strong colors and objectionable odors even when obtained by expelling, lipids from cottonseed and palm fruit have to be processed to obtain an edible product. In general terms, bland neutral oil is produced from crude oils by taking them through refining, bleaching, and deodorization steps.

There is a significant difference between cooking and salad oils. The term "salad oil" is applied to oils that remain substantially liquid in a refrigerator, that is, at about 40°F. In the standard method of evaluating salad oils, the cold test, described by the American Oil Chemists' Society (3), the oil sample in a sealed 4-oz bottle is placed in an ice bath at 32°F. If it remains clear after 5.5 hours, it meets the criteria of a salad oil or a suitably winterized oil. Well winterized oils remain clear for periods much longer than 5.5 hours. From a practical standpoint, the number of hours required for the oil to cloud is of less significance than the character and the amount of the crystals deposited after a prolonged period of storage at 32°F or at refrigeration temperature.

Crystallization inhibitors, such as lecithin, oxystearin, and polyglycerol esters, have been used for extending the cold test. Oxystearin is prepared by bubbling air through cottonseed oil hydrogenated to an IV of about 35. The fat is kept at a high temperature (about 200°C) during the bubbling operation. Characteristics of the final product will have to be within the following limits specified by the Food and Drug Administration regulations:

Acid number	15 maximum
Iodine number	15 maximum
Saponification number	225–240
Hydroxyl number	30–45
Unsaponifiable material	0.8 percent maximum
Refractive index	60 ± 1 at 48°C

Oxystearin may be used in salad oils at levels up to 0.125 percent based on the weight of the oil. Other crystal inhibitors, such as disaccharide esters (4), monosaccharide esters (5), and polysaccharide esters of hydroxy fatty acids (6) have been patented. The extent of commercial utilization of these additives is not known.

Sunflower seed, safflower seed, and corn oils may need dewaxing before they can meet the criteria of a salad oil. Soybean and rapeseed oils conform to the definition of a salad oil, but because they contain significant amounts of highly unsaturated and unstable linolenic acid, they are partially hydrogenated and winterized to produce cooking and salad oils. Soybean oil is used in the preparation of salad dressings without hydrogenation. Cottonseed oil, because of its high content of palmitic acid, has to be winterized to obtain a salad oil. Oils meeting the criteria of a salad oil cannot be prepared either from peanut oil because of noncrystallinity of its higher melting fraction, or from palm oil because of its high content of palmitic acid. High oleic safflower oil (6a), which contains about 80% oleic acid, is a good cooking oil, but its availability is still limited. Olive oil, which is used as both a cooking oil and a salad oil in some parts of the world, is unique because of its natural flavor relished by the users. The use of this oil is restricted by its high cost.

Manufacture of naturally flavored salad and cooking oils, such as olive oil, requires nothing more than expression of the oil from the oil-bearing materials and clarification of the oil by filtration. Among oils in their natural state, olive oil, especially virgin olive oil, is popular in many parts of the world, particularly in Mediterranean countries and the United States, because of its characteristic flavor. Commercial olive oil is probably a blend of oils from a number of different sources. The quality and the flavor of olive oil vary considerably from one season and one locality to another, and to turn out a product of more or less uniform characteristics, the packer must have access to a variety of oils. The annual consumption of olive oil in the United States averaged about 2.0 billion pounds per year during 1959–1972. Since 1973 the consumption has decreased to about 1.5 billion pounds per year. In the United States, production of olive oil fell from about 2.0 billion pounds per year during 1934–1960 to about 0.7 billion pounds during 1960–1972. Since 1974, production has been about 1.0 billion pounds per year.

There is a substantial market for cold pressed peanut, sesame, and safflower seed oils in many European, Asian, and South American countries. In the United States, use of such oils is becoming more popular in recent years because of nutritional considerations. However, there are no reliable statistics available for the quantities of such oils consumed on a world basis.

Consumption of all types of food fats and oils has been gradually increasing all over the world. Fats and oils consumed as part of foods can be conveniently divided into two broad categories: "visible" fats that are used in the preparation of food and "invisible" fats that are present in the food naturally. Per capita consumption of visible fats in the United States has increased from 45 lbs in 1961 to about 57 lbs in 1976. Table 5.1 shows the per capita consumption and total production of cooking and salad oils in the United States.

For many years cottonseed oil was the main source of cooking and salad oils in the United States. However, since World War II soybean oil has become the most prominent oil, accounting for well over half of all oils going into food products. Quantities of cooking and salad oils derived from various vegetable oils are shown in Table 5.2.

In the last decade palm oil has grown in prominence as a food fat both

Table 5.1   Total production (in million pounds) and per capita consumption (pounds per person) of cooking and salad oils in the United States

Year	Cooking and Salad Oils	
	Total	Per Capita
1961	1661	9.2
1962	2021	11.0
1963	2066	11.1
1964	2249	11.9
1965	2398	12.5
1966	2464	12.7
1967	2474	12.7
1968	2665	13.5
1969	2863	14.4
1970	3125	15.5
1971	3215	15.7
1972	3513	17.0
1973	3737	18.5
1974	3851	18.4
1975	3856	18.2
1976	4234	19.9

Source: Fats and Oils Situation, Agricultural Research Service, United States Department Of Agriculture.

in the United States and the world. To be converted to a cooking oil, palm oil must be fractionated to remove the solid fraction. The liquid palm oil fraction is used extensively in the snack industry. Production of palm oil rose from 1.4 billion pounds in 1968 to an estimated 3.9 billion pounds in 1976. During the same period imports of this oil into the United States increased from a mere 103 million pounds to 532 millions pounds. Palm oil imports into the United States in 1977 were estimated to be about 700 million pounds. Although sunflowerseed oil is a major commodity in Europe, only in recent years has it become a source of edible oil in the United States. Similarly, domestic production of safflower seed oil is gradually increasing. Both sunflower seed and safflower seed oils are becoming popular because of their high content of linoleic acid, which is nutritionally important. Some of the analytical characteristics of commercially important cooking and salad oils are shown in Table 5.3.

Table 5.2   Salad and cooking oils derived from various vegetable oils in the United States (in million pounds)

Year	Soybean	Cottonseed	Corn	Peanut	Safflower	Olive
1960	887	752	247	28	—	51
1961	1019	818	210	76	—	59
1962	1437	817	210	46	6	58
1963	1317	764	202	55	27	33
1964	1638	906	241	50	20	67
1965	1564	915	239	53	9	44
1966	1860	746	217	115	12	49
1967	1912	625	231	133	42	54
1968	2036	541	242	156	22	63
1969	2244	513	248	124	19	58
1970	2471	527	246	139	12	62
1971	2658	442	248	160	11	62
1972	2985	441	259	148	22	67
1973	2848	595	286	127	29	60
1974	3149	545	276	98	17	53
1975	3032	432	280	100	22	48
1976	3349	380	294	147	22	62

Source: Fats and Oils Situation, Agricultural Research Service, United States Department Of Agriculture.

Stability of Salad Oils and Cooking Oils

By proper processing, most oils can be made bland. Complete removal of the "corny" flavor of corn oil is rather difficult. However, this flavor of corn oil is preferred by most consumers. During storage or during cooking, bland oils undergo both physical and chemical changes as they are sensitive to both heat and light. In addition, trace metals like copper and iron, and other proxidants of unknown nature catalyze such chemical changes. Primary change brought about by these factors is the reaction of the unsaturation sites of the oil with oxygen. At low levels of oxidation, the degree and type of change, generally known as "flavor reversion," is characteristic of each oil. For example, soybean oil develops a flavor that is described as beany or grassy. This flavor has been attributed to the formation of 2-pentylfuran (7) and 3-*cis*-hexenal (8). Reversion flavors of sunflowerseed and safflowerseed oils are described by the term "seedy." Similarly, rapeseed and palm oils develop peculiar flavors. Palm oil reversion flavor is particularly objectionable. These "reversion flavors" are observed long before other oxidative flavors are formed. With many of these oils, the source of the "reversion flavors" is not known.

Table 5.3 *Analytical characteristics of typical commercial samples of salad and cooking oils*

	Salad Oils			Cooking Oils		
	Olive Oil	Corn Oil	Soybean Oil	Winterized Cottonseed Oil	Cottonseed Oil	Peanut Oil
Iodine number	85	125	130	112	108	95
Refractive index at 60°C	1.4546	1.4598	1.4600	1.4577	1.4572	1.4550
Free fatty acids, percent as oleic	1.5	0.05	0.02	0.02	0.03	0.03
Smoke point (°F)	—	450	450	450	440	440
Keeping quality, AOM[a] hours to peroxide value of 125	20	10	8	10	10	12
Color, Lovibond	—	40Y–4.0R	15Y–1.8R	20Y–2.0R	20Y–2.5R	25Y–2.0R
Cold test (hours to cloud at 32°F)	>24	Does not cloud	Does not cloud	20	<1	<1
Cloud point, ASTM (°F[b])	22	12	14	24	38	40
Solid point, ASTM (°F[b])	14	8	10	20	27	34
Titer (°C)	23.0	18.5	21.5	33	36.0	31.3

[a] Active oxygen method.
[b] Modified by examination of sample at intervals of 2°F rather than 5°F.

321

The initial product of reaction of oils with oxygen is the hydroperoxide, which has no odor or flavor. When it decomposes it forms a number of diverse compounds, such as aldehydes, ketones, acids, esters, alcohols, hydroxy compounds, hydrocarbons, polymers, and epoxides. The breakdown products are the cause of off flavors observed in oils and their products. Many of the compounds are present only in parts per billion, and not all of these compounds are objectionable. Many of the related esters, alcohols, acids, and ketones are responsible for the desirable flavors of tomato juice, ripened banana, olive oil, butter, and other foods. Similary, the flavor of fried foods has been attributed to 2,4,-decadienal, which is also a component of off-flavored soybean oil. Obviously, the proportions in which these compounds are present is more important in determining whether a given flavor is favorable, rather than the types of compounds (9).

Shelf stability of salad oils is dependent on many factors such as the conditions of processing, the presence and the amount of natural antioxidants, the type of container, and the temperature of storage. However, it can be generally stated that oils containing linolenic acid, whether hydrogenated or unhydrogenated, are less stable than other oils.

In the case of cooking oils, both thermal and oxidative degradation are important. Thermally-caused oxidative changes result in polymer formation as well as other types of alteration of the oil. These changes are important both from nutritional and aesthetic standpoints. Therefore, they have been extensively studied and discussed (10–19). Unfortunately, many of the nutritional studies have been conducted on fats and oils degraded under conditions far different from practical cooking operations. This has caused considerable uncertainty about the nutritional qualities of cooking oils that have undergone thermal and oxidative deterioration.

Both volatile and nonvolatile decomposition products of different cooking oils have also been studied (20–24).

Quality Evaluation of Salad Oils and Cooking Oils

Oxidation of oils and fats from the stage of obtaining the oil seed to the stage of consumption of the finished oil is a real danger to their quality. Therefore, many attempts have been made to measure the extent of such oxidation. The most important quality criterion in the case of cooking and salad oils is their blandness, which is changed by oxidation. The most widely used method of determining blandness is organoleptic evaluation, but this method is subjective and is influenced by many variables. Therefore, several objective methods to evaluate the quality of oils and fats have been proposed. These methods are generally based on the fact that oxidative degradation produces both volatile and nonvolatile chemical compounds with functionalities that can be measured by physical or chemical methods. However, for any method to be useful, the results must correlate with the organoleptic properties of the oils and fats in question. A second but equally important requirement is that a method should be capable of predicting the storage stability of the oils and the fats as well as the products derived from them.

4.1 KREIS TEST (25)

This test depends on the color produced by oxidized oils and fats with phloroglucinol. Recently, an improvement in reliability has been claimed by use of resorcinol instead of phloroglucinol (26, 27). However, correlation with organoleptic evaluations is poor.

4.2 PEROXIDE VALUE (28)

This is the most widely used method to determine the degree of oxidation. The primary oxidation products of oils and fats are hydroperoxides, the quantity of which can be quantitatively measured by determining the amount of iodine liberated by its reaction with potassium iodide. The peroxide content is expressed in terms of milliequivalents of iodine per kilogram of oil or fat. However, when these hydroperoxides start breaking down to produce flavor compounds, correlation of blandness with degree of oxidation almost disappears because of the dynamic free radical system created by such breakdown and the nonreactive nature of most of the breakdown products with potassium iodide.

4.3 CARBONYL VALUE (29)

This method measures volatile carbonyl compounds that result from the breakdown of hydroperoxides. These carbonyls are distilled and converted to hydrazones, and the absorption is measured at 480 nm. This method suffers from lack of reliable correlation with organoleptic properties of oxidized oils and fats.

4.4 BENZIDINE AND ANISIDINE VALUE (30, 31)

This method uses the reaction of nonvolatile α,β-unsaturated aldehydes with these reagents. Absorption readings are made at 350 nm in 1-cm cells. Originally, benzidine was used as the reagent. Later, because of the carcinogenicity of benzidine, anisidine was recommended. This method of determining the degree of oxidation is extensively used in European countries. Results thus far indicate that the method is more useful in determining the quality of crude oils and the efficiency of processing procedures than the quality of finished oils during storage.

An extension of this method is the "totox" value (32), which is the sum of benzidine or anisidine value and two times the peroxide value.

4.5   PENTANE VALUE (33, 34)

One of the compounds produced by the autoxidation of salad and cooking oils is pentane. The quantity of pentane is measured by a GLC procedure. Correlation of this value with organoleptic properties of the oil has not been unequivocally established.

4.6   THIOBARBITURIC ACID TEST (35, 36)

The method, developed initially for measurement of malondialdehyde, is used to measure the oxidative products of polyunsaturated fatty acid moieties; it lacks sensitivity, however (37).

All the preceding methods serve as useful tools under a given set of conditions but seem to fail in their universal applicability to the quality of oils and fats.

4.7   VOLATILE PROFILE METHOD (38–41)

This is a direct gas chromatographic method to examine the volatiles in oils, fats, and their products. Under control conditions, the method has good reproducibility. When the results are statistically treated, good correlation with organoleptic evaluation is obtained.

4.8   ACCELERATED TESTS

Study of the storage stability of salad oils is very time consuming and requires several months to complete. From a commerical standpoint, it is necessary to complete this determination in a shorter period of time. Therefore, partially successful attempts have been made to develop accelerated tests using light- and temperature-susceptible properties of oils and fats. Moser et al. (42) have designed an apparatus using fluorescent light to accelerate the aging of oils and fats. Radtke et al. (43) have studied the influence of light of varying intensity and wavelength on the oxidative deterioration of salad oils. They observed that photochemical action was dependent on the wavelength and increased with decreasing wavelength to a much greater extent than could be predicted from energy considerations.

From these light studies it may be possible to reduce the time of the storage stability test by exposing oils to light of short wavelength, followed by an analysis of the volatiles using the volatile profile method.

In the case of cooking oils during storage, degradation similar to that observed in salad oils occurs. In addition, during cooking more severe types of breakdown are caused by the high temperature used for cooking, the presence of moisture, oxygen, and food materials. These changes are reflected by increases in color, free fatty acids, polymers, polar materials, foaming, and so on. Even today, there are no reliable methods to predict the performance of a cooking oil and establish a discard point for the used oil. However, several empirical methods have been suggested to determine the extent of damage to the oil caused by the frying process.

4.9 FREE FATTY ACID

This is probably the most widely used characteristic of oil quality control. Well-deodorized cooking oils generally have a free fatty acid level less than 0.05%. During use, there is a buildup of free fatty acids. In the initial stages of cooking, free fatty acids are produced by oxidative breakdown, but in later stages hydrolysis of the fat caused by the presence of moisture in the food causes free fatty acids to increase. This is a dynamic process as some of the free fatty acids so produced are lost through oxidation and steam distillation from the food. Further, free fatty acids catalyze hydrolysis of the cooking oil. An increase in free fatty acid decreases the smoke point of the oil. When significant amounts of free fatty acids have accumulated in the oil, smoking becomes excessive and the quality of the food decreases; the oil then has to be discarded.

4.10 COLOR

Oils used for cooking darken because of the formation of oxidative materials, including polymers, and the presence of oil-soluble products from the fried food. Up to a certain point the increase in color can be measured with a Lovibond "Tintometer." Generally speaking, such a darkening of the oil without corresponding increase in other degradative products will only make the food look darker but does not significantly affect the taste of the food. Color alone as a measure of the extent of degradation of the cooking oil is highly unreliable.

4.11   OXIDIZED FATTY ACIDS

Many methods have been proposed to measure the amount of oxidized lipids in a cooking oil. Oxidative degradation will affect both the appearance and the taste of the cooked food.

A method (44) proposed by the "German Society for Fat Research" follows oxidative degradation by determining the petroleum ether-insoluble oxidized fatty acids. Under strictly controlled condition, the results are reproducible. One disadvantage of this method is that considerable quantities of such fatty acids are soluble in petroleum ether.

Another method (45) is the "saponification color number." This method is based on the fact that oxidized oils and fats give yellow to brown solutions on saponification with alcoholic potassium hydroxide.

4.12   POLYMERS

Sahasrabudhe and Bhalerao (46) have described a method for isolation and determination of polymers by assessing the fraction not included in a urea complex. There is an increase in this fraction with increased oxidative degradation. This method cannot be used directly on the oil.

Gel permeation chromatography (47, 48) will permit a direct analysis of the degraded oil for polymers. Ordinary liquid chromatography (49) has also been proposed for such an analysis. With the advent of high pressure liquid chromatography, the time required for such analyses has been significantly shortened (50).

A GLC method (51, 52) that determines the amount of nonelutables under conditions used for fatty acid distribution analyses is promising for following the degradation of cooking oils.

Additives for Salad and Cooking Oils

Salad and cooking oils have varying amounts of natural antioxidants, mostly tocopherols. With the introduction of synthetic antioxidants, it has been customary to supplement the natural antioxidants with synthetic ones. Most commonly used antioxidants are butylated hydroxy anisole (BHA), butylated hydroxy toluene (BHT), tertiary butyl hydroquinone (TBHQ), propyl gallate (PG), and ascorbyl palmitate. Many other antioxidants have been suggested, but their use is limited. Citric acid is added primarily to chelate trace metals which otherwise accelerate the oxidative process.

Sherwin and Luckadoo (53) found that the finished oils from safflower, sunflower, soybean and cottonseed crude oils that had been treated with antioxidants had higher oxidative stability and responded better to further antioxidant treatment than did those from untreated crude oils. List et al. (54) observed that phenolic antioxidants did not improve the flavor stability of finished oil from unbleached sunflowerseed oil. On the other hand, treatment of the bleached sunflowerseed oil improved its stability. Study of the effect of TBHQ and citric acid (55) on crude palm oil showed that a finished oil with a better keeping quality could be obtained by such a treatment. Applications and status of antioxidants have been reviewed recently by Sherwin (56).

Most of the evaluations of antioxidants have been done using the active oxygen method (AOM). Unfortunately, AOM stability does not correlate with room temperature stability, which is the criterion used for salad oils. Usefulness of synthetic antioxidants in cooking oils is uncertain because these antioxidants are steam distilled or degraded at frying temperatures. However, usefulness of these antioxidants in fried foods has been amply demonstrated.

Present knowledge in this area seems to indicate that the effectiveness of synthetic antioxidants in storage stability of salad and cooking oils is somehow dependent on the history of crudes from which they are made and the thoroughness of various processing steps.

Other additives have been suggested to improve the stability and performance of salad and cooking oils. Cunningham et al. (57) observed that addition of a mixture of poly(oxyethylene)-20-sorbitan monoleate (Tween 80) and decaglycerol decaoleate improved the stability of cooking oils. Incorporation of small amounts of cystine into salad oils by heating the mixture of oil and amino acid to 140°C imparted greater stability to the oil against oxidation (58). Chang and Morne (59) found that cooking oils could be stabilized with α-sitosterol. Many phosphoric acid derivatives have been proposed as preservatives. Polyoxyethylene sorbitan esters have been proposed as antispattering agents (60) in cooking oils.

Nutrition-Oriented Salad and Cooking Oils

The U.S. diet, as well as diets of many industrialized nations, are rich in calories contributed by oils and fats. Technological innovations are, therefore, directed toward finding low-calorie substitutes that do not affect the aesthetic or health values of food products. Sucrose polyesters with six or more hydroxy groups esterified with unsaturated fatty acids have been reported to have characteristics of conventional cooking and salad oils (61, 61a). These polyesters are not absorbed in the digestive system and, therefore, do not contribute any calories. However, there are a number of unanswered questions on the effect of these polyesters on other nutrients. Erickson (62) has reported that addition of as little as 0.5% of a monocarboxylic ester of plant sterol to cooking and salad oils will impart hypocholesteremic properties to the oil.

6.1 SALAD DRESSINGS

Salad dressings are divided into two broad categories: spoonable (includes mayonnaise) and pourable (includes French dressing). Both production (Table 5.4) and consumption (Table 5.5) of mayonnaise and other salad dressings have increased tremendously in many parts of the world, es-

Table 5.4 Shipments of salad dressings, mayonnaise, and related products (millions of gallons)

Year	Total	Salad Dressings	Mayon-naise	Sandwich Spreads	French Dressings	Other
1954	106.9	49.5	41.3	7.1	7.0	2.0
1955	111.6	49.8	44.1	7.2	7.4	3.1
1956	114.9	50.4	46.0	6.9	8.1	3.4
1957	212.4	54.0	47.7	7.2	8.0	4.5
1958	125.5	54.6	50.4	6.9	8.6	5.0
1959	134.0	56.7	45.3	6.9	9.4	6.4
1960	136.5	54.7	55.4	6.7	9.4	10.3
1961	140.6	56.9	57.2	6.8	9.5	10.2
1962	144.6	58.0	58.8	6.6	9.9	11.3
1963	160.3	62.7	66.3	6.9	10.4	9.5
1964	167.2	63.8	70.0	6.8	10.4	16.2
1965	174.8	65.9	71.7	6.9	10.7	19.6
1966	183.4	68.3	75.9	7.2	10.5	21.5
1967	191.0	69.4	79.0	7.5	10.8	24.3
1968	201.3	71.0	84.6	7.2	10.5	28.0
1969	210.7	71.9	88.2	7.8	11.2	31.6
1970	218.4	72.5	92.6	7.5	11.5	34.3
1971	235.0	74.2	101.1	7.4	12.2	40.1
1972	240.4	73.0	97.5	9.4	14.9	45.6
1973	251.3	71.7	104.1	10.2	17.0	48.2
1974[a]	262.7					
1975[a]	274.5					
1976[a]	286.8					

Source: Department of Commerce and Judges Almanac.

[a] Estimate from Association of Dressing and Sauces (no available breakdown).

Table 5.5   *Per capita consumption of salad products*[a]

Year	Per Capita Consumption (in pints)
1960	6.1
1961	6.1
1962	6.2
1963	6.8
1964	7.0
1965	7.2
1966	7.5
1967	7.7
1968	8.0
1969	8.3
1970	8.5
1971	9.1
1972	9.2
1973	9.6
1974	9.9
1975	10.3
1976	10.7

Source: Department of Commerce, Association of Dressings and Sauces.

[a] Includes sandwich spreads, pourable dressings, and spoonable dressings.

pecially in the United States in the past 25–50 years. The principal reason for this increase has been their widespread adoption as spreads for bread and in other food preparations. They have largely supplanted butter or margarine in sandwiches. To a commercial sandwich maker, they are attractive in a number of ways. They have considerable flavor of their own that blends very well with more expensive ingredients to form a delicious filling. They are also less perishable than butter or margarine and are more easily stored, handled, and spread.

6.2   MAYONNAISE

Composition.   The official definition adopted in 1950 by the Food and Drug Administration of the United States, describes mayonnaise as follows: Mayonnaise is the semisolid food prepared from edible vegetable oil, egg yolk, or white eggs (fresh or frozen or dried), any vinegar diluted with water to an acidity, calculated as acetic acid, of not less than 2.5% by weight, lemon and/or lime juice, with one or more of the following: salt, sweetener, mustard, paprika, monosodium glutamate, and other suitable food seasonings. The finished product contains not less than 65% of edible vegetable oil.

Sweetener may be sucrose, dextrose, corn syrup, invert sugar syrup, nondiastatic maltose syrup, glucose, or honey.

Mustard, paprika, other spices, or any spice oil or spice extract can be used, but not turmeric, saffron, spice oil, or spice extract, which imparts to the mayonnaise a color simulating the color imparted by egg yolk. Both mustard and oleoresin paprika impart a yellow color but, because of the small amounts used, do not cause any variance from the standard of identity.

Monosodium glutamate, because of its nutritional implications, has come under the Food and Drug Administration's scrutiny, and its continued use is doubtful. Any suitable harmless food seasoning or flavoring can be used provided it is not an imitation and does not impart to the mayonnaise a color simulating the color imparted by egg yolk.

Acidity, calculated as acetic acid, contributed by lemon and/or lime juice, should not be less than 2.5% by weight of the product. When an optional acidifying ingredient such as citric acid is used, the label on the package shall bear the statement "citric acid added." As of January 1, 1978, malic acid has been permitted as an acidifying agent by the Food and Drug Administration of the United States.

Calcium disodium EDTA or disodium EDTA is permitted to protect the flavor.

Physically, mayonnaise consists of an internal or discontinuous phase of oil droplets dispersed in an external or continuous aqueous phase of vinegar, egg yolk, and other ingredients; it is an oil-in-water emulsion. Consistency of the emulsion depends to a large extent on the ratio of its aqueous and oil phases and the amount and type of egg solids. Stability of the emulsion is not only dependent on the ingredients, but is also greatly affected by the type of equipment and how it is operated during manufacture. Table 5.6 gives a general composition for the product. Mayonnaise is a pale yellow creamy product. The color is derived mainly from the egg rather than from the oil.

Oil.   Edible vegetable oil used in the product may contain not more than 0.125% by weight of oxystearin to inhibit crystallization. Although mayonnaise may legally contain as little as 65% oil, the usual content is 75–82%. Salad oil is the preferred oil in the manufacture of mayonnaise. Any salad oil, such as winterized cottonseed oil, safflower oil, corn oil,

Table 5.6   Approximate composition of mayonnaise

Ingredient	Percent by Weight
Oil	75.0–80.0
Vinegar (4.5% acetic acid)	10.8–9.4
Egg yolk	9.0–7.0
Sugar	2.5–1.5
Salt	1.5
Mustard	1.0–0.5
White pepper	0.2–0.1

and unhydrogenated and partially hydrogenated winterized or unwinterized soybean oil, can be used in its preparation (63, 64). However, unhydrogenated soybean oil is the most commonly used oil. Oils with large quantities of saturated acids, such as palm oil, or oils that solidify at refrigeration temperatures, such as peanut oil, are seldom used as they will tend to break the emulsion in cold temperatures. Oil levels 1–2% higher than the range given in Table 5.6 are used when stiffer body is needed. However, if the oil content exceeds 85 percent, the emulsion tends to be unstable.

Vinegar.   Vinegar has a dual purpose in mayonnaise. The acetic acid acts as a preservative against microbial spoilage, including some types of yeasts (65). It also serves as a flavor ingredient when used at the proper level. However, excessive quantities of the acid will impair the flavor. Water in vinegar forms the continuous phase in the emulsion. Generally, the concentration of acetic acid is about 3.5% of the volatiles present in vinegar. Vinegar may contain high quantities of trace metals that can be detrimental to the stability of the dressing. Therefore, the quality of the vinegar used is important.

Egg.   The main function of the egg solids in mayonnaise is to serve as an emulsifying agent. The mayonnaise emulsion is built around egg yolk, which is one of nature's perfect emulsions. One of the factors influencing the stiffness and the stability of the emulsion is the amount and kind of egg yolk that goes into it. Egg yolk is a complicated chemical entity with a high lipid content. The lipids of the egg yolk form about 65% of the total solids. The composition (66) of the lipids in egg yolk is as follows: glycerides 40.3%, phospholipids 21.3%, and cholesterol 3.6%, based on the total solids. Of the fatty acids, 30% are saturated and the rest are unsaturated. Lipid components of egg yolk have a profound effect on flavor, color, and stability of mayonnaise and other dressings. Egg white is a very complex system of proteins that aids in emulsification by forming a gel structure on coagulation by the acid component (67).

Mustard.   Mustard is generally used in mayonnaise in the form of a flour. There are two varieties of mustard, white and black. White mustard is hot to taste but practically odorless. The brown variety, on the other hand, has a sharp odor. Therefore, the two varieties are blended in various proportions to achieve a desired level of flavor and pungency (68). Mustard contains a glucoside that, when hydrolysed, releases the pungent oil of mustard, allyl isothiocyanate. This imparts the bite to mayonnaise. According to Corran (69), in addition to its flavor contribution, the nature, the origin, and the method of addition of mustard influence the emulsion. He stresses that mustard is a very efficient emulsifier and is very effective if incorporated with egg yolk.

Because of the variation in the essential oil content and immiscibility of the mustard flour with the vegetable oil, Cummings (70) recommends the use of oil of mustard in place of mustard flour. If oil of mustard is used, the advantage of mustard flour as an emulsifying adjunct and as a possible contributor of color in salad dressings may be lost.

Other Spices and Seasonings.   Of the other optional ingredients permitted, paprika adds to the flavor, and salt and sugar act both as flavor ingredients and as stabilizers.

Quality Control and Stability Measurements

Mayonnaise may be classified as a semiperishable product. It is sufficiently stable to keep for reasonable lengths of time without refrigeration. Mayonnaise gradually becomes thinner with age. Thinning and separation of the phases of mayonnaise are greatly accelerated by mechanical shock. Separation of phases can also be caused by exposure to low temperatures. Gray (71, 72) studied the breakdown of mayonnaise and concluded that the higher the water content, the greater is the amount of egg solids needed to stabilize the emulsion. Other causes of breakdown listed by the same worker are (*a*) rapid addition of the oil, (*b*) unregulated agitation during emulsifying, (*c*) high temperature storage, and (*d*) agitation during transit.

Other common forms of spoilage in mayonnaise are due to oxidative degradation of various components, especially vegetable oil and egg lipids. Microbial spoilage is a rare occurence as the product is well protected by the high acid content. But molds and yeast and, to a lesser extent, Lactobacilli, may sometimes find conditions favorable for growth (73).

Vegetable oils used must be of good quality. Poor oils will produce inferior quality products with shortened shelf life. Similarly, the quality of the egg, both microbiologically and chemically, should be high; otherwise, both flavor and emulsion stability problems can be encountered. Mustard flour must be free from impurities. Strict microbiological control of raw materials is very important. During processing, control must be exercised in maintaining correct temperature and emulsification procedures. Proper sequential addition of ingredients and mixing times are important for emulsion stability (74). For good emulsification, the oil globules should be even and should be 1–4 μ in size (75). It is reported that freshness of mayonnaise can be retained substantially longer by processing in the absence of oxygen (76). Processing under nitrogen increased the shelf life of mayonnaise from 48 days to over 240 days (77). Specific gravity of the product with proper amount of inert gas is about 0.90–0.92, and that of the ungassed product is about 0.94. Amount of entrapped gas has to be controlled below 6%; otherwise, shrinkage during transit will cause deterioration of the product (77). Tryptophan or tryptophan derivatives (78), as well as L-cystine (79), have been found to improve the storage of mayonnaise.

Various instruments have been proposed for testing the important properties of consistency, stability, and flavor quality of mayonnaise and other dressings. Consistency is measured by a device that is a modified form of the Gardner mobilometer (80), which was originally designed for testing paints, varnishes, and enamels. The mobilometer is in reality a special form of viscometer in which a weighted, perforated plunger is allowed to fall through the sample. Consistency of the product is reported in seconds. Kilgore (81) has devised a simple method that is particularly suitable for control work, since it involves nothing more than dropping a pointed rod or a "plummet" into a sample from a definite height and noting the depth of its penetration. Penetration correlates inversely with viscosity. Brookfield Helipath viscometer can be used to measure viscosities. These consistency measurements do not totally describe the attributes of body and texture of mayonnaise or salad dressing.

Ability of mayonnaise emulsion to resist mechanical shock is commonly evaluated by testing samples under simulated shipping conditions.

As the quality of oil plays a major role in the flavor stability of mayonnaise and other dressings, only the best grades of oil should be used. The quality of flavor is generally judged by the subjective method of tasting. However, Dupuy et al. (38) have developed a gas chromatographic method to evaluate the flavor quality objectively.

Manufacture

Because the emulsion is an oil-in-water type, preparation of a stable emulsion is difficult. Thus major manufacturers of these products have developed proprietary techniques to achieve this goal. Effect of various factors on the stability of mayonnaise emulsions is not well understood. Therefore, production of stable mayonnaise has remained an art to some degree. Although egg solids are the backbone of a stable emulsion, processing conditions whose interactions are only partially understood play an important role.

A process for making mayonnaise in a batch mixer is to mix egg yolks, sugar, salt, spices, and a portion of the vinegar followed by gradually beating in the oil and then thinning out by mixing the remainder of the vinegar. This method of mixing is said to give a product of better consistency than is obtained by adding all the vinegar at the beginning of the operation.

Gray et al. (82) have made a study of the effect of the method of mixing on the consistency of the emulsion and have recommended that one-third of the vinegar be added initially and the balance to be added at the end. On the other hand, Brown (83) recommended that the mixture of eggs and other ingredients be as stiff as possible when the oil is beaten in. The stiffness can be controlled during the addition of the oil by adding small portions of vinegar. This method of preparation has been claimed to produce very small oil droplets, thus making the emulsion highly stable.

The temperature of the oil and other materials during mixing also influences the body characteristics of mayonnaise. A thin product results if the operation is carried out with materials that are warm. Gray and Meier (84) suggest a temperature of 60–70°F. The inconvenience of attempting to operate at temperatures below 60°F was not worthwhile as the initial superiority of the low temperature processed product is lost with slight aging of the product. Brown (83), however, recommends a temperature of 40°F.

The equipment most popular in the manufacture of mayonnaise and salad dressings is the Dixie mixer and the Charlotte colloid mill connected suitably with valves and pumps. A colloid mill is a mechanical device in which the product passes between a high-speed rotor (3600 rpm) and a fixed stator. Product enters the area at a low velocity and is subjected to a high shear with reduction of the particle size. Clearance between these two parts determines (*a*) the amount of shear imposed and the viscosity of the final product and (*b*) throughput of the mill. Weiss (68) recommends an opening within a range of 25–40 mils. The mixer is used to prepare a coarse emulsion that is then passed through the colloid mill to give the creamy texture.

A second system is the AMF system (85), which consists of a premix feed tank and two mixing stages. In the premix tank, slow agitation is maintained to keep the ingredients well dispersed but not to thicken the emulsion. The mix containing all the ingredients except a portion of the vinegar from the premix tank is converted to a coarse emulsion in the first stage of mixing and is then pumped to the second stage. At this stage, the balance of the vinegar solution is added. This mixer rotates at about 475 rpm, which is much slower than a colloid mill, but because of its teeth design the mixing is very intense. It is claimed that this system produces much smaller and more uniform droplets of oil than the colloid mill does and thus makes a highly stable emulsion.

Other equipment suggested are the Girdler CR mixer (86) and the Sonalator (87), which is an ultrasonic homogenizer. Rees (88) has made a comparative study of homogenizers and colloid mills for producing dressings and concluded that homogenizers can be used for certain type of dressing, but not for mayonnaise. Automation techniques for the salad dressing industry have been discussed by Joffe (85). Finberg (89) has reviewed advanced techniques for making mayonnaise and other salad dressings. Continuous systems that can improve productivity and have built-in versatility to permit production of different dressings have been described by Potter (90), Carlson (91), Lipschultz and Holtgrieve (92) and Täubrich (93). McKenzie and Ziemba (94) have described a streamlined production line intended to safeguard quality of dressings.

Spoonable and Pourable Salad Dressings

Among the pourable dressings, only French dressing has a standard identity. Standards of identity set by the Food and Drug Administration are essentially same as those for mayonnaise with the following exceptions: (a) there is a lower level of vegetable oil—not less than 30%; (b) starch pastes can be used as thickeners, (c) not less egg-yolk-containing ingredients than is equivalent in egg yolk solids content to 4% by weight of liquid egg yolk is permitted (d) gums are used as emulsifiers; and (e) dioctyl sodium sulfosuccinate should be used at a level not to exceed 0.5% of the weight of gum used.

Preferred oils are the same as those used in mayonnaise. However, a process for making stable salad dressings containing mixtures of liquid triglycerides that have an iodine value greater than 75 and solid triglycerides that have an iodine value not exceeding 12 has been patented (95). Quality of the oil is as important in these dressings as in mayonnaise. Cooked or partly cooked starchy paste prepared with a food starch, tapioca flour, wheat flour, rye flour, or any two or more of these is used. Many of these starches are modified to suit the product needs, such as acid stability. Gums that are polysaccharides or their derivatives are not truly soluble in water but are water-dispersible colloids. Gums that are permitted under the standard are gum acacia (also called *gum arabic*), carob bean gum (also called *locust bean gum*), guar gum, gum karaya, gum tragacanth, extract of Irish moss, pectin, propylene glycol ester of alginic acid, cellulose gum (sodium carboxymethyl cellulose, methoxy content not less than 27.5% and not more than 31.5% on a dry weight basis), hydroxypropyl methyl cellulose, or any mixture of two or more of these. However, the amount of such emulsifiers should not exceed 75 ppm by weight of the finished product. Most recently, the Food and Drug Administration has approved xanthan gum in dressings. Xanthan gum is a high-molecular-weight polysaccharide. This gum has been reported to stabilize emulsions and improve the overall quality of dressings (96). Further, it is claimed that the thickening efficiency of this gum permits a reduction in the quantity of stabilizers used in the product.

Large numbers of other emulsified and two-phase pourable dressings are available on the market. These are generally based on flavor appeal and use suitable ingredients to incorporate such flavors.

Low-Calorie Salad Dressings

Because of the calorie consciousness of the U.S. population, as well as the people of other industrialized nations of the world, several low-calorie dressings, such as imitation French, and imitation mayonnaise have been formulated in recent years. Although these dressings have reduced calories than the foods for which they are substituted, they do not qualify as low-calorie foods, which are defined as foods that will provide no more than 40 cal per serving with a maximum of 0.4 cal/gram (97). Use of proper stabilizers to improve the quality of reduced-calorie dressings has been discussed by Meer and Gerard (98). The advantages of using apo-carotenal, a naturally occurring carotenoid that can be synthesized, as a coloring agent in nonstandardized dressings and spreads has been discussed by Finberg (99). Freeze–thaw stable salad dressings have been prepared using cooked starch mixtures made from freeze-resistant starch and unwinterized oil (100). Use of highly esterified sugar esters to reduce the caloric content of dressings has been described by Mattson and Volpenhein (101). A stable low-calorie cream-type salad dressing containing only 10–12% of fat and 5–20 cal per tablespoon has been claimed by Szczesniak and Engel (102). A patent issued to Spitzer et al. (103) has claimed the use of agar–agar and methocel mixture to produce a low-calorie dressing. An emulsion base prepared by homogenizing fats and oils, like sunflower, coconut, lard, and cottonseed, with a solution containing proteins from whey, whole milk, sunflowerseed globulins or rapeseed has been used for preparing low-calorie dressings (103). Combination use of carrageenan and carob bean or guar gum to prepare low-calorie dressings has been patented (105). Weidemann and Reinicke (106) have prepared a dietetic salad dressing with only 15–35% oil. In addition to protection provided by the low pH, pasteurization of the product at 95°C for 40 min gives the product longer shelf life.

Heat-Stable Dressings

Improvement in the shelf stability of dressings can be expected if they can be heated to high temperatures without losing body or color. Further, such heat-stable dressings can be incorporated into canned meat, fish, vegetable salads, and other foods. A new microcrystalline cellulose has been found to impart such stability (107), even when the product is retorted at 240°F in the presence of food acids. Similarly, xanthan gum has been shown to produce a product with no visible signs of oil separation and with a viscosity that remains practically unchanged over a wide range of temperatures.

A dry preparation of the ingredients that may be reconstituted to form a salad dressing simply by the addition of water has been patented (108). Several such preparations are being marketed by different companies in the United States.

References

1. G. P. Lensack, *Food Eng.*, **12**, 97–100 (1969).
2. ICI United States, Inc., Bulletin 222-8.
3. American Oil Chemists' Society Method Ccll-53.
4. F. J. Baur and E. S. Lutton (to Procter & Gamble Co.); U.S. Pat. 3,158,490 (1964).
5. F. J. Baur (to Procter & Gamble Co.), U.S. Pat. 3,211,558 (1965).
6. F. R. Hugenberg and E. S. Lutton (to Procter & Gamble Co.), U.S. Pat. 3,353,966 (1967)
6a. G. Fuller, G. O. Kohler, and T. H. Applewhite, *J. Am. Oil Chemists' Soc.*, **43**, 477–478, (1966).

7. S. S. Chang, T. H. Smouse, R. G. Krishnamurthy, B. D. Mookherjee, and B. R. Reddy, *Chem. Ind.,* 1926–1927 (1966).

8. G. Hoffmann, *J. Am. Oil Chemists' Soc.,* **38,** 1–3 (1961).

9. S. S. Lin, T. H. Smouse, and R. R. Allen, paper presented at the American Oil Chemists' Society Spring Meeting, Mexico City, 1974.

10. N. R. Artman, *Adv. Lipid Res.,* **7,** 245–330 (1969).

11. H. W. Schultz, E. A. Day, and R. O. Sinnhuber, *Symposium on Foods: Lipids and Their Oxidation,* The AVI Publishing Co., Westport, Conn., 1962.

12. Anon., *B. I. B. R. A. Bulletin 10,* 4–8 (1971).

13. R. Guillaman, *Rev. fr. Corps. gras,* **18,** 445–456 (1971).

14. J. P. Freeman, *Food Process Mark. (Lond.),* **38,** 303–306 (1969).

15. R. J. Sims and H. D. Stahl, *Baker's Dig.,* **44,** 50–52 (1970).

16. U. Shimura, *J. Jpn. Oil Chemists' Soc. (Yukagaku),* **19,** 748–756 (1970).

17. E. Yuki, *J. Jpn. Oil Chemists' Soc. (Yukagaku),* **19,** 644–654 (1970).

18. W. W. Nawar, *J. Agr. Food Chem.,* **17,** 18–21 (1969).

19. H. Schmidt, H. L. Masson, and R. A. Rybertt, *Revista de Agroquimia y Technol. de Alimentos,* **9,** 423–427 (1969).

20. B. R. Reddy, K. Yasuda, R. G. Krishnamurthy, and S. S. Chang, *J. Am. Oil Chemists' Soc.,* **45,** 629–631 (1968).

21. K. Yasuda, B. R. Reddy, and S. S. Chang, *J. Am. Oil Chemists' Soc.,* **45,** 625–628 (1968).

22. R. G. Krishnamurthy and S. S. Chang, *J. Am. Oil Chemists' Soc.,* **44,** 136–140 (1967).

23. T. Kawada, R. G. Krishnamurthy, B. D. Mookherjee, and S. S. Chang, *J. Am. Oil Chemists' Soc.,* **44,** 131–135 (1967).

24. D. M. Lee, *Bull. Brit. Food Manuf. Ind. Res. Assoc.,* No. 80 (1973).

25. V. C. Mehlenbacher, *Analysis of Fats and Oils,* Gerard Press, New York, 1960.

26. G. Valentinis and B. Romani, *Bull. Lab. Chim. Prod.,* **18,** 240 (1967).

27. M. Loury and L. Garber, *Rev. fr. Corps. gras.,* **15,** 301 (1968).

28. American Oil Chemists' Society Method, Cd 8-53.

29. S. S. Chang and F. A. Kummerow, *J. Am. Oil Chemists' Soc.,* **32,** 341–344 (1955).

30. U. Holm, K. Ekbom, and G. Wode, *J. Am. Oil Chemists' Soc.,* **34,** 606–609 (1957).

31. G. R. List, C. D. Evans, W. F. Kwolek, K. Warner, B. K. Boundy, and J. C. Cowan, *J. Am. Oil Chemists' Soc.,* **51,** 17–21 (1974).

32. G. Johansson and V. Persmark, *Oil Palm News* (10/11), 3 (1971).

33. R. G. Scholz and L. R. Ptak, *J. Am. Oil Chemists' Soc.,* **43,** 596–599 (1966).

34. C. D. Evans, G. R. List, R. L. Hoffmann, and H. A. Moser, *J. Am. Oil Chemists' Soc.,* **46,** 501–504 (1969).

35. G. A. Jacobson, J. A. Kirkpatrick, and H. E. Goff, *J. Am. Oil Chemists' Soc.,* **41,** 124–128 (1964).

36. J. A. Fioriti, M. J. Kanuk, and R. J. Sims, *J. Am. Oil Chemists' Soc.,* **51,** 219–223 (1976).

37. B. Tsoukalas and W. Grosch, *J. Am. Oil Chemists' Soc.,* **54,** 490–493 (1977).

38. H. P. Dupuy, S. P. Fore, and L. A. Goldblatt, *J. Am. Oil Chemists' Soc.,* **50,** 340–342 (1973).

39. A. E. Waltking and H. Zaminski, *J. Am. Oil Chemists' Soc.*, **54,** 454–457 (1977).

40. H. W. Jackson and D. J. Giacherio, *J. Am. Oil Chemists' Soc.*, **54,** 458–460 (1977).

41. J. L. Williams and T. H. Applewhite, *J. Am. Oil Chemists' Soc.*, **54,** 461–463 (1977).

42. H. A. Moser, C. D. Evans, J. C. Cowan, and W. F. Kwolek, *J. Am. Oil Chemists' Soc.*, **42,** 30–33 (1965).

43. R. Radtke, P. Smits, and R. Heiss, *Fette, Seifen, Anstrichm.*, **72,** 497–504 (1970).

44. M. Ahrens, G. Guhr, J. Waibel, and S. Kroll, *Fette, Seifen, Anstrichm.*, **79,** 310–314 (1977).

45. U. J. Salzer and J. Wurziger, *Fette, Seifen, Anstrichm.*, **73,** 705–710 (1971).

46. M. R. Sahasrabudhe and V. R. Bhalerao, *J. Am. Oil Chemists' Soc.*, **40,** 711–712 (1963).

47. E. G. Perkins, R. Tanbold, and A. Hsieh, *J. Am. Oil Chemists' Soc.*, **50,** 223–225 (1973).

48. M. Unbehend, H. Scharmann, H. J. Strauss, and G. Billek, *Fette, Seifen, Anstrichm.*, **75,** 689–696 (1973).

49. K. Aitzetmuller, *Fette, Seifen, Anstrichm.*, **75,** 256–260 (1973).

50. G. Billek and G. Guhr, paper presented at American Oil Chemists' Society Meeting, New York, 1977.

51. J. A. Thompson, M. M. Paulose, B. R. Reddy, R. G. Krishnamurthy, and S. S. Chang, *Food Technol.*, **21,** 87A–89A (1967).

52. A. E. Waltking and H. Zaminksi, *J. Am. Oil Chemists' Soc.*, **47,** 530–534 (1970).

53. E. R. Sherwin and B. M. Luckadoo, *J. Am. Oil Chemists' Soc.*, **47,** 19–23 (1970).

54. G. R. List, C. D. Evans, and H. A. Moser, *J. Am. Oil Chemists' Soc.*, **49,** 287–292 (1972).

55. D. Carter and M. Pike, *Oil Palm News*, No. 19, 1 (1975).

56. E. R. Sherwin, *J. Am. Oil Chemists' Soc.*, **53,** 430–436 (1976).

57. R. G. Cunningham, R. D. Dobson, and L. H. Going (to Procter & Gamble Co.), U.S. Pat. 3,415,658 (1968).

58. H. Enci, S. Okumura, and S. Ota (to Aginomoto Co., Inc.), U.S. Pat. 3,585,223 (1971).

59. S. S. Chang and P. E. Morne (to Swift and Co.), U.S. Pat. 2,966,413 (1960).

60. E. R. Purves, L. H. Going, and R. D. Dobson (to Proctor & Gamble Co.), U.S. Pat. 3,415,660 (1968).

61. J. E. Thompson and J. B. Martin (to Proctor & Gamble Co.,), U.S. Pat. 3,634,397 (1972).

61a. R. W. Fallat, C. J. Glueck, R. Lutmer, and F. H. Mattson, *Am. J. Clin. Nutr.*, **29,** 1204 (1976).

62. B. A. Erickson (to Procter & Gamble Co.), Canadian Pat. 928,140 (1970).

63. CPC International, British Pat. 1,473,208 (1977).

64. F. J. Baur (to Procter & Gamble Co.), U.S. Pat. 3,027,260 (1962).

65. M. H. Joffe, "Mayonnaise and Salad Dressing Products," The Emulsol Corporation, Chicago, 1942.

66. R. H. Forsythe, *Cereal Sci. Today*, **2,** 211 (1957).

67. W. Flückinger, *Fette Seifen Anstrichm.*, **68,** 139–145 (1966).

68. T. J. Weiss, *Food Oils and Their Uses,* The AVI Publishing Co., Westport, Conn., 1970.
69. J. W. Corran, *Food Manuf., 9,* 17 (1937).
70. D. Cummings, *Food Technol., 18,* 1901–1902 (1964).
71. D. M. Gray, *Oil Fat Ind., 4,* 410 (1927).
72. D. M. Gray, *Glass Packer, 2,* 311 (1929).
73. R. B. Smittle, *J. Food Prot., 40,* 415 (1977).
74. A. McKenzie and J. V. Ziemba, *Food Eng., 36,* 96–98 (1964).
75. D. R. Beswick, *Food Technol. N.Z., 4,* 332–339 (1969).
76. G. T. Muys and J. A. Schaap (to Unilever Ltd.), British Pat. 1,130,634 (1968).
77. R. D. McCormick, *Food Prod. Dev.,* 15–18 (February–March 1967).
78. Anon., Kyowa Hakko Kogyo Co., Ltd. (Jap.), British Pat. 1,155,490 (1969).
79. H. Enei, A. Mega, O. Ayako, S. Okumura, and S. Ota (to Ajinomoto Co.), British Pat. 1,152,966 (1969).
80. H. A. Gardner and A. W. VanHeuckeroth, *Ind. Eng. Chem., 19,* 724–726 (1927).
81. L. B. Kilgore, *Glass Packer, 4,* 65–67, 90 (1930).
82. D. M. Gray, C. E. Maier, and C. A. Southwick, *Glass Packer, 2,* 397–400 (1929).
83. L. C. Brown, *J. Am. Oil Chemists Soc. 26,* 632–636 (1949).
84. D. M. Gray and C. E. Maier, *Glass Packer, 4,* 23–25, 40 (1931).
85. M. H. Joffe, *Food Eng., 28,* (5), 62–65, 100 (1956).
86. J. P. Bolanowski, *Food Eng., 39* (10), 90–93 (1967).
87. O. C. Samuel, *Food Process. Marketing,* 81–84 (1966).
88. L. H. Rees, *Food Prod. Devel.,* 48–50 (1975).
89. A. J. Finberg, *Food Eng., 27,* (2), 83–91 (1955).
90. S. E. Potter, *Food Process.,* 43–44 (1970).
91. V. R. Carlson, *Food Eng., 42* (12), 54–55 (1970).
92. M. Lipschultz and R. E. Holtgrieve, *Food Eng., 40* (11), 86–87 (1969).
93. F. Täubrich, *Fette, Seifen, Anstrichm., 65,* 475–478 (1963).
94. A. McKenzie and J. V. Ziemba, *Food Eng., 36,* (11), 96–98 (1964).
95. C. H. Japikse (to Procter & Gamble Co.), U.S. Pat. 3,425,843 (1969).
96. L. R. Stanislav and J. K. Sheets, *Food Prod. Devel.* (10), 56 (1971).
97. M. T. O'Brien, *Food Prod. Devel.* (9), 13–14 (1977).
98. G. Meer and T. Gerard, *Food Process.* (5), 170–171 (1963).
99. A. J. Finberg, *Food Prod. Devel., 4,* 46–47 (1971).
100. A. Partyka (to National Dairy Products Corp.), U.S. Pat. 3,093,485 (1963).
101. F. H. Mattson and R. A. Volpenhein (to Procter & Gamble Co.), U.S. Pat. 3,600,186 (1971).
102. A. S. Szczesniak and E. Engel (to General Foods Corp.), U.S. Pat. 3,300,318 (1967).
103. J. C. Spitzer, L. S. Nasareisch, J. L. Lange, and H. S. Bondi (to Carter Products, Inc.), U.S. Pat. 2,944,906 (1960).
104. J. Kroll, G. Mieth, M. Roloff, J. Pohl, and J. Bruecker, German (East) Pat. 106,777 (1974).

105. U. Steckowski (to Carl Kühne KG), German Pat. 2,311,403 (1974).

106. H. Weidemann and H. P. Reinicke, German Pat. 1,924,465 (1970).

107. C. T. Herald, G. E. Raynor, and J. B. Klis, *Food Process. Mark.*, **11**, 54–55 (1966).

108. M. H. Kimball, C. G. Harrell, and R. O. Brown (to Pillsbury Mills, Inc.), U.S. Pat. 2,471,435 (1949).

6
Miscellaneous Oil and Fat Products

In addition to the major classes of oil and fat products described in the preceding chapters, a number of other materials are important commercially. Fats and oils are used in a quite extraordinary range of industrial products. In some of these applications the ability of fats and oils to convert to polymers is the critical factor as in linoleum, shade cloth, oil cloth, factice, patent leather, putty, brake linings, or core oils. Lubricity is the important factor in other uses such as lubricating oils, greases, cutting oils, plasticizers, or leather and textile treating compounds. Surface activity is of dominant importance in most uses of soaps, fatty nitrogen compounds, or fat-based synthetic detergents. A combination of emollient, surface-active, and lubricating properties is important in cosmetics; a blend of high melting point and fuel value is important in candles; and in other products such characteristics as viscosity, resistance to oxidation, low volatility, or compatibility with modifying agents may be critical.

In general, the manufacture of these miscellaneous or specialty products is not standardized to nearly the same degree as that of more widely used fatty products. Often their preparation is more of an art than a science, and the method of manufacture and the materials used vary widely from one manufacturer to another. Because of this, and also because of the great diversity of minor uses for fats and oils, it is hardly possible to review this field as comprehensively as those previously covered. However, an attempt is made here to indicate the nature and the composition of some of the more important secondary products.

In many specialty uses a fatty oil is employed because of its ability to dry or polymerize. In many applications where polymerization does not take place, the fatty oils compete to some extent with mineral oils. Since the latter are stable toward oxidation and also are relatively cheap, they

343

are generally preferred except where the distinctive properties of the fatty oils or fatty acids confer on these materials particular advantages.

About 6 billion pounds of fats and oils are used annually in inedible products in the United States, and about 4 billion pounds of these are consumed in the miscellaneous industrial products described in this chapter. Table 6.1 summarizes the volumes of fats and oils used in inedible products in the United States during the last two decades. Very little statistical information is available on fats and oils consumption for the three-quarter billion pounds or so described as "other inedible products," and some of the volume information supplied subsequently represents educated estimates.

Drying Oil Products

1.1 Linoleum (1)

From 1900 until the mid-1950s linoleum was the dominant flexible floor covering. Primarily because of high labor costs compared with competitive floor coverings based on vinyl and phenolic resins or asphalt, linoleum has been largely replaced in the market place. From a technical standpoint the manufacture of linoleum is quite unique, and a review of processing is in order. The manufacture of linoleum is a relatively old industry. The first linoleum factory was built in England in 1864, and by 1874 linoleum was being made in the United States (2). Empirical processing methods developed over the years, but ultimately much of the technology was placed on a sound technical basis. Despite the traditional rather unscientific treatment of the constituent oils and resins in linoleum manufacture, high production requirements for a material with intricate inlaid designs led to extremely ingenious and altogether revolutionary improvements in the machinery for linoleum manufacture.

In its simplest form linoleum consists of one-third binder (oxidized oil, usually linseed, natural gums, and rosin), one-third inorganic fillers (pigments and ground limestone), and one-third organic fillers (ground cork and wood flour).

The essential material for the manufacture of linoleum is a specially prepared, highly oxidized and polymerized drying oil that has been compounded with rosin and usually other natural or artificial resins, to yield a stable, resilient, yet thermosetting product known to the industry as "cement." The cement is mixed with ground cork or wood and pigments or extenders and pressed onto a backing of burlap or other coarse fabric. In so-called inlaid linoleum, the design is made up of separate blocks or

Table 6.1 Consumption of fats and oils in inedible products in the United States[a]

Item	Consumption (Millions of Pounds)		
	1959	1966	1976
Soap	957	840	976
Paint or varnish	476	343	321
Fatty acids	1150	1950	1978
Feed	505	892	1464
Linoleum	63	32	(w)[c]
Resins and plastics	186	227	122
Lubricants and similar oils	151	192	200
Other inedible products[b]	577	763	640
Totals	4046	5278	5700

[a] U.S. Department of Commerce, Current Industrial Reports, M20K, 1962, 1967, 1977.

[b] Includes cosmetics, drying oils packaged for sale, glazing and caulking compounds, pharmaceuticals, polishes and waxes, printing inks, rubber and rubber products, sulfonated oils, synthetic organic detergents, toilet preparations, and other products.

[c] Consumption for linoleum in 1976 withheld to avoid disclosing figures for individual companies.

other figures of cement, which are pigmented with different colors. This method of manufacture causes the design to extend some distance into the linoleum, so that it is not readily worn away. Linoleum is flexible and resilient, relatively nonconducting to heat, and very durable.

True linoleums are to be distinguished from the much cheaper and less durable felt-base floor coverings, which are made by impregnating a felt-like material with asphalt and applying and baking a thick coating of enamel onto the upper surface. The print paints for felt-base floor covering are high in solids, quite viscous, and thixotropic, so that rather intricate multicolor patterns can be applied without any appreciable flow that would blur the pattern during baking. The manufacture of felt-base floor coverings has been described briefly elsewhere (3). In 1948 the United States production of linoleum and felt-base floor coverings amounted to 75 and 280 million square yards, respectively. By 1958, the corresponding figures were approximately halved (4), and the yardage was reduced to a low level by 1972 (5).

The first step in the manufacture of linoleum is the oxidation and polymerization of a suitable drying oil or mixture of drying oils until a relatively dry and nontacky solid is produced. Linseed oil is the most suitable raw material, although soybean and fish oils and tall oil fatty acid esters are used also. Tung oil or other conjugated acid oils tend to polymerize without extensive oxidation and hence are unsuitable. The process is primarily one of oxidative polymerization rather than thermal polymerization; it is essential for the oil to take up a large amount of oxygen before any considerable increase in its viscosity occurs. In the older processes, this limited the temperature for treating the oil to about 48–52°C; higher temperatures, for example, 60°C or above, were said to yield inferior products.

Originally, polymerized oil for the manufacture of linoleum was made by the "scrim" process, wherein lengths or scrims of light cotton fabrics were suspended in a warm room provided with air circulation and periodically (about every 24 hours) flooded with boiled linseed oil. After about 3–6 months the fabrics became coated with a layer of polymerized oil about 1 in thick, which was then stripped off.

The scrim process was largely replaced by the "smacker" process in which the oil, heated to about 50°C, was stirred in a horizontal jacketed drum equipped with a horizontal agitator with radial arms or paddles under a strong current of air. Before the smacker, the oil was previously oxidized to some degree by other means. Older practices included air blowing or spraying at a low temperature. A much more effective process made use of intensive mechanical agitation of oil and air in a machine of the turbo-gas-diffuser type (6, 7). In such an apparatus, the oxygen content of linseed oil can be increased from about 11 to 17–18% before the oil becomes highly viscous. Much higher temperatures (well above 95°C) can be employed, and the reaction is completed within a few hours, or a small fraction of the time required for attainment of a comparable state of oxidation by the conventional air-blowing or spraying processes.

In the smacker, the oil is often mixed with a small quantity of whiting or other extender. The product from the smacker is cooled and, after solidification, is cut up into chunks that are then stored or "stoved" for several days at about 35–40°C. During this period the product becomes still stiffer and more crumbly, and swells to form a honeycomb structure from the generation of gases within the mass. Swelling will occur without the presence of whiting or other carbonate and, in improperly processed material, may fail to take place in the presence of carbonate; hence it cannot be attributed to the generation of carbon dioxide from the latter.

In the second step in the preparation of linoleum cement, the polymerized oil is fluxed with rosin or other resins. The resin mixture originally recommended consisted of equal parts of rosin and kauri gum, but gradually the latter resin was omitted in favor of an all-rosin system (8). Approximately 1 part of resin is used to 4 parts of polymerized oil. The oil and resins are heated together at about 135–150°C until the desired degree of reaction and thickening has occurred. If kauri or other fossil gum is used, it is not previously "run," as in varnish making, to make it completely compatible with oil, and although reaction takes place between the polymerized oil and the rosin, to yield a particular variety of thermoplastic resin, fossil gums or hard synthetic resins appear simply to stiffen the product mechanically.

The final step in the development of the body of the finished material takes place after the linoleum is formed, when it must again be cured by "stoving" or "seasoning" at a moderately elevated temperature (65–95°C) in large ovens for a period of several days or weeks.

Linoleum cement is not merely another form of infusible drying oil–resin polymer, similar to that in an ordinary dried varnish film. In the manufacture of linoleum the separate processes of heat polymerization and oxidation polymerization are reversed from their usual order (8), as in the formation of protective coatings. It should be particularly noted that linoleum cement is unlike a hardened paint or varnish film in that it is fusible and thermosetting. This property is important in its relation to the mechanism of linoleum manufacture. Fusibility in the cement is particularly essential in the manufacture of inlaid linoleum, as this process requires the welding together of many separate blocks of materials and the reworking of much scrap. In a few large modern plants inlaying was carried out by entirely automatic high-speed machinery (2).

The chemistry of linoleum cement manufacture is even less well understood than that of ordinary polymerization processes involving a drying oil. The reaction occurring between the rosin and the polymerized oil in the fluxing operation has never been fully explained. The monofunctional character of the rosin is considered important (8). Presumably, it modifies the polymerization of the glycerides in the oil by reducing total functionality, to preserve the heat-convertible nature of the product. Although rosin itself is not an altogether essential component, some such acidic material appears to be indispensable; a peptizing effect of the rosin acids on the drying oil gel has been suggested. The chemical aspects of linoleum-making have been discussed at length by de Waele (9), whose theories, although old, do not appear to have been greatly improved on.

From a practical standpoint, the success of linoleum cementmaking largely depends, as stated previously, on the somewhat difficult matter of maintaining a high degree of oxidation polymerization in relation to heat polymerization.

The consumption of fats and oils in floor covering manufacture in the United States decreased from the maximum of about 167 million pounds in 1948 to about 63 million pounds in 1959 (10), 21 million pounds in 1967 (11), and perhaps half that volume in 1972 based on the dollar volume (5).

1.2   OILED FABRICS

Oiled fabrics are another class of drying oil-based products that have been largely supplanted by petrochemical-based plastic materials.

There are two classes of oiled fabrics: that exemplified by ordinary "oilcloth," in which the oil forms a continuous coating on one side of the fabric, and that in which the entire fabric is impregnated with oil. Fabrics of the first class are used as coatings for walls, tables, shelves, and other surfaces that are not walked on or otherwise subjected to severe abrasion. Those of the second class are coated to make them water repellant, and are used chiefly for raincoats, machine and instrument covers, and so on. Both classes, but particularly the second class, must be pliable.

Oilcloth is made by applying successive coats of paints and varnishes to a cotton backing, with the final coat designed to impart a high gloss to the surface. The paints and varnishes must be compounded to produce a coating that will withstand moderate flexing without cracking. Hence they cannot be made exclusively with highly drying oils but must also contain plasticizers. Shade cloth is another type of coated fabric in which oils or long oil alkyd resins are used as binders for pigments that provide varying degrees of opacity for light control in windows. Flexibility retention over long periods of time under the polymerizing effect of sunlight is essential to prevent cracking as shades are rolled up and unrolled.

Linens, silks and synthetics, such as polyamides and polyesters, are the fabrics that are usually oiled. The oiling operation consists merely of thoroughly impregnating the fabric with unpigmented drying oils and allowing each application of the oils to dry thoroughly. Strongly drying oils, such as perilla or tung oils, are less suitable for this purpose than oils that yield softer and more pliable films, such as linseed and fish oils.

At one time the cordage used in fishing nets and so forth was usually oiled, and the oils most used for the purpose consisted of linseed oil and fish oils, but now synthetics have largely replaced the oiled natural fiber products.

1.3   PUTTY AND OTHER SEALING OR CAULKING MATERIALS (12, 13)

A variety of sealing materials and cements have a drying or semidrying oil base. Ordinary window putty is the most common example. About 18 million pounds of this material was factory produced in the United States in 1958. Gradually the volume of putty decreased, partly because of replacement by the more resilient glazing and caulking compounds and partly because of construction changes such as metal storm windows with rubberlike mastic seals in place of the older wood construction.

Putty is usually composed simply of a thick, plastic paste of whiting (calcium carbonate) and linseed oil. Putty hardens and thus is not satisfactory for sealing where there may be slight relative movement of components. Glazing compounds, mastic products that are usually based on heavy blown soybean oil, have replaced putty to a considerable degree. The glazing compounds develop a heavy surface skin with good painting characteristics, but the bulk of the sealant remains plastic in character for long periods of time.

Some sealing materials with a fatty oil base are treated to produce rubberlike materials. Others are more nearly like heavy paints or varnishes. The stearine pitch remaining from the distillation of fatty acids may be placed in this category of products; it is used in the manufacture of wall board, floor tiling, electrical insulation, industrial paints, and other products requiring a relatively high-grade pitchlike material.

1.4   Rubberlike Materials

Unsaturated fats and oils, and particularly the drying oils, are capable of polymerizating to form various elastic, rubberlike materials.

The oldest materials of this class are the factices, which are oils polymerized with the aid of sulfur. The action of sulfur in this case is entirely similar to its action in the vulcanization of rubber; that is, it modifies the structure of the polymer by providing a ready means of cross-linking through sulfur or disulfide bonds.

There are two varieties of factice: white and brown. White factice is prepared by reacting a relatively saturated oil, such as castor oil or rapeseed oil, with liquid sulfur monochloride. The reaction takes place readily in the cold and is quite complex (14, 15). There is no particular relation between the unsaturation of the oil and the amount of reagent absorbed. Substitution also appears to take place, since in some cases, particularly in the treatment of highly unsaturated fish oils, there is a considerable evolution of hydrogen chloride gas.

White factice is a light-colored, compressible, but more-or-less crumbly material, which has been used as an extender or modifier for rubber. An important use for this material is in the manufacture of erasers, where it confers the degree of friability essential in this product.

Brown factice is made by first blowing and then bodying a drying oil until thickened, followed by reaction with about 5–30% of its weight of sulfur in a closed vessel, with stirring at a temperature of about 120–165°C, for 1–2 hours. Depending on the percentage of sulfur, the temperature, and the reaction time, the product varies in consistency from a dark, viscous, and sticky semisolid to a hard and relatively brittle solid. The characteristics desired depend on the specific use to which the factice is to be put. The chemical reactions taking place in the manufacture of brown factice are undoubtedly even more complex than those involved in making white factice, but again they may be compared to those occurring in the vulcanization of rubber.

Brown factice has been used not only as a rubber extender, but also to modify the properties of drying oil products, such as varnishes and linoleum.

Although somewhat rubberlike, none of the factices possesses the combination of elasticity and tensile strength of vulcanized natural rubber or its synthetic substitutes because of the polyfunctionality of glycerides, and the consequent production of an extensively cross-linked structure rather than one of long linear chains with occasional cross-linking characteristic of elastomers.

An essentially linear polymer of higher molecular weight can be constructed from fatty materials; for example, from dimerized linoleic acid esterified with a glycol or other dihydric alcohol. Hence by special techniques it has been found possible to make a fatty oil-based product that resembles rubber to a large degree. This product was developed under the name of Norepol (16, 17). It was manufactured on a trial scale during the rubber shortage created by World War II and would probably have found considerable use had soybean oil, the preferred raw material, remained in good supply in excess of edible oil requirements.

The first step in the manufacture of this material involved conversion of the polyfunctional glycerides to a monofunctional form by reaction of the oil with methyl alcohol or other monohydric alcohol, to yield monoesters. The latter were then heat polymerized, the reactive esters principally forming dimers, and the nonreactive esters remaining as monomers. The latter were removed from the mixture by fractional distillation, and the monohydric alcohol attached to the dimeric residue was replaced with ethylene glycol. This product was then further polymerized, after which it was vulcanized and compounded to yield the finished rubberlike material.

The rubberlike products so far manufactured from fat-based materials lack the tensile strength and the abrasion resistance of natural rubber and some other synthetic products, but they are satisfactory for most purposes where these properties are not of prime importance. They are particularly suitable materials for gaskets, stoppers, bumpers, tubing, electrical insulation, and the like, and for the rubberizing of fabrics. Their resistance to oxidation is good. About 1.3 million pounds of fats were used in their manufacture in 1962 (10).

1.5   CORE OILS (12)

Core oils are used as binding agents for the sand cores of hollow metal castings. The cores are prepared by mixing approximately 50 parts by volume of sand with 1 part of oil, molding the mixture in a wooden form, and baking at a temperature of about 200–235°C until a hard, coherent mass is formed. The sand is usually dampened to make it more easily molded, and a water-soluble binder such as gelatinized starch or dextrin is sometimes added to assist in maintaining the core in the proper form until it becomes hardened through heat polymerization of the oil.

The core should have sufficient mechanical strength to retain its form during the casting operation but should not, of course, be so hard as to be difficult to break and remove from the finished casting. It must have a certain degree of permeability to permit the escape of gases evolved during casting. Test specimens baked and tested in an ordinary cement tester should have tensile strengths of the order of 200 psi.

The oils used in core making include linseed, soybean, tung, fish, and esterified tall oils, and frequently they are combined with highly unsaturated polymeric petroleum resins. The oils are usually bodied so that the finished core oil has a viscosity of A–G, and they are used without driers. About 9 million pounds of fats were used in their manufacture in 1962 (10).

1.6   BRAKE LININGS

Original equipment automotive brake linings are commonly based on phenolic resins as binders. In contrast, replacement brake linings are almost invariably based on linseed oil because they are more flexible than linings from phenolic resins and will conform to the changed radius of worn brake drums without cracking when brakes are relined.

Typically the lining compound is formed by mulling a low-acidity vacuum-bodied M-37 linseed oil with lime, asbestos, sulfur, and solvent. The uncured or "green" lining compound is formed and baked at 50–60°C for several hours to remove the bulk of the solvent following which the temperature is raised and the lining is baked for about 5 hours at 150–160°C. The chemical reactions by which the highly polymerized linseed oil is cross-linked through sulfide or disulfide bonds are probably similar to the action of sulfur in vulcanization.

When brakes are applied vigorously at high speeds, friction between the lining and the shoe can increase the temperature precipitously. Under these conditions braking power decreases, or as it is usually described, the brakes "fade." Linings based on linseed oil tend to fade somewhat more than the phenolic based linings.

1.7 Concrete Protection and Curing

Concrete Protection. Concrete roadways and bridges deteriorate when exposed to alternate freezing and thawing, and this surface scaling and spalling is accentuated when salt is applied as a deicing agent. In the 1920s and 1930s boiled linseed oil dissolved in an equal volume of kerosene or mineral spirits was used to protect roads in a number of northern states (18). When air entrainment of concrete was introduced for road construction, the use of linseed oil was discontinued, since it was shown in laboratory studies that properly air entrained concrete had excellent freeze–thaw resistance, even in the presence of salt.

In the highway building boom started in the United States in the 1950s following World War II, air-entrained concrete was specified for open road surfaces, and conventional non-air-entrained concrete was usually specified for bridges because of its considerably greater strength. Many states in the snow belt adopted a "bare pavement" policy with increased use of deicing chemicals (usually sodium and calcium chlorides). Severe deterioration of bridge decks became common, but serious deterioration of air-entrained pavement surfaces also occurred. Both linseed oil solutions and linseed oil emulsions came into use for reducing damage to concrete highways, bridge decks, and parking ramps caused by wintertime application of salt (19–23). In a test of 110 coatings for concrete, Snyder (24) found that linseed oil solutions were best. Most test systems (epoxy resins, polyvinyl butyral, hydrocarbon resins, polyvinylidene chloride, acrylics, silicone resins, rubber solutions, lubricating oil, asphalt, etc.) rated poor in performance, usually with moderate scaling in fewer than 27 freeze-thaw cycles, whereas concrete coated with linseed oil solutions passed an excellent 200-plus freeze–thaw cycles.

To explain the marked increase in low-temperature deterioration of concrete in the presence of salt it is generally postulated that there is a gradient of salt concentration in concrete from high levels at the surface to levels approaching zero at some interior depth. As the temperature lowers from the freezing point to subzero temperatures freezing exerts its expansive forces at different levels in the concrete structure. Whereas unsalted pavements undergo a freeze–thaw cycle only at the freezing point of water, there will be freeze–thaw cycles at some depth of the salt-treated concrete whenever the temperature rises or falls in the temperature range below the freezing point of water. When linseed oil solutions or emulsions are applied to concrete, they penetrate to a depth of several millimeters (25), probably with formation of a relatively impermeable membrane that minimizes salt penetration.

Concrete Curing. When concrete is freshly laid, it is important that moisture be retained if the concrete is to attain design strength. Surface wear and deterioration are closely related to moisture retention during curing. Polyethylene sheeting is effective as a curing membrane for flat highway surfaces, but it is difficult to lay the polyethylene and maintain an effective barrier in windy weather. Liquid membrane curing agents based on resins or waxes were developed originally as a replacement for wet burlap, and they are widely used in field laying of concrete. Linseed oil solutions in mineral spirits (26) and linseed emulsions (21) proved effective for concrete curing, and they provide carryover protection against freeze–thaw cycles, whereas the resin and wax membrane curing agents provide no long-term protection. Water-soluble compositions based on maleic and fumaric acid adducts of linseed oil (27, 28) proved highly effective as curing agents that also provide carryover protection of the concrete against deterioration due to salt during freezing weather.

Despite the hugh potential for vegetable oil-based coatings for concrete the annual volume is only in the several million pound per year range.

1.8   HARDBOARD

Linseed oil and other drying oils are used both as internal binders (integral oil) and for external treatment (tempering oil) of the reconstituted wood board products known as *hardboard*. Approximately 10 million pounds of drying oils are used annually in hardboard manufacture.

The typical wet process for manufacture of hardboard involves steps akin to those involved in paper manufacture. Logs are debarked and chipped, the resultant chips are pressure cooked for a few minutes at 175–275 psi, and the pressure-cooked chips are reduced to a pulp in disc refiners. The pulp is suspended in recycle white water from the fourdrinier together with a small amount of make-up water. By recyling of process waters, integral oil that is not picked up on the board at the forming stage on the fourdrinier can be absorbed by the fiber at the pulp stage. After adjustment of consistency the pulp is treated with 1–1.5% of integral oil, an emulsion of drying oil and resins, and the treated pulp is applied continuously from a headbox to the screen of a forming machine or a fourdrinier. Over a distance of 10 ft or so the white water is largely removed leaving a wet sponge like board varying from $\frac{1}{2}$–1 in in thickness depending on the thickness desired in the final pressed board. A separate overlay of finer pulp containing a higher percentage of integral oil may be applied to the partially drained board through a second head box to provide a smoother veneerlike surface.

The wet board is cut to length and dried in huge ovens from which the boards emerge at an exit temperature of 130°C and a few percent moisture. The oven residence time is varied depending on whether $\frac{1}{8}$ or $\frac{1}{4}$ inch board is being manufactured. The dried boards are trimmed to press size, commonly 4 by 16 ft, and pressed between metal platens in a hydraulic press at a pressure of 4000 psi and temperatures of 210–240°C. The platens may be smooth, or they may be embossed with a variety of simulated wood grain textures, knots, wood imperfections, and panel separation grooves. Hot boards from the press are roller coated with approximately 1% by weight of tempering oil, and the process is concluded by baking for several hours at 135–150°C. The final section of the oven is not heated, but air saturated by bubbling through water is passed through this section to humidify the boards to their equilibrium moisture content of approximately 3%. Rehumidifying prevents subsequent board warpage.

There is a one-of-a-kind installation is which hardboard is made by dry processing. Cooked chips are separated into fibers, phenolic resin is added, and the treated fiber is dried. The resin-treated fibers are deposited on an endless belt in a 0.5-ft-thick layer, and the belt moves under a series of reduction rolls that evens the distribution of the fiber bed and reduces the thickness to about 25% of the original depth. Finally the fiber mat is cut, pressed, tempered, and baked as for the wet-process board.

Tung oil is preferred for both integral and tempering oils, because it heat cures more rapidly than other drying oils. Most board is now based on linseed oil because of the high price and short supply of tung oil. Integral and tempering oils may have the same composition, but more commonly they differ in both composition and drier levels. Integral oil strengthens the board and improves dimensional stability. The primary function of the tempering oil is to improve paintability, but the tempering oil has a favorable effect on other physical properties such as tensile strength, moisture resistance, toughness, machineability, and nailing chracteristics.

1.9   TANNING OILS

In the operation of oil tanning, the skins are impregnated with suitable drying oil and stored at a warm temperature (e.g., 35–45°C) until considerable oxidation of the oil has taken place. The excess of oil is then removed by scouring. Tanning occurs as the result of chemical combination of aldehydes or other products of oxidation with the collagen of the skins. In the process a sufficient quantity of oil or of long-chain glyceride fragments is fixed to make the finished leather very soft and pliable.

Oil tanning is employed in the tanning of furs and also for making chamois and other washable leathers for gloves and other materials.

The oils used in tanning consist mostly of marine oils, of which cod oil appears to be particularly preferred. The particular desirability of marine oils is attributed to the fact that they contain substantial proportions of both highly unsaturated and moderately unsaturated or saturated acids. Since the fatty acids are more or less evenly distributed in the glyceride molecules, the average molecule will thus contain both highly unsaturated acids, to make it oxidizable and reactive with the leather, and saturated or monoethenoid acids, to remain unoxidized and contribute a lubricating action.

Lubricants and Plasticizers

2.1 LUBRICATING OILS

Approximately 200 million pounds of fats and oils are used in lubricants (Table 6.1), but the major consumption is in greases, and only limited quantities are used directly for lubrication.

Fatty oils were once extensively used as lubricants for machinery; now they are largely supplanted by mineral oils. The latter do not hydrolyze and become acid and corrosive in use, and they are cheaper than fatty oils. Fatty oils have certain special advantages, however, that have ensured their continued use. About 20 million pounds of fats were used in the manufacture of lubricant oils in 1962 (10). The principal virtue of these oils as lubricants is their superior ability to cling to metal surfaces in the form of very thin films. Actually, this property appears to be due largely to the surface activity conferred by the small amount of free fatty acids occurring in the oils (29). The free acids are polar and tend to become adsorbed in layers of molecular dimensions at the metal–oil interface. The interposition of such films is effective in preventing metal seizure under conditions of extreme pressure or under other conditions tending to displace gross films of lubricants between the bearing surfaces. Fatty oils are also less easily displaced from metal surfaces by water than are mineral oils and hence are valuable ingredients for lubricants designed for the cylinders and the valves of steam engines.

Sulfurized fatty oils are sometimes added to lubricants for the preparation of cutting oils and extreme-pressure lubricants. It is thought that the sulfur prevents contact between closely opposed ferrous metal surfaces through the formation of an intermediate film of iron sulfide.

Fatty oils are used in some quantity for the lubrication of very light machinery and delicate mechanisms, such as watches, clocks, scientific instruments, spindles, looms, and sewing machines. Suitable oils are those sufficiently saturated to be free from gumming tendencies, yet liquid at ordinary temperatures, such as lard oil and neatsfoot oil. Olive oil belongs to the class of liquid, nondegumming oils, but it is not used to any extent as a lubricant because of its high price. Sperm oil, which is really a liquid wax for the most part rather than a triglyceride oil, was a most important lubricant for delicate equipment in the past, but the sperm whale was placed on the endangered species list and importation of sperm oil into the United States was banned in 1970. The ban of sperm oil stimulated research on esters as replacements (30–32).

Castor oil is more viscous than ordinary oils and hence is suitable for lubricating fairly heavy machinery. Its viscosity changes only slightly with temperature, and it has a low cold-test or solidifying point (ca. $-20°C$); hence it was once considered more or less essential for the lubrication of piston-type airplane engines. In this and similar applications it has been replaced by the newer low-cold-test mineral oils and especially by synthetic esters, which have become the lubricants of choice for jet aircraft engines and are now being used as long-life automotive lubricants (33, 34).

There is some production of blended oils, comprised of mineral oils containing a fatty oil. "Blown" rapeseed oil or rapeseed oil that has been thickened by polymerization is a particularly common ingredient of such products. It not only contributes surface-active properties to the blend, but also increases the viscosity of the mineral oil. Blends containing about 5–25% of fatty oil are particularly used for machinery operating under heavy loads, or at high temperatures, such as marine and diesel engines. The blends are less easily washed away from metal surfaces by water and steam than are ordinary mineral oils, and hence they are preferred for the lubrication of steam engines. Low-acid lard and tallow oils are also much used in compounding.

Lubricant demands continue to become more stringent, for instance, military specifications for aircraft require a maximum pour point of $-54°C$ ($-65°F$), a minimum flash point of $218°C$ ($475°F$), and a high viscosity index coupled with good oxidative resistance (35). Di-2-ethylhexyl sebacate (DOS) with added antioxidants and special compounding materials meets these requirements, and azelaic esters as well as pelargonic esters also find use as lubricants that combine good high-temperature stability and fluidity at low temperatures.

In a review of fat-based synthetic lubricants, Matthews (34) describes four major product types:

1. Simple monobasic acid esters such as methyl stearate, methyl oleate, methyl 12-hydroxystearate, butyl oleate, or hexyl stearate are used in lithium-base greases, textile fiber lubricants, mold release agents, rolling oils, and cutting oils.
2. Dibasic acid esters, such as dioctyl azelate or di-2-ethylhexyl sebacate, find application in compressors, gas tubines, hydraulic fluids, instruments, turbo and jet aircraft engines, and two- and four-cycle engines. They are characterized by high viscosity indexes and low pour points, and compared with mineral oils they have markedly improved thermal and oxidative stability.
3. Glycol esters, such as dipropylene glycol dipelargonate or triethylene glycol dicaprylate-caprate, have good overall properties, but they lack oxidative and thermal stability for the most demanding applications.
4. The most important fat-based esters are the monobasic acid esters of branched polyols, such as neopentyl alcohol, neopentyl glycol, trimethylolpropane, pentaerythritol, and dipentaerythritol. The most important chemical feature of these esters is the absence of hydrogen atoms on the highly-branched carbon atom of the alcohol portion of the molecule. This structure provides high thermal and oxidative stability.

There has been vigorous research on alternative fat-derived ester lubricants. Acrylic acid addition to linoleic acid in the presence of a catalyst produces a C_{21} dicarboxylic acid whose esters have potentially attractive properties as lubricants (36). Mono-, di-, and tricarboxystearic acids can be synthesized by hydroformylation followed by oxidation of oleates, linoleates, and linolenates. Several esters combine desirable high viscosity index, low temperature fluidity, and in some cases fair to good antiwear properties (37). Saturated cyclic acids can be prepared from linseed oil by heat treatment in the presence of alkali followed by hydrogenation and subsequent isolation of the cyclic fraction. Esters of these cyclic acids have low pour points, thus suggesting potential for lubricants (38). In one ambitious study Magne et al. (39) investigated 61 fat-derived compounds as lubricants and lubricant additives, particularly fatty esters of mono- and polyhydric alcohols, but also including some fatty amides and imidazolidines. Substituents such as chloro, phosphato, phosphorodithio,

thio, epithio, mercapto, and sulfuryl were evaluated for their effects on antiwear and extreme pressure lubricant performance, with the epithio (thirane) group showing the most consistent effect in enhancing extreme pressure and antiwear performance.

2.2 LUBRICATING GREASES (40, 41)

Many mechanical devices are constructed in such a manner that they cannot be lubricated conveniently or efficiently by a liquid oil, but require a semisolid, plastic lubricating grease. More than 90% of these greases consist of lubricating oils thickened by the addition of various soaps.

Lubricating greases are prepared by heating a mixture of lubricating oil and soap to a high temperature to form a homogeneous solution and then cooling the mass with continuous agitation. Soap separates on cooling in the form of curd fibers, neat soap, or other crystalline material and, as it is intimately dispersed in the oil, forms a plastic solid with the latter. The characteristics of the finished greases depend on the mineral oil used, the relative proportions of oil and soap, the fatty materials going to make up the soap, the metal used for forming the soap, the water content of the composition, and the method of solidifying and working the grease. In addition, certain minor constituents or additives, including glycerol from soaps formed in situ from a fat, may greatly influence the physical structure of the product. The diverse lubrication requirements of modern machinery make the technology of these products highly involved. Only a bare outline of the subjects can be presented here.

Annual production of lubricating greases in the United States is about 1.5–1.7 billion pounds, and the corresponding consumption of fatty materials is about 150 million pounds. The principal fatty raw materials (over 60%) are inedible tallows and greases together with the corresponding hydrogenated glycerides plus a considerable volume of hydrogenated fish oils and minor quantities of other oils. About one-fourth of the total quantity of fatty material is used in the form of the separated fatty acids, including commercial oleic and stearic acids and an increasing volume of hydrogenated castor oil fatty acids.

The metallic components of the soap, in order of importance, are sodium, calcium, lithium, aluminum, barium, and lead. Calcium and sodium soap greases are the two types most widely used and account for about 70% of the production. Lithium soap greases, largely lithium 12-hydroxystearate, represent about 10–15% of the market. Lead soaps are sometimes incorporated in certain types of grease for their lubricating value, rather than for their stiffening effect. In addition to mineral oil and soaps, greases may also contain such ingredients as asphalt, petrolatum, mineral wax, rosin, dyeing agents, perfuming substances, and inorganic materials, including graphite, organobentonites, and other nonabrasive solids.

Extreme-pressure additives are not needed for general-purpose industrial uses, but they are required under heavy loads as in steel rolling mills or in many gear applications. Lead soaps and various additives containing sulfur, chlorine, and phosphorus are used.

If a fat rather than a fatty acid is used in forming the soap, and if saponification is carried out directly in the mineral oil, as is often the practice, glycerol from the fat will remain in the grease. In some greases, particularly in calcium soap greases, considerable unreacted fat may remain, as well as substantial amounts of free alkali; a minor proportion of a fatty oil is occasionally added. A small amount of water is commonly associated with the soap to maintain a smooth, buttery type of cup grease structure.

The consistency of a lubricating grease depends much more on the amount of soap incorporated than on the viscosity of the mineral oil. Often a series of greases of widely varying consistency is prepared from a single mineral oil by simply using different proportions of soap. Thus, for example, ordinary calcium-base cup grease may contain from about 7% soap in the No. 00 grade to about 35% soap in the No. 6 grade. This is the ordinary range of soap content in most greases, although considerably less soap may be used where the object is merely to thicken a fluid oil. Very stiff, high-melting greases designed for service at high temperatures, such as locomotive driving journal compounds, may contain 50 or even 60% soap. The consistencies of lubricating greases vary from fluid soap-thickened oils to bricks that must be cut with a knife.

The body characteristics of lubricating greases are naturally related in an intimate way to the crystal or gel structure of the soap phase. As a result of microscopic study of greases, a classification of these materials has been proposed on the basis of the average length of their soap fibers, as shown in Table 6.2. The suitability of greases for specific uses can be correlated with fiber length. Thus long-fiber greases are particularly adapted to use in gear boxes, since they are readily worked between the

Table 6.2   Classification of greases (**42**)

Type	Fiber length (μ)	Character
Long-fiber	≥ 100	Fibrous, ropy
Medium-fiber	10–100	Clinging, slightly rough
Short-fiber	1–10	Slightly rough, short
Microfiber	<1	Smooth, unctuous

moving parts by the action of the gears, and they tend to take up the slack in worn gears and eliminate noise. Moreover, they are easily retained within a gear housing. On the other hand, these greases are not good lubricants for ball and roller bearings, since they do not cling well to the balls and rollers. Short-fiber greases are the most suitable for locomotive journals and open-type roll bearings, whereas cup greases are always of the microfiber type.

Long-, medium-, and short-fiber greases are made with sodium soaps, with the fiber length generally tending to become greater as the fat stock becomes higher in iodine number and less saturated. The presence of water, glycerol, or other polar compounds also facilitates the production of long fibers.

Calcium, aluminum, and lithium greases are usually classified as "microfiber" products, although actually the stiffening agent in these consists largely of a liquid crystalline phase and, in some cases at least, crystals or crystallites appear to be absent altogether.

In addition to the consistency at ordinary temperatures, the factors of resistance to high temperatures and to water are important. Table 6.3 summarizes the approximate maximum continuous operating temperatures of various greases. Sodium greases have a high melting point (usually above 150°C) but comparatively poor water resistance, whereas calcium and aluminum greases are characterized by good water resistance but a low melting point (below 95°C). The increasing popularity of lithium and barium greases is due in large part to their combination of good heat and water resistance.

Sodium greases are commonly made by saponifying the fat with caustic soda in open kettles in the presence of a portion of mineral oil, commonly in the SAE 20–30 range, although higher-viscosity oils may be used for special effects, heating to drive off excess water and produce a relatively dry soap, and then working in the remainder of the mineral oil and cooling. A few parts per million of silicone is commonly incorporated in the mixture to minimize foaming during saponification. Calcium soaps are processed similarly, using hydrated lime as the saponifying agent, and carrying out the reaction in a closed kettle under pressure at a moderately elevated temperature (ca. 150°C). In both cases it is preferred to have most of the fat consist of glycerides rather than free fatty acids, as the presence of glycerol liberated by saponification contributes to the desired physical properties of the finished grease. On the other hand, preformed soaps, usually of stearic acid or 12-hydroxystearic acid, are used in making both aluminum and lithium soaps. The soap is dissolved in the heated oil at 150–200°C, followed by cooling and working of the mixture in such a manner as to obtain the finished product in the proper physical condition.

Table 6.3   Maximum continuous operating temperature of soap greases (**43**)

Grease Type	Maximun Temperature (°C)	Limiting Factor
Hydrous lime soap	60–80	Dehydration of soap
Hydrous sodium soap	70–100	Dehydration of soap
Aluminum stearate	80–100	Phase transition of soap
Lime soap (stabilized)	90–120	Phase transition of soap; volatility and oxidation of oil
Barium soap + barium acetate	110–120	Phase transition of soap; volatility and oxidation of oil
Sodium soap	120–160	Phase transition of soap; volatility and oxidation of oil
Lime soap + calcium acetate	120–150	Phase transition of soap; volatility and oxidation of oil
Special barium and strontium soaps	140–170	Phase transition of soap; volatility and oxidation of oil
Lithium soap	130–160	Phase transition of soap; volatility and oxidation of oil

Metallic soaps in nonaqueous solvents exhibit a highly complex phase behavior; hence the consistency, the stability, and other physical properties of greases depend to a very large degree on the procedure for cooling and working. Details of these operations cannot be discussed here, but information is available in the literature (40).

Continuous methods have been devised and successfully used for the manufacture of calcium, lithium, and other greases. Because of closer control over the variables just mentioned, they are claimed to yield products of better and more uniform quality than the older batch methods.

Combinations of clays such as bentonite with fatty quaternary compounds, such as di(hydrogenated tallowyl)dimethylammonium chloride, yield a unique series of a gel-forming agents (44). These oganobentonite products marketed under the trade name Bentone can be used to make a variety of greases with graded properties depending on the type of quaternary ammonium salt used with the bentonite and the amount of organobentonite in the finished grease.

A variety of extreme pressure and antiwear additives are used in both lubricating oils and greases. Many of these are based on sulfurized, chlorinated, or sulfochlorinated animal fats (45). Sulfurized oils are prepared by controlled heating of animal fats and sulfur and typically contain 10–17% sulfur; sulfochlorinated products are produced by simple addition of sulfur monochloride to fats; and chlorinated products, containing from 8–12% chlorine, are obtained by direct addition of chlorine gas to heated fats to yield a combination of addition and substitution products. Under conditions of boundary lubrication or extreme pressure these additives presumably form metal sulfides or chlorides, thus permitting sliding contact between surfaces with sluffing of the metal salt rather than the welding and scarring that can occur when there is metal–metal contact.

Soaps of epiminostearic acid have been investigated as grease thickening additives for both paraffin oil and diester-based products (46). The lithium greases had the unusual advantage of needing no corrosion preventative, but the aziridine group did not lead to enhancement of either wear or extreme pressure properties.

2.3   CUTTING OILS

Cutting oils are used for the lubrication of tools for cutting, machining, stamping, and drawing of metals. In some cases the oils are used as such, and in others they are employed in the form of an emulsion with water. The cutting fluid not only performs a lubricating function, but also cools the tool and washes away chips and particles of detached metal.

For many years the standard cutting oil of the nonemulsifying type has been lard oil. The No. 1 or No. 2 grades with a free fatty acid content of 15–25% are the ones usually used. Ordinary mineral or paraffin oil is quite inadequate as a substitute for lard oil, but satisfactory products containing a minimum quantity of fatty oil are now prepared by compounding a sulfurized fatty oil with mineral oil. Oils of this type are used principally for thread cutting and similar applications where a heavy cut is taken at a relatively low speed.

For high-speed cutting, where frictional heat is generated rapidly, an aqueous emulsion-type cutting liquid is usually employed. The nonaqueous base for a liquid of this type may consist substantially of soap, a sulfonated oil, or a mixture of mineral oil with sulfonated oil or soap.

Wire-drawing compounds, stamping compounds, and so on may logically be classed with these materials. They usually have a soap base but require a fatty oil, such as blown fish oil, in addition to mineral oil to give them the desired properties.

2.4 OILS FOR LEATHER TREATMENT (47)

The fibers of finished leather must be lubricated with a thin coating of oil to give them good strength and flexibility. The liquid oils that are most used in leather treatment are neatsfoot oil and fish oils, particularly cod oil. Tallows and greases are the plastic fats usually employed. There is also some use of waxes, such as wool grease, beeswax and carnauba wax, as well as paraffin wax and mineral oils. In a 1971 review of leather tanning, particularly the use of surface-active agents, Filachione (48) cites a U.S. consumption of 40 million pounds of fats in the tanning industry.

Light leathers are usually oiled by the so-called fat-liquoring process, in which the skins are tumbled in a drum containing a dilute emulsion of oil and water stabilized by a sulfated or sulfonated oil. The sulfonated oil serves to carry the oil to the interior of the skins, where it is deposited by subsequent breaking of the emulsion. Heavier leathers are oiled by the hand stuffing process, in which the damp leather is smeared with a plastic grease or "dubbin," consisting typically of a mixture of tallow and cod oil. The high-melting fat in this composition is not absorbed but serves merely as a vehicle for the liquid oil and is removed at the conclusion of the operation. If it is desired to cause the leather to absorb higher-melting fats or waxes, it may be "drum stuffed," by tumbling the wet leather in a drum with the melted fat. Other methods of incorporating the oil are by "burning in," or applying the hot melted fat by hand to the dry leather, and dipping the leather in the melted fat. The latter methods are principally employed when it is desired to impart some degree of waterproofness to the material.

In all methods of oiling leathers the distribution of oil is assisted by the action of water. One function of the water appears to be to cause the leather fibers to spread and separate. The presence of water also favors the spreading of oil films on the fibers in some manner not yet understood. In the methods mentioned earlier in which oil or fat is applied to the dry leather, the oiled leather is later drummed with warm water to distribute the fat.

The grain surface of wet vegetable-tanned leather is "oiled off" by the application of a thin oil coating before the leather is dried to lubricate the grain and to permit the leather to dry without the concentration of tanning liquors at the surface. Some leathers, such as sole leather, receive little or no oiling except in this operation.

In the lubrication of leathers, the chemical reactivity of the fatty oils is important; hence mineral oils are not suitable substitutes. The action of leather-treating oils is not purely physical; some degree of chemical combination apparently occurs between the leather and the oil since the latter is not readily removed after it is once incorporated even on treatment with solvents.

Originally patent leather was manufactured by daubing successive coats of linseed oil pigmented with carbon black or some other pigment until a substantial film accumulated, and finally the coating was buffed to a brilliant luster; now urethane oils that combine good hardness with a high degree of flexibility have supplanted linseed oil.

Soaps and fat-based detergents including quaternary ammonium compounds (49) are used in scouring, fat liquoring, and dyeing of leather.

2.5   Textile Lubricants and Softening Agents (50, 51)

Several thousand specialty chemical products, exclusive of dyes and coloring materials, are used in the manufacture of textiles, and about 40% of these are fat derived (51). Detergents are by far the most widely used class of textile specialties. Soaps or synthetic detergents are used repeatedly in all phases of textile processing from scouring of natural fibers to removal of impurities before wet processing or mechanical finishing to cleaning after the various treatments. Detailed discussion of detergents uses in the textile industry is beyond the scope of this chapter.

Textile lubricants and softening agents are used for rendering textile fibers pliant and enabling them to move smoothly over one another in combing, spinning, weaving, and similar operations. Tallow, lard, castor oil, coconut oil, palm oil, and various hydrogenated oils are used widely as yarn lubricants to protect and strengthen the yarn during weaving and minimize abrasion. Any lubricants or softening agents applied to the fibers during processing must, of course, be removed from the finished textiles. Fatty oils or fatty derivatives are much more easily removed than are mineral oils, and in addition are superior lubricants; hence they are used almost exclusively for this purpose. It is difficult to separate the lubricating from the softening function; hence they are considered together. About 11.7 million pounds of fats were used in these applications in 1962 (10).

Many textile-softening agents and lubricants are surface active and require a proper balance of hydrophilic and hydrophobic groups. Softening agents and lubricants consist mainly of emulsions of oils, fats and waxes, soaps, sulfonated oils, sulfated alcohols, and fatty acid condensation products. At present, oil emulsions are usually fatty oil–mineral oil blends with a dispersing agent. Soaps and mixtures of soaps and fats are used extensively with bleached cotton and rayon knit goods.

Turkey red oil, which is manufactured by sulfuric acid treatment of castor oil, dates back to the 1870s, when it proved of value in various dyeing processes. Subsequently it was found that turkey red oil and reaction products of various unsaturated oils with sulfuric acid imparted a desirable softness and fullness to fabrics.

The use of such sulfonated oils derived from olive oil, castor oil, tallow, and lard has decreased sharply, and sulfonated sperm oil, which was used formerly, is no longer available since the sperm whale was placed on the endangered species list. Sulfonated oils have been largely replaced by various anionic and nonionic synthetic detergents. Sulfated fatty alcohols, however, are valuable softening auxiliaries that also provide surface lubricating actions. The nonionic fatty acid or alcohol condensation products with ethylene oxide are among the more useful softening agents.

Cationic softeners based on quaternized fatty amines have become extremely important, and the volume has expanded rapidly since the early 1960s. Distearyldimethylammonium chloride, which is used as the base for many home laundry fabric softeners, is also the base for a number of commercial textile softeners (52). Among the fatty nitrogen compounds used in textile operations are tertiary amines, fatty acid–amine condensates, imidazoline salts, ethoxylated amines, and a variety of quaternized products (53, 54).

Unmodified oils such as neatsfoot oil, lard oil, and inedible olive oil are also used to a limited extent as textile lubricants. Under certain conditions some oils have a reverse effect and induce interfabric friction, producing during compression a crunchy feel and a sound known as "scroop" that simulates the rustling crackle of silk that is rubbed against itself. Scroop finishes are based on dispersions of fatty alcohols or fatty acid esters in protein emulsions, for example, stearates of oxyethylated butanol in casein solutions (55).

2.6 PLASTICIZERS (56)

A plasticizer is a substance incorporated into a plastic or an elastomer to increase its flexibility, workability, or distensibility. In general, plasticizers are high-boiling polar liquids that may be regarded as nonvolatile solvents for the polymeric material. From 1974 to 1976 about 1.3-1.6 billion pounds of plasticizers were used annually (57), mainly with polyvinyl chloride, cellulosics, and rubber.

Lacquers are a combination of "nitrocellulose" (more properly cellulose nitrate), plasticizer, and solvent. The principal plasticizers include phthalate esters, blown castor oil, blown linseed and soybean oils, epoxidized vegetable oils, and coconut oil–based alkyd resins.

Vinyl resin plasticizers include (a) phthalate, adipate, sebacate, and azelate diesters, (b) phosphate esters, particularly tricresyl phosphate, (c) epoxidized glycerides and epoxidized fatty acid or tall oil fatty acid esters, and (d) polyesters, especially adipic acid–glycol polyesters plus some sebacates and azelates.

To function as a plasticizer, a compound must have a proper balance of polar functional groups to hydrocarbon structure and also proper balance of size of groups. This is apparent if we look at the compatibility with polyvinyl chloride of simple fatty esters, RCOOR′, containing a total of 20 carbon atoms. Unsymmetrical esters are incompatible, but symmetrical products with the ester functionality near the center of the molecule are compatible.

Ethyl stearate	Incompatible
Butyl palmitate	Incompatible
Hexyl myristate	Incompatible
Octyl laurate	Compatible
Decyl caprate	Compatible
Lauryl caprylate	Compatible
Myristyl caproate	Incompatible
Palmityl butyrate	Incompatible
Stearyl acetate	Incompatible

Until the 1950s fats and their simple derivatives were used to only a minor extent as plasticizers because of their limited compatibility with synthetic polymers. The pioneering work of Swern and his coworkers (58–61) focused attention on the commercial potential of epoxidation of oils, and the first commercial process was developed shortly (62). The introduction of the oxirane group into oleic or linoleic acid esters yields symmetrical structures of the type needed for good compatibility.

Polyvinyl chloride (PVC) is a hard horny material that has limited utility alone but provides a broad array of products when properly plasticized. Certain plasticizers can be used over a broad range with PVC resins. These are termed primary plasticizers, and phthalate esters such as dioctyl phthalate (DOP) are by far the most important. If very stiff vinyl objects are desired a low level of plasticizer is used, for example, 25 phr (parts per hundred parts of resin); for ultraflexible plastic materials 100 phr or higher might be used.

Some plasticizers cannot be used individually over a wide range of plastic compositions. At higher levels of use a clear seemingly compatible sample might be obtained, but on aging incompatibility is evident by a greasy exudation, loss of transparency, stickiness, and stiffening. Such plasticizers are termed secondary plasticizers; they cannot be used alone, but they can be used with a more compatible primary plasticizer to lower costs, impart some desired characteristic such as resistance to extraction, or provide a heat- and light- stabilizing function.

Epoxidized oils are secondary plasticizers. Actually, they are used more for their stabilizing action than for their plasticizing ability. Polyvinyl chloride resins deteriorate when exposed to heat and light; they become brittle and darken, eventually becoming black. As deterioration is initiated hydrogen chloride is released that accelerates the deterioration; epoxidized products act as scavengers for the acid and minimize the autocatalytic deterioration:

$$-CH_2-CH-CH_2-CH- \xrightarrow[\text{light}]{\text{heat}} -CH=CH-CH_2-CH- + HCl$$
$$\quad\quad\; | \quad\quad\quad\; | \quad\quad\quad\quad\quad\quad\quad\quad\quad\quad\quad\; |$$
$$\quad\quad\; Cl \quad\quad\quad Cl \quad\quad\quad\quad\quad\quad\quad\quad\quad\quad\; Cl$$

Polyvinyl chloride

$$HCl + -CH-CH- \rightarrow -CH-CH-$$
$$\quad\quad\quad\quad \backslash \; / \quad\quad\quad\quad\; | \quad\; |$$
$$\quad\quad\quad\quad\quad O \quad\quad\quad\quad\; OH \; Cl$$

Oxirane group   Chlorohydrin

Metal soaps such as barium, cadmium, and lead laurates are used also as resin stabilizers, and they, too, are effective by removing acid:

$$(RCOO)_2Ba + 2 HCl \longrightarrow 2 RCOOH + BaCl_2$$

Epoxy plasticizers are produced from oils such as soybean oil or from unsaturated fatty acid esters of monohydric alcohols by oxidation of the double bond by hydrogen peroxide in the presence of acetic acid or formic acid. The peroxy acid is the actual oxidizing agent.

$$\overset{\displaystyle O}{\overset{\displaystyle \|}{CH_3C}}-OH + H_2O_2 \overset{H+}{\rightleftharpoons} \overset{\displaystyle O}{\overset{\displaystyle \|}{CH_3C}}-OOH + H_2O$$

$$-CH=CH- + \overset{\displaystyle O}{\overset{\displaystyle \|}{CH_3C}}-OOH \rightarrow -CH-CH- + \overset{\displaystyle O}{\overset{\displaystyle \|}{CH_3C}}-OH$$
$$\quad\quad\quad\quad\quad\quad\quad\quad\quad\quad\quad\quad \backslash \; / $$
$$\quad\quad\quad\quad\quad\quad\quad\quad\quad\quad\quad\quad\quad O$$

Net reaction: $-CH=CH- + H_2O_2 \rightarrow -CH-CH- + H_2O$
$$\quad\quad\quad\quad\quad\quad\quad\quad\quad\quad\quad\quad\quad\quad \backslash \; /$$
$$\quad\quad\quad\quad\quad\quad\quad\quad\quad\quad\quad\quad\quad\quad\quad O$$

Several methods are used commercially for manufacture of epoxidized oils (63).

Preformed Peracetic Acid Process (64).   Peracetic acid solutions with concentrations as high as 40% can be produced by reaction of acetic acid with concentrated hydrogen peroxide solutions in the presence of sulfuric acid or polystyrenesulfonic acid resins. Neutralization or separation of the acid catalyst yields relatively stable peracetic acid solutions which can be transported, stored, and used to epoxidize oils. Although preformed peracetic acid is useful for laboratory preparations, it has never achieved substantial commercial significance.

In situ Process.   In the *in situ* method hydrogen peroxide, acetic acid, and sulfuric acid are added to oil in a stirred reactor. As peracetic acid forms it reacts with double bonds to form epoxides. When formic acid is used in place to acetic acid it is not necessary to use a mineral acid to catalyze peracid formation (62). The epoxidation reaction is highly exothermic and efficient cooling is required.

In one commercial process (65) the spent aqueous layer is mixed with fresh oil to utilize available active oxygen, and the partially epoxidized oil is treated with fresh hydrogen peroxide, acetic acid, and sulfuric acid. In such processing almost all the hydrogen peroxide is used for oxidation of double bonds.

During processing the oxirane ring may be opened to yield several byproducts and reduce the oxirane content of the finished product. Starting with soybean oil having a typical iodine value of 134 the theoretical oxirane content is 7.8%, but cleavage of oxirane groups to form hydroxyacetate, hydroxysulfate, and vicinal glycol structures reduces the practical oxirane content of the product to about 6.0–6.5%.

In situ Resin Process.   Solid polystyrene sulfonic acid ion-exchange resins can be used as catalysts with less ring cleavage and higher conversions of unsaturation to oxirane than can be obtained using sulfuric acid. At the conclusion of the reaction the catalyst is removed by filtration, and the epoxidized product is separated from the aqueous layer by centrifuging. To obtain reasonable reaction rates, it is necessary to use 12–16% of resin catalyst based on the starting oil. The catalyst can be reused for six to eight consecutive runs, corresponding to an average of about 2% resin per batch. The catalyst swells and degrades with continuing use leading to lowered oxirane yields and difficulty in separating the resin catalyst from the reaction mixture. The oxirane content of oils epoxidized by the resin process is normally somewhat higher than that produced using sulfuric acid as the catalyst.

Philips Process (*Union Carbide*) (66, 67).   A unique synthetic method makes use of peroxy intermediates of the process for manufacture of acetic acid from acetaldehyde. In the conventional vapor-phase oxidation process for manufacture of acetic acid, ethylene is first oxidized to acetaldehyde in the presence of a metal catalyst:

$$CH_2{=}CH_2 + O_2 \longrightarrow CH_3CHO$$

Oxidation of acetaldehyde yields peracetic acid, which converts acetaldehyde to acetic acid or acetic anhydride depending on reaction conditions and provisions for removal of the water of reaction:

$$CH_3CHO + O_2 \longrightarrow CH_3COOOH$$
$$CH_3COOOH + CH_3CHO \longrightarrow 2\,CH_3COOH$$
$$or \quad (CH_3CO)_2O + H_2O$$

In the Philips process the initial air oxidation of acetaldehyde is carried out at a low temperature:

$$2\,CH_3CHO + O_2 \longrightarrow \underset{\underset{\displaystyle OH}{|}}{CH_3CHOOCOCH_3}$$

Acetaldehyde monoperoxyacetate (AMP)

Vapor-phase pyrolysis of the intermediate, AMP, in an inert solvent such as ethyl acetate yields acetaldehyde, which is recycled, and a solution of peracetic acid in the solvent. (*CAUTION:* AMP is explosive.)

$$\underset{\underset{\displaystyle OH}{|}}{CH_3CHOOCOCH_3} \xrightarrow{\text{heat}} CH_3CHO + CH_3COOOH$$

The peracetic acid solution is an effective epoxidation reagent. In effect, the Philips process uses oxygen from air as the oxidizing agent for epoxidation rather than preformed hydrogen peroxide.

The annual volume of epoxidized oils and fatty esters has risen to 125–150 million pounds, about 10% of the total plasticizer volume, and use of these products has expanded to the point that epoxy plasticizers are present in most plasticized systems (Table 6.4).

Table 6.4 Production of epoxy ester plasticizers (in millions of pounds) (57)

Year	Epoxidized Soybean Oil	Octyl Epoxy-tallate	Other Epoxies	Total
1962	37.7	14.6	1.7	55.0
1964	38.4	17.0	2.6	58.0
1966	59.2	11.5	15.9	86.6
1968	71.1	—	30.4	101.5
1970	69.3	23.1	2.7	95.1
1972	85.1	31.3	—	116.5
1974	126.9	14.9	12.1	153.9
1976	91.4	—	26.0	117.4
1977 (68)	—	—	—	126

Epoxidized soybean oil is by far the most important of the epoxy plasticizer–stabilizers, comprising about 75% of the market, but other products are used for special purposes. Epoxidized linseed oil has higher oxirane content and better compatibility than epoxidized soybean oil and is used in applications demanding good permanence or superior resistance to extraction as in vinyl liners of bottle caps or medical tubing.

Experimentally produced lots of epoxidized sunflower oil have better vinyl resin compatibility than epoxidized soybean oil, as would be expected from the combination of higher oxirane and lower saturated acid contents (69). Sunflowers are becoming an important oil-seed crop in the United States, and, if economics are favorable, epoxidized sunflower oil could become an important factor in the plasticizer market.

Epoxidized alkyl oleates and tallates are used for applications demanding excellent low-temperature flexibility together with generally good permanence and stabilizing value. In such applications they compete with octyl adipates, azelates, and sebacates.

Approximately 200 lb of plastics was used per automobile in 1977, and automotive design engineers project that plastic usage will hold steady or even increase despite a move toward smaller more energy-efficient automobiles. To provide more luxurious seating, manufacturers are aiming at 1 lb of plasticizer to 1 lb of resin in automotive vinyl upholstery. Requirements for minimizing cold cracking and fogging have led to specialty products such as epoxidized glycol dioleates or ditallates, isodecyl epoxytallate, and linear alkyl epoxy esters.

Dialkyl azelates and sebacates are more compatible than fatty acid esters and are used where there is need for outstanding low-temperature flexibility and low volatility. These esters are compatible with a broad range of polymers, including nitrocellulose, ethyl cellulose, cellulose acetate-butyrate, and most synthetic rubbers. High cost limits adipates, sebacates, and azelates to specialty applications where other lower-cost plasticizers cannot provide the desired combination of properties.

Polymeric plasticizers are prepared by esterification of adipic, sebacic, and azelaic acids with an excess of a glycol such as propylene glycol, butylene glycol, or 1,4-butanediol. The resultant polyesters containing free hydroxyl groups are commonly terminated with a fatty acid such as lauric acid or pelargonic acid, and in some cases the fatty acid component may approach 50% of the weight of the polyester. The higher cost sebacic and azelaic polyesters are superior to adipic acid–based products in providing a combination of low migration, low volatility, and good resistance to extraction by oils and solvents.

Typical polyvinyl chloride resins are copolymers of vinyl chloride and vinyl acetate, and external plasticizers are used to provide the broad range of properties required for applications as diverse as a rigid washing machine rotor and a flexible plastic shower curtain. The use of fatty derivatives as internal, permanently bound plasticizers has been suggested (70). Vinyl stearate, vinyl epoxystearate, octadecyl acrylate, and similar substances that polymerize with appropriate commercial monomers such as vinyl chloride, vinylidene chloride, acrylonitrile, and vinyl acetate permit the building of the plasticizer into the high-molecular-weight polymer molecule. The concept is attractive, but commercialization was not achieved because each internally plasticized polymer tends to fit into a rather specific market and must be plasticized with external plasticizers to fit needs of other products.

A variety of N-disubstituted amides have been proposed as plasticizers. Product types include ester amides from reaction of fatty acids with diethanolamine (71), N,N-dimethylamides (72), and other N,N-disubstituted amides (73). The various products as plasticizers for polyvinyl chloride exhibit fair to good low-temperature flexibility, but as a group they tend to be borderline in compatibility and heat stability, and most would be too costly for commercialization.

Miscellaneous Nondrying Oil Products

3.1 ILLUMINANTS AND FUELS

Fatty oils were once used widely as burning oils for illumination, but they have been entirely replaced by cheaper petroleum products, except for a few highly specialized uses, and in isolated regions where petroleum is not available. The early development of the American fatty acid industry by companies such as Procter and Gamble and Emery Industries in the late nineteenth century and early twentieth century is usually associated with the use of stearic acid in candles for lighting uses.

Because they are of ornamental value and are used for religious purposes and for special occasions, there is still a considerable world manufacture of candles. Although the basic materials for candles are paraffin wax and beeswax, candles made from these materials alone are not sufficiently high melting to maintain their shape well in hot weather or to burn without excessive dripping. Stearic acid has long been the standard hardening agent for candles, and since the advent of the hydrogenation process there has also been considerable use of highly hydrogenated fats for this purpose, either as such or more usually in the form of their fatty acids (74).

The amount of stearic acid or other fatty hardener used in candles depends on the type of candle and the other ingredients; it may be as little as 5% or as much as 50% or more.

A combination of greater home use of candles for ornamental purposes, including candles made by the popular "do-it-yourself" kits, coupled with an increasing popularity of candles for restaurant lighting has greatly expanded the market. During the last decade demand for stearic acid for candle manufacture increased from 3.5 million pounds to an estimated 10–12 million pounds (74).

Palm oil and other fatty oils have been used to a limited extent as diesel engine fuels. There is a considerable amount of literature on the pyrolysis of fats to produce light motor fuels; and, during periods when petroleum was not available, plants have actually been operated in China. In general, calcium or other soaps appear to be better materials for pyrolysis than the fat itself, although the latter may be cracked under pressure (75, 76). "Napalm," the material used on a very wide scale during World War II and the Vietnam War as a gelling and thickening agent for gasoline in the manufacture of incendiary bombs, consisted of mixed aluminum soaps of naphthenic, oleic, and coconut oil fatty acids (77, 78). A viscous gel, which is stable over a wide temperature range, forms rapidly using simple agitation when the soap is added to gasoline at ambient conditions.

3.2   COSMETIC AND PHARMACEUTICAL OILS

Cosmetics.   In cosmetics, as in many other products, petroleum fractions have taken over many of the traditional functions of fatty oils. However, oils, fats and fatty acids are still important materials in the cosmetic and pharmaceutical industry. Barnett (79) classified cosmetic materials derived from fats and oils as follows:

1. Triglyceride esters—castor, sesame, peanut, safflower, soybean, and coconut oils and tallow.

2. Fatty acids—lauric, palmitic, myristic, stearic, oleic, linoleic, ricinoleic, hydroxystearic, arachidic, behenic, erucic, and arachidonic acids.

3. Fatty alcohols—lauryl, cetyl, stearyl, oleyl, ricinoleyl, linoleyl, lanolin, and tallow alcohols.

4. Soaps—sodium; potassium; ammonium; mono-, di-, and triethanolamine; mono-, di-, and triisopropylamine; and amino glycol salts of fatty acids.

5. Detergent—alkyl sulfates from coconut oil fatty acids, amide sulfonates, ester sulfonates, N-acyl sarcosinates, alkylolamides, amines, and alkyl β-aminopropionates.

6. Cationic antiseptic and softener rinses—quaternary ammonium compounds, morpholinium compounds, and pyridinium compounds.

7. Alkyl fatty acid esters—isopropyl and butyl myristate, palmitate, stearate, oleate, and linoleate.

8. Polyhydric alcohol esters—propylene glycol, glycerol, sorbitol, and sorbitan fatty acid esters.

9. Ethoxylated fatty alcohols—polyethylene glycol mono- and difatty acid esters.

10. Ethoxylated fatty alcohols—polyethylene glycol ethers of cetyl, stearyl, oleyl, and lanolin alcohols.

11. Ethoxylated sorbitan esters.

12. Branched-chain high-molecular-weight alkyl esters—hexadecyl-myristate:

$$C_{13}H_{27}COO\ CH_2\ CH\!-\!C_8H_{17}$$
$$|$$
$$C_6H_{13}$$

13. Lanolin-derived fatty acids and fatty alcohols—this wax ester yields normal fatty acids (even-numbered, C_{10}–C_{26}), isofatty acids (even-numbered, C_{10}–C_{28}), anteiso fatty acids (odd-numbered C_9–C_{31}), and fatty alcohols (aliphatic, sterol, and triterpenoid).

14. Lanolin derivatives—lanolin fatty acid amine soaps, lanolin fatty alcohol esters of acetic acid and fatty acids, and ethoxylated ethers.

Fats in the form of almond oil, olive oil, palm oil, hydrogenated lard, and so on are used in the treatment type of creams, ointments, and lotions, since they are more readily absorbed by the skin than mineral oils. Vanishing creams, foundation creams, powder bases, and the like, are basically similar to brushless shaving creams; that is, they consist of an emulsion with a base of incompletely saponified stearic acid.

In the early cosmetics, natural fats were the dominant emollient materials, but they have been replaced in many applications by mineral oils that are lower in viscosity and more stable to both moisture and oxygen. Lower alkyl fatty acid esters, such as isopropyl myristate and palmitate and butyl stearate, have also replaced fatty oils in many types of cosmetic product. These esters are alcohol soluble, highly stable to oxidation, and much less viscous than natural oils, and they may be rubbed into the skin without leaving the residual greasiness characteristic of natural oils. The myristate is somewhat favored because it is lower melting and will remain fluid at lower temperatures than palmitates and stearates.

Combinations of monoglycerides with soap or other emulsifiers yield stable oil-in-water creams and lotions. Usually the products are based on so-called "monoglycerides" of 35–50% monoglyceride content prepared by glycerolysis of fats or partial esterification of fatty acids, but distilled products containing 90% or more of monoglyceride are used also to emulsify fats or mineral oils. Other partial fatty esters containing varying proportions of free hydroxyl groups are based on polyols such as ethylene and polyethylene glycols, propylene glycol, sorbitol, mannitol, or sucrose. Depending on the hydroxyl content, these emulsifiers vary from water repellant to strongly hydrophilic, and hence they find use in a wide range of cosmetic preparations.

Castor oil is used in hair dressings because of its property of mixing freely with alcohol. It is also used as a lacquer plasticizer in certain nail polishes.

Stearic acid in the form of zinc, calcium, and magnesium salts is widely used in face, bath, and talcum powders. Many cosmetic preparations contain fats in the form of soaps, sulfonated oils, and various emulsifying agents.

The following fats, oils, fatty acids, and waxes are officially listed in the United States Pharmacopeia (USP) and the National Formulary (NF): Oleum Amygdalae Expressum USP (expressed oil of almond), Oleum Chaulmoograe USP (chaulmoogra oil), Oleum Gossypii Seminis USP (oil of cottonseed), Oleum Lini USP (linseed oil), Oleum Maydis USP (corn oil), Oleum Morrhuae USP (cod liver oil), Oleum Olivae USP (olive oil), Oleum Ricini USP (castor oil), Oleum Sesami NF (sesame oil), Oleum Tigilii NF (croton oil), Acidum Oleicum USP (oleic acid), and Acidum Stearicum USP (stearic acid), Adeps (lard), Adeps Benzoinatus (benzoinated lard), Adeps Lanae (wool fat), Adeps Lanae Hydrosus (hydrous wool fat or lanolin), Oleum Theobromatis (cocoa butter), Sevum Preparatum (prepared suet), Cera Alba (white wax), Cera Flava (yellow wax), and Cetaceum (spermaceti). These are all USP materials.

As pointed out previously, natural fats were the dominant emollient materials in early cosmetics, but they were largely replaced by high-purity hydrocarbon oils about the turn of the present century. Kroke (80) points out that this trend is being reversed in Europe, where hydrocarbon oils are being replaced by synthetic triglycerides and liquid waxy esters derived from natural products. To a large extent this change in formulation reflects a concern for dermatologic and toxicologic safety.

In the United States similar concern for safety, particularly against trace materials that might be teratogenic or carcinogenic, is leading to replacement of hydrocarbons in cosmetics and toiletries by fat-derived materials that are viewed as "natural" products.

Johnson (81) reviewed the use of fatty acids in cosmetics and toiletries, and he cites estimates of $7.6–$9.8 billion as the value of the total U.S. market in 1976. Fatty acid consumption was 31 million pounds in 1971, and the 1977 volume is estimated to be 40 million pounds.

Pharmaceuticals. In pharmaceutical preparations, oils and fats are used as emollients and as carriers for medicinal substances to be applied to the skin. Castor oil and croton oil are used as cathartics, and chaulmoogra oil is used in the treatment of Hansen's disease (leprosy). The use of cod liver oil and other fish liver oils as sources of vitamins A and D is, of course, well known. Now synthetic vitamin A preparations are available cheaply and in large quantity, and they have largely supplanted the natural products. Oils, such as corn oil, sesame oil, and cottonseed oil, are employed as carriers for vitamin concentrates and other fat-soluble substances.

Cocoa butter has the property of remaining firm at ordinary temperatures but melting at the temperature of the body. Hence it is a standard base for suppositories. Water-soluble suppository bases with high melting points and good release of medicinal components are prepared from combinations of polyethylene glycol with polyoxyethylene sorbitan monopalmitate (82) or polyoxyethylene stearate with water, waxes, and sulfosuccinates (83). Blends of saturated fats and alkyl esters yield suppository bases useful for climates ranging from temperate to tropical depending on the proportion of ingredients (84).

Antibiotics such as penicillin, streptomycin, the tetracyclines, and neomycin are produced by submerged culture aerated fermentations in tanks with capacities as large as 25,000 gal or more. Unsaturated oils, especially soybean oil, as well as unsaturated fatty acids, are used to control the voluminous foam. The effect is much more than that of a simple antifoam agent; the oils or the fatty acids act as a nutrient and are consumed in the fermentation process, and they markedly increase the yield of antibiotic (85).

Fatty acids, soaps, fat-based detergents, and fatty esters are used in a wide variety of pharmaceuticals as emulsifiers, stabilizers, solubilizing agents, and dispersing agents. Comprehensive treatment is beyond the scope of this chapter, but a few applications are indicated. Calcium, aluminum, and sodium soaps are used as tablet lubricants by incorporation into the final tablet composition to facilitate compression without sticking to the punches and dies of the tableting machines (86). Monoglycerides (87, 88), polyethylene glycol esters of fatty acids (89), polyoxyethylene sorbitan monoesters of fatty acids (90), and polyoxyethylene sorbitan monoesters of fatty acids (91) act as solubilizers, stabilizers, or emulsifiers for vitamins, hormones, estrogens, antiseptics, antitussives, and other pharmaceutical materials. Similar nonionic fat derivatives are used as solubilizers or emulsifiers of drugs for parenteral administration. Carriers for such injectable products must be stable, nontoxic, nonirritating, antipyretic, and acceptable to the Food and Drug Administration (92). An interesting type of injectable product is the fat emulsion used as a concentrated source of calories for intravenous feeding. Monoglycerides, sorbitan esters, polyoxyethylene esters and similar products are used as emulsifiers together with egg phosphatides (93–95).

Some fatty acids and fatty acid derivatives have antimicrobial activity (96, 97). Common sodium soaps have been recognized as sanitizing agents for many years, and although they are not the most active germicides, they are by far the safest. Sodium and especially zinc undecanoates are used as fungicides, particularly for athlete's foot fungus. Fatty amides and more importantly fatty quaternary ammonium salts have general bactericidal properties. Quite surprisingly, high-purity monolaurin shows pronounced antimicrobial activity (97) and is marketed in a proprietary formulation.

Fats and fatty acids are used as protective coatings for pills and tablets to mask unpleasant taste, to protect ingredients from moisture or oxidation, or to delay release of ingredients. Reviews of literature and patents on fatty materials as coatings for medicinals are available (98).

3.3 TINNING OILS

In the mid-1940s hot dipping was the principal method for producing tin plate used to make "tin" cans. By 1965 most tin plate in the United States was manufactured by electroplating, and only 1% was produced by hot tinning. Outside the United States hot-dipped tin plate still makes up a large part of the total.

In tin plating a thick layer of oil is placed over the molten tin in the exit portion of a two-compartment plating bath. Cold-reduced steel from the rolling mill is degreased to remove rolling lubricant, acid pickled to remove scale, frequently using an amine as a corrosion inhibitor to prevent attack on the base metal, rinsed in cold water, and fed into the inlet compartment of the bath through a flux of zinc chloride. The steel sheet emerges from the bath of molten tin through a series of rolls immersed in the oil bath. The oil bath is recirculated through a jacketed reservoir to prevent overheating, and the temperature is maintained at approximately 240°C, slightly above the melting point of tin. The oil protects the tin bath from oxidation, absorbs metal oxides and flux residues, maintains the tin in a molten condition as the metal sheet emerges from the metal bath, and protects the hot freshly tinned sheet from oxidation. Finally, sawdust or bran is used to remove the thin oil film from the tinned sheet. In terne plating, from a tin–lead bath, a coating of oil serves a similar purpose, although some terne plating is carried out in a single bath with only a flux covering the molten metal.

For years crude palm oil was used to the practical exclusion of other oils in tinning because it was relatively cheap compared to other oils, had a composition near the optimum for the purpose, and usually contained a substantial quantity of free fatty acids that promote wetting of the sheet steel by the tin. The oil used must not be very unsaturated, else it will polymerize too rapidly at the high temperature (240–285°C) of the bath. On the other hand, a certain degree of heat degradation in the oil is desirable since it produces free fatty acids and possibly other compounds that assist in dissolving metallic oxides and promote uniform wetting of the metal by the oil. Tallow with added free fatty acids is an excellent substitute for palm oil. Other materials that have undergone successful mill tests include a mixture of dimerized linoleic acid and its monohydric alcohol ester, partially hydrogenated tallow fortified with a controlled amount of free fatty acids, and a mixture of the monohydric alcohol esters of a marine oil with hydrogenated glycerol esters.

At peak production, 30–35 million pounds of fat were used annually in the United States for manufacture of tin plate. Now hot dipped tin plating is limited to applications that have a heavier tin layer than the usual 0.001-in thickness produced by electroplating or to products in which there is demand for the higher brilliance of hot-dipped tin compared with the lower-luster electroplated products

3.4 HYDRAULIC OILS

Castor oil is used as a base for fluids for hydraulic systems, and particularly those exposed to low temperatures such as automobile brake systems and those used for various purposes in airplanes. Castor oil has certain advantages over mineral oils and other fatty oils; it has a relatively low cold test, is compatible with polar liquids of low viscosity, and has little effect on rubber gaskets. Esters of dibasic acids such as adipates and sebacates compete with castor oil in the hydraulic fluid market together with tall oil esters. Several million pounds of fatty materials are used annually.

3.5 INSECTICIDES AND FUNGICIDES

Fatty oils are used in insecticide sprays in the form of soaps or other surface-active materials, as wetting and spreading agents. Oils as such are used to promote adhesion of the spray in the "inverted-spray" method for the control of the codling moth. The oils employed are fish oils, principally because of their low cost.

In some cases fats or fatty derivatives may themselves be toxic agents for insects. Soaps have been found to be effective against some insects, although in practice they are used only as adjuncts to other and more potent insecticides. Certain fatty amines are effective against houseflies, and dioctyl amine has been suggested as a substitute for pyrethrum in household insect sprays (99). Sesame oil is used as an addition agent to pyrethrum-based insect sprays because of the synergistic action of its unsaponifiable materials (100).

Fatty oils, particularly fish oil, are used in the preparation of sticky, weatherproof, tree-banding compounds, used to prevent infestation of trees by larvae from the ground.

Commercial stearic acid with a very low content of unsaturated acids is used in the manufacture of a fungicide for fruit trees (101).

Commercial Fatty Acids and Their Derivatives (102, 103)

Although the characteristics and the uses of the commercially available fatty acids are mentioned in various other places in this book, it appears desirable to recapitulate the data pertaining to this important class of fatty materials, as well as certain derivatives that are produced and marketed in sufficient quantity to be well-recognized commercial products. Production of animal, vegetable, and marine fatty acids, together with tall oil fatty acids, has increased sharply over the last 20 years. Spearheaded by increased demand for fatty acids in soaps, fatty nitrogen compounds, and surfactants, the production of fatty acids has approximately doubled during each of the past two decades (Table 6.5).

Table 6.5   Production of fatty acids in the United States

| | Millions of Pounds | | |
	1958	1966	1976
Saturated acids	218	332	584
Unsaturated acids	137	196	416
Tall oil acids	105	337	387
Total	460	865	1387

4.1 PRODUCTION OF FATTY ACIDS

Raw Materials. Aside from tall oil (104), which, of course, is not a typical fatty material, the major raw materials for fatty acid production are inedible tallow and grease. Coconut oil is an important source of fatty acids for both surfactants and plasticizing-type alkyd resins. The other primary raw material for fatty acid processing is vegetable oil soapstock (Table 6.6).

Stearic and Oleic Acids by Pressing. For years the products commonly known as oleic and stearic acids have been made by fractional crystallization and pressing of the mixed fatty acids from inedible tallow

Table 6.6   Raw materials for fatty acid production

| | 1960 | | 1965 | | 1976 | |
Material	Millions of Pounds	Percent	Millions of Pounds	Percent	Millions of Pounds	Percent
Tallow and grease	385	65	540	74	834	81
Coconut oil	85	15	55	8	107	10
Vegetable foots	75	13	75	10	34	3
Other	40	7	55	8	59	6
Total animal and vegetable	585	100	725	100	1034	100
Tall oil	675		1025		944	
Grand total	1260		1750		1978	

Source: U.S. Department of Commerce, Current Industrial Reports, Fats and Oils, M20K (1961, 1966, 1977).

(105). Fats are split at atmospheric pressure by the time-honored Twitchell process, which is still used to some extent, but more commonly fats are hydrolyzed in batch autoclaves or continuous countercurrent splitters (106). Melted fatty acids are cascaded into rectangular aluminum trays and cooled slowly to a final temperature of 1–3°C to develop large, well-defined crystals. The solidified cakes of fatty acids are wrapped in burlap or cotton cloth, stacked in vertical hydraulic presses, and pressure is applied slowly to separate the liquid fraction (oleic acid or red oil) from the solid material (single-pressed stearic acid). After remelting the solid acids and cooling to room temperature, a second pressing yields "double-pressed stearic acid," and finally "triple-pressed stearic acid" is produced by pressing the recrystallized solid at still higher temperatures. The oily fractions, hot press oil, from these higher temperature separations are quite similar in composition to the starting fatty acids, and they are recycled back to the feedstock tank.

The panning–pressing operation merely produces a gross separation of the saturated and unsaturated acids in the mixture; hence these acids are by no means pure or even fairly pure compounds. Actually, commercial stearic acid contains slightly more palmitic acid than stearic, and the ratio between the two acids is about 55:45. In addition to these two saturated acids, stearic acid also contains 2–8% unsaturated acids, consisting principally of oleic, with a small amount of linoleic acid. Commercial oleic acid, also known as *red oil*, usually contains not more than about 70% actual oleic acid, and the impurities consist of the above-mentioned saturated acids and linoleic acid in approximately equal proportions, plus 2–5% of unsaponifiable matter.

Stearic and oleic acids are made in so-called saponified and distilled grades, according to whether the fatty acid stock is taken directly from the splitting equipment or distilled before pressing. Considerable purification of the acids is obtained by distilling; on the other hand, saponified acids are usually made from higher-grade materials than distilled acids and are not necessarily inferior in quality.

Stearic acid is single-, double-, or triple-pressed, according to the hardness and the degree of unsaturation desired. Average melting points of single-, double-, and triple-pressed stearic acids are respectively about 52–53, 53–54, and 54–55°C. Iodine numbers of the three grades average about 12, 7, and below 5, respectively, whereas average titers are approximately 53, 54, and 55°C. Good-quality stearic acid is white or only slightly yellowish in color; has a characteristic tallowy, but not strong, odor; and is pulverulent and only slightly waxy in consistency. It shrinks markedly upon solidification.

Commercial oleic acid or *red oil* from panning and pressing of tallow is normally colored to some degree; the color of the best products is similar to that of refined but unbleached vegetable oils, for example, 35 yellow and 8–10 red on the Lovibond scale, but much is a deep red or reddish brown. The color depends to a large degree on the extent to which the product has been contaminated with iron. With prolonged heating in ordinary carbon steel containers, it will become almost black; special light grades with a Lovibond color in the neighborhood of 15 yellow and 1.5 red are shipped in aluminum drums or tank cars.

The iodine value of commercial oleic acid is usually about 88–93 and often is quite close to 90, which is the iodine value of pure oleic acid. However, this fortuitous result occurs because considerable amounts of linoleic acid and saturated acids are present in approximately equal proportions. The titer is ordinarily a little below that of pure oleic acid, or about 10°C. Special grades are available in which the saturated acids have been reduced by recrystallization to lower the titer to about 3–5°C and to increase the iodine value to 93–95.

Stearic and Oleic Acids by Solvent Separation.   At one time panning and pressing accounted for almost the total production of stearic and oleic acids, but this process with its extensive hand labor has been replaced almost entirely in the United States by lower-cost, continuous solvent separation methods. The Emersol Process (105, 107, 108) is by far the most important solvent crystallization method.

Animal fatty acids are dissolved in 90% methanol to yield a 25–30% solution, and the solution is pumped continuously through a scraped film crystallizer cooled to about −15°C. The crystallized fatty acids are removed on a rotary vacuum filter, washed with cold 90% methanol, and discharged from the filter. After stripping the solvent from the filter cake, the melted solid fatty acids are discharged from the still for flaking, beading, or other packaging forms of stearic acid. Similarly the filtrate containing the liquid acids is stripped in a solvent recovery still, and the liquid acids are removed for distribution as commercial oleic acid.

The Armour process, or as it is sometimes called, the Armour-Texaco process, is a similar low-temperature crystallization of mixed fatty acids using acetone as a solvent (109, 110).

Application of the solvent crystallization process to other fatty acid stocks yields a variety of liquid and solid acids with compositions that differ from the traditional tallow-based products. For example, fractionation of mixed cottonseed acids yields solid acids that have a very high content of palmitic acid and liquid acids high in polyunsaturated acids.

Stearic and Oleic Acids by the Henkel or Hydrophilization Process.   The

Henkel process is a water-emulsion separation of solid and liquid fatty acids in the presence of a surfactant (111–117). Mixed fatty acids such as tallow acids are cooled slowly in a scraped blade heat exchanger to yield a slurry of saturated acids and unsaturated acids. Treatment of the mixture of solid and liquid acids with an aqueous surface-active material leads to emulsification of the liquid acids. The crystals of solid acids are "hydrophilized" so that they leave the oil phase and enter completely into the aqueous phase. The dispersion is separated in a centrifuge to yield a lighter oily fraction of liquid acids and a heavier aqueous layer that is a suspension of saturated acids. Finally, the suspension is heated, and the molten saturated acid fraction is separated from the aqueous phase. A one-step separation of tallow acids yields a stearic acid fraction with an IV of 13–18 depending on such process conditions as concentration of emulsifier, temperature, and ratio of aqueous phase to incoming mixed acids. A second-stage separation of the solid fraction at higher temperature yields a stearic acid fraction with an iodine value of 4–6.

A variety of surfactants including fatty alcohol sulfates, alkylbenzene sulfonates, soaps, alkane sulfonates, cationics, and nonionics can be used with alkyl sulfates being preferred. A 1–2% concentration of an inorganic salt in the aqueous phase improves the separation. Magnesium sulfate is preferred.

During the last 5 years several American and Canadian companies selected the Henkel process over low-temperature crystallization for manufacture of commercial oleic and stearic acids (118). Energy requirements of the aqueous emulsion separation technique are lower than those of the low-temperature solvent crystallization systems (119). The same separation principles can be applied to fats and fatty alcohols.

Distilled Fatty Acids.   Industrial demands for higher-purity materials of known and constant composition and the development of new processing techniques have combined in recent years to make available many grades of fatty acids other than the traditional "stearic acid" and "red oil." Batch or fractional distillation of fatty acid stocks with or without prior hydrogenation provides a versatile approach to a wide range of fatty acids (120).

Distilled vegetable oil fatty acids, usually derived from the soapstock produced by alkali refining, are available as such and in hydrogenated forms (121) of very high titer. Soybean oil fatty acids are in particular demand for the manufacture of oil-modified alkyd resins. Fatty acids from fish oils or from vegetable oils such as cottonseed or soybean are distilled to produce fractions high and low in iodine value, and coconut or palm kernel oils are similarly distilled to separate fractions of high and low molecular weight. Dehydrated castor fatty acids are used primarily in surface coatings, and 12-hydroxystearic acid derived from hydrogenated castor oil finds major use in lubricating greases.

By distillation of saturated acid mixtures or methyl esters through efficient fractionating columns, it is possible to obtain cuts consisting substantially of a single fatty acid or ester. For a number of years, C_8–C_{18} saturated fatty acids and their methyl esters containing 90–99% of a single component and relatively free of unsaturated and unsaponifiable contaminants have been available commercially.

A relatively recent development is a new line of "stearic acids" of exceptionally high functional purity, that is, with an iodine value not in excess of 0.5 and a content of unsaponifiable matter not exceeding 0.2%. One of the results of such high functional purity is extremely low color combined with remarkable color stability. The "stearic acids" are available with varying percentages of true stearic acid, up to a maximum of 98%.

Another recent development is a commercial grade of purified oleic acid containing about 88% oleic acid and not more than 3% of polyunsaturated acids. The purified grades are useful in preparing chemical derivatives, such as alkyl epoxystearates, and for cleavage to azelaic and pelagonic acids.

In addition to the more or less special fatty acids mentioned earlier, the mixed fatty acids from almost any common oil or fat are available commercially. In a few special cases, such as that of large-scale continuous soap manufacture, plants are set up so that the processing step utilizing fatty acids is immediately preceded by stages of fat splitting and distillation. Usually, however, fatty acids are produced and used in different establishments.

The recent trend has been toward the production of fractionated fatty acids of high purity comprising relatively close cuts with respect to chain length, and, more particularly, degree of unsaturation, and containing less nonfat impurities than older products. In many applications, including production of fatty nitrogen derivatives, organic syntheses, and the manufacture of polymers and synthetic surface-active agents, mixed acids are frequently less satisfactory than products which consist predominantly of one acid; in many others, absence of appreciable color, taste, or odor is essential. There is every indication that this trend will continue and that the fatty acid industry, once dependent almost entirely on mixtures, will make increasing use of purified fatty acids of known and reproducible composition.

Hot fatty acids are corrosive, and variants of the 300 series of stainless steels are used for construction of fat splitting autoclaves, stills, piping, and tankage to minimize metal contamination, which can lead to poor initial color of fatty acids or poor stability during further processing (122).

Fatty Acid Methyl Esters.   Methyl esters are replacing fatty acids in some applications (123, 124). Most commonly, methyl esters are made by methanolysis of triglycerides with an excess of methanol, but direct esterification is used also, particularly to prepare low-grade esters from acidulated soapstocks. Methyl esters have several advantages over fatty acids. Under normal vacuum distillation conditions methyl esters distill about 30°C below the boiling point of fatty acids, and for this reason less energy is needed for fractionation than for the corresponding acids. Methyl esters are more stable than acids that decarboxylate to some extent at distillation temperatures, and they are less corrosive, thus permitting processing in carbon steel equipment, whereas fatty acids require more expensive stainless steel or other corrosion resistant equipment.

Methyl esters are preferred over acids for production of fatty alcohols, alkanolamides, and lower esters such as isopropyl or butyl esters.

Tall Oil Fatty Acids.   Tall oil is not a typical fatty raw material, but it is the source of approximately 400 million pounds per year of specialty fatty acids (Table 6.5). Crude tall oil is a byproduct of the sulfate or Kraft process for pulp production from pine wood. Acidification of the black liquor soaps from the pulping process yields crude tall oil, a mixture of about 50% fatty acids; 40% rosin acids; and 10% unsaponifiable components, including sterols, hydrocarbons, waxes, and higher alcohols. Production of tall oil exceeded 1.6 billion pounds in 1978 in the United States (125). The bulk of the crude tall oil is fractionally distilled to provide tall oil rosin and tall oil fatty acids containing varying amounts of rosin and not less than 90% of fatty acids. Typically, Type I low-rosin tall oil fatty acids contain less than 1% rosin acids, 50% oleic acid, 7% conjugated linoleic acid, 42% unconjugated linoleic acid, and 2% saturated fatty acids. Tall oil fatty acids have broad uses for inedible products (126, 127).

4.2   USES OF FATTY ACIDS (102)

Among the manufactured products in which purified fatty acids are used in large quantities are soaps, synthetic surface-active agents, fatty nitrogen derivatives, plasticizers, polymers, alkyd resins, lubricating greases, rubber tires and other rubber products, candles, crayons, cosmetics, polishes and buffing compounds, matches, mold lubricants, and waterproofing and water-repellent compositions. During the past two decades the consumption of fatty acids has increased significantly for virtually all end

Table 6.7 Consumption of fatty acids

Use	1956 (102)		1965 (102)		1973 (128)	
	Millions of Pounds	Percent	Millions of Pounds	Percent	Millions of Pounds	Percent
Surfactants	42.6	13.2	74.1	15	392[a]	33[a]
Soaps	43.5	13.5	54.3	11		
Chemicals	57.9	17.9	103.7	21	—	
Fatty nitrogen compounds	—	—	—	—	214	18
Resins and plastics	37.5	11.6	69.1	14	48	4
Rubber	30.0	9.3	49.4	10	119	10
Lubricant	19.9	6.2	29.6	6	60	5
Paint and varnish	21.8	6.7	24.7	5	119	10
Textiles	—	—	—	—	60	5
Food additives	—	—	—	—	12	1
Cosmetics	—	—	—	—	12	1
Other	69.7	21.6	88.9	18	154	13
Total consumption	322.9	100	493.7	100	1190	100

[a] Surfactants and soaps combined.

uses (Table 6.7), but the fastest growing segments of the market are soaps and synthetic detergents, fatty nitrogen derivatives, and rubber applications.

Soaps. There is a considerable manufacture of water-insoluble, heavy-metal soaps, usually "stearates" or "oleates," which are prepared in most cases by the double decomposition of a sodium or ammonium soap and a solution of a metal salt.

The water-insoluble metal soaps are valuable as metal carriers and combine many of the properties of the alkali soaps, including a waxy nature and an ability to form gels with nonaqueous solvents.

The most common metal soaps are aluminum and zinc stearates. Aluminum soaps are widely used in the manufacture of lubricating greases and in waterproofing compunds. Zinc stearate is an internal lubricant and mold release agent for rubber and in compounding white or pastel shades. It is a common ingredient of face powders and many cosmetic and pharmaceutical specialties. Other metallic salts of purified stearic acid are used as die lubricants in powdered metal working and in wire drawing of aluminum. Stearates are excellent paint thickeners and retard setting. The zinc soaps of odd-chain fatty acids, such as zinc undecenoate, have been found to be effective nontoxic fungicides. There is also a large production of calcium stearate, as well as smaller amounts of soaps of magnesium, barium, copper, mercury, nickel, cobalt, lead, cadmium, and chromium. The heavy-metal soaps used as paint dryers, especially cobalt and manganese soaps, are now largely naphthenates and to a lesser degree resinates or tall oil products, rather than fat derivatives. Metal salts of fatty acids, such as the barium and cadmium salts, are used in conjunction with epoxy esters to improve the heat and light stability of polyvinyl chloride.

Alkali metal soaps have been largely replaced by synthetic surfactants for laundering, textile scouring, dish washing, and general cleaning applications, but soap has retained its dominance in toilet soaps. Increasingly, fatty acids are used instead of glycerides for soap manufacture. Among the advantages of fatty acids compared with fats and oils are faster reaction with alkalies without requirement for boiling, lower heat requirements, and superior quality because of the considerable purification effected by the fatty acid distillation. Fats are split in continuous fat splitters (129–131), and the crude fatty acids are distilled under vacuum to remove a small forerun of volatile materials and a still residue consisting chiefly of unreacted glycerides that is returned to the splitter. The distilled fatty acids are blended if desired, neutralized continuously with sodium hydroxide or sodium carbonate (132, 133); and the neat soap is dried, milled, perfumed, colored, and formed into bars.

Jungermann has reviewed several alternative continuous soapmaking processes that are still in use (134). The first process of this type was pioneered by Procter and Gamble following development by Mills (130) of a high-pressure fat-splitting process. The Mazzoni process features neutralization of fatty acids with either sodium hydroxide or sodium bicarbonate using a unique proportioning system consisting of two duplex pumps driven by a single motor but with variable speed adjustment to accommodate output requirements. A pH electrode in the circulating soap transmits a signal to the control unit to adjust the stroke of the alkali pump so that a constant pH is maintained. The Armour process combines a Colgate–Emery fat splitter, a distillation step with removal of an odor cut, and neutralization by the DeLaval process in which a viscosity-sensing device controls the neutralization step. In common, these continuous soap manufacturing processes combine high output, excellent product uniformity, and great versatility.

Liquid soaps are prepared from soybean fatty acids or blends of coconut acids and unsaturated acids by neutralization with potassium hydroxide. These soaps are generally standardized at a 25–35% concentration for application by liquid dispensers as hand soaps or hospital scrub soaps.

Fatty Nitrogen Derivatives (53, 54, 135). The volume of commerical nitrogen derivatives of fatty acids has increased from about 35 million pounds per year in 1950 to approximately 400 million pounds in 1978, and new uses for these versatile fatty acid derivatives are developed on a very regular basis.

Fatty acid amides are used as antiblock agents in plastic and wax formations to reduce surface tack, as mutual solvents to increase the compatibility of synthetic resins with waxes, in printing inks as antioffset and antiscratch agents, as thickeners and stabilizers for emulsions, as dye solubilizers and "builders" in synthetic detergents, as intermediates in the synthesis of permanent textile water repellents, and in the preparation of nonionic surface-active agents by reaction with ethylene oxide.

Fatty acid nitriles are used primarily as intermediates for manufacture of amines, but they have some direct use as low-temperature plasticizers for polyacrylonitrile and vinyl polymers, as yarn lubricants, and as intermediates in the synthesis of textile water repellents.

Fatty acid amines are insoluble in water, but they can be converted to a water-soluble form by salt formation or quaternization. Fatty amines are used in the rubber industry as reclaiming, mold release, and accelerating agents; in the textile industry for preparing cationic softening agents, dyeing assistants, and antistatic agents; in mining for froth flotation and concentration of minerals; in metal products as corrosion inhibitors; in the preparation of cationic surface-active agents that have emulsifying, germicidal, and algaecidal properties; and in the preparation of nonionic surface-active agents by reaction with ethylene oxide.

A number of other fatty acid-based nitrogen derivatives are important in addition to the simple fatty amides, nitriles, and amines and their salts or quaternary derivatives. The most important of these are the alkanolamides. They are prepared by reaction of fatty acids, especially coconut acids, with monoethanolamine, diethanolamine, and isopropanolamine.

Alkanolamides are nonionic agents that are used widely as detergent additives to stabilize foam, improve detergency, and increase viscosity. They also find use as fuel oil additives, corrosion inhibitors, latex stabilizers, antistatic compounds, and dye-leveling agents. The properties of diethanolamides depend on the ratio of reactants, and "Kritchevsky amides" (136, 137), prepared by using a ratio of 2 moles of diethanolamine per mole of fatty acid, contain a water-soluble mixture of amides, amino esters, amido esters, piperazine or morpholine derivatives, and unreacted diethanolamine. These products are highly effective foam stabilizers with pronounced wetting and detergent properties.

Diethanolamides prepared from methyl esters have properties that are quite different from products derived from fatty acids (123, 138). These "superamides" contain about 90% diethanolamide compared with a purity of about 60% for the acid-based products.

Reaction of fatty acids with polyamines such as ethylenediamine or diethylenetriamine yields substituted amides and imidazolines depending on reaction conditions (139, 140). The simplest product is derived from ethylenediamine, but a number of products are available commercially using other polyamines:

$$RCOOH + H_2NCH_2CH_2NH_2 \rightarrow RCONHCH_2CH_2NH_2 + H_2O$$

$$\Big\downarrow Heat$$

$$RC \begin{array}{c} N-CH_2 \\ | \\ NH-CH_2 \end{array} \quad + H_2O$$

2-Alkylimidazoline

These products are used as fungicides, anticorrosion agents for lubricants, petroleum deemulsifiers, detergents, antistatic agents, and textile softeners.

Virtually all amphoteric or ampholytic surfactants are derived from fatty nitrogen compounds. For example, N-alkyl-β-aminopropionic acids are made by addition of fatty amine to acrylate esters followed by hydrolysis of the resulting esters (141–143).

$$RNH_2 + CH_2{=}CHCOOCH_3 \rightarrow RNHCH_2CH_2COOCH_3$$

$$\Big\downarrow H_2O$$

$$RNHCH_2CH_2COOH + CH_3OH$$

In a closely related addition reaction amines are cyanoethylated by reaction with acrylonitrile, following by reduction to yield the corresponding diamines (144). These N-alkyl-1,3-diaminopropanes have uses generally similar to those of the simple amines, but they are characterized by exceptional cationic surface activity:

$$RNH_2 + CH_2{=}CHCN \longrightarrow RNHCH_2CH_2CN \xrightarrow{H_2} RNHCH_2CH_2CH_2NH_2$$

A similar reaction sequence starting with fatty alcohols yields "ether amines", primary amines containing the solubilizing effect of an ether oxygen group.

$$ROH + CH_2{=}CHCN \longrightarrow ROCH_2CH_2CN \xrightarrow{H_2} ROCH_2CH_2CH_2NH_2$$

Recently another product has assumed major commercial importance. Reaction of tertiary amines with hydrogen peroxide yields amine oxides. As with other fatty amine derivatives, the amine oxides find their major outlet in detergent products as foam stabilizers.

$$R_3N + H_2O_2 \longrightarrow R_3N \rightarrow 0 + H_2O$$

Finally, all the fatty nitrogen derivatives described previously that contain primary or secondary amino or amido groups with active hydrogen atoms will react with ethylene oxide or propylene oxide to yield a series of nonionic surface active agents whose properties vary with the molar ratio of alkylene oxide to nitrogen compound:

$$RNH_2 + (x + y)CH_2{-}CH_2{-}RN \diagup \begin{matrix} (CH_2CH_2O)_xH \\ \diagdown \\ (CH_2CH_2O)_yH \end{matrix}$$
$$\diagdown O \diagup$$

Kilheffer and Jungermann (145) developed rather convincing evidence that x and y are approximately equal. Coconut fatty amine adducts are especially important, and products range from 5–10-mole adducts up to 20-mole adducts or higher.

Dibasic Acids.   The dibasic acids, sebacic and azelaic, are produced commercially from fats, the former by high-temperature alkaline cleavage of castor oil and the latter by oxidative cleavage of oleic acid with ozone or other oxidizing agents (146). These dibasic acids find important outlets in the manufacture of plasticizers, synthetic lubricants, and high-molecular weight polymers.

Reductive ozonolysis of unsaturated fatty acids has been proposed. The products from methyl oleate are pelargonaldehyde and methyl azelaaldehydate. These aldehyde products have been suggested for synthesis of low-temperature plasticizers for vinyl resins (147), intermediates for nylon-9 synthesis (148), and as starting materials for several types of coatings (149, 150), but no commercial interest has been shown because of the relatively expensive processing sequence.

Undecanedioic acid from oxidation of castor fatty acids (151) and the C_{13} dibasic acid brassylic acid from ozone oxidation of erucic acid have been produced in pilot scale levels, but no commercial interest has developed. 9(10)-Carboxystearic acid from addition of carbon monoxide to oleic acid by way of the Koch reaction (152) or the oxo reaction (153) has also been offered experimentally, but has not been commercialized. Several million pounds per year of a C_{21} dibasic acid marketed under the name "Diacid" is produced by addition of acrylic acid to the linoleic acid of tall oil fatty acids in the presence of a catalyst (36). The acid has value in high-solids specialty soaps for use in lubricants and latex emulsions.

Mixtures of C_{19} dicarboxylic acids can be prepared by the Koch reaction of oleic acid, tall oil fatty acids, or partially hydrogenated tall oil fatty acids with carbon monoxide in the presence of sulfuric acid catalyst. Under the best conditions, light-colored, heat-stable C_{19} dicarboxylic acids are obtained in 83% overall yield at 96% purity, containing 75% tertiary and 25% seconary isomers (153a).

When linoleic acid (or its esters) is heated to 250–300°C in an inert atmosphere, dimers are formed. The reaction is a Diels–Alder addition reaction involving one molecule of ordinary nonconjugated linoleic acid and one molecule of thermally conjugated linoleic acid or ester (154). The resulting dibasic acids, known commercially as "dimer acids," are used in making polyamides, polyesters, and polyurethane foams (155). Dimer acids are now produced commercially by heating oleic acid or especially tall oil fatty acids with a clay catalyst (156, 157).

In a typical manufacturing process tall oil fatty acids are heated for 6–8 hours at 230°C with 4% of montmorillonite clay. The crude product is bleached with phosphoric acid and clay, and the filtered product is fractionated in a wiped-film evaporator to yield an approximately 60:40 ratio of dimer acids to monomeric acid. About 50 million pounds of dimer acids are produced annually in the United States.

During the dimerization process some of the unsaturated fatty acids rearrange to yield branched chain and cyclized monomeric acids. The monomer fraction consisting of a mixture of normal straight chain and rearranged acids is marketed as such, or it may be hydrogenated and separated into liquid and solid components (158). The liquid fraction, termed *isostearic acid,* is unique in its low melting point. Its titer is below 10°C, whereas the conventional stearic acid from tallow segregation has a titer of 55°C, and stearic acids from hydrogenated vegetable oils have titers of about 70°C. The combination of low melting point and high oxidative stability associated with saturated acids leads to usage of isostearic acid esters in lubricants and cosmetics.

Fatty Alcohols.   Fatty alcohols are obtained either by sodium reduction of fatty esters or by hydrogenolysis of fatty acids, their esters, or metallic salts (159, 160). The former method is more costly, but double bonds are essentially unaffected. A commercial plant for production of unsaturated and saturated fatty alcohols by the sodium reduction process operated briefly in the United States in the late 1950s and early 1960s, but limited markets for unsaturated alcohols coupled with a relatively noncompetitive position compared with saturated alcohols by hydrogenolysis stopped the venture. Synthetic fatty alcohols with carbon chain-length in the C_{10}–C_{14} detergent range are produced by oxidation of aluminum alkyls from polymerization of ethylene using triethyl aluminum as a catalyst.

About 300 million pounds of fatty alcohols are produced annually; the most important use is in the production of sodium alkyl sulfates, an important detergent ingredient. Reaction of fatty alcohols with ethylene oxide (161) or propylene oxide (162) yields nonionic detergents. Ethoxylated long-chain alcohols are the most important nonionic surfactants, with a 1976 volume of more than 0.5 billion pounds (163). Fatty alcohols with a carbon chain length of 16–22 carbon atoms have been proposed for evaporation control from reservoirs in hot and arid areas, but they are not sufficiently hydrophilic for effective spreading into monolayers; 5-mole ethylene oxide adducts (74) are about optimum. Fatty alcohols are converted to vinyl ethers by reaction with acetylene. These monomers are readily polymerized by ionic catalysis, yielding interesting surface coating and waterproofing agents.

Fatty Acid Esters.   Other important fatty acid derivatives are esters with monohydric and polyhydric alcohols (164). The latter are by far the more important. Esters of fatty acids with glycerol, ethylene glycol, propylene glycol, and pentaerythritol are widely used in the surface coatings industry, especially when the fatty acids are polyunsaturated. Partial esters of polyhydric alcohols with fatty acids are important emulsifiers. When monoglycerides, such as glyceryl monooleate or monostearate, are acetylated, acetylated monoglycerides are obtained (164). These range from liquids to solids melting as high as 60°C, depending on the degree of acetylation and the fatty acid. Acetostearins have an unusual combination of properties in that they are relatively sharp melting, flexible, nongreasy fatty materials with extremely high elongations, in some cases over 800%.

Food use of fatty esters continues to grow, and various products function as dough conditioners, bread softeners, emulsifiers, coating agents, release agents, whipping agents, starch complexing agents, cloud inhibitors, stabilizers, and defoaming agents. Plastic fats have long been considered indispensible for bread baking, and indeed, the substitution of oils for lard or hydrogenated shortenings in conventional bread formulas leads to low loaf volume and poor crumb color and texture. However, bread of excellent functional quality is now made using soybean oil as a shortening in combination with monoglycerides and polysorbates (165), and soybean oil can be used similarly in sweet rolls and cakes (166).

Simple esters such as amyl, butyl, ethyl, hexyl, methyl, and nonyl fatty acid esters, especially of lower-molecular weight fatty acids, are used as synthetic flavoring materials.

Mono- and diglycerides are the most important ester-type food additives; others include polyglycerol esters, acetyl tartaric derivatives of mono- and diglycerides, monoglyceride citrate and lactate, lactylic esters of fatty acids, sucrose and sorbitan mono- and partial esters, polyoxyethylene esters, polyoxyethylene sorbitan monoesters, and calcium or sodium stearoyl-2-lactylate. Polyglycerol esters (167) and other food emulsifiers (168) have been reviewed recently.

Synthetic Fats and Fatty Acids (169–172)

There are a number of alternative approaches to the manufacture of synthetic fatty acids from petroleum raw materials. Fineberg (170) lists a number of processes that have been used at least to a limited extent commercially or that have the potential for commercialization (Table 6.8).

Fatty Acids by Oxidation of Paraffins.   Historically, the first synthetic fatty acids were manufactured by oxidation of aliphatic hydrocarbons in Germany in the years during and immediately preceding World War II. At the peak of production they were apparently produced at a rate approaching 200 million pounds yearly.

Paraffins prepared from water gas by the Fischer–Tropsch process were employed as the raw materials. The crude paraffins, consisting of a mixture of members of widely varying molecular weights, were fractionally distilled, and those containing about 19–28 carbon atoms were retained for the next step. It was necessary to employ paraffins with hydrocarbon chains approximately twice the length of the fatty acids desired.

The selected paraffin stock was subjected to an oxidizing pretreatment involving mixing with about 0.1% powdered potassium permanganate or other chemical oxidizing agent, heating to about 150°C for a brief period,

Table 6.8   Synthetic acid processes **(170)**

Process	Commercial	Potential
Oxidation of wax—straight chain	x	
Oxidation of aldehydes	x	
Oxidation of alcohols	x	
Oxidation of olefins	x	
Hydroformylation of olefins (oxo)	x	
Carbonylation of olefins (Koch)	x	
Ozonolysis of unsaturates	x	
Pyrolysis and alkaline cleavage of castor oil	x	
Dimerization–oxidation of alcohols	x	
Fermentation		x
Carbonylation of alcohols	x	
Addition of acids to olefins	?	
Oxidation of diols		x
Addition of HCN to olefins		x
Addition of acrylonitrile to olefins		x
Addition of acrylates to diolefins		x
Telomerization of acrylates, formates, and acetates, with dienes		x

and then blowing with air at about 100°C. The pretreatment served to initiate the formation of peroxides, which must be present for the subsequent stage of oxidation to proceed rapidly.

After pretreatment, a catalyst consisting of manganese soaps amounting to 0.5–1.0% of the weight of the stock was added, and the stock was blown with air at about 100°C in towers of special design until 30–50% of the paraffins had been oxidized to acids. About 24 hours of oxidation time was required. In the course of oxidation, rupture occurred more or less at random along the hydrocarbon chains. Hence a mixture of fatty acids of different chain lengths was formed, in addition to a great variety of other decomposition products.

The fatty acids and other acidic substances were separated from a high proportion of unsaponifiable material by saponifying the mass with sodium carbonate and discarding the water-insoluble material remaining after saponification. The crude fatty acids obtained by splitting the resulting soaps with a mineral acid were purified by extraction with various solvents, followed by fractional distillation, which serves to select the fatty acids of medium molecular weight that are desired in the final material. Table 6.9 lists a typical analysis and fatty acid composition.

Although synthetic fatty acids were produced primarily for use in soaps

Table 6.9   Synthetic fatty acids from hydrocarbon oxidation

Solidification point	26.2°C
Odor	Practically none
Acid value	244.2
Saponification value	247.2
Average molecular weight	229.8
Iodine value	4.9
Hydroxyl value	3.7
Naphtha insoluble (%)	0.0
Unsaponifiable matter (%)	0.29
Fatty acid composition (wt. %)	
C_8	0.4
C_9	2.1
C_{11}	7.3
C_{12}	13.9
C_{13}	16.3
C_{14}	15.2
C_{15}	12.4
C_{16}	8.1
C_{17}	7.3
C_{18}	3.5
C_{19-21}	7.3
C_{22}	1.2

and other detergents, appreciable quantities of selected fractions were esterified with glycerol or a monohydric alcohol and used in the manufacture of margarine. The following analytical data have been reported for a synthetic glyceride product (173): melting point, 36.5%; iodine value, 11; saponification value, 234; odor and flavor, none; keeping time, over 2 years; appearance, like butterfat; composition; C_{11} acids, 18%; C_{12} acids, 18%; C_{13} acids, 20%; C_{14} acids, 8%; C_{16} acids, 15%; C_{10}, C_{18}, C_{19}, and C_{20} acids, 3%; and esters of high boiling point, 11%.

The German synthetic fatty acid program was deeply involved with political consideration and was propagandized accordingly. Consequently, it has been difficult to give it a proper scientific and economic evaluation. The best disinterested opinion, however, appears to be that the synthetic methods are wholly uneconomic on any other than an emergency basis, and that there is considerable doubt concerning the acceptability of the synthetic fats as a food product. Although they may be freed of certain impurities, including hydroxy acids, keto acids, dibasic acids, and lactones, by suitable processing, there is a considerable content of branched-chain acids that is very difficult to remove. There is evidence that some of these are toxic on long ingestion (174).

Following World War II the Soviet Union and eastern European socialist countries continued to develop the field of paraffin oxidation. In the United States the plentiful supply of fats and oils discouraged research on synthetic fatty acids from petroleum intermediates; in contrast, shortage of food fats in eastern Europe focused attention on synthetic fatty acids from paraffins to satisfy the requirements for fatty acids for soap and other nonedible products. Fineberg (172) reports that eastern Europe produces over a billion pounds of synthetic fatty acids annually and that there are approximately 20 plants in the eastern European socialist countries. A typical analysis of Russian synthetic fatty acids is shown in Table 6.10.

Sonntag and Zilch (169) have reviewed the Russian processes. Selected petroleum-based paraffin fractions are oxidized by air at 100°C to an acid value of approximately 70 in the presence of a permanganate catalyst. Low-molecular-weight materials are removed by water washing, and acids are separated from unsaponifiable constituents by alkali treatment. The unsaponifiables are recycled, and after acidification of the soap solution the crude fatty acids are fractionated.

The first reaction product is thought to be a hydroperoxide that decomposes to primary and secondary alcohols and ketones (169); continuing oxidation yields acids.

Fatty Acids from Olefins by Oxo Reactions.   The reaction of a mixture of carbon monoxide and hydrogen (water gas) with the double bonds of unsaturated hydrocarbons in the presence of cobalt and other metals is known as the "oxo" or hydroformylation reaction. The carbonyl products formed by the primary reaction can be reduced to alcohols or oxidized

Table 6.10   Typical analysis of Russian synthetic fatty acids from oxidation of paraffins (128)

Chain Length	Weight Percent
$C_5–C_6$	3
$C_7–C_9$	10
$C_{10}–C_{16}$	40
$C_{17}–C_{20}$	25
C_{21} and higher, including still residues	22

to carboxylic acids (171):

$$RCH{=}CH_2 \xrightarrow[H_2]{CO} \underset{\underset{CHO}{|}}{RCH_2{-}CH_2} + \underset{\underset{CHO}{|}}{RCH{-}CH_3} \rightarrow \underset{\underset{CH_2OH}{|}}{RCH_2{-}CH_2} + \underset{\underset{CH_2OH}{|}}{RCH{-}CH_3}$$

$$\underset{\underset{COOH}{|}}{RCH_2{-}CH_2} + \underset{\underset{COOH}{|}}{RCH{-}CH_3}$$

Terminal linear olefins available by cracking of straight-chain hydrocarbons or by Ziegler chemistry yield linear acids with the carboxyl group in the terminal position together with a smaller quantity of acids with the carboxyl group in the 2-position. Reaction of branched chain olefins from dimerization or trimerization of propylene or isobutylene yields branched-chain carboxylic acids.

Fatty Acids from Ziegler Intermediates.   In the process sequence that is basic for the production of polyethylene, ethylene and other unsaturated hydrocarbons polymerize in the presence of aluminum triethyl to yield long-chain aluminum alkyls whose size is controlled by the amount of ethylene. For synthesis of fatty alcohols or fatty acids, an average chain length in the range of 10–16 carbon atoms is desirable.

$$(x + y + z)CH_2{=}CH_2 + (CH_3CH_2)_3Al \rightarrow CH_3(CH_2CH_2)_yCH_2{-}Al \overset{CH_3(CH_2CH_2)_xCH_2}{\underset{CH_3(CH_2CH_2)_zCH_2}{\Big\langle}}$$

The intermediate Ziegler alkyls can be decomposed in a variety of ways to yield olefins, fatty alcohols, and possibly fatty acids.

$$(RCH_2CH_2)_3Al \xrightarrow{[O]} (RCH_2CH_2O)_3Al \xrightarrow{H_2O} RCH_2CH_2OH \xrightarrow{[O]} RCH_2COOH \quad (1)$$

$$\xrightarrow{[O]} (RCH_2COO)_3Al \xrightarrow{H_2O} RCH_2COOH \quad\quad\quad\quad (2)$$

$$\xrightarrow{CO_2} (RCH_2CH_2COO)_3Al \xrightarrow{H_2O} RCH_2CH_2COOH \quad\quad (3)$$

Direct reaction of the Ziegler intermediate with oxygen (reaction 1) yields fatty alcohols. The average chain length covers a range centered about the desired size. For example, when an average of four ethylene units are added per ethyl group in aluminum triethyl, the weight percent

*Table 6.11   Chain distribution of fatty
alcohols produced by reaction of four
ethylene units per aluminum triethyl
bond* (175)

Alcohol Chain Length	Weight Percent
2	0.5
4	3.2
6	9.2
8	15.9
10	19.5
12	18.5
14	14.3
16	9.3
18	5.2
20	2.6
20+	1.8

of fatty alcohols peaks around the expected 10-carbon chain length, as shown in Table 6.11.

Oxidation of the resultant alcohol yields straight-chain fatty acids containing an even number of carbon atoms. Alternatively, it has been suggested (reaction 2) that direct oxidation of the aluminum alkyls will yield fatty acids directly. Another direct route to fatty acids from Ziegler intermediates has been proposed by way of direct carboxylation of the aluminum alkyls using carbon dioxide (reaction 3). The reaction with carbon dioxide to form acids is more complex than the carboxylation of Grignard reagents.

In each of the preceding methods for synthesis of fatty acids by way of Ziegler intermediates, the expensive aluminum alkyl catalyst is destroyed; hence these methods are not competitive for fatty acid production and separation of fatty acids from natural sources.

In contrast, olefins can be made at low cost by Ziegler chemistry because the aluminum catalyst remains effective. Ethylene polymerization is carried out typically at about 100°C and < 1000 psi. At higher temperatures (176) the growing alkyl group may be displaced by ethylene with formation of α-olefin:

$$RCH_2CH_2Al\!\!\begin{array}{c} {}^{R'} \\ {}_{R''} \end{array} + CH_2{=}CH_2 \rightarrow RCH{=}CH_2 + CH_3CH_2Al\!\!\begin{array}{c} {}^{R'} \\ {}_{R''} \end{array}$$

The growth and displacement reactions may occur separately or simultaneously, and an increasingly larger fraction of ethylene displaces alkyl groups rather than adding to them as the temperature is raised.

The relatively low cost Ziegler olefins will yield terminal and 2-carboxylic acids by way of the oxo process as described previously. Chemical oxidation of the olefins using ozone or nitric acid produces acids containing one less carbon than the starting olefin. Alternatively, the olefin can be oxidized directly to acids by air oxidation, but this reaction is difficult to control and yields mixed acids because of the random attack of oxygen or oxygen-induced free radicals on chain methylene groups.

Fatty Acids from Olefins by the Koch Reaction.   Polymerization of propylene or isobutylene to the dimer or trimer stage using acidic catalysts yields branched-chain olefins that are intermediates in the manufacture of high-octane gasoline components. These olefins are carboxylated by way of the Koch reaction (177, 178) by reaction with carbon monoxide in the presence of sulfuric acid:

$$
\begin{array}{c}
\underset{\displaystyle |}{CH_3} \\
R-C=CH_2 + H_2SO_4 \longrightarrow
\underset{\displaystyle \underset{\displaystyle CH_3}{|}}{R-\overset{+}{C}-CH_3} + HSO_4^- \\
\downarrow CO \\
\underset{\displaystyle \underset{\displaystyle \overset{+}{C}=O}{|}}{R-C-CH_3} \\
\downarrow H_2SO_4 \\
\underset{\displaystyle \underset{\displaystyle OSO_2OH}{|}}{\underset{\displaystyle \underset{\displaystyle C=O}{|}}{R-C-CH_3}} \xrightarrow{H_2O}
\underset{\displaystyle \underset{\displaystyle COOH}{|}}{R-C-CH_3}
\end{array}
$$

Commercial branched-chain fatty acids in the C_5–C_{13} range are sold as "Neo-Acids" or "Versatic Acids."

Fatty Acids from Aldehyde Condensation.   2-Ethylhexanoic acid is prepared from butyraldehyde by aldol condensation followed by hydrogenation and oxidation of the dimer:

$$CH_3CH_2CH_2CHO \xrightarrow{OH^-} CH_3CH_2CH_2CH—CHCHO \xrightarrow{-H_2O}$$

$$\underset{OH \quad CH_2CH_3}{|\qquad\quad|}$$

$$CH_3CH_2CH_2CH=C—CHO \xrightarrow[Ni]{H_2}$$

$$\underset{CH_2CH_3}{|}$$

$$CH_3CH_2CH_2CH_2CH—CHO \xrightarrow{(O)}$$

$$\underset{CH_2CH_3}{|}$$

$$CH_3CH_2CH_2CH_2CH—COOH$$

$$\underset{CH_2CH_3}{|}$$

Economics of Synthetic Fatty Acids.   Fineberg (170, 172) cites production of 100 million pounds of petrochemical acids (exclusive of acids by paraffin oxidation in the Soviet Union and China), and he describes additional capacity of 340 million pounds from plants now under construction or planned in the near future. He also analyzes the economics of coconut fatty acids against petrochemical acids from oxidation of oxo process tridecyl alcohol, alkaline oxidation of alcohol, or oxo carbonylation of α-olefins. At 1979 pricing, synthetics could be competitive with natural fatty acids based on a coconut oil price of $0.55 per pound, but the differences between the costs of the natural and synthetic acids are relatively close.

Fats in Feeds

Although not a manufactured or industrial product in the usual sense, fats in feeds represent a tremendous outlet for fats, mainly inedible animal fats, such as tallows and greases and their simple derivatives. Vegetable oil foots are also sources of feed additives.

The major role of fats in feeds is as a caloric supplement, as fats have the highest caloric density of all feed materials. As caloric supplements, fats compete with corn, and the price of the fats (on a caloric basis) must be competitive. Fats, however, have certain advantages over corn and can command a slightly higher price than would be calculated on a caloric basis only. The technology of utilization of fats in poultry and other livestock feeds has been reviewed (179).

At present, fats in feeds represent the largest single use of inedible animal fats. During the past two decades the consumption of fats in feeds has roughly quadrupled from 400 million pounds per year in 1956 to 892 million pounds in 1966 and 1464 million pounds in 1976. The consumption is expected to increase.

References

1. P. O. Powers, "Linoleum," in *Kirk-Othmer Encyclopedia of Chemical Technology*, 1st ed., Vol. 8, Interscience, New York, 1952, p. 392.
2. G. A. O'Hare, *J. Am. Oil Chemists' Soc.*, **25**, 105 (1948).
3. G. A. O'Hare, *J. Am. Oil Chemists' Soc.*, **27**, 530 (1950).
4. U.S. Department of Commerce, *Census of Manufactures*, MC58 (2)-39D (1961).
5. U.S. Department of Commerce, "Hard-Surface Floor Coverings," SIC3996, *Census of Manufactures*, MC72 (P)-39 D-4 (1974).
6. G. A. O'Hare and W. J. Withrow, *Ind. Eng. Chem.*, **29**, 101 (1947).
7. E. Hazlehurst, U.S. Pat. 2,446,652 (1948).
8. A. B. Miller and F. D. Snell, *Ind. Eng. Chem.*, **25**, 1307 (1933).
9. A. DeWaele, *Ind. Eng. Chem.*, **9**, 6 (1917).
10. U.S. Dept. of Commerce, *Current Industrial Reports*, M20K (62)-13 (1963).
11. U.S. Dept. of Commerce, *Current Industrial Reports*, M20M (67)-13 (1968).
12. S. S. Gutkin, *J. Am. Oil Chemists' Soc.*, **27**, 538 (1950).
13. C. T. Rairdon, in *Kirk-Othmer Encyclopedia of Chemical Technology*, 2nd ed., Vol. 4, Interscience, New York, 1964, p. 28.
14. H. P. Kaufmann, J. Baltes, and P. Mardner, *Fette, Seifen, Anstrichm.*, **44**, 337 (1937).
15. H. P. Kaufmann, E. Gundsberg, W. Rottig, and R. Salchow, *Ber.*, **B70**, 2519 (1937).
16. J. C. Cowan and D. H. Wheeler, *J. Am. Chem. Soc.*, **66**, 84 (1944).
17. J. C. Cowan, W. C. Ault, and H. M. Teeter, *Ind. Eng. Chem.*, **38**, 1138 (1946).
18. C. E. Morris, *J. Am. Oil Chemists' Soc.*, **38**, 24 (1961).
19. R. S. Yamasaki, *J. Paint Technol.*, **39**, 394 (1967).
20. C. E. Morris, *The Use of Linseed Oil for the Protection of Portland Cement Concrete Surfaces*, American Road Builders Association Technical Bulletin No. 257 (1965).
21. W. L. Kubie and J. C. Cowan, *J. Am. Oil Chemists' Soc.*, **44**, 194 (1967).
22. W. L. Kubie, L. E. Gast, and J. C. Cowan, "Linseed Oil for the Preventive Maintenance of Concrete," Highway Research Board, Maintenance Practices, 1967, No. 254, 1968, p. 61.
23. W. E. Grieh and R. Appleton, *Public Roads*, **33** (1), (1964).
24. M. J. Snyder, *National Highway Research Report No. 16*, National Academy of Sciences–National Research Council (1965).
25. L. E. Gast, W. L. Kubie, and J. C. Cowan, *J. Am. Oil Chemists' Soc.*, **48**, 807 (1971).
26. C. H. Scholer and C. H. Best, *Highway Res. News*, **16**, 70 (1964).
27. A. E. Rheineck and R. A. Heskin, *J. Paint Technol.*, **40** (525), 450 (1968).
28. A. E. Rheineck and R. A. Heskin, *J. Paint Technol.*, **42** (544), 299 (1970).

29. G. L. Clark, B. H. Lincoln, and R. R. Sterrett, *Proc. Am. Petroleum Inst. III*, **16**, 68 (1935).

30. T. Perlstein, A. Eisner, and I. Schmeltz, *J. Am. Oil Chemists' Soc.*, **51**, 335 (1974).

31. R. R. Mod, F. C. Magne, and G. Sumrell, *J. Am. Oil Chemists' Soc.*, **54**, 589 (1977).

32. E. W. Bell, L. E. Gast, and F. L. Thomas, *J. Am. Oil Chemists' Soc.*, **54**, 259 (1977).

33. Anon., *J. Am. Oil Chemists' Soc.*, **53**, 664A (1976).

34. D. M. Matthews, *J. Am. Oil Chemists' Soc.*, **56**, 841A (1979).

35. Military Specification Mil-L-23699, amendment dated June 10, 1964.

36. B. F. Ward, Jr., C. G. Force, and A. M. Bills, *J. Am. Oil Chemists' Soc.*, **52**, 219 (1975).

37. E. J. Dufek, W. E. Parker, and R. E. Koos, *J. Am. Oil Chemists' Soc.*, **51**, 351 (1974).

38. J. P. Friedrich, *J. Am. Oil Chemists' Soc.*, **44**, 244 (1967).

39. F. C. Magne, R. R. Mod, G. R. Sumrell, R. E. Koos, and W. E. Parker, *J. Am. Oil Chemists' Soc.*, **52**, 494 (1975).

40. C. J. Boner, *Manufacture and Application of Lubricating Greases*, 2nd ed., Reinhold, New York, 1964.

41. R. E. Lee, Jr. and E. R. Boozer, "Lubrication and Lubricants," in *Kirk-Othmer Encyclopedia of Chemical Technology*, 2nd ed., Vol. 12, Interscience, New York, 1967, p. 557.

42. B. B. Farrington and W. N. Davis, *Ind. Eng. Chem.*, **28**, 414 (1936).

43. R. G. Larson and A. Bondi, "Lubrication and Lubricants," in *Kirk-Othmer Encyclopedia of Chemical Technology*, 1st ed., Vol. 8, Interscience, New York, 1952, p. 495.

44. J. W. Jordan, B. J. Hook, and C. M. Finlayson, *J. Phys. Colloid Chem.*, **54**, 1196 (1950).

45. W. Ramney, *Lubricant Additives*, Noyes Data Corp., Park Ridge, NJ, 1973.

46. A. Eisner, T. Perlstein, G. Maerker, and L. Stallings, *J. Am. Oil Chemists' Soc.*, **48**, 811 (1971).

47. F. O. Flaherty and R. L. Stubbins, "Leather," in *Kirk-Othmer Encyclopedia of Chemical Technology*, 2nd ed., Vol. 12, Interscience, New York 1967, p. 303.

48. E. M. Filachione, *J. Am. Oil Chemists' Soc.*, **48**, 334 (1971).

49. T. E. Nestler and G. L. Royer, *J. Am. Leather Chemists' Assoc.*, **40**, 40 (1945).

50. E. L. Valko, in *Man Made Fibers, Science and Technology*, H. F. Mark, S. M. Atlas, and E. Cernia, Eds., Vol. 3, Wiley, New York, 1968, pp. 499, 533.

51. S. Cohen, in E. S. Pattison, Ed., *Fatty Acids and Their Industrial Applications*, Chapter 11, Marcel Dekker, New York, 1968.

52. R. R. Egan, *J. Am. Oil Chemists' Soc.*, **55**, 118 (1978).

53. R. A. Reck, *J. Am. Oil Chemists' Soc.*, **56**, 796A (1979).

54. C. W. Glankler, *J. Am. Oil Chemists' Soc.*, **56**, 802A (1979).

55. S. A. Kaplan, U.S. Pat. 2,483,917 (1949).

56. H. A. Sarvetnick, *Polyvinyl Chloride*, Reinhold, New York, 1969.

57. U.S. International Trade Commission (formerly U.S. Tariff Commission), *Production of Selected Synthetic Organic Chemicals*, S. O. C. Series C/P-75-12.

58. D. Swern, T. W. Findley, and J. T. Scanlan, *J. Am. Chem. Soc.*, **66**, 1925 (1944).

59. T. W. Findley, D. Swern, and J. T. Scanlan, *J. Am. Chem. Soc.,* **67,** 412 (1945).

60. D. Swern, G. N. Billen, and J. T. Scanlan, *J. Am. Chem. Soc.,* **68,** 1504 (1946).

61. D. Swern, *Chem. Rev.,* **45,** 1 (1949).

62. W. D. Niederhauser and J. E. Koroly, U.S. Pat. 2,485,160 (1950).

63. J. G. Wallace, "Epoxidation," in *Kirk-Othmer Encyclopedia of Chemical Technology,* 2nd ed., Vol. 8, Wiley, New York, 1965.

64. W. R. Schmitz and J. G. Wallace, *J. Am. Oil Chemists' Soc.,* **31,** 363 (1954).

65. A. W. Wahlroos, U.S. Pat. 2,813,878 (1957).

66. B. Philips and P. S. Starcher, British Pat. 735,974 (1955).

67. B. Philips, F. C. Frostick, and P. S. Starcher, U.S. Pat. 2,804,473 (1957).

68. Anon., *Modern Plastics,* **54,** (9), (1977).

69. W. M. Budde, private communication, 1978.

70. W. S. Port, E. F. Jordan, Jr., W. E. Palm, L. P. Witnauer, J. E. Hanson, and D. Swern, *Ind. Eng. Chem.,* **47,** 472 (1955).

71. F. C. Magne, R. R. Mod, and E. L. Skau, *J. Am. Oil Chemists' Soc.,* **40,** 541 (1963).

72. R. R. Mod, F. C. Magne, and E. L. Skau, *J. Am. Oil Chemists' Soc.,* **45,** 385 (1968).

73. L. W. Mazzeno, Jr., F. C. Magne, R. R. Mod, E. L. Skau, and G. Sumrell, *Ind. Eng. Chem., Prod. Res. Devel.,* **9,** 42 (1970).

74. M. Weiss, N. J. R. Rosberg, N. O. V. Sonntag, and S. Eng, *J. Am. Oil Chemists' Soc.,* **56,** 849A (1979).

75. C. Chang and S. Wan, *Ind. Eng. Chem.,* **39,** 1543 (1947).

76. M. L. Mandlekar, T. N. Mehta, V. M. Parekh, and V. B. Thosar, *J. Sci. Ind. Res. (India),* **B5,** 45 (1946).

77. L. F. Fieser, G. C. Harris, E. B. Hershberg, M. Morgana, F. C. Novello, and S. T. Putman, *Ind. Eng. Chem.,* **38,** 768 (1946).

78. K. J. Mysels, *Ind. Eng. Chem.,* **41,** 1435 (1949).

79. G. Barnett, *Soap Chem. Spec.,* **36,** 141 (1960).

80. H. P. Kroke, *J. Am. Oil Chemists' Soc.,* **55,** 444 (1978).

81. D. H. Johnson, *J. Am. Oil Chemists' Soc.,* **55,** 438 (1978).

82. J. E. Goyan and M. Wruble, U.S. Pat. 2,975,099 (1961).

83. H. M. Gross and C. H. Becker, *J. Am. Pharm. Assoc., Sci. Ed.,* **17,** 498 (1953).

84. Belgian Pat. 633,957 (1963).

85. M. Goldschmidt and H. Koffer, *Ind. Eng. Chem.,* **42,** 1819 (1950).

86. W. A. Strickland, Jr., *Drug Cosmetic Ind.,* **85,** 318 (1959).

87. G. Y. Brokaw and W. C. Lyman, Jr., U.S. Pat. 2,976,251 (1961).

88. A. F. Woodhour and T. B. Stim, U.S. Pat. 3,149,036 (1964).

89. J. L. Kanig, L. Chavkin, and L. Lerea, *Drug Cosmetic Ind.,* **75,** 180 (1954).

90. E. Ullmann, *Mitt. Deutsch Pharm. Ges.,* **27,** 1 (1957).

91. E. G. Rippie, D. J. Lamb, and P. W. Ramig, *J. Pharm. Sci.,* **53** (11), 1346 (1964).

92. E. L. Parrott, *Drug Cosmetic Ind.,* **96,** 320 (1965).

93. W. S. Singleton, J. L. White, L. L. Pitrapani, and M. L. Brown, *J. Am. Oil Chemists' Soc.,* **39,** 260 (1962).

94. E. B. McQuarrie and H. P. Andersen, *Am. J. Clin. Nutr.,* **16,** 23 (1965).

95. A. J. Wretland, U.S. Pat. 3,169,094 (1965).

96. J. J. Karbara, in *Pharmacological Effects of Lipids*, J. J. Kabara, Ed., American Oil Chemists' Society, Champaign, Illinois, 1979, p. 11.

97. J. J. Kabara, *J. Am. Oil Chemists' Soc.*, **56,** 760A (1979).

98. E. Stempel, *Drug Cosmet. Ind.*, **98** (1), 44 (1966); ibid., **98** (2), 36 (1966).

99. A. W. Ralston, J. P. Barrett, and E. W. Hopkins, *Oil Soap*, **18,** 11 (1941).

100. H. L. Haller, E. R. McGovran, L. D. Goodhue, and W. N. Sullivan, *J. Org. Chem.*, **7,** 177 (1942).

101. H. W. Thurston, Jr., *Agr. Chem.*, **5,** 28, 99 (1950).

102. E. S. Pattison, Ed., *Fatty Acids and Their Industrial Applications*, Marcel Dekker, New York, 1968.

103. J. C. Kern, *J. Am. Oil Chemists' Soc.*, **56,** 716A (1979).

104. L. G. Zachary, *J. Am. Oil Chemists' Soc.*, **54,** 533 (1977).

105. K. T. Zilch, *J. Am. Oil Chemists' Soc.*, **56,** 739A (1979).

106. N. O. V. Sonntag, *J. Am. Oil Chemists' Soc.*, **56,** 729A (1979).

107. L. D. Myers and V. J. Muckerheide, U.S. Pat. 2,293,676 (1942).

108. L. D. Myers and V. J. Muckerheide, U.S. Pat. 2,421,157 (1947).

109. R. H. Potts and G. W. McBride, *Chem. Eng.*, **57** (2), 124, 172 (1950).

110. V. J. Muckerheide, in K. S. Markley, Ed., *Fatty Acids, Their Chemistry, Properties, and Uses*, Part 4, Chapter 24, Interscience, New York, 1967.

111. W. Stein, *J. Am. Oil Chemists' Soc.*, **45,** 471 (1968).

112. W. Stein and H. Hartmann, U.S. Pat. 2,800,493 (1957).

113. W. Stein and H. Hartmann, British Pat. 925,674 (1955).

114. W. Stein and H. Hartmann, U.S. Pat. 3,737,444 (1973).

115. H. Waldemann and W. Stein, German Pat. 1,107,659 (1961).

116. H. Waldemann and W. Stein, U.S. Pat. 3,052,700 (1962).

117. H. Hartmann and W. Stein, U.S. Pat. 3,733,343 (1973).

118. N. O. V. Sonntag, *J. Am. Oil Chemists' Soc.*, **56,** 861A (1979).

119. E. Fritz, *J. Am. Oil Chemists' Soc.*, **56,** 744A (1979).

120. R. Berger and W. McPherson, *J. Am. Oil Chemists' Soc.*, **56,** 743A (1979).

121. R. C. Hastert, *J. Am. Oil Chemists' Soc.*, **56,** 732A (1979).

122. E. E. Rice, *J. Am. Oil Chemists' Soc.*, **56,** 754A (1979).

123. R. D. Farris, *J. Am. Oil Chemists' Soc.*, **56,** 770A (1979).

124. N. O. V. Sonntag, *J. Am. Oil Chemists' Soc.*, **56,** 751A (1979).

125. R. L. Logan, *J. Am. Oil Chemists' Soc.*, **56,** 777A (1979).

126. R. Herlinger, in E. S. Pattison, Ed., *Industrial Fatty Acids*, Chapter 4, Reinhold, New York, 1959.

127. L. G. Zachary, H. W. Basak, and F. J. Eveline, Eds., *Tall Oil and Its Uses*, McGraw-Hill, New York, 1965.

128. A. G. Johanson, *J. Am. Oil Chemists' Soc.*, **54,** 848A (1977).

129. M. H. Ittner, U.S. Pats. 1,918,603 (1933), 2,139,589 (1938), 2,221,799 (1940), 2,435,745 (1948), 2,458,170 (1949), and 2,480,471 (1949).

130. V. Mills, U.S. Pats. 2,156,863 (1939), 2,159,397 (1939), and 2,233,845 (1941).

131. A. C. Brown, U.S. Pat. 2,486,630 (1949).

132. A. B. Herrick and E. Jungerman, *J. Am. Oil Chemists' Soc.*, **40**, 615 (1960).

133. K. N. Richardson, *Soap, Perfumery, Cosmet.*, **38**, 137 (1965).

134. E. Jungermann, *J. Am. Oil Chemists' Soc.*, **56**, 827A (1979).

135. S. H. Shapiro, in E. S. Pattison, Ed., *Fatty Acids and Their Industrial Applications*, Chapter 5, Marcel Dekker, New York, 1968.

136. W. Kritchevsky, U.S. Pats. 2,089,312 (1937) and 2,096,746 (1937).

137. H. L. Sanders, D. E. Libman, and Y. D. Kardish, *Soap Chem. Spec.*, **32** (1), 33 (1956).

138. A. Cahn, *J. Am. Oil Chemists' Soc.*, **56**, 809A (1979).

139. H. L. Morrill, U.S. Pat. 2,508,415 (1950).

140. F. Bailles and C. Paquot, *Oleagineux*, **21** (7), 44 (1966).

141. A. F. Isbell, U.S. Pat. 2,169,467 (1952).

142. D. Aelony, U.S. Pat. 2,811,549 (1957).

143. D. L. Andersen, U.S. Pat. 2,816,920 (1957).

144. M. Cooperman, U.S. Pat. 3,222,402 (1965).

145. J. Kilheffer and E. Jungermann, *Anal. Chem.*, **32**, 1178 (1960).

146. R. G. Kadesch, *J. Am. Oil Chemists' Soc.*, **56**, 845A (1979).

147. E. H. Pryde, D. J. Moore, J. C. Cowan, W. E. Palm, and L. P. Witnauer, *Polym. Eng. Sci.*, **6**, 60 (1965).

148. E. H. Pryde, *J. Am. Oil Chemists' Soc.*, **48**, 349 (1971).

149. P. R. Sampath and A. E. Rheineck, *J. Paint Technol.*, **41**, 17 (1969).

150. A. E. Rheineck and P. R. Lakshmanan, *J. Am. Oil Chemists' Soc.*, **46**, 452, 456 (1969).

151. T. R. Steadman and J. O. H. Peterson, *Ind. Eng. Chem.*, **50**, 59 (1958).

152. E. T. Roe and D. Swern, *J. Am. Oil Chemists' Soc.*, **37**, 661 (1960).

153. K. Buchner, O. Roelen, J. Meis, and H. Langwald, U.S. Pat. 3,043,871 (1962).

153a. N. E. Lawson, T. T. Cheng, and F. B. Slezak, *J. Am. Oil Chemists' Soc.*, **54**, 215 (1977).

154. J. C. Cowan, *J. Am. Oil Chemists' Soc.*, **31**, 529 (1954).

155. D. E. Floyd, *Polyamide Resins*, 2nd ed., Reinhold, New York, 1966.

156. J. C. Cowan, *J. Am. Oil Chemists' Soc.*, **39**, 534 (1962).

157. E. C. Leonard, *J. Am. Oil Chemists' Soc.*, **56**, 782A (1979).

158. D. V. Kinsman, *J. Am. Oil Chemists' Soc.*, **56**, 823A (1979).

159. G. R. Wilson, *J. Am. Oil Chemists' Soc.*, **31**, 564 (1954).

160. J. A. Monick, *J. Am. Oil Chemists' Soc.*, **56**, 853A (1979).

161. A. N. Wrigley, A. J. Stirton, and E. Howard, Jr., *J. Org. Chem.*, **25**, 439 (1960).

162. A. N. Wrigley, F. D. Smith, and A. J. Stirton, *J. Am. Oil Chemists' Soc.*, **36**, 34 (1959).

163. U.S. International Trade Commission, *Synthetic Organic Chemicals, U.S. Production and Sales of Surface-Active Agents, 1976* (September 1977).

164. M. W. Formo, *J. Am. Oil Chemists' Soc.*, **31**, 548 (1954).

165. D. I. Hartnett and W. G. Thalheimer, *J. Am. Oil Chemists' Soc.*, **56**, 944 (1979).

166. D. I. Hartnett and W. G. Thalheimer, *J. Am. Oil Chemists' Soc.*, **56**, 948 (1979).
167. R. T. McIntyre, *J. Am. Oil Chemists' Soc.*, **56**, 835A (1979).
168. J. L. VanHaften, *J. Am. Oil Chemists' Soc.*, **56**, 831A (1979).
169. N. O. V. Sonntag and K. T. Zilch, in E. S. Pattison, Ed., *Fatty Acids and Their Industrial Derivatives*, Chapter 17, Marcel Dekker, New York, 1968.
170. H. Fineberg, *J. Am. Oil Chemists' Soc.*, **54**, 842A (1977).
171. N. E. Bednarcyk and W. L. Erickson, *Fatty Acids, Synthesis and Application*. Noyes Data Corp., Park Ridge, NJ, 1973.
172. H. Fineberg, *J. Am. Oil Chemists' Soc.*, **56**, 805A (1979).
173. A. Krautwald, *Deutsch Gesundheitsw.*, **3**, 354 (1948).
174. K. Thomas and G. Weitzel, *Deutsch Med. Wachschr.*, **71**, 18 (1946).
175. M. F. Guatreaux, W. T. Davis, and E. D. Travis, in *Kirk-Othmer Encyclopedia of Chemical Technology*, 3rd ed., Vol. 1, Wiley, New York, 1978, p. 740.
176. K. Ziegler, *Angew. Chem.*, **64**, 23 (1952).
177. H. Koch, U.S. Pat. 2,831,877 (1958).
178. E. F. Jason and E. K. Fields, U.S. Pat. 3,076,842 (1963).
179. W. C. Ault, R. W. Riemenschneider, and D. H. Saunders, *U.S. Department of Agriculture, Utilization Report No. 2* (1960).

7
Analytical Methods

Introduction

1.1 IMPORTANCE AND CLASSIFICATION OF ANALYTICAL METHODS

For our purposes the analytical methods used in the fats and oils, fatty acid, fatty chemical, and related edible- and inedible-product industries can be conveniently classified into five groups. Except for the first group, which consists largely of unofficial procedures rather than methods, the second to the fifth groups fall logically into a decreasing order of method "standardization."

The first group consists of procedures used to identify fats and oils with particular respect to their source and origin. Fats and oils may be simply classified into 11 categories that comprise the 10 groups outlined in Volume 1, Chapter 5, pages 282–287 plus a "miscellaneous" group that includes rarer fats and oils whose glycerides contain fatty acids with epoxy, acetylenic, cyclopentenoid, or other unusual functional groups. Occasionally, identification of fats and oils is required by source and origin, that is, the land animal, milk fat (animal), oil-seed vegetable, tall oil vegetable, or fish animal nature of the fat or oil is required to be established. In addition, it is sometimes necessary to determine the specific animal fat or the particular oilseed or oil pulp vegetable origin of a sample. Although no routine standard methods are available, a number of general procedures are used that vary from fat to fat and oil to oil.

The second (and largest) group consists of important "official" methods. Certain procedures of the American Oil Chemists' Society (AOCS) (1), the American Society for Testing Materials (ASTM) (2), the Association of Official Analytical Chemists (AOAC) (3), and other organizations are customarily used for routine analysis of fats and oils, finished fatty acids, and other derivatives for sale or purchase, occasionally for "in-process" control to establish compliance with certain quality specifications. Other methods apply to vegetable source materials such as cottonseed, peanuts, soybeans, tung fruit, castor beans, flaxseed and safflower seed, and to oilseed byproducts like cake, meal and meats, linters and hull fibers, soya flour, and castor pumice.

Most of the methods in the second group involve the determination of a component through chemical reaction at a characteristic functional group, others involve the determination of certain physical properties, and a few require the evaluation of the performance of the product under prescribed conditions. Among the conventional "official" methods for fats and oils are those used for the determination of moisture, insoluble impurities, unsaponifiable matter, iodine value, acid value, ash and peroxide value. The conventional "official" methods applied to fatty acids are not necessarily identical to those used for fats and oils but include methods for the determination of moisture, acid value, titer, unsaponifiable matter, color, iodine value, saponification value, ash and others. Additional "official" methods are used for fat-based products, such as soap, soap products, synthetic detergents, fatty alkyl sulfates, various grades of glycerol, sulfonated and sulfated oils, soapstock, fatty nitrogen chemicals, mono- and diglycerides, "dimer" acids, and numerous other kinds of products.

A third group of analytical methods, now becoming increasingly important and more widely used, was developed specifically for the purpose of monitoring and controlling the levels of objectionable or toxic components that occasionally occur in fat and oil products. Certain contaminants are potentially derived from the fat and oil raw materials or from their degradation products, others from inadvertent contamination from the environment, and still others, perhaps, by contamination from the extraction solvents, the metal equipment or the processing chemicals used in the manufacture of the products.

Typical of methods used to detect contamination from raw material sources are those used to determine trace metal content in vegetable oils (from the soil in which the plants were grown), those used to detect both gossypol (4) and cyclopropenoids (5) in cottonseed derived products, and those used for polyethylene in animal fats (6) and for methyl ketones in milk fats (7).

The possibility of environmental contamination necessitates analytical methods such as those used for the detection and control of (a) "chick edema factor" (presumably chlorodibenzodioxin residues from chlorinated pesticides) (8), (b) aflatoxins, which are highly toxic carcinogenic metabolites produced during certain mold decay in peanut, other nut, or cottonseed oils (9, 235–239), (c) nitrosamines (10, 10a), which are powerful animal carcinogens found occasionally in minute quantities in detergent compositions, and edible vegetable oils [N-nitrosodimethylamine (NDMA) has been found in 21 of 61 samples at a level of 1 μg/kg to 23 μg/kg, and N-nitrosodiethylamine (NDEA), in 18 of 61 samples at a level of 1–27.8 μg/kg (11), in margarine (NDMA in 15 of 107 samples at 1–5.8 μg/kg, and NDEA in 33 of 107 samples at a level of 1–7.5 μg/kg), in beer (12), and in certain other products], (d) thioglucosides in *Crambe* (13), and (e) a pungent factor that occurs in mustard seed and is capable of generating objectionable allyl isothiocyanate by hydrolysis (14).

"In-process" contamination may result accidentally in fats and oils, fatty acids, and other fat-derived products from trace or small quantities of metals such as nickel, tin, and copper used as process catalysts; from traces of soap left in fats and oils from alkalies used in the refining process; from soluble iron through metal corrosion; from particulate iron or steel due to abrasion of moving parts such as agitator shafts; from heating fluids such as Dowtherm® A or Therminol® 66 through leakage from heating unit jackets; from inefficient removal of bleaching clays; and from the presence of toxic substances in ethoxylated derivatives (14a). Fortunately, efficient technology exists that may be used to remove or minimize these objectionable materials in most fat and oil products. Reliable, precise, reproducible, reasonably rapid and thoroughly evaluated analytical methods are available and continue to be in increasing demand. They are essential for monitoring the quality of the finished products. In many cases, analytical methodology and new sophisticated equipment designed to apply these procedures efficiently are in a constant state of reevaluation and refinement, and we can expect continuing improvement in sensitivity and precision for many of the important methods.

A fourth group of analytical methods was designed for the qualitative and quantitative detection of relatively small quantities of nontoxic components that have a somewhat advantageous or beneficial effect on the quality or the performance of the finished products. Typical examples include methods for the determination of β-tocopherol in wheat germ oil (15) and of vitamin A in butterfat (16); identification and assay of several ferulate natural antioxidants in rice bran oil (17); analysis of sesamol and related compounds in sesame oil (18); determination of n-eicosanyl ferulate in linseed and rapeseed oils (19); determination of dimethylpolysiloxanes, which are used as approved trace additives in fats and oils for antifoam control (20); and a gel permeation chromatographic method for the determination of synthetic antioxidants of the phenolic type (BHA, BHT, and TBHQ) in edible fats and oils (21).

The optimization of desired quality or advantageous performance with the level of responsible component in many products is achieved initially by establishment of the relationship, and then afterwards, by routine quality control to assure that optimization is achieved during production. Both of these steps in product development require appropriate and satisfactory analytical methods.

Finally, a fifth group of so-called procedures generally involving complicated separations and sophisticated instrumental techniques of analysis can be applied to the determination of small or trace quantities of components that ordinarily have little or no effect on the quality or the performance of the finished products. Aside from some academic or scientific interest, the information obtained from such analyses is of minor importance, and these procedures are rarely used for quality control purposes; thus few, if any, of them have attained the status of standard or "official" methods of analysis. Representative of these procedures are the determination of sterols in many fats and oils (22, 23), triterpene alcohols in fat and oil unsaponifiables (22–24), traces of natural hydrocarbons in virgin olive oil (25), and furanoid fatty acids in cod liver oil (26). As new trace components are discovered and their effects on the quality or the performance of the products established, or as further information on existing components is obtained, our overall dependence on reliable and sensitive analytical methods is increased.

1.2   ANALYTICAL METHODS DEVELOPMENT

The principal developers of standard or "official" analytical methods are the industrial producers in the fats and oils and related industries, the professional societies and associations such as the AOCS, the ASTM and the AOAC, and occasionally governmental regulatory agencies such as the FDA.

When standard methods are deemed necessary, professional societies such as the AOCS customarily undertake a development program involving technical committees that conduct round-robin collaborative evaluation of a suggested or candidate method with selected standard samples. The scope, the reproducibility, and the reliability of the method are ascertained. The AOCS publishes information on the evaluation of test methods (27) and recommends procedures for method evaluation and writing and approval of methods (28).

The AOAC has prepared a statistical manual that describes the planning of collaborative studies and the interpretation of the collaborative test data (28b). A number of the professional societies assign to developed methods a tentative status for some period of time to establish the proven usefulness and reliability before an "official" status is assigned.

1.3   THE SMALLEY PROGRAM FOR ANALYTICAL CHEMIST AND LABORATORY ACCREDITATION

With worldwide production of fats and oils at near 100 billion pounds per year, and with much of this raw material traded on the international market, it is readily apparent that there is a continuing need for certification of referee analytical chemists to assure that quality of purchased or exchanged goods is established in conformity with purchase specifications. The Smalley program, operated by the AOCS since 1915, provides a satisfactory means for assuring reliable fat and oil analysis. Certification of analytical chemists at their organization laboratories is provided through a series of test analyses on specific fat and oil products. Chemists are accredited for specific fat and oil analyses if their test analyses on standard samples conform within rigid precision requirements. Analytical chemists may qualify for Smalley certification in analysis of a single product or in a whole series of fat and oil products.

Methods and Procedures

2.1   PROCEDURES FOR IDENTIFICATION OF COMMON FATS AND OILS

The identification of most common fats and oils that are not in admixture is considerably easier than it was 15–20 years ago, although the establishment of the origin of a specific sample is by no means a routine and simple task. The common fats and oils have been classified with respect to the general ranges of fatty acid distributions in their glycerides (see Chapter 6, Volume 1 for details of the fatty acid distributions tentatively adopted by the Food and Agriculture Organization/World Health Organization Codex Alimentarius Committee on Fats and Oils for common fats and oils). The use of the standard GLC fatty acid analysis, as typified by AOCS Method Ce-1-62, affords a clue to the classification group to which the fat or oil is ordinarily assigned. Unofficially, these comprise 11 groups of fats and oils, consisting of the 10 groups outlined in Chapter 5, Volume 1, pages 282–287, plus a "miscellaneous" group including fats and oils whose glycerides contain substantial quantities of fatty acids with unusual structure (epoxy, acetylenic, cyclopentenoid, furanoid, etc.). The GLC analysis, taken alone, is rarely definitive enough for identification purposes, particularly since many vegetable oils are assigned to the oleic–linoleic group, and most of them vary widely with recent genetic changes, thus affording broad latitude in their fatty acid composition.

Recourse is usually made to the determination and the interpretation of several physical and chemical characteristics for the fat and oil and, further, in some cases to the analysis for several characteristic trace components such as sterols, tocopherols, triterpenes, ferulates, and pigments before identification can be reasonably certain. Since only approximately 4700 of about 250,000 species of plants known to botany are analytically characterized today (1981), one can appreciate that it would, of course, be impossible to accomplish the identification of samples of all of the known vegetable oils.

Table 7.1 presents a summary of information that is helpful in establishing the identification of 37 common and 11 rarer fats and oils where little or no background information is on hand concerning the unknown sample. Used in conjunction with GLC fatty acid distribution data, and with information from several physical and chemical characteristics methods, these are usually sufficient for the positive identification of about 60 fats and oils. Section I of the AOCS Official Methods (1) provides a tabulation of the physical and chemical characteristics of 72 common fats and oils of animal and vegetable origin, including 8 marine oils and 12 waxes, which is also very helpful.

An ASTM method (28a) recommends that drying oils be converted to methyl esters and analyzed by the GLC technique as an aid to identification of the drying oil type. Tabulations of fatty acid distributions of 11 common drying oils are included for purposes of comparison.

The identification and analysis of fat and oil mixtures is usually extremely difficult. In the simple case of two component mixtures where one of the components has a unique and distinctive constituent capable of direct analysis, such as the use of GLC for 12:0 fatty acid in the identification of coconut oil in soybean oil, the problem is quite simple. On the other hand, the analysis of vegetable oil mixtures from the oleic–linoleic group is very difficult, and the identification of beef tallow in lard (hog fat) by fatty acid homolog comparison is virtually impossible.

2.2 "OFFICIAL" OR STANDARD METHODS OF ANALYSIS FOR FATS AND OILS, SOURCE MATERIALS, BYPRODUCTS, FATTY ACIDS, AND VARIOUS OTHER FAT AND OIL PRODUCTS

Raw Materials and Byproducts. Some of the important analytical methods that have been developed and which are applied to the determination of quality of some source materials for vegetable oils are included in Table 7.2; several methods applicable to byproducts resulting from the conversion of vegetable source materials to finished product fats and oils are included in Table 7.3.

Table 7.1. Several characteristic constituents useful in the identification of the origin of fats and oils

Fat or Oil	Characteristic Constituents[a]	Analytical Method	Reference
Land animal, general	Branched-chain, odd-numbered, 17:0, 17:1 FAG; relatively high amounts of *trans* isomers; large cholesterol content in sterols	Isolated *trans* isomers Sperry–Webb Method	29 30
Butterfat	Low-molecular-weight fatty acids, 4:0 (2.8–4), 6:0 (1.4–3) FAG; SV 210–250	Reichert-Meissel value 22–34 Polenske value 2–4 Kirschner value 20–26	31 31 31
	Lactones (~0.011)	Various methods[b,c]	32a–f
	Methyl ketones (0.006–0.022)	Steam distillation	7
	Vitamin A 6–i2 μg/g, β-carotene 2–10 μg/g	UV spectrophotometric or colorimetric method	33
Lard (hog)	20:2, 20:4 (0.3–1) total 20–24s (1–2) FAG; high-melting palmitodistearin; specific triglycerides: SPS (~30), OOP (~40)	Boemer test Differential thermal analysis	34 35
Tallow (beef)	18:1 (40–50), 16:0 (24–37), 18:0 (14–29), 18:2 (1–7), traces of 17:0, 17:1, 18:3, 20:1, 20:4, FAG; about 5% *trans* acids; SV-193–202	Isolated *trans* isomers	29
Fish animal, general	Unhydrogenated: 20:0, 20:1, 20:4, 20:5, 22:1, 22:5 and 22:6 FAG in most oils		
Menhaden	17:0 (0.4–1), 22:6 (3–12) FAG; IV 150–165, SV 192–199		
Whale	Fatty acids vary with species; SV 185–202: IV 110–135, UNS (<2)		
Sperm	About 76% wax esters, 22.6% triglycerides, 0.16% hydrocarbons, 1.2% nonpolar compounds; 20:1 (~16), 22:1 (8–10) FAG; SV 120–150; UNS (17.5–44); SG 10°C 0.844–0.881		

413

Table 7.1. (Continued)

Fat or Oil	Characteristic Constituents[a]	Analytical Method	Reference
Cod liver	22:6 (8–19) FAG; Furanoid fatty acids (~1) FAG		28
	Vitamin A (3M–30M USP/g)		33
Tall oil vegetable, general	Rosin [crude TO (~40), dist'd. TO (20–35)]; in TOFA 16:0 + 18:0 (3), 18:1 (~48), 18:2 (~46), 18:3 (nil) FAG; UNS (~1.7); stigmasterol (~0.9) and stilbenes (0.005)	Linder–Persson method	36
			37
			24, 25
		UV spectrophotometric method	38
Common vegetable, general	Phytosterols in UNS	Liebermann–Burchard method	39
		Vegetable oil sterols by GLC	39a
Coconut	Medium-molecular-weight fatty acids 12:0 (44–52), 14:0 (13–19) FAG; UNS (0.2–0.4); SV 250–264; RI 40°C 1.448–1.450; IV 7.5–10.5; sterols (0.15–0.2), β-sitosterol 69–75%; tocopherols (~0.008), α-tocopherol 99% in crude; refined apparently has no α-, β-, or γ-tocopherols but has δ- (~0.00024) and α-tocotrienol (~0.0020); lactones (~0.01), C₁₀ δ-type (~0.005)	Reichert-Meissl value (6–8)	31
		Polenske value (15–18)	31
		TLC method	24, 25
		Column chromatography	41, 71
Palm kernel	Medium-molecular-weight fatty acids 8:0 (3–5), 10:0 (3–7), 12:0 (40–52), 14:0 (14–18), 16:0 (7–9), 18:1 (11–19) FAG; SV high 245–255; UNS (0.4–0.6); five sterols in UNS plus cholesterol; lactones, total (0.0039), C₁₂ δ-type (0.0009); triterpene alcohols (0.07), six in UNS	Reichert-Meissl value (5–7)	31
		Polenske value (10–12)	31
		Kirschner value (0.5–1)	31
		TLC method	25
		Column chromatography	78
Babassu	8:0 (~5), 10:0 (6–7), 12:0 (~45), 14:0 (~15), 16:0 (~8.5), 18:1 (~16) FAG; lactones, total (~0.0044), C₁₀ δ-type (~0.0022)	TLC method	24, 25
		Polenske value (10–12)	31
		Reichert-Meissl value (5–7)	31
		Column chromatography	78

414

Oil/Fat	Composition and characteristics	Method	Ref.
Cocoa butter	16:0 (24–27), 18:0 (32–35), 18:1 (33–37) FAG; SV 190–200, IV 35–40, UNS (~0.4); phospholipids (0.05–0.13); sterols (~0.3), β-sitosterol is ~59%; polymorphism in triglycerides	TLC method	25
		X-Ray diffraction	42
		Differential scanning calorimetry	43
Kokum butter	Melting point 39–43; IV 23–37; 2-oleodistearin (~40), mp 23, 37, and 44°C; 76 mole % S_2U in triglycerides	X-Ray diffraction and "thermal analysis"	44
		Possible HPLC method	45, 46
Cottonseed	16:0 (20–23), 18:0 (1–3), 18:1 (23–35), 18:2 (42–54), 20:0 (0.2–1.5) FAG; gossypol pigments; cyclopropenoid fatty acids (0.5) in crude, trace in refined; phosphatide in crude (0.7–0.9); sterols (0.26–0.31), β-sitosterol 89–93%; tocopherols (crude ~0.09, refined ~0.05) α- and γ- about equal	UV spectrophotometric method	47
		Halphen test	48
		Spectrophotometric and HBr methods	49, 50
		TLC method	51
		TLC and GLC methods	25
		Silica gel TLC method	52
Peanut	16:0 (6–9), 18:0 (3–6), 18:1 (53–71), 18:2 (13–72), 20:0 (2–4), 22:0 (1–3), 24:0 (1–2.5) FAG; sterols (0.18–0.25), β-sitosterol 70–76%; pigments β-carotene and lutein	TLC method	40
		TLC method	25
		Spectrophotometric method	53
Olive	18:1 (65–85) FAG; high squalene (0.2–0.7); 2-monoglycerides from pancreatic hydrolysis never have more than 1.6% 16:0; sterols (0.1–0.2), β-sitosterol 91%; tocopherols (~0.01), α-tocopherol 88%; triterpene alcohols (~0.16), six in UNS; erythradiol in olive husk oil.	Alumina chromatography	54
		Pancreatic hydrolysis	56, 57
			56
		TLC method	25
		Silica gel TLC method	40, 58, 59
		TLC method	24, 25
		Crismer test (68.5–71)	55, 55b
Palm	16:0 (32–47), 18:0 (2–8), 18:1 (40–52), 18:2 (5–11) FAG; high β-carotene (0.05–0.2); sterols (0.03–0.1), β-sitosterol ~74%; tocopherols (0.003–0.11), α-tocopherol predominates	Spectrophotometric method	60
		TLC method	25
		Various methods	15, 15a

415

Table 7.1. (Continued)

Fat or Oil	Characteristic Constituents[a]	Analytical Method	Reference
Sunflower	16:0 (3–6), 18:0 (1–3), 18:1 (14–43), 18:2 (44–75), FAG; UNS (0.301.3); sterols (0.542–0.584), β-sitosterol 62–75%; methyl sterols (~0.11)	Silica gel TLC method TLC method TLC method	40 25 25, 26
Sesame	16:0 (7–9), 18:0 (4–5), 18:1 (37–49), 18:2 (35–47) FAG; sesamol (trace), sesamine (0.4–1.1) and sesamoline (0.3–0.65); UNS (≤1.8); sterols (0.43–0.55), β-sitosterol ~62%; methyl sterols (~0.42), phosphatide content crude (~0.1)	Baudouin test Villavecchia test Fluorescence method TLC method TLC method	61 62 63 25 25, 26
Corn	16:0 (8–12), 18:0 (2–5), 18:1 (19–49), 18:2 (34–62) FAG; phosphatide content crude (1–2); UNS (<2); sterols (0.58–1), β-sitosterol 66%; tocopherols (~0.01), unique γ-tocopherol 69–70%; dihydro-β-sitosterol ferulate (small)	 TLC method Various methods Silica gel TLC method	51 51 25 15 40, 65
Safflower, High linoleic	16:0 (3–6), 18:0 (1–4), 18:1 (13–21), 18:2 (73–79) FAG; sterols (~0.35), β-sitosterol ~52%; triterpene alcohols (0.066); tocopherols (~0.03), α-tocopherol ~84%; AOM value (97°C) 8 hours	TLC method TLC method Silica gel TLC method Active oxygen method	25 24, 25 40, 58 66
High oleic	16:0 (4–8), 18:0 (4–8), 18:1 (74–79), 18:2 (11–19) FAG; sterols (~0.38), β-sitosterol ~52%; triterpene alcohols (~0.060); AOM value (97°C) ~29.3 hours	TLC method TLC method Active oxygen method	25 24, 25 66
Tobacco seed, West Bengal	16:0 (~8.4), 18:0 (~3.8), 18:1 (~15.7), 18:2 (~71.2) FAG		
Poppyseed, Russian	18:2 (68.2–80.3) FAG		

Teaseed	18:1 (74–87), 18:2 (7–14), saturated fatty acids (7–11) FAG; characteristic red color vs. olive with Ac_2O, $HCCl_3$, and H_2SO_4	Fitelson test	68
Kapok	16:0 (10.5–10.8), 18:0(4.9–8.6), 18:1(46.1–56.6), 18:2 (27.7–34.6) FAG; no gossypol; cyclopropenoid fatty acids	Halphen test Spectrophotometric and HBr methods	48 49, 50
Rice bran	16:0 (12–18), 18:0 (1–3), 18:1 (40–50), 18:2 (29–42), FAG; UNS (3–5); active lipase; sterols (~0.75), β-sitosterol about 49%, campesterol ~28%, and stigmasterol ~15%; tocopherols (~0.04); contains three ferulate antioxidants (0.006–0.027)	TLC method Silica gel TLC method	25 40 69–71
Sorghum	16:0 (6–10), 18:0 (3–6), 18:1 (30–47); 18:2 (40–55) FAG; SV 181–191, IV 108–122		
Buckwheat seed	18:1 (32.4–38.5), 18:2 (30.8–34.7), 23:0 (~1.4), 24:0 (~4.7) FAG		72
Rapeseed High erucic type	22:1 (40–55) FAG: contains trace of S compounds; UNS (0.5–1.2); SG 25/25°C 0.9123; sterols (0.35–0.50), *Brassica* and *sinapsis* oils contain ~10% brassicasterol plus ~56% β-sitosterol and ~25% campesterol, but no stigmasterol; triterpene alcohols (0.05), 13 in UNS; AOM value ~19.2 hours; pigments 6–7 ppm β-carotene and neolutein A and B; polymorphism in triglycerides	Coin and silver benzoate tests TLC method TLC method Active oxygen method Spectrophotometric method X-Ray diffraction Phase transitions Crismer test (76–82)	79 25, 56, 74 24, 25 66 64 75 75, 76 55, 55a
Low erucic type	22:1 (0–5) FAG; SG 25–25°C 0.9159; AOM value, 35–50 hours	Active oxygen method Crismer test (68–69)	66, 67 55, 55a
Nonerucic type	22:1 (0–trace) FAG; SG 26/25°C 0.9171	Crismer test (67–69)	55, 55a

Table 7.1. (Continued)

Fat or Oil	Characteristic Constituents[a]	Analytical Method	Reference
Crambe	18:3 (5.5–7), 20:1 (2.9–4.6), 22:0 (1.6–2), 22:1 (52–68), 24:1 (0.8–2.9); *Crambe* has high C_{22} and low C_{20} fatty acid distribution FAG		77
Linseed	16:0 (4.7), 18:0 (2–5), 18:1 (12–34), 18:2 (17–24), 18:3 (35–60) FAG; UNS (~1); IV 165–204; RI correlates with IV; sterols (~0.4), β-sitosterol 46% and campesterol 29%; triterpene alcohols (~0.15), five in UNS	TLC method TLC method	25 24, 25
Soybean[d]	16:0 (7–11), 18:0 (2–6), 18:1 (15–33), 18:2 (43–56), 18:3 (5–11) FAG; crude has phosphatides (1.5–2.5), lecithin (0.5–0.8); UNS (~0.7); randomized mp −7 to +5.5°C; triterpene alcohols (~0.06), no 24-methylenecycloartanol; sterols (~0.42), ~55% β-sitosterol, ~20% stigmasterol, and ~20% campesterol; tocopherols crude ~0.17, refined 0.06–0.1, α~12%, γ~58%, and δ ~30%; AOM value ~11.5 hours	TLC method TLC method GC method Various methods[c] Silica gel/TLC method Active oxygen method	80, 81 24, 25 56 82 15 40 66
Perilla	16:0 (~7), 18:0 (~2), 18:1 (~13.5), 18:2 (~14.7), 18:3 (~65.2) FAG; IV 192–208; drying time 2.45 hours vs. 4.3 hours for linseed oil	Drying time test	83
Hempseed	18:1 (7–14), 18:2 (46–69), 18:3 (16–28) FAG; IV 140–175		
Wheat germ	16:0 (11–16), 18:0 (1–6), 18:1 (8–30), 18:2 (44–63), 18:3 (4–10), FAG; UNS (~3.2); sterols high (1.3–1.7), β-sitosterol ~67%, campesterol ~22%; tocopherols (0.18–0.45), β-tocopherol	TLC method Silica gel TLC method Gas chromatography	25 40 84 85

418

	(~0.04) is unique, also α-tocopherol (~0.06); dihydro-γ-sitosterol ferulate (small)	Browne test	86
Tung	18:3 conjugated (~69 + ?) plus a rare C_{18} di-hydroxydienoic acid FAG	Worstall test	86
			86
Oiticica, Brazilian	16:0 (~7), 18:0 (~5), 18:1 (~6), licanic acid (73–83), hydroxyacids (~4) FAG; RI 25°/D 1.512–1.516; UNS (0.8); triterpene alcohols (~0.02), 11 in UNS	TLC method	24
Castor	Ricinoleic (83–89), dihydroxystearic (0.4–1) FAG; IV 81–91; UNS (0.5); SG 25/25°C 0.945–0.965; acetyl value 144–150; sterols (~0.2), β-sitosterol ~44%, and stigmasterol ~25%	Acetyl value	87
		TLC method	25
	Unbodied dehydrated castor oil[c]: saturated fatty acids (~0.5), OH acids (~5), 18:1 (~7.5), 9,12-octadecadienoic (~65), 9,11-octadecadienoic (~22) FAG; SV 188–195; IV 125–145; RI 25°C 1.4805–1.4825; gel time at 600°F 145 min; gel time "set-to-touch" 2.5 hours		87a
			87b
Rarer vegetable			
Indian ironweed	Vernolic acid (~78) FAG		88
Boleko (Isano)	Isanic acid (~50) FAG		89
Sterculia Java olive	Sterculic acid (~72) FAG		90
Cardamine impatiens	Dihydroxy acids (~25) FAG		91
Kamala species	Kamlolenic acid (~70) FAG		92
Puramnia tariri	Tariric acid (~90) FAG		93
Santalum album	Ximenynic (santalbic) acid (~95) FAG		94
Okraseed	(−)-Epoxyoleic acid (~3) FAG		95–97

Table 7.1. (Continued)

Fat or Oil	Characteristic Constituents[a]	Analytical Method	Reference
Euphorbia lagas-cae	Vernolic acid (58–62) FAG		98
Chaulmoogra	Chaulmoogric acid (~22.5) FAG		99
Gorliseed	Gorlic acid (~15) FAG		100

[a] Figures in parentheses are percentages of total oil except where otherwise noted. FAG—figures in parentheses are percentage of the fatty acids in the glycerides by an analytical technique on interchanged triglycerides using AOCS Method Ce 2-66 followed by GLC separation according to Method Ce 1-62. UNS—unsaponifiable matter by a method such as AOCS Method Ca 6a-40 (general) or Ca 6b-53 (marine oils).

[b] Methods for assay of lactones in butterfat includes Ellis and Wong's use of aluminum oxide and gas chromatography (32g), Kinsella, Patton, and Dimick's use of silicic acid column chromatography and TLC (32c), and steam distillation, column chromatography, and other separative methods (32a–f).

[c] Methods include Emmerie–Engel Method (31a), Quaife, Harris, and Biehler modification of Emmerie–Engel Method (31b), colorimetric γ-tocopherol method (31c), Emmerie–Engel method with molecular distillation (31d), the Weisler method (31e), and a specific TLC-GLC method for tocotrienols and tocopherols in palm oil (15a).

[d] Determination of tocopherols and sterols in soya sludges and residues by gas chromatography (AOCS Method Ce 3-74).

[e] ASTM D 961-75, *Annual Book of ASTM Standards*, Part 20, "Paint, Varnish, Lacquer and Related Products," 1976.

The primary source materials for animal fats are hogs, cattle, fish, whales (exclusive of the United States) and other marine animals. The operations of converting land animals into animal fat, namely, slaughtering, cutting and trimming, rendering, and occasionally refining depend on technology and methodology for the optimization of sanitation, cleanliness, and preservation of food quality of the products as well as procedures for quality control based on analytical methods.

With respect to the rendering and refining operations for both lard and tallow production, the analytical methods are largely those of the AOCS, that is, free fatty acid (Method Ca 5a-40), saponification value (Method Cd 3-25), acid value (Method Cd 3a-63), peroxide value (Method Cd 8-53), iodine-value (Method Cd 1-25), and soap in oil (Methods Cc 15-60 and Cc 17-79).

Greases are secondary source materials of animal origin with somewhat inferior quality that are used largely for industrial products. The various grades are supported by official methods such as those of the AOCS for free fatty acid (Method Ca 5a-40), titer (Method Cc 12-57), "MIU," and FAC color (Method Cc 13a-43).

Commercial Fats and Oils: General.   Most of the important ASTM and AOCS "official" and standard analytical methods for commercial fats and oils are listed in Table 7.4. (Additional "official" methods suited for the assay of objectionable components in fats and oils are given in Section 2.3.) A brief treatment of the physical and chemical background, and applicability and limitations of each method in Table 7.4 is appropriate.

SAMPLING.   Since no analytical evaluations are meaningful unless it is known that the sample is representative of the entire lot under examination, both the ASTM and AOCS give directions for the proper sampling of fats and oils in bulk and in containers. AOCS Method C 1-47 (revised in 1978) gives directions for sampling animal and vegetable fats and crude and refined vegetable and marine oils with core samplers, bomb or zone samplers, triers for solid fats, and "oil thieves." The method provides guidance for the size and the number of samples to be taken in multipackage lots.

INSOLUBLE IMPURITIES.   Appearance and the property of being "free of foreign matter" are important criteria of any fat or oil. Most of the appraisals are based on visual observations such as settled matter in clear oils and the occurrence of bones, paper, sediment, dirt, or other matter in animal fat and similar matter in vegetable oils. AOCS Method Ca 3-46

Table 7.2   AOCS "official" or standard analytical methods for vegetable-source materials of oil products

	AOCS Method No.
Cottonseed	
Sampling	Aa 1-38
Foreign matter	Aa 2-38
Moisture and volatile matter	Aa 3-38
Oil content	Aa 4-38
Nitrogen-Ammonia-Protein	Aa 5-38
Conversion table for nitrogen–ammonia-protein	Aa 5-38
Free fatty acids	Aa 6-38
Residual lint	Aa 7-38
Peanuts (groundnuts)	
Sampling	Ab 1-39
Moisture and volatile matter	Ab 2-39
Oil content[a,b]	Ab 3-39
Protein	Ab 4-39
Free fatty acids	Ab 5-39
Soybeans	
Sampling	Ac 1-45
Moisture and volatile matter	Ac 2-41
Oil content	Ac 3-44
Protein	Ac 4-41
Free fatty acids	Ac 5-41
Tung fruit	
Sampling	Ad 1-48
Moisture and volatile matter	Ad 2-52
Oil content	Ad 3-52
Physical analysis	Ad 4-52
Tung kernels, oil content	Ad 5-52
Tung fruit, hulled, oil content	Ad 6-52

affords a determination of extraneous matter gravimetrically by extraction with kerosene of the residue from the moisture and volatile matter determination of AOCS Ca 2b-38 or Ca 2d-25.

Commercial Fats and Oils: Test Methods on Physical Properties.   The test methods based on physical properties of fats and oils are important for identification and characterization and afford information about the fatlike character of the products and their response to changes in temperature.

Table 7.2.   *(Continued)*

	AOCS Method No.
Castor beans	
Moisture and volatile matter	Ae 2-52
Oil content	Ae 3-52
Free fatty acids	Ae 4-52
Flaxeeed	
Sampling	Af 1-54
Moisture and volatile matter	Af 2-54
Oil content	Af 3-54
Safflower seed	
Oil content	Ag 1-65
Sunflower seed	
Moisture and volatile matter	Ai 2-75
Oil content[b,c]	Ai 3-75
Protein	Ai 4-75

[a] Kuck and St. Angelo (29a) offer a suggestion on an improved method of grinding peanuts so that oil losses are minimized.

[b] Srinivasan (29b) found the CPMG sequence best in pulsed nuclear magnetic resonance (NMR) estimation of oil content of sunflower seed and peanuts, with moisture content showing little or no effect on results.

[c] Robertson and Morrison (29c) evaluated wide-line NMR for the analysis of oil content of sunflower seed. The evaluation of variables in sunflower seed analysis (sample size, oil composition, etc.) may expedite the adoption of this method as an official method for domestic trading in sunflower seed.

MELTING POINT.   Conventional melting point determinations of fats and oils are little used. The melting points of commercial fats and oils are never sharp and the actual melting change occurs over a range of temperatures. The existence of polymorphic forms of many fats and oils, which melt at different temperatures, further complicates the situation. Fats and oils are usually mixtures of mixed and symmetrical triglycerides and nonoil components that melt entirely uncharacteristically compared to pure organic compounds. Most of the melting point determinations that are used depend critically on the specific conditions, such as the rate of heating and other factors. From among the many determinations only two are assigned "official" status by AOCS and both of these are somewhat empirical. AOCS Method Cc 1-25 (Reapp. 73) is a capillary tube melting

Table 7.3   AOCS "official" or standard analytical methods for oilseed byproducts.

	AOCS Method No.
Cakes, meal, and meats	
Sampling	Ba 1-38
Moisture and volatile matter	Ba 2-38
Oil content	Ba 3-38
Protein	Ba 4-38
Ash	Ba 5-49
Acid-insoluble ash	Ba 5a-68
Crude fiber	Ba 6-61
Urease activity	Ba 9-58
Protein dispersibility	Ba 10-65
Nitrogen solubility index	Ba 11-65
Trypsin inhibitor activity	Ba 12-75
Linters and hull fiber	
Sampling	Bb 1-38
Oil content	Bb 2-38
Cellulose yield	Bb 3-47
Soya flour	
Sampling	Bc 1-50
Moisture and volatile matter	Bc 2-49
Oil content	Bc 3-49
Nitrogen and protein	Bc 4-49
Ash	Bc 5-49
Crude fiber	Bc 6-49
Screen test	Bc 7-51
Castor pomace	
Moisture and volatile matter	Bd 2-52
Oil content	Bd 3-52

Table 7.4   AOCS, AOAC, and ASTM "official" and standard analytical methods for commercial fats and oils for determination of "nonobjectionable" components used for quality control or "finished-goods" assay

Method Title[a]	Method Number		
	AOCS (1)	AOAC (3)	ASTM (2)
Sampling	C 1-47	—	D 1466
Insoluble impurities	Ca 3-46	—	—
Melting point			
Capillary method	Cc 1-25	28.011	—
Wiley method	Cc 2-38	28.009	—
Softening point	Cc 3-25	—	—

Table 7.4. *(Continued)*

Method Title[a]	Method Number		
	AOCS (1)	AOAC (3)	ASTM (2)
Slipping point, modified Bailey–Whitney method	Cc 4-25	—	—
Cold test	Cc 11-53	23.116	—
Titer test	Cc 12-59	28.012	D 1982
Congeal point	Ca 14-59	—	—
Flow test	Cc 5-25	—	—
Cloud test	Cc 6-25	—	—
Consistency–penetration test	Cc 16-60	—	—
Liquid and solid fatty acids	Cd 6-38	28.044	—
Solid fat	Cd 10-57	—	—
Specific gravity			
Bottle method	Cc 10-25	28.003	—
Fryer–Weston method	Cc 10b-25	—	—
Drying oils, etc.	To 1b-64	—	D 1963
Refractive index	Cc 7-25	28.006	—
Viscosity, bubble-time method	Tq 1a-64	—	D 1545
Color			
Wesson (Lovibond) method	Cc 13b-45	—	—
FAC method	Cc 13a-43	—	—
Gardner	Td 1a-64	—	D 1544
Spectrophotometric method	Cc 13c-50	—	—
Iodine value			
Wijs method	Cd 1-25	28.020	D 1959
Hanus method	—	28.018	—
Thiocyanogen value	Cd 2-38	28.024	—
Acid value	Cd 3a-63	—	—
Free fatty acid	Ca 5a-40	28.029[b]	—
Saponification value	Cd 3-25	28.025	—
Oxirane	Cd 9-57	—	—
Hydroxyl and *acetyl* values	Cd 13-60	28.016	D 1957
	Cd 4-40	—	—
Peroxide value	Cd 8-53	28.022	—
α-Monoglycerides	Cd 11-57	28.139	—
Polyunsaturated acids, *ultraviolet spectrophotometric*	Cd 7-58	28.045	—
Enzymatic method	Cd 15-78	28.071	—
Reichert-Meissl, Polenske, and Kirschner values	Cd 5-40	28.036	—
Mole percent butyric acid in fat	—	28.038	—
Diene value	Th 1a-64	—	D 1961
Conjugated dienoic acids	Ti 1a-64	—	D 1358

Table 7.4 (Continued)

Method Title[a]	Method Number		
	AOCS (1)	AOAC (3)	ASTM (2)
Methyl ester preparation, BF_3 method	Ce 2-66	28.053	D 1983[c]
H_2SO_4 method	—	28.052	—
Fatty acid homologues	Ce 1-62	28.057	D 1983
Docosenoic (erucic) acid, GLC	—	28.066	—
Crismer test, Fryer and Weston modification	Cb 4-35	—	—
Refining Loss, crude oils of peanut, coconut, corn, soybean, and cottonseed	Ca 9a-52	—	—
Soybean oil, extracted	Ca 9b-52	—	—
Refining Loss, degummed, hydrated, and extracted soybean oil	Ca 9c-52	—	—
Degummed, expressed soybean oil	Ca 9d-52	—	—
Solvent extracted cottonseed oil	Ca 9c-52	—	—
Bleach Test, refined cottonseed oil	Cc 8a-52	—	—
Refined soybean and sunflower oils	Cc 8b-52	—	—
Tallows and greases	Cc 8d-55	—	—
Modification for soybean oil	Cc 8e-63	—	—
Neutral oil and loss	Ca 9f-57	—	—
Break test	Ca 10-40	—	D 1969
Smoke, Flash and Fire Points,			
Cleveland cup	Ce 9a-48	—	D 1310
Closed cup	Cc 9b-55	—	D 1393
Fat stability, AOM	Cd 12-57	—	—
Foots in raw linseed oil, gravimetric method	—	—	D 1966

[a] Methods listed in italic print are obsolete, approximate, or known to be inaccurate to some extent and should not be used where alternate methods are available.

[b] Also the official method of National Cottonseed Products Association.

[c] Includes alternate BF_3–diethyl ether method.

point method designed to be carried out on dry and filtered fats or oils in sealed capillary tubes. The melting point is defined as the temperature at which the sample becomes perfectly clear and liquid when the agitated bath is heated at the rate of about 0.5°C/min. AOCS Method Cc 2-38 (Reapp. 73) is a modified AOAC method for Wiley melting point. In this technique prechilled and solidified cylindrical disks of fat or oil are melted in alcohol/water mixtures in a test tube by heating at a 22°C increase per 10 min with air-bubble agitation. The temperature at which the disk becomes completely spherical is considered the melting point (if the water bath temperature is not more than 1.5°C above the sample melting point). A second determination is required.

Applications of sophisticated melting point determinations in oleochemical analysis include the use of the Boemer test to detect tallow in pork fat based on the observed large difference in the melting points of the glycerides and fatty acids of unhydrogenated pure hog fat and the small differences for tallow (AOCS Method Cb 5-40), and the use of differential thermal analysis to detect the various triglycerides of lard (35).

SOFTENING POINT.   This is generally defined as the temperature at which the fat softens or becomes sufficiently fluid to slip or run. AOCS Method Cc 3-25 (Reapp. 73) is an open-tube melting point determination applicable to fats like coconut oil, stearin, hydrogenated fats, and hard tallows, less satisfactory for lard, soft tallow, and animal greases, and unsatisfactory for lard compounds and mixtures of hard and soft fats or emulsions. The dry, filtered samples are solidified in at least three (1-mm inside diameter) capillaries at 4–10°C for 16 hours, and then heated, attached to a thermometer in a water bath 8–10°C below the softening point of the samples at the rate of 1°C/min and at 0.5°C/min near the softening point. The average temperature at which fat rises in all the tubes is termed the softening point.

SLIPPING POINT.   AOCS Method Cc 4-25 is a modified Bailey–Whitney method for the determination of the slipping (softening) point of the sample, not necessarily constant for the fat type itself. It is not applicable to samples that have been melted or otherwise modified but can be used for lard, lard substitutes, butter, margarine, emulsions, and fatty substances in their natural state. Three brass cylinders packed with fat are heated in a saturated brine bath at the rate of 1°C/min and 0.5°C/min near the slipping point until fat rises in the brass cups. The average temperature is the slipping point.

COLD TEST.   AOCS Method Cc 11-53 (Reapp. 73) is designed to measure the resistance of a normal, refined, dry oil to crystallization at 0°C. This reflects the oil's "winterization." The sealed sample is cooled in an ice bath at 0°C from 25°C for 5.5 hours. Samples that pass the test remain clear, brilliant, and limpid. (AOAC Method 23.116 is identical).

TITER.   As a logical extension of one of the oldest test methods applied to fats and oils, AOCS Method Cc 12-59 (Reapp. 73) is applicable to normal animal and vegetable fats and oils and marine oils. It is carried out on the fatty acids of the fat or oil triglycerides and is defined as the temperature at which solidification occurs under the conditions of the test. The method is based on determining the temperature at which there is a change in the cooling curve as crystallization begins and the heat of solidification is evolved. The titer test critically depends on the complete saponification of the fat or oil, reacidulation of soaps to fatty acids, and care in the maintenance of test conditions. Although directions are specific in the titer test, the determination is always carried out best with an experienced analyst.

The fatty acid sample is first dried at 130°C and, without reheating again, is cooled in a special titer apparatus with agitation at a regulated rate of 100 complete up and down motions per minute for a stirrer moving a distance of 38 cm. For titers below 35°C, the cooling bath contains ethylene glycol and dry ice; for titers above 35°C ice water is the cooling medium. The bath temperature at the start should be about 10°C above the actual titer point. The cooling is continued until the falling temperature shows a slight rise or is constant for 30 secs. In practice, it is best to repeat the titer test on a second dry sample and to report the data only when agreement is reached on duplicate tests within 0.2°C.

Unless it is known that the sample is a natural unaltered fat or oil, the information obtained from the titer test can be quite misleading. Many industrial and edible oils are blended mixtures or are altered chemically (selectively, partially, or completely hydrogenated) or physically (*cis* to *trans* isomerized) with a considerable change in the titer of the component fatty acids. Accordingly, the interpretation of titer data must be made with care, and it is always best to support this with other information such as fatty acid homologue distribution obtained from GLC (AOCS Methods Ce 2-66 and Ce 1-62).

CONGEAL POINT.   The congeal point is the solidification temperature of the fat or oil under the test conditions in contrast to titer, which is that of the fatty acid components of the glycerides. AOCS Method Ca 14-59 (Reapp. 73) is applicable to fats and fat mixtures which congeal or solidify within the temperature range of 25–45°C. Obviously, the fat or oil must be dry and free of fine suspended matter. Ice or cold water is used as a cooling bath at 15 ± 0.5°C for congeal points below 35°C, and at 20 ± 0.5°C for congeal points at 35°C or above. The cooling is carried out in two stages, the first in a liquid cooling bath with agitation at 250 rpm to below the cloud point, followed by cooling in an air bath at 20°C during the congealing change. The empirical conditions for carrying out this test are such that two single determinations in one laboratory should not differ by more than 1.52°, and single determinations performed in two different laboratories should not differ by more than 2.38°. Like the titer test data, interpretation of congeal point data must be carried out with care.

FLOW TEST.   This is an empirical test relating to the lowest temperature at which the sample will flow under prescribed conditions. AOCS Method Cc 5-25 is applicable to neatsfoot and lard oils, grease oils, and similar products, but not to the harder fats and fatty mixtures. The dried sample is cooled in a special sized test jar for 2 hours with a 15% brine/chopped ice bath. The jar is then transferred to a holder in a room or air bath at 5–8°C above the expected flow point; and, as the sample warms up, the temperature at which the solidified fat flows from one end of the jar to the other when inverted is observed. The method is highly empirical and yields results that should agree within ± 0.8°C between different operators.

CLOUD TEST.   This is an empirical indication of the temperature at which a cloud is formed in the dry sample indicative of the first stage of crystallization. AOCS Method Cc 6-25 is applicable to all normal animal and vegetable fats. The sample is cooled in a bottle immersed in a water bath with stirring until it is about 10°C above the cloud point. The temperature at which the thermometer immersed in the oil is no longer visible when viewed horizontally through the sample and bottle is termed the *cloud point*.

CONSISTENCY–PENETRATION METHOD.   A number of arbitrary test methods have been devised to measure the firmness of plasticized fats by measuring the distance a dropped standard weight of defined shape will penetrate the fat. These tests range from relatively simple ones such as drop tests (a standard spatula dropped down the center of a vertical pipe) to the AOCS Method Cc 16-60 (Reapp. 73) employing a specially fabricated penetrometer and applicable to plastic fats and solid fat emulsions such as shortenings, margarine, butter, and similar products. It is readily apparent that the consistency–penetration measurement is temperature dependent.

The standard penetrometer consists of a 45-gram aluminum cone mounted on a 47.5-gram shaft with a support to grip and release the penetrating device, a platform to support the sample, spirit level and adjustments to maintain the penetrometer in a level position, and a gauge graduated to allow reading the depth of penetration in 0.1-mm units. In use, the penetrometer is mounted above the surface of the tempered fat with the cone tip just touching the clean surface. In release the tip penetrates the fat by its weight for a controlled period of 5 secs, and the depth of penetration is read from a scale graduated in tenths of millimeters and recorded on a gauge. The temperature of the fat and the penetration is recorded and several readings are taken on clean surfaces at least 1 in apart. The cone is cleaned between each measurement. The coefficient of variation within a laboratory is 3.5 and between laboratories is 4.0.

LIQUID AND SOLID FATTY ACIDS.   AOCS Method Cd 6-38 (Reapp. 73), called the modified *Twitchell method*, is designed to measure the content of liquid (oleic, linoleic, linolenic, etc.) and solid acids (stearic, palmitic, etc.) of the glycerides of high C_{18} fatty acid oils. It is based on the insolubility of lead salts of saturated acids and solubility of unsaturated acids in 95% ethanol. It is an adaptation of Twitchell's modification (101) of the classical lead soap separation technique.

The method is not applicable to fats of the lauric acid group, or the milk fat group nor the high-molecular-weight group containing rapeseed or mustard seed oils, nor to tall oil containing rosin. The method is an approximation only; no account is taken of the unsaponifiable matter, and it is recognized that the lead soaps of saturated acids are slightly soluble under these conditions.

In practice, the fat or the oil is saponified, and the potassium soaps are acidulated and treated with lead acetate in alcohol. After 2 hours at 15°C the insoluble lead soaps are removed by filtrating, dried, and weighed. Isooleic acids may be estimated by attributing the entire IV of the solid acids to be due to them. The method had utility before 1940, but is rarely used since the development of the efficient GLC fatty acid distribution analysis for specific fatty acids.

SOLID FAT INDEX.   The solid fat index (Cd 10-57 Reapp. 73) is an empirical measure of the solid fat content. It is calculated from the specific volumes at several temperatures utilizing a dilatometric technique. The measurements for shortening and margarine are commonly run at 10, 21.1, 26.7, 33.3, and 37.8°C. This method is much used to follow the course of ester–ester interchanges, particularly those of lard, cocoa butter, and edible oils in the margarine, shortening, and hard butter segments of the food industry.

Madison and Hill (102) have developed a method and instrumentation for the accurate and precise measurement of solid fat in commercial shortenings or hydrogenated oils by a transient NMR technique. The incorporation of a tempering step at 26.7°C for all samples before measurement has improved the precision of the measurements (± 0.2% solids).

SPECIFIC GRAVITY. Although specific gravity and density are general physical characteristics used in the classification and identification of fats and oils, neither are highly definitive for characterization except for a very few high-density oils like castor and hydrogenated castor oils. The great majority of the common vegetable oils have specific gravities, measured at 25/25°C, in the range of 0.914–0.924, whereas castor oil has a specific gravity of 0.945–0.965.

The AOCS lists three methods for the determination of specific gravities of fats and oils. Method Cc 10-25 (Reapp. 73) is for the determination of the specific gravity of liquid oils at 25/25°C or 25/60°C using a specific gravity bottle. Most frequently, the specific gravities of vegetable oils are reported and given at 25/25°C.

Method Cc 10b-25 (Reapp. 73) is the Fryer–Weston technique for solid fats and waxes at 25/25°C. This is the "float-sink" method in which dry, air-free, filtered, and resolidified pellets of the fat are suspended in successively denser aqueous alcohol solutions at 25°C. The solutions range from 0.940 to 1.000 (at 25°C) in increments of 0.01 units. At the point when the pellet first begins to float, and the addition of a single drop of alcohol causes the sample to sink, the specific gravity of the solution is determined by means of a Westphal balance at 25°C. At this point, its specific gravity is identical with that of the sample.

AOCS Method To 1b-64 (Reapp. 73) is a specific gravity procedure for natural and synthetic drying oils and derivatives. Directions are given for oils that have viscosities up to 40 stokes (with Leach type pycnometers), and also for more viscous oils with viscosities over 40 stokes (using wide-mouth specific gravity bottles such as the Hubbard [ASTM Specification D-70)].

REFRACTIVE INDEX. The refractive indices of fats and oils offers some identification opportunities among highly unsaturated, conjugated, hydroxy-substituted, or uniquely structured members. Measurement of refractive index is convenient for following the progress of catalytic hydrogenation of fats and oils; it offers the advantage of speed and simplicity. (Refer to Chapter 3, Volume 1, pages 217–221).

AOCS Method Cc 7-25 describes a determination of the refractive index of all normal oils and liquid fats with any standard refractometer equipped with Abbé or Butyro or any other standard scale, and the conversion to n_D values for the specified temperatures. The constant temperature bath must be controlled to $\pm 0.1°C$. The method suggests 40 and 60°C as the reference temperatures.

AOCS Method Tp 1a-64 is intended for drying oils (and fatty acids). The method is essentially similar to that used for normal fats and oils, except that an Abbé refractometer is specified, and the temperature of measurement is 25°C for oils liquid at that temperature (65°C for solid fatty acids).

VISCOSITY.   Except for ricinoleic acid oils such as castor or hydrogenated castor or blown, heat-bodied, or other polymerized oils, most common fats and oils have similar viscosities. Although they differ from oil to oil, they do not differ very widely within a general class. (Refer to Chapter 3, Volume 1, pages 178–185.) This physical property is of some value in connection with handling fats and oils in bulk, especially in the design of pumping, mixing, and transferring systems. The viscosity of oils can be measured with the usual instruments suitable for transparent liquids; ASTM D 88 is a Sayboldt viscosimeter method. AOCS Method Tq 1a-64 (Reapp. 73) and ASTM 1545 are bubble-time methods applicable to natural and synthetic drying oils (and fatty acids) or any transparent liquids covered by the range of the viscosity standards (except silicones or water dispersions of gums, etc.). The viscosity is determined in "bubble-seconds" at 25°C by timing and/or comparison with a graded series of standards. The apparatus is arranged so that a uniform sized bubble travels a distance of 73 mm up a tube held in a perfectly vertical position. The method includes a table that converts bubble-seconds to "approximate" stokes and also relates this to equivalent Gardner–Holdt letters, a scale of viscosity long used in the paint and protective coatings industries. Today, stokes are the preferred scale. A series of reference standards for comparison are arranged in three series: *light* series, 15 tubes, 0.22–8.0; *heavy* series, 14 tubes, 10–200; and *very heavy* series, 7 tubes, 250–1000. Despite the inaccuracy of this method for measurement of the viscosity of liquids with a bubble time of less than 4 secs, it has reasonably satisfactory precision. Evaluated on liquids ranging from 4.46 to 440.2 stokes, the precision from 10 laboratories was: within laboratory (68%), 2.5%; within laboratory (95%), 4.9%, between laboratory (68%), 4.5%, and between laboratory (95%), 9.0%.

Determinations of viscosity by kinematic viscosity methods are the most accurate and require relatively simple and inexpensive equipment. Suspended-level viscosimeters or modified Ostwald viscosimeters that have a range of calibrated bores for testing liquids of varying viscosities are used to measure the efflux time for a volume of sample at a specified temperature. Time required is measured in seconds and then converted into stokes. Kinematic viscosity, in stokes, is equal to viscosity, in poises, divided by density of the fluid, in grams per cubic centimeter. The centipoise and the centistoke are often used in preference to poise and stoke, for convenience of expression.

COLOR.   The colored substances that occur in natural fats and oils are principally pigments of vegetable origin, such as β-carotene, chlorophyll, lutein (xanthophyll), lycopene, gossypol, and others. Animal fats generally have less color pigments than vegetable oils, but similar pigments do occur in them as a result of the vegetable diet of the animals. A number of other fat and oil components contribute to color; for instance, γ-tocopherol on oxidation generates chroman-5,6-quinone, an intensely dark red quinone, which can be a source of color in cottonseed and soybean products. Molds and other microorganisms, as well as oxidation and conjugation changes during storage and processing, can cause color development in fats and oils.

Some crude vegetable oils have relatively intense colors. Palm oil, which contains 0.05–0.20% carotenoids, is orange. The colors of refined vegetable oils are not objectionable; the pale and attractive shades are more or less characteristic for each oil and are a distinct asset for marketing in the edible oil area. The slightly greenish coloration of olive oil is unique. The brilliant golden yellow carotene-based hues of the common vegetable oils, like cottonseed, corn, safflower, and soybean, vary somewhat and contribute much to eye appeal. The deep yellow color of butter is a distinctly natural one and has often been associated with relatively high levels of desirable carotene and vitamin A. Attempts are made to duplicate it in man-made margarines.

Many methods have been devised for the measurement of the color of animal and vegetable fats and oils, but it has been only relatively recently that there has been an application of sound scientific principles to these measurements. One of the earliest and most widely used methods for measuring color of fats and oils was the Wesson colorimeter technique using standard red and yellow Lovibond glasses in suitable combination as described in AOCS Method Cc 13b-45 (Corr. 77), and applicable to all normal fats and oils providing no turbidity is present in the sample. Originally, the Wesson-type colorimeter was constructed and employed, later the Stevenson colorimeter (102a), and more recently various Lovibond tintometers and subsequent improved models were used to make the color comparisons. The color glasses were standardized to conform to AOCS

Tintometer color scale tolerances. It was essential that suspended matter be filtered off before color matching. The method, which is still in use, gives directions for combining the yellow and red glasses for matching crude and raw coconut oil colors, for tallows, greases, dark oils, refined oils, and refined and bleached oils. Comparison matching is done with 133.35-mm (5.25-in) or 25.4-mm (1-in) tubes of oil sample. Although this method was used broadly, its use is declining.

Another early color measurement for fats and oils, presumably designed to fill the gap not covered by the Wesson method, was the so-called FAC standard colors method, outlined in AOCS Method Cc 13a-43 (Reapp. 73). This was (and still is) applicable to (a) animal fats, some of which have light odd colors with green casts, (b) predominantly yellow fats, (c) dark-colored fats with red casts, and (d) dark green cast fats or, in short, to all types that are not adequately described by Wesson standards. This is a relatively large group of fats and oils. The FAC standards are 26 permanent colors numbered from 1 to 45 in odd numbers, divided for use into five groups: *light-colored* fats, 1–9; *predominantly yellow* fats, 11–11C; *dark fats (red cast)*, 13–19; *very dark fats (green)*, 21–29; and *very dark fats (red)*, 31–45. The FAC color standards are arranged in four disks in the following order: disk 1: 1, 5, 9, 11A, 11B, 11C, 13, 17, 21; disk 2: 3, 7, 11, 11A, 11B, 11C, 15, 19, 23; disk 3: 25, 29, 33, 37, 41, 45; and disk 4: 27, 31, 35, 39, and 43.

Another widely used method for color measurement of drying oils primarily (also for fatty acids and derivatives) employs the Gardner scale of 1963 glass standards and is outlined in AOCS Method Td 1a-64 (ASTM D 1467 for drying oils, ASTM D 555 for rosin oils, and ASTM 803 for tall oil). Unlike the Lovibond system, the color comparison is not carried out additively, and the color measurement is reported as between or equal to one of two successive intensities of standard, thus 5, 5+, 6– or 6. The standards are 18 glass colors of specified chromaticity coordinates. This method has widespread use in drying oils, and also for paint and protective coating raw materials and for fatty acids and derivatives. The paint industry can be expected to continue to express the color of much of its raw materials on the Gardner scale.

A spectrophotometric index method for the measurement of the color of cottonseed, soybean and peanut oils, written as AOCS Method Cc 13c-50 (Corr. 77), is recommended only for those oils but appears to have broader eventual applicability. Measurements at 460, 550, 620 and 670 nm at 25–30°C on the filtered samples are taken with a $NiSO_4$-calibrated spectrophotometer. The photometric index $= 1.29A_{460} + 69.7A_{550} + 41.2A_{620} - 56.4A_{670}$, where A is the absorbance. This method represents an attempt to measure the color of fats and oils on a scientific basis with the elimination of human judgment.

The platinum–cobalt scale (previously called the *Hazen scale*) is another system for the measurement of color of fat and oil products, glycerine, and it is especially suited for the measurement of very light colors. It has little or no applicability to natural fats and oils (see later in this section, under commercial fatty acids "Color").

ODOR.   Those fats and oils used in cosmetics, pharmaceuticals, toiletries, and in the food industry usually have specifications that involve some restriction on the odor of the product. In the past, this was usually included in the product description with the words "bland" or "characteristic." Today, emphasis on clean and fresh smelling products has progressed to the point where the consumer expects bland and wholesome aromas even after prolonged storage. Thus the importance of odor evaluation is increasing. Producers serving these odor-sensitive product markets are alert to the need for processes and techniques for reliably producing such products, for storing them in a protective environment, for fortifying them if necessary, and for reliable methods for evaluating odors.

The odor of a fat or oil may be attributed to two entirely different causative factors, that is, a natural and inherent odor, and a "derived" odor, or one that the oil is capable of achieving if certain chemical changes are permitted to occur.

The natural or inherent odor of a fat or oil is the faint, bland, somewhat oily and characteristic type of fugitive odor typical of high quality material. The characteristic odors of well refined and preserved olive, peanut, and coconut oils are all unmistakable. Fresh animal fats such as lard and beef tallow have immediately distinguishable odors. Butter, cocoa butter and linseed oil are further examples of fats and oils with uniquely characteristic odors. If fats and oils are refined properly from quality raw materials, and if they are preserved and fortified with care, there is little reason to be concerned about any adverse effects of their natural odors. Actually, good-quality fats and oils are prized in all sorts of foods, and their attractive and appetizing odors contribute much to the enjoyment of eating.

Bacterial action, enzyme-catalyzed hydrolysis, and oxidation by the twin processes of autoxidation or photooxidation are the chemical changes that may contribute a whole series of more or less unpleasant odors ranging from slightly "off flavor" to fouly rancid. It is principally because of the marked tendency for these changes to occur, particularly autoxidation, that the measurement of odor or aroma becomes necessary with some fat and oil products.

Panel techniques, such as those applied to edible fats and toiletries, are applicable to all fats and oils where subjective reaction to odor must be evaluated. Organoleptic evaluations of odor have been, and probably will continue to be, the most effective means to resolve unsatisfactory product problems. A method for odor evaluation is presented by Konigsbacher et al. (103). When the contaminating odor in fats and oils is slight or very fugitive, storing samples in closed glass jars for prolonged periods is useful in concentrating the odor principle for evaluation purposes.

In the past two decades remarkable progress in GLC analysis has been achieved. This has been particularly true in the field of odor principle identification, including those initial odors developing in the early stages of autoxidation of fats and oils. No more positive contribution to the present art of odor evaluation can be offered than scientific methods for separating and quantitatively identifying the one or more constituents responsible for a particular odor.

Commercial Fats and Oils: Test Methods for Chemical Properties.   The test methods for chemical properties are a fundamental group dependent upon characteristic functional group reactivity. Depending on the uniqueness of the chemical reaction involved, all have a high degree of specificity.

IODINE VALUE.   Probably no analytical test method in all of oleochemistry has had the universal widespread use as has the measurement of unsaturation in fats and oils by determination of iodine value. Von Hübl, in 1884, probably first suggested that the addition of halogen be used for analytical purposes; his reagent was iodine and mercuric chloride used in about 100% excess in 95% ethanol (103a). Since 1898 a great many other innovations have been developed, continuing until recently. Mixed halogens such as iodine monobromide (in the Hanus method) and iodine monochloride (in the Wijs method) were quickly applied; their use continues today with the analytical results reported for iodine value in terms of centigrams of iodine absorbed per gram of fat or oil sample. The iodine value, thus, is a measure of the amount of unsaturated fat or oil and, when used and interpreted intelligently in conjunction with other data, such as polyunsaturated acid content, is useful in establishing the proportion of unsaturated fats, but not the kind. Used in conjunction with data obtained from GLC on methyl esters, it can be specific in the assay of certain unsaturated fat fractions.

All determinations of iodine value of fats and oils are applications of analytical iodimetry. An excess of the reagent is added to the dry, filtered sample, allowing the mixture to react with or without a catalyst for about 30 mins at about 25°C, treating the excess reagent with potassium iodide to convert it to equivalent iodine, and titrating the latter with standard sodium thiosulfate reagent, using starch as indicator.

The Wijs method employs iodine monochloride as reagent and is an official method of AOCS [Method Cd 1-25 (Reapp. 73)]. The method gives appropriate directions for the preparation of the iodine monochloride reagent, but today most laboratories find it convenient to purchase the reagent from laboratory supply houses. The reagent is reasonably stable if stored in the dark, in a dry area, and not above about 30°C. The method provides exact directions for sample size to ensure that the required 100–150% excess of reagent is available.

The Hanus method uses iodine monobromide as reagent and is the official method of the AOAC (104). The reagent is used in 150% excess of that absorbed in acetic acid solution. The scope of the Hanus method is limited "to all normal fats and oils not containing conjugated systems." As a consequence of exhaustive research in the area of halogen addition to fatty materials during the last 50 years, it is apparent that both the Wijs method and the Hanus method are subject to severe limitations and that strict adherence to method directions must be maintained. The addition of bromine to oleic acid itself at normal room temperature is accompanied by about 8–10% anomalous substitution that results in the evolution of hydrogen bromide and the formation of bromo derivatives halogenated in other than the 9- and 10-positions of the fatty acid chain (105). It is obvious that the excess of reagent must be controlled in both the Hanus and Wijs methods to eliminate the possibility of competing reactions of substitution. It is now generally appreciated that the addition of halogenating agents to unsaturated fats depends critically on (a) the halogenating agent, (b) the time, temperature, solvent, and presence or absence of a catalyst, and (c) the structure of the unsaturated fatty acid component of the glycerides, such as proximity of the carboxyl group to the double bond, degree of conjugation, and alkyl branching. Ponzio and Gastaldi (106) demonstrated that the double bond must be at least three methylene groups removed from the carboxyl group for absorption of halogen to be approximately theoretical. Fortunately, natural fats and oils are essentially free from fatty acid components with these interfering structures. Accordingly, both the Hanus and Wijs methods, when properly applied and carried out, have served for well over 80 years in the analysis of most of the fats and oils encountered. However, the use of iodine value for new and uniquely structured fats and oils must be undertaken with care.

For following the course of plant operations, such as hydrogenations, the Wijs method may be carried out in as little as 3 min, compared to the official method time of 30 min if the addition reaction is catalyzed by mercuric acetate (107, 108); however, the disposal of mercury-containing laboratory residues presents environmental problems.

A comprehensive treatment of the mechanism, the stereochemistry, and the theory of halogen addition to monoethenoid fatty acids, and the application of this reaction for analytical purposes, is available, including details regarding 15 major test methods for measuring unsaturation in fatty acids and their derivatives which involve halogen addition (109).

UNSATURATION BY HYDROGENATION.   The distinct limitations of halogen addition for the analysis of unsaturation in fats and oils has prompted attention to catalytic hydrogenation methods. Although this method has not been applied to process control operations, there are some applications in fat research that are useful. For accurate quantitative catalytic hydrogenation of olefinic double bonds on a micro and semimicro scale, a number of satisfactory procedures are available using relatively simple apparatus. The methods involve hydrogen measurement manometrically or volumetrically. Many variations in apparatus design permit contact with reactants, introduction of sample, equilibrium, and ease of operation. The so-called Brown hydroanalyzer (110), an all-glass apparatus for quantitative hydrogenation using platinum borohydride complexes as catalyst, is useful. For details on the determination of "hydrogen number," refer to the review of Polgar and Jungnickel (111).

THIOCYANOGEN NUMBER.   The addition of thiocyanogen to the double bonds of unsaturated fats and oils was carried out in the past for the purpose of determining the content of oleic, linoleic, and linolenic acids when iodine values were also determined. It was assumed that thiocyanogen added to the double bond of oleic acid, to one of the two bonds of linoleic, and to two of the three double bonds of linolenic acid. Later, it was determined that the absorption of thiocyanogen by linoleic and linolenic acids was less than was first assumed. The determination was outlined in AOCS Method Cd 2-38 (Reapp. 73) with thiocyanogen numbers defined as equivalent to iodine numbers, or in centigrams of iodine absorbed per gram of sample. The procedure is rarely used for this purpose today since GLC methods have largely displaced it.

ACID VALUE AND FREE FATTY ACID.   These two characteristics define the free fatty acid content of the fat or oil on different bases. The acid value is defined as the milligrams of potassium hydroxide required to neutralize the free carboxyl groups in 1 gram of the sample. Free fatty acid is the content of equivalent oleic acid in the sample on a percentage basis. Specifically, free fatty acid, when expressed as percent oleic acid, is about one-half (0.504) the acid value. Both characteristics indicate the quality of the oil with respect to acidity, and have been used to follow the course of fat or oil refining.

The standard test method for the determination of acid value is AOCS Cd 3a-63 (Reapp. 73). The method depends on the room temperature neutralization of the acids in the sample with aqueous standard $0.1N$ potassium hydroxide using isopropanol–toluene as the solvent and phenolphthalein as indicator. The titration is complete when the pink end color of the indicator persists for 30 secs. The method gives directions for the proper choice of sample size and also provides an alternate procedure for the determination of acid value of dark colored fats or oils with the use of a pH meter. Under these conditions fats and oils are not subject to the competing reaction of saponification.

AOCS Method Ca 5a-40 (Reapp.) describes the technique for determining the free fatty acid. This is conducted in 95% ethanol and the sample is titrated with either $0.1N$, $0.25N$, or $1.0N$ standard aqueous sodium hydroxide depending on the free fatty acid content. A sample size table is included. Phenolphthalein is the indicator and the pink end color must persist for 30 secs.

In AOAC method 28.029, also an official method of the National Cottonseed Products Association, the sample weight is chosen so that the titration volume represents the free fatty acid. For crude oils, 7.05 grams is the sample weight with $0.25N$ NaOH; for refined oils 56.4 grams of oil is used with $0.1N$ NaOH. For crude oils, the titration is a direct indication of free fatty acid as oleic acid; for refined oils, the titration multiplied by 0.05 is free fatty acid as oleic acid.

Chapman (112) recommends the addition of urea to the usual BF_3–methanol or BCl_3–methanol esterification reagents to eliminate transesterification of triglycerides to methyl esters and permits the direct GLC analysis of free fatty acids of vegetable oils.

Although the practice should be discouraged, there has been a tendency to express free fatty acid in the palm oil industry in terms of palmitic acid and free fatty acid in the coconut oil industry in terms of lauric acid. Occasionally, when the determination of total acidity of a fat and oil is determined, the use of methyl orange as an indicator is useful in determining the mineral acid acidity, whereas the phenolphthalein titration affords total acidity. The difference in the titrations represents the free fatty acid acidity.

SAPONIFICATION VALUE.   Saponification value is defined as the milligrams of potassium hydroxide required to react *completely* with all the reactive groups in 1 gram of sample. As this test is carried out in alcoholic alkali at the reflux temperature for 30–60 min, it is apparent that alkali is consumed not only for the saponification of all the triglycerides, diglycerides and monoglycerides in the sample, but also for the neutralization of all the free fatty acids in the sample and also for other reactive esterlike components, such as lactones. However, since waxes react very sluggishly under the conditions of the saponification value determination, the analysis cannot be used reliably for fats and oils containing substantial quantities of waxes. The difficulty is that there is uncertainty in determining that saponification is complete because of insolubility of the hydrolyzed long-chain alcohols in the saponification medium.

The saponification value and certain related values are measures of the average molecular weight or, more correctly, the *equivalent weight*, of fatty materials. The saponification equivalent of a fat or other ester is the number of grams of fat or ester saponified by 1 mole (56.104 grams) of potassium hydroxide. The ester value of the fat or oil is number of milligrams of potassium hydroxide required to saponify the neutral oil in 1 gram of sample, exclusive of that required to neutralize any free fatty acids and, hence, is equal to the saponification value minus the acid value. The following relation holds between the saponification value and the saponification equivalent: saponification value × saponification equivalent = 56,104. As a carryover from fatty acid usage and terminology, there is a tendency to replace the terms "saponification value," "ester value," and "acid value" with the equivalent terms "saponification number," "ester number," and "acid number."

AOCS Method Cd 3-25 describes the saponification value method applicable to all normal fats and oils. A sample of about 4–5 grams is saponified with 50 ml of alcoholic potassium hydroxide (about 40 grams per liter of low carbonate grade) for 60 min "until saponification is complete" as judged by the clarity of the reaction medium. A blank is run simultaneously. The blank and sample test runs are washed down and back titrated with standard aqueous 0.5N HCl with phenolphthalein indicator until the pink color disappears:

$$\text{Saponification value} = \frac{28.05 \text{ (titration of blank — titration of sample)}}{\text{weight of sample}}$$

AOCS Method Cd 3b-76 is a method for the determination of the saponification value of vegetable oil deodorizer distillates and sludges. It differs from the standard determination for fats and oils in that a 2-gram sample is employed.

AOCS Method Tl 1a-64 is intended for industrial drying oils (and fatty acids) and differs from the standard method for normal fats and oils only in sample size and in the time permitted for completion of saponification.

HYDROXYL AND ACETYL VALUES.   Analyses for the assay of hydroxyl value (milligrams of potassium hydroxide equivalent to hydroxyl groups in 1 gram of sample) and acetyl value (milligrams of potassium hydroxide equivalent to acetyl groups in 1 gram of sample of acetylated fat) are used mainly with certain fatty oils, diglycerides, and monoglycerides.

AOCS Method Cd 13-60 (Reapp. 73) and ASTM D 1957 are applicable to the determination of the hydroxyl value of fatty oils such as castor and hydrogenated castor oils and to mono- and diglycerides and hydroxy acids. The method depends on the back titration of the acetic acid derived from both the excess acetic anhydride by hydrolysis and from acetic acid as by product in the acetylation of the hydroxyl groups in the sample with pyridine–acetic anhydride reagent. In the reaction it is necessary to correct for free fatty acid acidity if present, and this is accomplished by running a simultaneous acid value determination.

The published method includes a table specifying the sample weights required for the range of hydroxyl values of the sample. Although the calculations required are not difficult, hydroxyl values are not easy to run. Precision for castor and dehydrated castor oils samples are: two single determinations in one laboratory should not differ by more than 2.4; and single determinations performed in two different laboratories should not differ by more than 3.1.

AOCS Method Cd 4-40 (Reapp. 73) is for the determination of hydroxyl and acetyl values of acetylated fats. This method requires the simultaneous determination of saponification value. The hydroxyl value is defined as the number of milligrams of potassium hydroxide equivalent to the hydroxyl content of the sample based on the weight of unacetylated fat.

Several other methods that involve acylation reactions that are applied for hydroxyl group assay include the use of 150–200% excess acetyl chloride in toluene–pyridine solvent (113), stearic anhydride as a stearoylating agent (114), 3,5-dinitrobenzoyl chloride in pyridine as a benzoylating agent (115), and acetic anhydride in ethyl acetate or pyridine with perchloric acid as a catalyst (116). All have advantages and disadvantages for certain applications.

OXIRANE OXYGEN.   The official method is based on the ring-opening reaction of epoxy (oxirane) groups with hydrogen bromide in acetic acid using crystal violet as indicator. AOCS Method Cd 9-57 is applicable to epoxidized soybean oils and other epoxidized oils. Interferences are cyclopropenoids such as those found occasionally in cottonseed oil (50), oxidized fats, which contain peroxides or hydroperoxides, α,β-unsaturated ketones, and soaps. The test method is subject to possible loss of HBr during the titration and the reagent is unstable on standing. The probability limits (95%) are: difference between duplicate determinations made within a laboratory should not exceed 0.08; and difference between average or duplicate determinations made in different laboratories should not exceed 0.19.

PEROXIDE VALUE.   A number of methods for the detection of peroxides, the initial evidence of the development of rancidity in unsaturated fats and oils, have been proposed and used. The peroxide value is defined as the content of reactive oxygen in terms of millimoles of peroxide or milliequivalents of oxygen per 1000 grams of fat (1 mmol = 2 meq). In the United States the AOCS Method Cd 8-53 (Reapp. 73) is used to determine peroxides in fats through its ability to liberate iodine from potassium iodide in glacial acetic acid solution. The method is sensitive and highly empirical. A blank determination on the reagents should be conducted.

In Europe other methods used for the determination of peroxides are those of Lea (117), or of Taffel and Revis (118). Less widely used methods for the determination of peroxides (119, 120) are the ferric thiocyanate method, the dichlorophenol–indophenol method, and the stannous chloride method.

α-MONOGLYCERIDES.   AOCS Method Cd 11-57 (Corr. 76) consists of two sections: Section I is for the determination of α-monoglycerides under 15%, and Section II is for the determination of α-monoglycerides over 15%. The analysis is carried out on a series of aliquots. Samples to be analyzed are dissolved in chloroform containing 5% dimethylformamide (to assure solution of any glycerine), and aliquots are oxidized with excess periodic acid. A solution of periodic acid in water extracts and oxidizes free glycerine and a solution of periodic acid in methanol oxidizes α-monoglycerides plus glycerine. The unused periodic acid is determined iodometrically by reaction with potassium iodide and titration with standard arsenite solution. The oxidation of periodic acid is specific to α-monoglycerides, which have the 1,2-vicinal glycol structure; β-monoglycerides are not oxidized.

Calculation of the α-monoglyceride content is based on the amount of periodic acid consumed in the oxidation of the α-monoglyceride plus glycerine corrected for the amount consumed in the oxidation of the free glycerine alone. In practice, the analysis is sensitive and should be carried out by an experienced chemist. There are reports that certain monoglyceride–emulsifier mixtures do not deemulsify to permit the separation of the aqueous and chloroform layers in this analysis.

Total monoglycerides, as well as di- and triglycerides, can also be determined by silica gel column chromatography using AOAC (1980) Method 28.133–28.138 for monoglyceride concentrates and 28.148–28.153 for shortening.

Wood, et al. (120a) efficiently separated 1-, and 2-monoglycerides as trimethylsilyl ether derivatives on packed and large bore capillary columns of diethyleneglycol succinate polyester (DEGS) and Apiezon *L*.

Test Methods for Minor Fat and Oil Constituents.   Test methods for fat and oil minor constituents assay those unavoidable components that, although not entirely objectionable, are generally undesirable, especially if they are present in greater than ordinary amounts. Test methods for these minor constituents include those for moisture, unsaponifiable matter, metal content, ash, rosin acids in oils and fatty acids, free glycerol, *trans* isomers, "MIU," moisture and volatile matter, soluble mineral matter as soap, and oxidized fatty acids. A discussion of the test methods for these materials is included in Section 2.3.

Test Methods for Various Fat and Oil Components.   Test methods for the various fat and oil components are important because they afford an indication of the composition of a substantial part of the fat or oil, usually as a consequence of the fatty acid composition of the triglycerides. Frequently, as a result of the information obtained from these test methods some indication of the performance of the products is obtained. For example, an idea of the nutritive value of the sample is obtained from the essential fatty acid content as determined by the *cis,cis*-polyunsaturated acids, and from determination of conjugated diene an indication of the film-forming or polymerization properties may be estimated.

REICHERT–MEISSL, POLENSKE, AND KIRSCHNER VALUES.   These test methods relate to the volatility and water solubility of the short-chain (C_4–C_{10}) fatty acid components of low molecular weight fats such as milk fats and those of the lauric acid group (coconut, palm kernel, babassu, etc.). The Reichert–Meissl value is a measure of the soluble volatile fatty acids (chiefly butyric and caproic), the Polenske value is an index of the insoluble volatile fatty acids (mostly caprylic, capric and lauric), and the Kirschner value is a measure of butyric acid. All are expressed in terms of milliliters of $0.1 N$ NaOH required to neutralize the fatty acids obtained from a 5-gram sample under the specific conditions of the method. All three test methods are described in AOCS Method Cd 5-40 (Reapp. 73). In determining these values it is first necessary to saponify the fat, acidulate the soaps, and to treat the aqueous suspension–solution empirically by distillation. In determining the Reichert–Meissl value, 5 grams of sample in a 300-ml distillation flask is saponified with 20 ml of a standard glycerol–soda solution and heated until completely saponified. After the addition of 135 ml of recently boiled distilled water and 6 drops of a dilute sulfuric acid sufficient to completely acidulate the soaps, exactly 110 ml of distillate is collected during the course of 30 min while preserving the distillate exit temperature below 20°C. After filtration through filter paper, exactly 100 ml of the filtrate is titrated with $0.1N$ NaOH standard solution and, after conducting a blank, the Reichert–Meissl value is calculated:

Reichert–Meissl value = 1.1 (titration of sample – titration of blank)

It can be assumed that this figure represents the equivalent butyric and caproic acids in the fat triglycerides.

The Polenske value is obtained by washing the insoluble acids on the filter paper with water and discarding the washings, followed by dissolving them with three 15-ml portions of 95% ethanol neutralized with phenolphthalein. The entire ethanol washings are then titrated with $0.1N$ NaOH and represent the approximate caprylic, capric, and lauric acid content of the triglycerides:

Polenske value = milliliters of $0.1N$ NaOH required

To determine the Kirschner value, another 100 ml of Reichert–Meissl distillate is neutralized with $0.1N$ Ba(OH)$_2$ solution just to a faint pink phenolphthalein endpoint in a closed flask to eliminate CO_2 absorption. Then, 0.3 gram of Ag$_2$(SO$_4$) is added as a fine powder and the flask and contents are shaken and permitted to stand for 1 hour with intermittent shaking. The mixture is filtered, and 100 ml of filtrate is acidulated with 10 ml of dilute sulfuric acid and diluted with 35 ml of distilled water. Distillation is carried out at the rate of 110 ml in approximately 20 min. After 110 ml has been collected, it is filtered again, and 100 ml is titrated with $0.1N$ NaOH to a pink phenolphthalein endpoint. A blank determination is run similar in all respects:

$$\text{Kirschner value} = \frac{A \times 121 \times (100 + B)}{10,000}$$

where A is the titration of the sample minus titration of the blank and B is milliliters of $0.1N$ Ba(OH)$_2$ required to neutralize the original 100 ml of Reichert–Meissl distillate. It is assumed that the barium salts of all the fatty acid components are water soluble, that the silver salts of acids higher than butyric are water insoluble, and that the Kirschner value represents butyric acid content that is readily and completely steam distilled under the conditions of the test. Table 7.5 includes Reichert–Meissl, Polenske, and Kirschner values for a series of common low-molecular-weight animal and vegetable fats and oils.

POLYUNSATURATED ACIDS.   AOCS Method Cd 7-58 (Reapp. 73) is an ultraviolet spectrophotometric method. The method depends on the measurement of conjugated components on a first sample in isooctane at the following wavelengths (nm): dienoic, 233; trienoic, 262, 268, 274; tetraenoic, 308, 315, 322; and pentaenoic, 346. A second sample is isomerized (conjugated) by one of two methods depending on the nature of the polyunsaturated acids present. Potassium hydroxide in ethylene glycol (6.6%) is used when the sample contains only linoleic and linolenic acids; the saponification–conjugation is carried out at $180 \pm 0.5°C$ for 25 min. When the sample contains linoleic, linolenic, arachidonic, or pentaenoic acids, the use of 21% KOH in ethylene glycol for 15 min at $180 \pm 0.5°C$ is preferred. Both saponification and conjugation operations are carried out carefully under a nitrogen blanket and blanks are run on the KOH–ethylene glycol solutions. The reaction solution is diluted with reagent methanol and measured at the wavelengths previously indicated. The calculations are somewhat complicated for a four component polyunsaturated acid system but are simplified if it is known that certain components are absent in the sample. If the iodine value of the sample is measured, an estimate of the oleic acid content may be made, and saturated acids may also be calculated by difference.

This method is applicable to determination of polyunsaturated acids from dienoic through pentaenoic in animal and vegetable fats containing only the natural *cis* isomers of the nonconjugated polyunsaturated acids, only small amounts of preformed conjugated material, and only small amounts of pigments whose absorption may undergo considerable change with alkali isomerization (refined vegetable oils appear to be preferred

Table 7.5   Reichert–Meissl, Polenske, and Kirschner values for common animal and vegetable fats and oils

	Butterfat	Coconut Oil	Palm Kernel Oil	Ordinary Fats and Oils
Reichert–Meissl value	22–34	6–8	5–7	<1
Polenske value	2–4	14–18	10–12	<1
Kirschner value	20–26	1–2	0.5–1	<0.5

candidates for analysis). The method is not applicable, or is applicable only with specific precautions, to modified drying oils, hydrogenated oils, or other fats containing *trans* isomers of the unsaturated fatty acids ("elaidinized fats"), to fish oils or similar fats containing acids more highly unsaturated than pentaenoic, to crude oils or unusual samples containing pigments whose absorption undergo changes during the alkali procedure causing interferences which are unknown and cannot be determined, or to fats and oils containing large quantities of preformed conjugated fatty acids.

A comparative analysis of 103 samples of corn oil grown in the United States "Corn Belt" during a period of 2.5 years in the early 1960s showed that the linoleic acid content was about 2.5 units lower than the data obtained from GLC fatty acid distribution analysis (121).

AOCS Method Cd 15-78 is an enzymatic method for the estimation of *cis,cis*-polyunsaturated fatty acids. It determines *cis*-,*cis*-1,4-methylene interrupted diene structures, —CH=CH—CH$_2$—CH=CH—, in the form of esters or triglycerides by means of the enzyme lipoxidase. The enzyme is unreactive with conjugated systems, *trans* double bonds or monoenoic esters. In use, the sample is saponified and treated with the enzyme in a buffered system, then transferred to the spectrophotometer and measured at 234 nm. For comparison, pure standard trilinolein is treated similarly, both with boiled enzyme solution (inactive) and with enzyme, and a standard curve is established relating absorbance to μg/ml. Two determinations made by the procedure in one laboratory should not differ by more than 1.6% and between two laboratories by more than 2.2%.

Other methods for the determination of essential fatty acids involving the use of lipoxidase oxidation (122, 123) are available that correlate well with the rat bioassay method (124).

DIENE VALUE.   This method attempts to measure the diene value of conjugated oils or of oils with unsaturated systems easily capable of generating them. The diene value is expressed, like the iodine value and the thiocyanogen value, in terms of equivalents of iodine. Maleic anhydride functions as a dienophile in the Diels–Alder reaction and adds to the conjugated systems of oils to form bicyclic anhydride compounds. In AOCS Method Th 1a-64 (Reapp. 73) and ASTM D 1961, the sample is refluxed with excess maleic anhydride–toluene reagent, the excess reagent is decomposed by boiling with water to hydrolyze the excess maleic anhydride, and then, after taking up in diethyl ether and extracting with water, titrating the combined water extracts with standard alkali. A blank is run on the maleic anhydride–toluene reagent. These methods are known to give low results compared to the spectrophotometric method because of incomplete Diels–Alder adduct formation, but they are still useful in giving an approximate idea of the latent and potential diene conjugation present in the sample. The reaction of oleic acid with maleic anhydride, as shown in the structural formula, in a nonDiels–Alder substitution reaction with double bond migration (125) cannot be a competitive reaction; if this were to occur to any degree, the result for conjugation would be higher.

$$RCH{=}CHCH_2CH{=}CHR'$$

$$\downarrow$$

$$RCH_2CH{=}CHCH{=}CHR' \; + \quad \text{(maleic anhydride)} \longrightarrow$$

$$CH_3(CH_2)_6CH{=}CH{-}CH(CH_2)_7COOCH_3$$

CONJUGATED DIENOIC ACID. AOCS Method Ti 1a-64 is a spectrophotometric method applicable to dehydrated castor oil (and the methyl and ethyl esters or free fatty acids thereof). The measurement is at 233 nm in isooctane solvent, and the results are reported as percent conjugated dienoic acid.

FATTY ACID HOMOLOGUE DISTRIBUTION. Fats and oils are not traded on specifications based on fatty acid distribution, but are increasingly being classified into groups on the basis of fatty acid composition (125a), and we can expect to see increasing reliance on compositional limitations in the future. No advance in oleochemical analytical chemistry has been more important than the development, within the last 25 years, of the rapid and convenient GLC method for fatty acid composition. In the short period of its use, this reliable and accurate method has already eliminated the time-consuming previous procedure involving fractional distillation of methyl esters and the tedious analysis of separated fractions by iodine value and other tests. Furthermore, it has largely eliminated the need for a number of somewhat unreliable analytical methods, including the thiocyanogen value, liquid and solid fatty acids, the diene value, and, to a considerable extent, the spectrophotometric method for polyunsaturated fatty acids. The convenience and the degree of separation for many fatty acid separations are constantly being improved, and there are literally hundreds of research papers on aspects of this analytical technique for numerous applications. It is virtually impossible to design a thoroughly comprehensive single analytical method that would be applicable to all the classes of fats and oils that occur in nature, nor would it even be desirable to do so. However, the methodology and the instrumentation that is currently available is remarkably versatile, and by varying the use of columns and techniques somewhat, a great deal of analytical latitude is already possible.

Although it is possible to use GLC with fatty acids (126–130), the present official methods are based on the use of methyl esters, and for the purpose of fat and oil analysis, this requires methods for quantitative conversion of fats and oils into methyl esters with no change in structure or composition. Fortunately, there are a large number of methods for preparing methyl esters. The AOCS recommends a boron trifluoride–methanol technique for "long-chain fatty acids," applicable to common fats and oils (and fatty acids) in AOCS Method Ce 2-66 (Reapp. 73). It is recognized that unsaponifiables are not removed and that, if present in large amounts, they may interfere with subsequent analyses. The procedure is known to be inapplicable to fats and oils that contain epoxy, hydroperoxy, cyclopropenyl, cyclopropyl, and possibly hydroxyl groups, and it is not intended for fats or oils containing these groups. Furthermore, when applied to the analysis of the low-molecular-weight acid groups, such as those found in milk fats, the method is not reliable because C_4–C_6 methyl esters are lost by evaporation. A large number of methods for methyl ester preparation are available; included are those employing boron trifluoride–methanol (131, 132), boron trichloride–methanol, dimethoxyacetone (133), and others (134). MacGhee and Allen (136) suggest the use of wet halogenated hydrocarbons in the esterification procedure to eliminate the loss by volatilization of acids lower than caproic (C_6). This procedure is applicable to the methyl esters of the short and medium chain fatty acids, such as those found in butterfat, and the lauric acid group.

AOCS Method Ce 1-62 (Reapp. 75), ASTM D 1983, and AOAC-IUPAC 28.057 are the general methods specified for polyester liquid phase or acid-washed Chromasorb W columns or the equivalent. The AOCS separation is effective with fatty acids having C_8–C_{24} carbon atoms but is not applicable to epoxy or oxidized fatty acids, to polymerized acids, and is somewhat unreliable for castor or hydrogenated castor fatty acids. Used strictly according to directions, two single determinations of major components ($>5\%$) performed in one laboratory should not differ by more than 1.0 percentage unit; and two single determinations performed in different laboratories should not differ by more than 3.0 percentage units.

Slover and Lanza (135) have developed the technique of glass capillary gas chromatography (GCGC) for the analysis of fatty acid distribution in fats and oils. Using a wall-coated SP 2340 capillary glass column in a semiautomated system, an improved direct separation of $18:2\omega6$-*cis,cis*, *cis,trans*, *trans,cis*, and *trans,trans* isomers was possible, but $18:1$ geometrical and positional isomers could be only partially separated. More than 1900 samples were analyzed with one column at temperatures up to 200°C with no appreciable column deterioration. We can look forward to enhanced utility for such methods and equipment.

Performance Test Methods for Fats and Oils

CRISMER TEST.   AOCS Method Cb 4-35 is the Fryer–Weston modification of the Crismer test, which is a method for determination of miscibility of the oil in a standard solvent of equal volumes of ethanol/amyl alcohol adjusted with water to incipient turbidity at 70°C with sweet almond oil. The results are reported as the temperature at which equal volumes of the dried oil and the standard solvent show a cloudiness on cooling in a stirred test tube. The method includes a correction for the free fatty acid of the oil determined as oleic acid in the free fatty acid method. The corrected Crismer values are more or less characteristic within a narrow range for a given kind of pure oil. Sahasrabudhe (55a) correlated the Crismer values of a whole series of Canadian rapeseed oils with the erucic acid content and found reasonable agreement in a linear relationship. Table 7.6 shows Crismer values for representative vegetable oils.

REFINING LOSS.   The AOCS lists five separate methods for the evaluation of refining losses of different crude vegetable oils. AOCS Method Ca 9a-52 (Rev. 79) is applicable to crude oils of peanut, coconut, corn, soybean (expeller and hydraulic), and cottonseed (expeller and hydraulic) oils. AOCS Method Ca 9b-52 (Reapp. 73) is applicable only to extracted soybean oil. AOCS Method Ca 9c-52 is applicable only to degummed hy-

Table 7.6   Crismer test data for a series of vegetable oils

Oil	Crismer Test	Reference
Olive	68.5–71.6	55
Zero erucic rapeseed[a]	68.2	55a
Zero erucic rapeseed[b]	68.6	55a
Low erucic rapeseed[c]	69.2	55a
High erucic rapeseed[d]	76.7	55a
High erucic rapeseed[e]	77.5	55a
High erucic rapeseed[f]	81.8	55a

[a] From *Brassica napus*, "Zephyr" variety, 0.5% erucic acid.

[b] From *Brassica campestris*, "Span" variety, 0.3% erucic acid.

[c] Commercial sample, "canbra" oil, 3.2% erucic acid.

[d] From *Brassica napus*, "Bronowski" variety, 20.1% erucic acid.

[e] Commercial sample, 26.2% erucic acid.

[f] From *Brassica campestris*, "Yellow sarson" type, 54.2% erucic acid.

draulic and extracted types of soybean oil. AOCS Method Ca 9d-52 (Reapp. 73) is applicable only to degummed, expeller type of soybean oil. AOCS Method Ca 9e-62 (Reapp. 73) is applicable to solvent extracted crude cottonseed oil and to the reconstituted whole mixture of crude cottonseed oil from prepressing and subsequent solvent extraction of cottonseed. It is not applicable to crude cottonseed oil extracted from prepressed cake by solvent. All these methods are specific detailed laboratory runs of actual refining operations designed to measure the yield (and losses). Each method gives the conditions of the refinings for each type of oil and specifies the caustic required, and the number of cups to be run, based upon the acid value for each sample. The oil refining apparatus holds 6, 12, 18 or 24 cups, as required. The refined oils are generally held for the determination of bleaching test, color, etc.

BLEACHING TEST.   The AOCS specifies four tests for bleaching and color evaluation on a laboratory scale. AOCS Method Cc 8a-52 (Reapp. 73) is applicable to refined cottonseed oil, Cc 8b-52 (Rev. 79) to refined soybean and sunflower oils, Cc 8d-55 (Reapp. 73) to tallows and greases for soap production, and AOCS Cc 8e-63 (Corr. 79) is a modified test for soybean oil. The three methods applicable to the vegetable oils specify the standard bleaching earth and the conditions and duration of bleaching, and those for soybean oil give directions for bleaching of normal colored oils and also for high chlorophyll oils. In the case of the animal fats, directions are given for both refining and bleaching operations. These tests can be conducted in the same apparatus used for the refining loss test.

NEUTRAL OIL AND LOSS.   AOCS Method Ca 9f-57 is an alumina chromatographic method for the determination of total neutral oil (triglyceride plus nonsaponifiables) of cottonseed, soybean, peanut, linseed and coconut oils. The method states that although it has been investigated with other oils, it is probably applicable to practically all natural animal and vegetable fats and oils. Ether–methanol is the solvent. For reproducibility, two single determinations should not differ by more than 0.15, and single determinations performed in two different laboratories should not differ by more than 0.30.

BREAK TEST.   AOCS Method Ca 10-40 (Rev. 79) is the modified Gardner method for the determination of "break" material in crude soybean oil. In practice, the test measures the material that separates under the test conditions and is rendered insoluble in carbon tetrachloride.

SMOKE, FLASH, AND FIRE POINTS.   Although fats and oils are organic materials and will burn, they are not considered especially hazardous materials. Flash and fire points are important because they permit fats and oils to be classified into groups for proper labeling during shipment (or storage). Except for the case where fats are contaminated with organic solvents, their flash points are all above 550°F, and they are relatively nonflammable. The determination of smoke point is of interest since certain oils function as lubricants, for example, and as pan oils in the baking industry; thus their smoking behavior is important.

AOCS Method Cc 9a-48 (Corr. 77) is the general method for smoke, flash, and fire points as carried out with the standard Cleveland open flash cup. The smoke point is the temperature at which the sample gives off a thin, continuous stream of bluish smoke when heated under the specified conditions of the test. The flash point is the temperature at which a flash appears at the point on the surface of the sample under the conditions of ignition and heating specified in the test. The fire point is that temperature at which application of the test flame causes burning for at least 5 secs.

The presence of free fatty acids affects the smoke, flash and fire points of fats and oils; the smoke points of common vegetable oils range from about 450°F (0.01% free fatty acid) to 200°F (100% free fatty acid), flash points from about 625°F (0.01% free fatty acid) to 380°F (100% free fatty acid), and fire points from 685°F (0.01% free fatty acid) to 430°F (100% free fatty acid). (See Chapter 3, Volume 1, p. 212.)

AOCS Method Cc 9b-55 (Corr. 77) is a flash point method suited for use with the Pensky–Martens closed-cup flash tester. It is applicable to animal, vegetable, and marine fats and oils that flash at temperatures below 300°F. These fats may or may not contain amounts of volatile flammable solvents.

FAT STABILITY (ACTIVE OXYGEN METHOD).   For routine testing of the oxidation stability of fats, the method first suggested by Wheeler (137), then modified and standardized by King et al. (138), and now officially AOCS Method Cd 12-57 (Reapp. 73) has wide utility. It is known as the *active oxygen method* (AOM) and is applicable to all normal fats and oils intended for human consumption. The AOM method was also known as the *Swift stability test.*

The sample is continuously aerated at 280°F (97.8°C), and, by periodic titration, the time required for a specific peroxide number to develop is determined. The test method suggests no criteria for specific fat stability, although it was originally designed for use with lard and it was suggested that a peroxide number of 20 (milliequivalents per kilogram) be taken as the onset of rancidity. There are no universally accepted values; standards vary from location to location. For comparative purposes, lard, hydrogenated lard shortenings, oleo oil and hydrogenated vegetable oil shortenings (from cottonseed and soybean oils) have values of 20, 40, 60, and 80, respectively, coordinating with the beginning of definite rancidity. Unhydrogenated olive and peanut oils have peroxide values of 50–60, whereas cottonseed, corn, sunflower, and soybean oils become rancid at peroxide values between 125 and 150. In a number of investigations the AOM test was continued for long periods and effluent air was evaluated organoleptically without regard to the peroxide value of the sample.

A simple test for oxidation stability is the *oven* or *Schaal test* (139, 140). Held in a loosely sealed container for sufficient time to develop peroxide content and odor or taste effects at 140–145°F, a sample can be easily evaluated. Like the AOM test, no standards of performance were assigned.

Oxygen uptake test methods are numerous, and in many cases they are quite effective. Method improvement over the years has made it possible for a "50-hour shortening," as termed by the AOM 97.8°C method, to be analyzed in 20 hours by a 110°C method (141), in 2 hours and 45 min by an oxygen bomb method (OBM) (142), and in 1 hour and 45 min by a modified bomb test (141). In a recent modification (141) a 10-gram sample of shortening is held at 135°C at an oxygen pressure of 110 psig until the oxygen absorption corresponds to a drop of 2 psig. This test is claimed to be more precise than the original AOM method.

Two relatively new methods are of interest because they evaluate oxidation stability of fats and oils and also provide an efficient means to evaluate new antioxidants. The Marco method (143) and simplifications of it (144) appear to be capable of satisfactory use. Methyl linoleate emulsified with β-carotene is followed colorimetrically at 50°C; the destruction of the color pigment requires only 1–3 hours. The hematin-catalyzed oxygen uptake method (145), which is carried out at 45°C with a Beckmann oxygen analyzer (146), also offers promise. Two other tests are available and have been used for the determination of fat stability to autoxidation:

1. The *Kreis test*, is highly empirical and nonspecific. It allegedly depends on the presence of epihydrinaldehyde, but this is unlikely. Although it has some qualitative utility, it cannot be recommended. A modification of the test, however, making it suited for colorimetry (147), has substantially improved it.

2. The *thiobarbituric acid* (TBA) *test* (148), related to the Kreis test, is much more sensitive and responsive at earlier stages of autoxidation. Products of oxidation of unsaturated fatty acids like linoleic acid are apparently responsible for the color reaction with TBA (149, 150). For example, malonic dialdehyde and methyl oleate hydroperoxide both give the color reaction. A comprehensive evaluation of the method by Marcuse and Johansson (151) using standard alkanals, alkenals, and 2,4-alkadienals indicate that all the aldehydes form a yellow pigment (absorption at 450 nm) with TBA, whereas only the 2,4-dienals, and to a lesser extent the alkenals, produce a red pigment (absorption at 530 nm). It appears desirable to include a measurement of both absorption maxima for the total aldehydic products of unsaturated fat autoxidation.

Other analytical techniques for assaying the course of autoxidation in unsaturated fats include change in refractive index (152); measurement of soybean oil rancidity by gas chromatography and calculation of the oxidation value (OV), as correlated with PV (153); color development when autoxidized fats are treated with concentrated alkali (through α-dicarbonyl condensations); and measurement of ultraviolet absorption as conjugation develops.

In the food industry the evaluation of the oxidation stability of a food fat is intimately related to the evaluation of flavor and odor. Evaluations of flavor are quite important and are almost always carried out on an organoleptic basis. Pohle et al. (154) evaluated the PV and TBA test on stored fats and shortenings and found that either test could be used to measure flavor development, but that the flavor score could not be so estimated. Significantly, within any one product, development of off-flavors is related to PV and TBA intensities, but the relative levels vary from product to product.

Commercial Fatty Acids: General.   Although many of the test methods for commercial fatty acids are similar to those used with fats and oils, there are also methods that are quite different. Furthermore, there are five "official" test methods that are carried out on the corresponding methyl esters, and in four of these the test method is not applicable to the fatty acid *as is*. Table 7.7 includes most of the important "official" AOCS and ASTM analytical methods for commercial fatty acids. (Section 2.3 includes additional "official" methods suitable for the assay of objectionable components occasionally found in commercial fatty acids.)

SAMPLING.   AOCS Method Ta 1a-64 (Reapp. 73) points out that copper, brass, bronze or other copper alloys should not come into contact with fatty acids as containers or as equipment. Otherwise, the directions applicable to sampling in AOCS Ta 1-64 (industrial oils and their derivatives) prevail for fatty acis.

Commercial Fatty Acids: Tests for Physical Properties

MELTING POINT. .   Although commercial fatty acids do not melt with the sharpness that pure organic compounds do, there are fractionated fatty acids of a high degree of purity, and various melting point methods are always required, especially in fatty chemical research. Neither AOCS or ASTM specify melting point methods for fatty acids, but the conventional methods are all satisfactory; in fact, even AOCS Method Cc 1-25, a fat and oil sealed capillary method, finds utility.

SOFTENING POINT.   A ring–ball method for the determination of softening point (often referred to as *melting point*), as in ASTM D 36, is intended for butuminous materials, but is also suited for fatty acid pitch.

TITER.   Titer, defined as a solidification point under somewhat empirical test conditions, is probably the oldest and most generally used characteristic for both fatty acids and fats and oils, although it is now apparent that its use is declining somewhat, at least for fatty acids. The general

Table 7.7   AOCS and ASTM "official" or standard analytical methods for commercial fatty acids

	Method No.	
Method Title[a]	AOCS (1)	ASTM (2)
Recommended practices for testing	S 1-64	D 1467
Sampling	Ta 1a-64	D 1466
Titer	Tr 1a-64	D 1982
Specific gravity	To 1a-64	D 1963
Bottle method	Cc 10b-25	—
Fryer–Weston method	Tp 1a-64	—
Refractive index	—	—
Viscosity, bubble-time method	Tq 1a-64	D 1545
Iodine value, Wijs method	Tg 1a-64	D 1959
Color		
Gardner	Td 1a-64	D 1544
Pt-Co scale	Td 1b-64	D 1209
Photometric index	Td 2a-64	D 1469
Thiocyanogen value	Cd 2-38	—
Acid number (value)	Te 1a-64	D 1980
Saponification number (value)	Tl 1a-64	D 1962
Hydroxyl number (value)	Cd 13-60	D 1957
Polyunsaturated acids		
Ultraviolet spectrophotometric method	Tj 1a-64	—
Enzymatic method	Cd 15-78	—
Diene value	Th 1a-64	D 1961
Conjugated dienoic acids	Ti 1a-64	D 1358
Fatty acid homologues (by GLC)	Ce 2-66	D 1983
Flash and fire points	Tn 1a-64	D 92
Heat stability (color after heating)	Td 3a-64	D 1981

[a] Methods indicated in italics are obsolete, approximate, or known to be inaccurate to some extent and should not be used where other methods are available.

test methods are AOCS Tr 1a-64 (Rev. 78) and ASTM D 1982. As applied to fatty acids, it is not necessary to saponify, acidulate, and dry the fatty acids, as is the case with titer determination of fats and oils. However, if it is suspected that the sample contains fat or monoglyceride, the saponification step must be included. Rarely, however, is this the case, and it is usually necessary only to filter and dry the sample before measurement, with care taken not to reheat more than once. Otherwise, the same restrictions and limitations apply as already indicated for fat and oil titers (see earlier in this section, under "Titer"). The precision is better for tallow fatty acids than for common vegetable fatty acids:

	Fatty Acids		
	Soybean, Linseed	Tallow	Coconut
At a confidence level of 95%, in the range of values of	17–26	39–44	21–24
Single determinations performed in one laboratory shall not differ from mean value by more than	2.9%	1.0%	1.1%
Single determinations performed in different laboratories shall not differ from the mean value by more than	7.0%	1.0%	2.4%

SPECIFIC GRAVITY.   Pure fatty acids present a number of trends in density and specific gravity that relate to chemical structure. Generally, specific gravity or density increases with unsaturation, although the differences are not well suited for purposes of identification. Furthermore, *cis* unsaturation appears to increase density more than does *trans* unsaturation, the density of liquid fatty acids appears to decrease as the double bond is farther removed from the carboxyl group, and an increase in the chain length of saturated fatty acids decreases the liquid density or specific gravity. For detailed discussion of the relation of specific gravity and density to structure and other factors, refer to Chapter 3, Volume 1, pages 186–192 or to the compilations of Singleton (155) and Pryde (156).

The method for the determination of the specific gravity of liquid fatty acids is usually AOCS Method To 1a-64 (Reapp. 73) or ASTM D 1963. This is a specific gravity bottle method that determines the ratio of the weight of a unit volume of the sample at 25 or 70°C to the weight of a unit volume of water at 25°C. It is applicable to all liquid fatty acids. For solid fatty acids the specific gravity may be determined by use of a Westphal balance as in AOCS Method Cc 10b-25 (Reapp. 73), which is the Fryer–Weston method designed for solid fats and waxes. It is generally determined at 25°C and reported as 25/25°C.

REFRACTIVE INDEX.   Refractive index is related in an orderly way to the density, molecular weight, and internal structural arrangement of fatty acids. Increases in both molecular weight and degree of unsaturation cause increases in refractive index. Other things being equal, conjugated unsaturation results in a higher refractive index than does unconjugated unsaturation. Hydrogenation causes the refractive index to decrease in proportion to the decrease in iodine value. For this reason, and primarily because refractive index can be determined almost instantly, it is often used to follow the course of fatty acid (and oil) hydrogenations. For more interpretive information relating refractive index with composition, refer to Chapter 3, Volume 1, pages 217–221 and to the discussion by Singleton (157).

AOCS Method Cc 7-25 is a general method for the determination of refractive index employing refractometers with either a Butyro- or Abbé scale, and it may be used, with appropriate conversion tables, to yield n_D values of refractive index. A more convenient method is AOCS Method Tp 1a-64 (for drying oils or fatty acids), which employs an Abbé refractometer and recommends a measurement temperature of 25°C for fatty acids liquid at that temperature and 65°C for solid fatty acids. The ASTM method for drying oils and fatty acids is D 555 and is identical with it.

VISCOSITY.   The viscosity differences between common fatty acids, except for blown, heat-bodied, or certain substituted fatty acids, are small and have little identification potential. As with viscosities of fats and oils, the information has value in design of pumping, transferring and mixing systems. The measurement of viscosity is important in many polymerization, heatbodying, and air-blowing reactions where fatty acids are used as raw materials. Increase in viscosity may be employed satisfactorily to follow the course of such reactions.

If the usual instruments suited for the determination of viscosity are constructed of materials resistant to fatty acid corrosion, they can be used for liquid fatty acids, all of which are transparent liquids. Sayboldt viscosimeters constructed with orifices of stainless steel are satisfactory in ASTM D 88.

Viscosity of natural and synthetic drying oils and fatty acids that are used in the protective coatings industry are customarily measured by Gardner–Holt bubble-type standards. Originally this was expressed in letter units with tables appended to convert to approximate stokes. The test methods are AOCS Tq 1a-64 (Reapp. 73) and ASTM D 1545, and they have been revised to provide for direct reading of results in stokes.

The determination of viscosity of fatty acids by kinematic viscosity methods is most accurate. In general, the same methods applicable to fats and oils (see earlier in this section, under "viscosity") can also be applied to fatty acids.

COLOR.   With the use of efficient fractional distillation technology, the color of fatty acids is frequently improved over that of the parent fats and oils; therefore, more lighter shade color test methods are usually required for fatty acids than for fats and oils. Older methods, such as the Gardner and Lovibond systems, still find use, although there appears to be an emphasis for the use in recent years of the platinum–cobalt scale and the photometric index systems of color definition. Generally, four test methods are used in the evaluation of color of fatty acids.

AOCS Method Cc 13b-45 (Corr. 77) is based on Lovibond standards and was originally designed for fats and oils. It has long been used to measure the color of fatty acids, but its use is sharply declining (see earlier in this section, under "Color").

The Gardner color standards, developed originally for the measurement of color of drying oils, later fatty acids and derivatives, can be expected to be used indefinitely because of the reluctance of the coatings industry to make changes in raw material standards. AOCS Method Td 1a-64 (Corr. 76) describes the test method said to be applicable to natural and synthetic drying oils, fatty acids, and oil derivatives "which do not differ appreciably from the standards." It is based on the use of 18 color standards supplied by either of two reliable supply houses. The ASTM recommends the use of the Gardner method in its Method D 1544. It defines the Gardner series in terms of CIE chromaticity coordinates, then gives preference to 1963 permanent glass standards for applicability to fatty acids used in protective coatings (ASTM Method D 1467). For drying oils, ASTM Method D 555 is used; for rosin oils, D 1131; and for tall oil, D 803, with the older iron–cobalt standards replaced by the 1963 standards. In 1953, the ASTM had adapted color standards using chloroplatinate in place of ferric chloride–cobalt chloride in tubes 1–8. The chloroplatinate standards were not sufficiently soluble to permit concentrations for tubes above 8. The 1953 change was made because of the better color stability of chloroplatinate over the iron–cobalt standards. They were replaced by the permanent 1963 glass standards.

AOCS Method Td 1b-64 (Reapp. 73) is the platinum–cobalt scale method for color of fatty acids and other light-colored liquids. (The terminology "Hazen scale" is rapidly disappearing for this method.) This scale was needed because of the availability of fatty acids of superior and fainter colors. Color matching is carried out on thicker layers than in the Gardner system. Although it could be accomplished, it is probably not desirable to generate a relationship between the Gardner and the platinum–cobalt systems based upon the CIE system of chromaticities or the use of color-difference devices developed by Hunter Associates Laboratory, Inc., and Gardner Laboratory. The ASTM recommends the platinum–cobalt color test method in its Method D 1209 for fatty acids in protective coatings.

The most recent test method for fatty acid color is the most scientific method since it eliminates human judgment; it is also precise and rapid. AOCS Method Td 2a-64 (Reapp. 73) is the photometric index method. The color of fatty acids is designated as a photometric index expressed as 100 × the absorbance at each of two wavelengths, 440 and 550 nm. It is being used increasingly in the specification literature of United States fatty acid producers.

Two test methods for the determination of color in fats and oils—AOCS Method Cc 13a-43 (Reapp. 73), the FAC standard color method for animal fats primarily, and AOCS Method Cc 13c-50 (Corr. 77), a photometric method used only for cottonseed, soybean, and peanut oils—are not used for the measurement of fatty acid color.

ODOR.   Although the need for organoleptic evaluations of fatty acid odor is, in general, less critical than for fats and oils, the demand for product stability, accompanied by a requirement for bland-smelling products, is increasing for fatty acids also. This is probably a result of their use in cosmetics, toiletries, and also in food, where their use in food additive derivatives continues to grow (see earlier in this section, under "Odor").

Commercial Fatty Acids: Tests for Chemical Properties

IODINE VALUE.   The AOCS Method Tg 1a-64 (Rev. 73) is the Wijs method employing iodine monochloride in glacial acetic acid (Wijs solution) and expresses iodine values as centigrams of iodine absorbed per gram of sample. Like the method used with normal fats and oils, the excess of reagent used in the test is critical. With fatty acids, sample weights are tabulated in the method that is to be used for normal fatty acids (100–150% excess reagent) and conjugated acids (115–135% excess reagent). The precision for this determination is:

	Fatty Acid	
	Soybean, Linseed	Tallow, Coconut
At a confidence level of 95%, in the range of values of	130–200	7–55
Two single determinations performed in one laboratory should not differ from the mean value by more than	±1.8%	±4.6%
Single determinations performed in different laboratories shall not differ from the mean value by more than	±2.1%	±5.2%

It cannot be assumed that the iodine value of fatty acids containing conjugated unsaturation is that due to the total unsaturation present (see
earlier in this section, under "Iodine Value" for a thorough discussion
of limitations of this method, which is equally applicable to the determination of iodine value of fatty acids).

UNSATURATION BY HYDROGENATION.   With fatty acids, the use of catalytic
hydrogenation is sometimes preferred where interfering substitution reactions are suspected in the determination of unsaturation by halogen
addition.

THIOCYANOGEN VALUE.   Similar to the case of its use on fats and oils, this
test method, as outlined in AOCS Method Cd 2-38 (Reapp. 73) (and
originally intended for use on fats and oils), is largely obsolete. Prior to
1939 it was much used in conjunction with iodine value determination to
estimate oleic, linoleic, and linolenic acid contents in many vegetable-
derived fatty acids, but the absorption of thiocyanogen is now known to
be less than that originally thought, and it was largely displaced in the
1960s by the ultraviolet spectrophotometric method (AOCS Method Tj
1a-64 (Reapp. 73), and the latter, in turn, has been largely supplanted by
the GLC method (AOCS Method Ce 2-66 or ASTM D 1983).

ACID NUMBER (ACID VALUE) AND FREE FATTY ACID.   Acid number is perhaps the most frequently used test method applied to fatty acids. There
is a continuing trend to use the term "acid number" in place of the term
acid value and, although the two terms are used interchangeably, the
former is perhaps preferred because it illustrates the additive nature of this
characteristic when used in conjunction with "saponification number"
(for saponification value), "hydroxyl number" (for hydroxyl value), and
so on. There is an increasing tendency for the use of acid number in the
specification lists of American producers. The term is defined as the
milligrams of potassium hydroxide required for the neutralization of the
free carboxyl groups in 1 gram of the sample.
    When the acid number is determined on samples free of inert or neutral
matter, the information obtained is of value in indicating the mean molecular (equivalent) weight of the fatty acids present. When only fatty
acids are present in the sample (fractionated fatty acids frequently meet
this requirement), the average chain length is revealed from the acid
number, and the technologist has a potential clue to the origin of the
sample. The acid number has its greatest value, however, when it is used
in conjunction with other supporting analytical data such as saponification

number, unsaponifiables, and homologue distribution (GLC analysis) for a more complete indication of the nature and composition of the sample.

AOCS Method Te 1a-64 (Reapp. 73) and ASTM D 1980 are the two standard test methods for the determination of acid number. They are applicable to all fatty acids and polymerized fatty acids. In use, a 5-gram sample of fatty acids is titrated in 75–100 ml of hot neutral alcohol with $0.5N$ NaOH standard solution until the first pink color of phenolphthlein indicator persists for 30 secs. Thus:

$$\text{The acid number, mg KOH per gram of fatty acids} = \frac{\text{titration} \times N \times 56.10}{\text{weight of sample}}$$

The precision of the method is:

	Polymerized Fatty Acid
At a confidence level of 95%, in the range of values of	162–195
Two single determinations performed in one laboratory shall not differ from the true mean value by more than	0.8%
Single determinations performed in different laboratories shall not differ from the mean value by more than	1.3%

The term "free fatty acid" apparently was devised by the early fatty acid industry to represent the free acidity of all fatty acids expressed as percent oleic acid. Since there is no official free fatty acid test method for fatty acids, it is obtained by determining the acid number by the standard method [AOCS Method Te 1a-64 (Reapp. 73)] and calculating the percent oleic acid from the data obtained. Numerically, free fatty acid, when expressed as percent oleic acid, is about one-half (0.504) the acid number. Occasionally, free fatty acid is expressed for coconut oil fatty acids based on lauric acid, in which case free fatty acid is about one-third (0.357) the acid number, and quite rarely, for palm fatty acids, based on palmitic acid, in which case free fatty acid is (0.456) the acid number. For the sake of international uniformity, both these practices should be discouraged, and it appears much more desirable to express all fatty acids in terms of the more acceptable acid number designation.

SAPONIFICATION NUMBER (VALUE) AND ESTER NUMBER (VALUE). Saponification number, interchangeable with saponification value, is the number of milligrams of potassium hydroxide required to react completely "with all the reactive groups" in 1 gram of a sample. It is obvious that since the test is carried out in alcoholic alkali at the reflux temperature for a period of 30–60 min, or "until the sample is completely saponified as judged by the clarity and homogeneity of the saponification medium," alkali is consumed for carboxylic acid groups in the fatty acids, and also for other esterlike groups in any oils or partial glycerides that may be present, as well as lactones and "estolides" and the like. However, natural waxes, which are the esters of long-chain alcohols and fatty acids, saponify slowly and incompletely under the saponification conditions to yield nonhomogeneous saponification media containing long-chain alcohols; therefore, mixtures of fatty acids and waxes cannot be submitted to the usual saponification number analysis.

Principally, saponification numbers are useful in providing information relating to the proportion of glycerides and acids that are present in a given sample. Used in conjunction with acid number this indication is of optimum usefulness; the combined data are indicative of the mean molecular weight of the acids in the sample. Short-chain acids such as lauric acid have high saponification numbers, whereas longer-chain acids such as stearic acid give correspondingly lower numbers. When saponification number is expressed as percent oleic acid, it is occasionally known as "total fatty acid." The term "saponification equivalent" is sometimes used and represents the number of grams of sample that will react under these saponification conditions with 56.1 grams of potassium hydroxide. Expressed this way, the numerical results are a measure of the average equivalent weight of the fatty acids.

The saponification number is identical to the acid number when the fatty acid is entirely free of esterlike contaminants. The difference between the saponification number and the acid number is called the *ester number* (or *ester value*) and for fatty acids commercially available today, this number is rarely more than a point or two, with the saponification number higher than the acid number. When applied to certain fatty acid derivatives, such as dimer acid esters, saponification number determinations are carried out in higher-boiling solvents such as ethylene glycol.

The AOCS recommends that Method Tl 1a-64 (Corr. 79), originally intended for industrial oils and derivatives, be used for fatty acids. The ASTM equivalent method is D 1962. In carrying out the procedure, the dry sample is saponified with 50 ml of alcoholic potassium hydroxide (27 grams of KOH per liter of 95% ethanol) until it is "completely saponified" and back titrated with standard $0.5N$ HCl until the pink color of phenolphthalein has disappeared. A blank is run, and the saponification number is calculated according to the following formula:

Saponification number, mg KOH/ gram sample

$$= \frac{56.10 \times N \text{ acid} \times (\text{titration blank} - \text{titration sample})}{\text{weight of sample}}$$

The precision of the test method is:

	Fatty Acids	
	Polymerized	Tallow, Coconut, Soybean, Linseed
At a confidence level of 95%, in the range of values of	167–197	195–275
Two single determinations performed in one laboratory shall not differ from the mean value by more than	± 5.5%	± 1.6%
Single determinations performed in different laboratories shall not differ from the mean by more than	± 5.7%	± 1.7%

The saponification number test is applicable to many types of normal esters; however, there are a number of hindered esters that saponify sluggishly to which it cannot be applied. For more specific details concerning the saponification of esters that react with difficulty, or to saponification of dark-colored substances, and micro and semimicro methods, refer to Hall and Shaefer (158).

HYDROXYL AND ACETYL NUMBERS.   The hydroxyl number is of use in the analysis of crude or partially esterified fatty acid–alcohol mixtures, the analysis of hydroxy-substituted fatty acids, (e.g., hydroxystearic acid), and the analysis of partially esterified polyhydroxy compounds, among others. The hydroxyl number is defined as the number of milligrams of potassium hydroxide equivalent to the hydroxyl groups in 1 gram of sample. AOCS Method Cd 13-60 (Reapp. 73) and ASTM D 1957 applicable to oils, mono- and diglycerides, and hydroxystearic acids, are used for hydroxyl number determination for many hydroxy-containing substances. The method is discussed earlier in this section, under "Hydroxyl and Acetyl Values."

The older acetyl number (value), as outlined in AOCS Method 4-40

(Reapp. 73) and intended for acetylated fats and oils, is rarely applied to fatty acids, as the utility and the need are completely satisfied by the hydroxyl number determination.

OXIRANE OXYGEN.   The AOCS does not recommend the test method involving ring opening of epoxide groups with hydrogen bromide in glacial acetic acid, which finds limited use in the analysis of epoxidized oils (Method Cd 9-57, Rev. 79). This method is sensitive and subject to too many interferences; for example, peroxides, hydroperoxides, cyclopropenoids, α,β-unsaturated compounds, and soaps interfere. If the sample is known to be free of these interfering substances, the analysis can be carried out if sufficient care is taken. In view of the need to standardize the reagent before each determination, the instability of the reagent, and the possible loss of HBr during the titration, this analysis has not been considered sufficiently satisfactory. It appears that a standard instrumental method of analysis for the oxirane functional group could be developed to fill this void (159).

Test Methods for Minor Fatty Acid Constituents.   The minor constituents that occur in commercial fatty acids are almost entirely those that fill the role of an unavoidable nuisance; that is, they are rarely of major consequence unless for some reason they occur in greater quantities than normal. The test methods for these minor constituents include those for moisture, unsaponifiable impurities, ash, moisture and volatile matter, *trans* isomers, chick edema factor, and others. For a discussion of these test methods, see Section 2.3 on methods for detection of objectionable components.

Test Methods for Various Fatty Acid Components.   A number of test methods for fatty acids relate to identification of substantial components that define the degree to which the products satisfy specific applications. Frequently, these refer to degree and type of unsaturation present. Of great importance is the determination of the homologues of saturated and unsaturated fatty acids. As a group, these tests are important in those instances where they are still used and have not been replaced by more effective methods.

POLYUNSATURATED ACIDS.   AOCS Method Tj 1a-64 (Reapp. 73) is applicable to the determination of polyunsaturated fatty acids from dienoic through pentaenoic and which contain only the natural *cis* isomers. The method is quite similar to AOCS Method Cd 7-58 (Reapp. 73), which describes the analysis of fats and oils, and it can also be used for methyl, ethyl and, presumably, partial glyceride esters. It is an ultraviolet spectrophotometric analysis.

The complete formation of fatty acids can be assured because use of either the 6.6% potassium hydroxide solution (for a 25-min isomerization) or a 21% solution (for a 15-min isomerization) provides sufficient alkali to form potassium salts with excess alkali left over for the actual isomerization of nonconjugated to conjugated components. Over the years, much use has been made of this analytical method, and, despite the many limitations known to be inherent in it, there has been reasonable acceptance. However, within the last two decades further limitations and disadvantages have become evident. Fatty acids that are derived from fats and oils by the normal processes of fat splitting and distillation are now known to contain small but variable quantities of *trans* isomers. Unsaturated fatty acids derived from fats and oils through alcoholysis are likely to contain small quantities of conjugated isomers since alkaline catalysts are employed in alcoholysis. Furthermore, the presence of *trans*-conjugated systems of double bonds, such as those present in many drying oil fatty acids, is known to cause interferences that limit the precision of the determination. Fortunately, the development of specific GLC has filled much of the gap in the need for analytical detection of unsaturated fatty acids of the polyunsaturated type.

DIENE VALUE.   The Diels–Alder adduct method of analysis for conjugated oils, as described in AOCS Method Th 1a-64 (Reapp. 73), is now known to give low results; therefore, this method affords only an approximation of the conjugated diene content of oils such as dehydrated castor oil and others. Although a number of attempts have been made to apply this method to fatty acids, it is probable that the reliability is even less for these materials. Specifically, the acidity titration of hydrolyzed excess maleic anhydride may be interfered with by the fatty acid. Under these circumstances it is best to assay for conjugated diene systems in drying oil fatty acids by alternate methods such as photometric analysis (AOCS Method Ti 1a-64 (Reapp. 73), or GLC techniques.

CONJUGATED DIENOIC ACID.   AOCS Method Ti 1a-64 (Reapp. 73) is an ultraviolet spectrophotometric method for the determination of dienoic conjugation, expressed as percent, in materials like dehydrated castor oil, dehydrated castor fatty acids and their methyl or ethyl esters. The analysis is carried out in isooctane solution at 233 nm. The observed precision for the method is:

Repeatability—results determined by the same operator and the same apparatus should not differ from the mean by more than ±0.39%.

Reproducibility—results in different laboratories by different operators should not differ from the mean by more than ±1.28%.

FATTY ACID HOMOLOGUE DISTRIBUTION.   This is an exceedingly important analytical method for both fatty acid analysis and the determination of the fatty acid distribution of fat and oil triglycerides.

Although recent developments (126–130) indicate that fatty acids may be used directly in GLC separation and analysis, this has not been developed for routine analysis; methyl esters are usually used. For the analysis, all that is required is that the fatty acid sample be converted quantitatively into methyl esters by esterification, with no changes in composition or structure of the fatty acid components. AOCS Method Ce 2-66 (Reapp. 73) advocates a boron trifluoride–methanol reagent equally suited for the direct esterification of fatty acids or the interesterification of common fats and oils. ASTM D 1983 and AOCS Method Ce 2-66 (Reapp. 73) describe the gas-liquid chromatographic methods.

Commercial Fatty Acids: Performance Test Methods

FLASH AND FIRE POINTS.   Fatty acids have greater volatility than do triglyceride fats and oils; therefore, it is to be expected that their flash and fire points would be substantially lower than those of comparable fats and oils. The smoke point characteristics of fatty acids are not especially important; the AOCS recommends no method for this determination on fatty acids. Typical flash points for common fatty acids are about 380°F, whereas fire points are about 430°F. These characteristics are important because they permit classification of fatty acids into groups to permit their proper labeling and handling during transportation. Like most organic compounds, stearic acid will burn when it is exposed to heat or flame. It is not considered a hazardous material; most classification of organic materials relegate it to the "slight" fire hazard category (160). However, it should be pointed out that certain mixtures of fine stearic acid dusts, and presumably other fine fatty acid dusts also, will flash like most other organic dusts.

The test methods used for determination of the flash and fire points of common fatty acids that flash above 300°F are AOCS Method Tn 1a-64 (Reapp. 73) and ASTM D 1959 using the Cleveland open cup. Occasionally, fatty acids are admixed with volatile solvents and their flash points are substantially lowered. AOCS Method Cc 95-55 (Corr. 77) is a closed-cup method suited for use with materials that flash below 300°F. See earlier in this section, under "Smoke, Flash, and Fire Points."

HEAT STABILITY.   The application of heat to fatty acids causes a number of changes before ultimate dehydration to fatty acid anhydrides or pyrolysis. Among the changes is the development of color, which is important if the fatty acid is intended for use in alkyd resin manufacture. Fatty acids intended for alkyd resins generally have heat stability specifications; the test methods designed to measure this performance are AOCS Method Td 3a-64 (Reapp. 73) and ASTM D 1981. Both methods specify that about 40 grams of sample be heated in a 1 × 6-in test tube at 205°C under nitrogen for a stated time interval depending on unsaturation. Fatty acids having an iodine value over 15 are heated for 1 hour, and fatty acids with an iodine value of less than 15 are heated for 2 hours. The AOCS Method is called "Color After Heating." Gardner colors are measured by the standard test method before and after the tests.

Fatty acids and derivatives are tested in a variety of heat tests for many end use applications. Time of heating, temperature, presence of additives, and the use of air or oxygen (which changes the test to one measuring oxidation stability) vary quite widely in these tests. In the Swift stability test or the AOM test, 97.8°C is the heating temperature [AOCS Method Cd 12-57 (Reapp. 73)]; in the Schaal oven test (139) the temperature is 63°C.

OXIDATION STABILITY.   Many of the test methods which are applied in the measurement of oxidation stability of fats and oils (see earlier, under "Peroxide Value" and "Fat Stability") are applicable to fatty acids.

Perhaps the most widely used test method that measures degree of autoxidation of a fatty acid (or fat or oil and derivatives) is the peroxide value, which measures the initial chemical evidence of oxidation, namely, hydroperoxides and peroxides. This test method is frequently coordinated with other test data such as organoleptic evaluations of taste or odor development. An improved organic polarographic determination of peroxide value is available (161, 162).

Eckey (163) describes a test that has occasionally been used to measure the oxidation stability of fatty acids. The test is based on the rate of oxygen absorption when a sample of fatty acid is held at elevated temperatures over a period of many hours or days. Degree and type of unsaturation in a fatty acid are the most important factors in determining the rate of oxidation, but the presence or absence of impurities, especially traces of metal prooxidants, may also greatly effect the rate of oxidation.

The Mackey test (164) is a method for the determination of the temperature of spontaneous combustion, usually for oleic acid and oleate esters, based on measurement of the rise of temperature when the sample is absorbed on cotton in a wire mesh cylinder through which air is circulating.

Fatty acids that are used in the manufacture of lubricating greases may be tested for oxidation stability by means of the oxygen bomb method of ASTM D 942. Modifications of the standard ASTM Method 525, designed originally to measure the induction period of gasolines, may be applied to the study of low molecular weight fatty acids. Note, however, that AOCS Method Cd 12-57 (Reapp. 73), designed to measure the oxidation stability of fats and oils, is not applicable to fatty acids.

Although fatty acids are much too volatile to be used as such in aircraft jet engine lubrication, many of their ester derivatives are applicable. For instance, USAAF Military Specifications MIL-L-7808F or MIL-L-9236 require strict synthetic ester performance requirements which involve heating the esters for long periods at elevated temperatures in air. The tests are evaluated on the basis of changes in acid number, saponification number, viscosity, and the corrosion rates on critical metals like lead suspended in the test medium.

ESTABLISHING ORIGIN OF FATTY ACIDS.   The identification of fatty acids by origin is exceptionally complex and is more difficult than that for fats and oils. During the processing of fatty acids many of the raw materials are completely or partially hydrogenated, and during distillation they are fractionated such that the pattern of fatty acid distribution characteristic of the parent raw material is obliterated. Moreover, the trace nonoil constituents that provide a means of identification in the case of some fats and oils are largely removed in distillation or other processing. Thus, for practical purposes, fatty acids cannot be easily identified as to source of fat or oil, but afford only what limited information can be deduced from their finished fatty acid homologue composition.

Fatty Nitrogen Chemicals.   Special test methods .vere developed by the AOCS during 1960–1964 to meet the need for reliable methods of analysis for fatty amines, amidoamines, quaternary ammonium chlorides, and fatty diamines. In cooperation with the ASTM, official method acceptance commenced in 1964. The appropriate official AOCS and ASTM Methods are included in Table 7.8.

AMINE VALUES.   Commercial fatty amines are mixtures of primary, secondary, and tertiary types, and it is usually necessary to assay for the

Table 7.8   AOCS and ASTM "official" analytical methods for fatty nitrogen chemicals

	Fatty Amines	Amido-amines	Quaternary Ammonium Chlorides	Fatty Diamines
Sampling	Ta-1c	Ta-1c	Ta-1c	Ta-1c
Moisture	Tb-2a	Tb-2a	Tb-2a	Tb-2a
	(D 2072)	(D 2072)	(D 2072)	(D 2072)
Color				
Gardner	Td-1a	Td-1a	Td-1a	Td-1a
Pt–Co.	Td-1b	—	Td-1b	—
Total amine value				
Potentiometric	Tf-1a	Tf-1a	—	Tf-1a
	(D 2073)	(D 2073)	—	(D 2073)
Indicator	Tf-1b	—	—	—
	(D 2074)	—	—	—
Amine value,	Tf-2a	—	—	Tf-2a
potentiometric	(D 2073)	—	—	(D 2073)
Primary, secondary,	Tf-2b	—	—	—
and tertiary with				
indicator				
Percent primary,	Tf-3a	—	—	—
secondary, and	(D 2083)	—	—	—
tertiary amines				
Primary amines by	Tf-4a	—	—	—
GC				
Percent nonamines	Tw-1a	—	—	Tw-1a
(in fatty amines	(D 2082)	—	—	(D 2082)
and diamines)				
Iodine value,	Tg-2a	Tg-2a	Tg-3a	Tg-2a
modified Wijs	(D 2075)	(D 2075)	(D 2078)	(D 2075)
	(D 2075)	(D 2075)	(D 2078)	(D 2075)
Ash	—	—	Tm-2a	—
	—	—	(D 2077)	—
Flash point, tag	—	—	Tn-2a	—
open cup				
Average molecular	—	—	Tv-1a	—
weight (of	—	—	(D 2080)	—
quaternary				
ammonium				
chlorides)				
pH (of fatty	—	—	Tu-1a	—
quaternary	—	—	(D 2081)	—
ammonium				
chlorides)				
Nonvolatiles (solids)	—	—	Tc-2a	—
	—	—	(D 2079)	—
Acid value and	—	—	Te-3a	—
amine value	—	—	(D 2076)	—

content of each to establish the quality of the product. The "total amine value" is defined as the number of milligrams of potassium hydroxide equivalent to the entire basicity in 1 gram of sample. It may be determined by one of two methods, a potentiometric method, somewhat time consuming but suited for all samples including dark colored ones, and an indicator method, for rapid assay of lighter colored samples.

AOCS Method Tf 1a-64 (Reapp. 73) or ASTM D 2073 is the potentiometric method applicable to fatty amines, and it is also recommended for fatty diamines and amidoamines. The sample is titrated in 90% chloroform/ 10% isopropanol solution with standard 0.5N HCl in 99% isopropanol using a pH meter and recording pH measurements each milliliter near the endpoint with construction of a titration curve. The endpoint is the midpoint of the inflection on the curve:

$$\text{Total amine value, mg KOH/gram} = \frac{\text{ml} \times N \times 56.1}{\text{weight of sample}}$$

The precision is:

	Amines	Amidoamines	Diamines
At a confidence level of 95%, in the range of values	75–268	143–156	305–406
Two single determinations performed in one laboratory shall not differ from the true mean by more than	±2.7%	±4.0%	±2.6%
Single determinations performed in two different laboratories shall not differ from the true mean by more than	±4.0%	±6.2%	±3.3%

AOCS Method Tf 1b-64 (Reapp. 73) or ASTM D 2074 is the indicator method recommended only for fatty amines. It involves titration of the sample in isopropanol with 0.5N or 0.2N HCl in 99% isopropanol using bromphenol blue as indicator (to the yellow endpoint). The indicator method has slightly better precision than the potentiometric method (for light-colored samples only). The precision is:

	Amines
At a confidence level of 95%, in the range of values of	75–267
Two single determinations performed in one laboratory shall not differ from the true mean by more than	±2.8%
Single determinations performed in two different laboratories shall not differ from the true mean by more than	±3.1%

The determination of separate primary, secondary, and tertiary amine values is carried out by a series of methods that involve complicated chemical separations illustrated by the following series of reactions:

$$RNH_2 \; + \; O{=}CH \text{（salicylaldehyde）} \; \rightarrow \; R{-}N{=}CH \text{（salicyl）} \; + \; H_2O \quad (1)$$

$$RNH_2 \; + \; (CH_3CO)_2O \rightarrow RNHCOCH_3 + CH_3COOH \quad (2)$$

$$R_2NH \; + \; (CH_3CO)_2O \rightarrow R_2NCOCH_3 + CH_3COOH \quad (3)$$

$$RNH_2 \; + \; SCN\text{（phenyl）} \; \rightarrow \; RNHCSNH\text{（phenyl）} \quad (4)$$

$$R_2NH \; + \; SCN\text{（phenyl）} \; \rightarrow \; R_2NCSNH\text{（phenyl）} \quad (5)$$

AOCS Method Tf 2a-64 (Reapp. 73) or ASTM D 2073 is a potentiometric method applicable to fatty amines and recommended by AOCS also for fatty diamines; the method is suited for most products even dark colored ones. AOCS Method Tf-2b-64 (Reapp. 73) is the indicator method, a rapid method for light colored samples, intended for fatty amines only. In both methods, the determination of total amine value by AOCS Method Tf-1a-64 (Reapp. 73) or Tf-1b-64 (Reapp. 73) is carried out first.

In both methods, a sample is reacted with salicylaldehyde, with which only primary amines react, according to reaction 1 to form a Schiff base, which does not react with HCl. A titration of the reaction mixture is carried out, and only secondary and tertiary amines titrate under these conditions. A third sample is reacted with acetic anhydride in the case of the potentiometric method, and only primary amines (reaction 2) and secondary amines (reaction 3) react. Titration of this reaction mixture with standard perchloric acid affords a titration of only the tertiary amines present. In the case of the indicator method, a third sample is reacted with phenyl isothiocyanate. Only primary (reaction 4) and secondary (reaction 5) amines react to form nonbasic thioureas. On titration of this reaction mixture with HCl only the tertiary amines will titrate. Thus, for both methods, with a series of three differential titrations, one can calculate the total amine value, and the primary, secondary, and tertiary amine values, all expressed as mg KOH/gram of sample. The precision for the potentiometric method is:

	Fatty Amines			Fatty Diamines		
	Primary	Second-ary	Tertiary	Pri-mary	Second-ary	Tertiary
At a confidence level of 95%, in the range of values of	0.2 5–257	9–105	0.8–1.2	67	161–197 136–198	6–11
Two single determinations performed in one laboratory shall not differ from the true mean value by more than	— ±2.4%	±6.3%	±55%	±4.3%	±6.4% ±4.3%	±33%
Single determinations performed in different laboratories shall not differ from the true mean value by more than	— ±2.6%	±8.3%	±57%	±4.4%	±7.7% ±4.3%	±33%

The precision for the indicator method is:

	Primary		Secondary		Tertiary
At a confidence level of 95%, in the range of values of	0.1	5–256	8–105	0.9–1.5	68
Two single determinations performed in one laboratory shall not differ by more than	±180%	±2.9%	±6.3%	±67.5%	±3.1%
Single determinations performed in different laboratories shall not differ by more than	±187%	±2.9%	±8.0%	±176%	±10.4%

AOCS Method Tf 3a-64 (Reapp. 73) or ASTM D 2083 is a calculation procedure for percent primary, secondary, and tertiary amines based on the data obtained in Methods Tf-1a or Tf-2a and the Percent Nonamine Method Tw-1a.

AOCS Method Tf-4a is a gas chromatographic method for the determination of homologues applicable to animal and vegetable primary amines containing 6–24 carbon atoms. Saturated and unsaturated primary amines are separated. The method is unsuited for undistilled primary amines unless they are of high purity. The separations are carried out on the trifluoroacetyl derivatives on 6 ft by ½-in outer-diameter columns packed with 15% ECNSS-S on 80/100 mesh GasChrom Q. The precision is:

1. Two single determinations of major components greater than 5% performed in one laboratory shall not differ by more than 2.5 percentage units.
2. Two single determinations of major components greater than 5% performed in different laboratories shall not differ by more than 3.6 percentage units.

AOCS Method Te 3a-64 (Reapp. 73) is a method applicable to fatty quaternary ammonium chlorides for the determination of acid value and free amine value of these products. The acid value is defined as the acidity expressed as equivalent to milligrams KOH per 1 gram of sample; the free amine value is the basicity equivalent to KOH expressed as the milligrams per 1 gram of sample. Twin samples are used. For the acid value, 5–20 grams of the sample in 100 ml of isopropanol is treated with phenolphthalein indicator. If the sample remains colorless, it is titrated with $0.1N$ NaOH to the pink endpoint; if the original is pink, it is titrated with $0.2N$ HCl to the bromphenol endpoint (sample contains free caustic plus free amines). A second sample is neutralized with the calculated amount of HCl and titrated with $0.1N$ NaOH to the phenolphthalein endpoint. A table of molecular weights of representative fatty quaternary ammonium chlorides enables an approximate calculation of percent free amine, free amine hydrochloride, free caustic and the acid, and free amine values.

PERCENT NONAMINES IN FATTY AMINES AND DIAMINES.   AOCS Method Tw-1a-64 (Reapp. 73) or ASTM D-2082 is a method for the determination of nonamine components (amides, nitriles, alcohols, and unsaponifiables derived from the original fatty acids or fatty alcohols) in fatty amines or diamines. These are determined with chromatographic columns packed with Dowex 50W-X4 ion exchange resin.

IODINE VALUE.   The iodine value is a measure of the unsaturation of the alkyl group or groups and is expressed in terms of centigrams of iodine absorbed per gram of sample (percent of iodine). For measuring the iodine values of fatty nitrogen derivatives such as fatty amines, diamines, and amidoamines, the normal Wijs method is modified with the use of a mercuric acetate catalyst. AOCS Method Tg 2a-64 (Reapp. 73) or ASTM D 2075 describes the method tò be used which includes a table for sample size in which the weights are correlated with the iodine value of the sample. The precision is:

	Amines	Amido amines	Amido amines	Diamines
At a confidence level of 95%, in the range of values of	3–81	2–3	73–79	8–71
Two single determinations performed in one lab shall not differ from the true mean value by more than	± 4.6%	±28%	± 9.5%	± 8.3%
Single determinations performed in different labs shall not differ from the true mean value by more than	±11.3%	±55%	±12.4%	±13.1%

For quaternary ammonium chlorides AOCS Method Tg 3a-64 (Reapp. 73) or ASTM D-2078 is suited. Since there is some evidence that the use of mercuric acetate catalyst gives somewhat erroneous results with quaternaries, the catalyst is omitted. Furthermore, sodium lauryl sulfate is added to the titration medium to prevent free iodine from being tenaciously held by the nonaqueous phase and thus gives a sharper endpoint.

The precision is, at a confidence level of 95%, in the range of values of 0.55–0.8:

1. Two single determinations performed in one laboratory shall not differ from the true mean value by more than ±28%.
2. Single determinations performed in different laboratories shall not differ from the true mean value by more than ±56%.

AVERAGE MOLECULAR WEIGHT OF FATTY QUATERNARY AMMONIUM CHLORIDES.   AOCS Method Tv-1a-64 (Corr. 72) or ASTM D 2080 is a method for the determination of the molecular weight of a fatty quaternary ammonium chloride that depends upon converting it to the acetate, titrating potentiometrically, and correcting for the content of nonquaternary compounds present. The determinations of acid and free amine values, percent free amine, and amine hydrochloride (Te-3a), ash (Tm-2a), and nonvolatiles (Tc-2a) are first carried out. The precision is, at a confidence level of 95%, in the range of 349–561:

1. Two single determinations performed in one laboratory shall not differ from the true mean value by more than ±3.1%.
2. Single determinations performed in different laboratories shall not differ from the true mean value by more than ±6.4%.

FLASH POINT OF FATTY QUATERNARY AMMONIUM CHLORIDE. AOCS Method Tm-2a-64 (Reapp. 73) describes the Tag open cup method for the determination of the flash point of fatty quaternary ammonium chloride products. Its purpose is to define whether the product flashes at 80°F or below, the limit of the ICC classification for the flammable category. The procedure may give erroneous values for samples with flash points over 150°F. The method is applicable to fatty quaternary ammonium chlorides and their isopropanol solutions that are liquids at temperatures 20°F below their expected flash points. The precision is, at a confidence level of 95%, in the range of values of 69–188:

1. Two single determinations performed in one laboratory shall not differ from the true mean value by more than ±5.5%.
2. Single determinations performed in different laboratories shall not differ from the true mean value by more than ±10.4%.

pH OF FATTY QUATERNARY AMMONIUM CHLORIDES. AOCS Method Tu-1a-64 (Reapp. 73) or ASTM D-2081 is a method for the determination of the pH in water applicable to fatty quaternary ammonium chlorides. The determination of solids by Method Tc-2a is accomplished first. A 5% solution is made from an about 50% aqueous standard isopropanol, and the pH is determined with a standard pH meter to ±0.1 unit. The precision is, at a confidence level of 95%, in the range of values of 6.4–8.5:

1. Two single determinations prepared in one laboratory shall not differ from the true mean value by more than ±4.0%.
2. Single determinations performed in different laboratories shall not differ from the true mean value by more than ±8.1%.

NONVOLATILES (SOLIDS) OF FATTY QUATERNARY AMMONIUM CHLORIDES. AOCS Method Tc-2a-64 (Reapp. 73) or ASTM D 2079 is designed to determine the nonvolatiles (solids) content of fatty quaternary ammonium chlorides by a vacuum oven technique. Two heatings and "volatilizations" are involved: 1 hour at 105 ± 3°C in an air oven, followed by 8 hours at 100 ± 3°C at 27–29 in Hg. The precision is, at a confidence level of 95%, in the range of values of 51–77:

1. Two single determinations performed in one laboratory shall not differ from the true mean value by more than ±2%.
2. Single determinations performed in different laboratories shall not differ from the true mean value by more than ±5%.

AMINE OXIDE METHODS.   Amine oxides may be determined by potentiometric titration with alcoholic HCl in isopropanol using methyl iodide to remove tertiary amine interference (165). Substitution of the methyl iodide by benzyl chloride is possible.

A titration procedure based on the Polonovsky reaction (reaction 1) (166) serves as a rapid method for the determination of free tertiary amine in amine oxides. It is necessary to remove excess hydrogen peroxide with formaldehyde (reactions 2 and 3) before reacting the amine oxide with acetic anhydride. This ultimately produces nontitratable N,N-disubstituted acetamide, and the free amine is titrated with HCl to a red thymol blue endpoint:

$$R_3\overset{+}{N}\!\!-\!\!\overset{-}{O} + (CH_3CO)_2O \rightarrow R\!\!-\!\!\underset{\underset{R}{|}}{N}\!\!-\!\!\overset{\overset{O}{\|}}{C}\!\!-\!\!R + CH_3COOH \tag{1}$$

$$HCHO + H_2O_2 \rightarrow HCOOH + H_2O \tag{2}$$

$$HCOOH + H_2O_2 \rightarrow H_2CO_3 + H_2O \tag{3}$$

A rapid and efficient TLC method (167) may be used for purposes of process control for determination of tertiary amine in amine oxide samples. For this purpose, the sample is run in comparison against a known sample of amine oxide, which is dissolved in isopropanol and spotted on a TLC plate that is a small microscope slide that has been dipped into a slurry of 50-gram silica gel and 150-ml chloroform. After the chromatogram is developed, it is sprayed with 2,7-dichlorofluorescein indicator. The area of the tertiary amine spot is determined, and the weight of tertiary amine is obtained by referring to the standard curve. Comparison with results of the potentiometric method show good agreement.

Miscellaneous Oil and Fat Products.   It is impossible within the intended scope of this treatment to cover the analytical methods for oil and fat products in the voluminous application areas in any degree of detail. The cosmetics, leather, soap and detergent, emulsifier, food additive, lubrication, cleaner and sanitizer, and ore flotation industries all have additional specific methods for many products.

Table 7.9 tabulates only the AOCS (official) methods for soap and soap products, soap-containing detergents, fatty alkyl sulfates, alkylbenzene sulfonates, glycerol, sulfated and sulfonated oils, soapstock, and lecithin. The supporting literature on analytical methods for oil and fat products is extensive and is growing daily. Surprisingly, useful information is still to be found in the 1921 compilation by Lewkowitsch (233), the first attempt to deal systematically with the analysis of fats and oils and their products. In 1960, Mehlenbacher (167h) authored a comprehensive coverage of the analytical chemistry of fats and oils. Two small-surveys by Metcalfe, one dealing with traditional analytical methods in the fatty acid industry (167a), the other with recent analytical developments in the field (167b), are excellent. Hummel has provided *Identification and Analysis of Surface Active Agents by Infrared and Chemical Methods*, (167c), and Boekenoogen's *Analysis and Characterization of Oils, Fats and Fat Products* (167d) is valuable. A fine collection of new analytical methods of particular value to the oleochemical researcher is in *Analysis of Lipids and Lipoproteins*, edited by Perkins (167e). From the various volumes on cosmetic products perhaps the recent series *Cosmetic Science and Technology* (167f) is best. The standard series of *Chemical Analysis* (323) has many volumes of particular serviceability for fat and oil analytical chemists, including Vol. 12, *Systematic Analysis of Surface Active Agents* (167g); Vol. 25, *Atomic Absorption Spectoscopy* (324); and Vol. 32, *Determination of Organic Compounds: Methods and Procedures,* (325). Milwidsky's *Practical Detergent Analysis* (228) and Longman's *The Analysis of Detergents and Detergent Products* (326) are also useful. Of more general utility are Snell and Biffen's *Commercial Methods of Analysis* (327) and the AWWA-APHA-WPCF treatise, *Standard Methods for the Examination of Water and Wastewater* (328). Two publications (329, 330) deal with microanalytical aspects of entomology, particularly in the food industry.

Only three of numerous methods for miscellaneous fat products are detailed here.

ACID VALUE OF FATTY ACIDS IN SOAP AND SOAP PRODUCTS.   AOCS Method Da 14-48 (Reapp. 73) is a method for the determination of acid value in soaps and soap products, defined as the milligrams of potassium hydroxide necessary to neutralize fatty (and rosin) acids in 1 gram of sample. It is applicable to total acids after their separation from the soap products. The method is carried out similarly to AOCS Method Ta 1a-64.

Table 7.9   AOCS "official" or standard analytical methods for soap and soap products, soap-containing synthetic detergents, fatty alkyl sulfates, and alkylbenzene sulfonates, glycerol, sulfonated and sulfated oils, soapstock, and lecithin

	AOCS Method
Soap and soap products	
Sampling	Da 1-45
Alcohol soluble and alcohol insoluble matter	Da 3-48
Free acid or free alkali, soda soaps	Da 4a-48
Free alkali and potassium carbonate, potash paste soaps	Da 5-44
Water insoluble matter	Da 6-48
Total alkalinity of alcohol insoluble matter	Da 7-48
Total anhydrous soap and combined alkali	Da 8-48
Chlorides	Da 8-48
Unsaponified plus unsaponifiable matter	Da 10-42
Titer test	Da 13-48
Acid value of fatty acids	Da 14-48
Iodine value	Da 15-48
Saponification value	Da 16-48
Borax	Da 17-52
Alkaline silicates	Da 18-42
Carbonates, gravimetric absorption method	Da 19a-42
Carbonates, volumetric evolution method	Da 19b-42
Phosphates, gravimetric method	Da 20a-48
Phosphates, electrometric method	Da 20b-57
Tetrasodium pyrophosphate	Da 21-48
Sulfates	Da 22-48
Free glycerol	Da 23-56
Sugars	Da 24-48
Starch	Da 25-48
Volatile hydrocarbons	Da 26-42
Combined sodium and potassium oxides	Da 27-48
Screen test	Da 28-39
Copper	Da 31-58
Soap-containing synthetic detergents	
Alcohol soluble and insoluble matter	Db 2-48
Free acid or free alkali	Db 3-48
Water insoluble matter	Db 4-48
Total alkalinity of alcohol insoluble matter	Db 5-48
Total anhydrous soap	Db 6-48
Chlorides	Db 7-48
Chlorides, potentiometric method	Db 7b-55
Saponification value	Db 8-48
Alkaline silicates	Db 9-48
Fatty matter	Db 10-48

Table 7.9 *(Continued)*

	AOCS Method
Fatty alkyl sulfates	
Sampling	Dc 1-59
Alcohol soluble matter	Dc 3a-59
Alcohol insoluble matter	Dc 3b-59
Ester SO$_3$	Dc 4-59
Combined alcohols	Dc 5-59
Alkalinity	Dc 6-59
Sodium sulfate	Dc 7-59
Unsulfated material	Dc 8-59
Aklylbenzene sulfonates	
Sampling	
Sodium alkylbenzene sulfonate by ultraviolet absorption	Dd 3-60
Natural oil (unsulfated material)	Dd 4-60
Glycerol	
Sampling	Ea 1-38
Ash	Ea 2-38
Acidity or Alkalinity	Ea 2-38
Sodium chloride	Ea 2-38
Total and organic residue at 175°C	Ea 3-58
Glycerol	Ea 6-51
Apparent Specific gravity	Ea 7-60
Color (APHA)	Ea 9-65
Sulfonated and sulfated oils	
Organically combined sulfuric anhydride	
Titration method	F 2a-44
Extraction–titration method	F 2b-44
Ash-gravimetric method	F 2c-44
Total desulfated fatty matter	F 3-44
Active ingredients	F 4-44
Inorganic salts	F 6-44
Total alkalinity	F 7-44
Total ammonia	F 8-44
Acidity, except in the presence of ammonium or triethanolamine soaps	F 9a-44
Acidity, dark-colored oils	F 9b-44
Acidity, in presence of ammonium or triethanolamine soaps	F 9c-44
Water-immiscible organic solvents	F 10-44
Soapstock	
Sampling	G 1-40
Total fatty acids, wet extraction method	G 2-53
Oxidized fatty acids	G 3-53

Table 7.9   *(Continued)*

	AOCS Method
Total fatty acids, coconut soapstock	G 4-40
Neutral oil	G 5-40
Titer test	G 6-40
pH	G 7-56
Total gossypol, cottonseed soapstock	Ba 8-78
Lecithin	
Petroleum ether insoluble matter	Ja 2-46
Acetone soluble and acetone insoluble matter	Ja 4-46
Phosphorus, total	Ja 5-55
Acid value	Ja 6-55

ORGANICALLY COMBINED SULFURIC ANHYDRIDE IN SULFONATED OR SULFATED OILS.   The AOCS lists three separate analytical methods for the determination of combined sulfuric anhydride applicable to sulfated oils (and one of which is applicable to sulfated and sulfonated oils).

AOCS Method F 2a-44 (Reapp. 73) is applicable to sulfated oils that split off their combined SO_3 on boiling with mineral acids and do not contain compounds that cannot be titrated in water solution with methyl orange as indicator. Only that portion of the S bound through O to C is determined; S bound directly to C, that is, in true sulfonates, is not determined. In this method, alkalinity by F 7-44 is determined first, then sufficient $1N$ H_2SO_4 is added to exactly neutralize it, plus 25 ml of excess; the mixture is refluxed for 1.5 hours. Titration to the methyl orange endpoint with $1.0N$ NaOH with a blank run on the reagents completes the analysis:

$$\text{Organically combined sulfuric anhydride, } \% = (T - B) \times 0.8007$$

where T = ml of $1.0N$ NaOH required by sample and B = ml of $1.0N$ NaOH required by the blank.

AOCS Method F 2b-44, an extraction–titration method, is equivalent in scope to F 2a-44 and is an alternate method to eliminate influence of materials such as sodium acetate or other partially titratables. It involves extraction of the acidulated sulfates with ether or ether (2 parts by volume) /95% ethanol (1 part by volume). After evaporation of the solvent, the titration is carried out as in Method F 2a-44.

AOCS Method F 2c-44 (Reapp. 73), an ash-gravimetric method, is used for the determination of sulfated and sulfonated oils and those containing sodium acetate or other partially titratable components. Two procedures are involved, one in the presence of ammonia (NH_3 is removed by volatilizing it off with NaOH before ashing), the other in the absence of NH_3. Both methods involve ashing as equivalent Na_2SO_4:

$$\text{Organically combined sulfuric anhydride, \%} = 1.1274 \times \% \text{ extract} - \text{ash}$$
$$1.1274 = \text{the molecular ratio } 2SO_3 : Na_2SO_4$$

METAL CONTENT OF DRYING OIL ADDITIVES.   The analysis of metal content in fat-derived products such as metallic stearates and paint driers is of importance. Frequently, the standard analytical methods for the determination of metals are applicable after the sample has been ashed. The drying oil and coatings industry uses a variety of metallic salts of fatty acids as drying oil additives and the determination of lead, manganese, copper, zinc, and other metals is frequently required.

The ASTM lists methods for the determination of calcium and zinc (D 2613), lead (D 2374), manganese (D 2375), and cobalt (D 2373) by modifications of the EDTA method.

Although metal stearates and paint dryer analyses involve the determination of relatively high quantities of metal content, the determination of trace quantities of metals is of importance when these metals represent an objectionable contaminant (see Section 2.3, on objectionable constituents).

2.3   ANALYTICAL METHODS FOR THE DETECTION OF OBJECTIONABLE OR TOXIC COMPONENTS THAT OCCASIONALLY OCCUR IN FATS AND OILS, FATTY ACIDS, OR OTHER FAT AND OIL PRODUCTS

Methods for the Detection of Adulteration.   One of the problems encountered by the oleochemical analytical chemist is the determination of adulteration of more expensive oils with cheaper fats or with inert diluents such as mineral oil, paraffin, or other components. Examples include the difficult task of establishing adulteration of vegetable oils with animal fats, for example, to establish compliance with kosher dietary requirements; to detect the adulteration of cocoa butter with cheaper oils such as cottonseed or soybean, of coconut oil with palm or soybean oils, of vegetable or even certain land animal fats with fish animal fats or oils; and the adulteration of several milk fats. The relatively few recognized procedures that are available are rarely entirely satisfactory, but those that are used usually depend on the identification and the determination of certain characteristic constituents, usually fatty acids from the glycerides or trace constituents of nonoil composition. Further confirmation is supplied by physical and chemical tests to substantiate and support the suspected degree of adulteration.

In a few cases, methyl ester analysis by GLC provides satisfactory identification of adulteration when applied to fats and oils. Coconut oil adulteration of soybean oil, cottonseed oil, or tallow can be detected by means of a 12:0 fatty acid assay. Castor oil adulteration can be ascertained by ricinoleic or hydroxystearic acid content and vegetable oil adulteration with animal fats by measurement of the 17:0 and 17:1 fatty acids. The results of the latter determination must be interpreted with care in view of the trace occurrence of heptadecenoic and heptadecanoic acids among the fatty acids from the glycerides of a very few vegetable oils, such as Indian sesame oil (168). The identification of peanut oil through its high content of arachidic acid (1.5–4.8%) in the modified Renard test (169) depends on saponification, conversion to lead saops, acidulation, and the separation by crystallization of arachidic acid (melting point 71–72°C) from 90% ethanol. In the modified Bellier test (170), peanut oil is detected in olive, cottonseed, or corn oils by the turbidity of the acidulated, saponified medium. On cooling slowly, if turbidity develops before 9°C (olive) or 13°C (cottonseed, corn, or soybean), peanut oil is present. A semiquantitative determination of fish or marine oils in vegetable oils (171) depends on a visual observation of insoluble polybromides produced by reaction with bromine in acetic acid and the chloroform solution of the sample compared to that from fish oil. Up to 15% and as little as 1% adulteration with fish oils is said to be detected if metallic salts are absent.

Examples of trace quantities of nonoil constituents that can provide a satisfactory basis for the instrumental analysis of binary (or multicomponent) mixtures include determination of the high squalene content of olive oil to detect a relatively high olive oil content in olive–vegetable oil mixtures (54), the specific occurrence of β-tocopherol only in wheat germ oil (127a), the rosin content in tall oil for the analysis of tall oil–vegetable oil or tall oil fatty acids–vegetable fatty acid mixtures, the cholesterol content in animal fat to determine animal fat adulteration in vegetable oils (172), the comparison of microcrystalline differences between cholesterol and phytosterols to determine vegetable oils in butterfat (173), the difference in sterol content of teaseed oil compared to olive oil as a means of colorimetric determination of the former in the latter by the Fitelson test (modified Liebermann–Burchard test) (174), and the use of sesamol and related compounds that occur in sesame oil as the basis for methods based on colorimetry (63) fluorescence and phosphorescence (64).

In addition to the identification of specific and unique fatty acids from the fat or oil glycerides and the occasional assessment of trace nonoil constituents, the triglyceride structure provides a basis for some present and future analytical utility. Since many fats and oils have characteristic triglyceride structural arrangements, it follows that any separation technique that is able to separate the existing isomers as such would be potentially capable of providing a satisfactory method of analysis. One technique that offers promise is high-performance liquid chromatography (HPLC). Using reverse-phase HPLC on a C_{18} μ-Bondapak column with acetonitrile–acetone solvent mixtures, Plattner et al. (45) separated the major triglycerides in coconut, corn, linseed, safflower, soybean, and sunflower oils and, more recently, carried out a separation of the triglycerides from *Vernonia anthelmintica* (175). Such separations offer large potential for the resolution of many of the problems in fat and oil adulteration.

An outstanding example of the use of specific triglyceride structure for the determination of adulteration of olive oil is the use of fractional extraction of the more organic solvent–soluble triglyceride components (olive oil has about 30% OLL and OOL triglycerides) with a methanol–acetone mixture at 0°C and examination of the fatty acids of the extracted glycerides by GLC. Olive oil adulteration with cottonseed, sesame, corn, soybean, or linseed oils is claimed to be readily determined by this method (176).

Foreign fats containing tristearin can be determined in lard by the Boemer number method (176), a thermometric method (difference between melting points of triglycerides and the fatty acids derived from them, which is large for pure pork fats and small for beef fats). Since lard contains a high percentage of saturated acids in the 2-position of the triglycerides in comparison to other animal fats, including beef fat, determination of the fatty acids in the 2-position can be used to detect the presence of lard in beef fats (176a).

Physical or chemical characteristics that may be employed for the detection of adulteration in fats and oils include the Polenske value, Reichert–Meissl value and/or Kirschner value for the detection of low-boiling milk fatty acids, saponification value for the detection of cocoa butter adulteration with paraffin, refractive index for adulteration of linseed oil, specific gravity for castor oil contamination with oils of similar iodine value, and Crismer values for determining adulteration in rapeseed oils (55a).

Analytical Methods for Detection of Constituents that are Objectionable Other than for Toxicity. The natural constituents in fat and oil products that are objectionable for reasons other than for toxicity include those that occur as fatty acid components of the glycerides and those that are nonoil in chemical nature. Among the former are cyclopentenoid fatty acids such as those found in chaulmoogra and gorliseed oils; erucic acid in certain rapeseed oils; and *trans*-unsaturated, polymerized, and oxidized oils. The analytical methods that find largest use in the separation and identification of these components are the GLC methyl ester method (ASTM D 1983 and AOCS Method Ce 2-66) and other separation techniques. Varying degrees of objection to nonoil minor constituents are manifest. The contaminants include moisture, nonsaponifiable matter, trace metals, ash, rosin and stilbenes in tall oil or its fatty acids, waxes, phosphorus and sulfur compounds, chlorophyll, β-carotene, and a few others.

MOISTURE.   Perhaps no series of analytical test methods are more extensive than those intended to measure moisture in fat and oil products. The AOCS lists 17 separate "moisture and volatile matter" test methods for fats and oils, and their source materials, fatty acids; sulfonated and sulfated oils; and an additional 12 methods for moisture in fats and oils, fatty acids, industrial oils, and derivatives.

The presence of moisture is universal; almost never are naturally derived products totally devoid of it, and fat and oil products are rarely delivered completely dry to customers. The solubility of water in common fats and oils is variable and is as high as 0.05–0.30% without physical evidence of its presence. Water is soluble in winterized cottonseed oil to the extent of 0.07% (30°F) and 0.14% (90°F) (177, 178). The solubility of water in fatty acids is much greater: 9.7% in caproic acid (46°C), 2.35% in lauric acid (43°C), and 0.92% in stearic (69°C) acid (178).

In view of the enormous volume of fat and oil products traded and exchanged internationally, the determination of moisture assumes considerable importance. Although fat and oil products are traded on a dry basis, and shipments are generally made under optimum conditions, it is recognized that it is virtually impossible to keep liquid, semisolid, or solid materials absolutely anhydrous despite their hydrophobic and water-repelling nature. Table 7.10 lists the moisture test methods recommended for vegetable-source materials, fats and oils, fatty acids, industrial oils, and derivatives. Only a few of these methods can be discussed here.

The moisture and "moisture and volatile matter" methods have been developed to provide accurate moisture assays in minimum test time. The older methods involve the evaporation of water by heating with or without use of vacuum for the quickest drying without damage to the solids by decomposition, charring, or loss of weight as the result of pyrolysis of portions of the oil. The choice of drying temperature is critical in these determinations. The Karl Fischer method of water analysis eliminates the need for heat and is rapid and convenient. It is based on a specific reaction

Table 7.10   AOCS "official" and standard analytical methods for the determination of moisture in vegetable source materials, fats and oils, fatty acids, fatty nitrogen chemicals, industrial oils, and other fat and oil products

	AOCS Method No.
Moisture and volatile matter	
Cake, meal, and meats	Ba 2-38
Castor beans	Ae 2-52
Castor pomace	Bd 2-52
Cottonseed	Aa 3-38
Flaxseed	Af 2-54
Peanuts (groundnuts)	Ab 2-49
Soya flour	Be 2-49
Soybeans	Ac 2-41
Sunflower seed	Ai 2-73
Tung fruit	Ad 2-52
Fats and oils	
Air oven method	Ca 2c-35
Hot-plate method	Ca 2b-38
Vacuum-oven method	Ca 2d-25
Fatty acids, hot-plate method	Tb 1a-64
Soap and Soap Products, air oven method	Da 2a-48
Soap-Containing Synthetic Detergents	Db 1-48
Sulfonated and Sulfated oils	F 1b-48
Moisture	
Alkylbenzene sulfonates	
Distillation method	Dd 2a-59
Karl Fischer method	Dd 2b-59
Fats and oils	
Distillation method	Ca 2a-45
Karl Fischer method	Ca 2e-55
Fatty acids, Karl Fischer method	Tb 2-64
Fatty acid sulfates, distillation method	Dc 2-59
Fatty nitrogen compounds, Karl Fischer method	Tb 2a-64
Glycerol, "distilled" glycerides, Karl Fischer method	Ea 8-58
Lecithin	Ja 2-46
Soap and soap products, distillation method	Da 2b-42
Sulfonated and sulfated oils, distillation method	F 1a-44

of water with the Karl Fischer reagent, a standard mixture of iodine, sulfur dioxide, pyridine, and methanol or glycol ether. As long as water is present, iodine is reduced to colorless hydrogen iodide in the complex sequence of chemical reactions:

$$C_5H_5N \cdot I_2 + C_5H_5N \cdot SO_2 + C_5H_5N + H_2O \rightarrow 2C_5H_5N \cdot HI + C_5H_5N \cdot SO_3$$

$$C_5H_5N \cdot SO_3 + ROH \rightarrow C_5H_5N \cdot HSO_4R$$

The ASTM gives a standard method and description of the use of Karl Fischer reagent for the analysis of water in many organic compounds (179). In use, the titration is continued until the appearance of excess free iodine is determined visually or electrometrically.

For fats and oils, air-oven, hot-plate, and vacuum-oven methods for the determination of moisture and volatile matter are available. AOCS Method Ca 2c-25 (Reapp. 73) is applicable to animal and vegetable fats and oils, but not to drying or semidrying oils or oils of the coconut oil group. It is not applicable to fats containing added monoglycerides. In practice, a 5-gram sample is dried for 30 min at $101 \pm 1°C$ in a tared aluminum moisture dish. Moisture and volatile matter is the loss in weight. AOCS Method Ca 2b-38 (Reapp. 73) is a hot-plate method applicable to all the ordinary fats and oils, including emulsions, such as butter and oleomargarine, and high-acid coconut oil. It is not applicable to certain abnormal samples, such as solvent-extracted fats and oils, that may contain residues from solvents with fairly high boiling points or to samples containing added monoglycerides. This is a sensitive method in which the temperature of the 5–20-gram sample is not allowed to exceed 130°C and in which the endpoint is determined by placing a clean dry watch glass above the tared beaker containing the sample. Absence of condensed water droplets from steam indicates that the drying is complete. Obviously, the method is unreliable if foaming of product causes a loss of fat by entrainment in the escaping water vapor. However, it has served satisfactorily for the determination of relatively large water and volatile matter contents.

AOCS Method Ca 2d-25 (Reapp. 73) is a vacuum-oven method generally suitable to all normal fats except those of the coconut oil group containing 1% or more of free fatty acids (for these oils use Method Ca 2b-38). It is not applicable to fats containing added monoglycerides. In use, a 5-gram sample in an aluminum moisture dish is dried to constant weight at not less than 20°C and not more than 25°C above the boiling point of water at the operating pressure, which is not more than 100 mm Hg. A table is appended in the method to give minimum and maximum oven temperatures for the various internal oven pressures.

AOCS Method Ca 2a-45 (Reapp. 73) is a distillation method used for all normal fats and oils including emulsions for the determination of moisture only, as differentiated from moisture and volatile matter. It cannot be used for samples that contain water-miscible volatile substances or to samples that contain added monoglycerides. It depends on the azeotropic distillation of water and toluene and on the ultimate separation of water and toluene after condensation in layers and volumetric measurement of the liberated water. The method specifies the sample size to be used consistent with the moisture range of the sample. The method has been used widely in the past, but today its use is declining because it is time consuming and relatively inaccurate for fat and oil samples that contain less than 0.5% moisture.

The most widely used method for the determination of moisture in fats and oils is AOCS Method Ca 2e-55, the modified Karl Fischer method. This procedure is probably one of the most widely used methods, together with the determinations of iodine value and acid and saponification values, in the entire realm of oleochemical analytical chemistry. The method is applicable to fats and oils that do not react with and are soluble in the reagents. All of the common fats and oils comply with this requirement. Actually, the method may be developed for the determination of water in reagent-insoluble samples if sufficient time is permitted for the water to leach out of the sample. The method gives directions for the preparation of Karl Fischer reagent, but it is common practice to employ previously prepared reagent, which keeps well if stored properly in a dry environment. In use, 5–20-gram sample, dissolved in 50 ml of chloroform, is treated with 25 ml of the pyridine–methanol–sulfur dioxide solution and titrated, either to a visual endpoint or electrometrically, with a standard calibrated iodine–methanol solution to the yellow to red visual endpoint, or the electrometric endpoint, consistent with a sudden increase of current that will persist for 1 min. A blank is also run. With average moisture levels of 0.52 and 0.57%, the precision is:

1. Within laboratory: coefficient variation is $\pm 2.7\%$.
2. Between laboratories: coefficient variation is $\pm 10.7\%$.

Although there is a hot-plate method for determination of moisture and volatile matter in fatty acids (AOCS Method Tb 1a-64), moisture is almost always determined in commercial fatty acid samples by the Karl Fischer method, AOCS Method Tb 2-64 (Reapp. 73). The method is similar to that carried out on fats and oils (AOCS Method Ca 2e-55), except that no precision data are given for fatty acids.

For the determination of moisture in nitrogen chemicals, such as fatty amines, fatty diamines, difatty secondary amines, and fatty quaternary ammonium chlorides, it is more advantageous to back titrate with water–methanol solution after the reaction of the water in the sample with excess Karl Fischer reagent. The Method is AOCS Tb 2a-64 (Reapp. 73). The procedures are slightly different for the three classes of nitrogen derivatives, that is, (*a*) fatty amines and diamines (*b*) difatty secondary amines, and (*c*) quaternary ammonium chlorides, with respect to choice of solvents used, but otherwise are similar in the order of steps used and in the calculations.

The precision is:

	Amine	Quaternary	Amidoamine
At a confidence level of 95%, in the range of values of	0.0–0.3	4–14	0.1–0.4
Two single determinations performed in one laboratory shall not differ from the true mean value by more than	±95%	±31%	±44%
Single determinations performed in two different laboratories shall not differ from the true mean value by more than	±110%	±34%	±85%

UNSAPONIFIABLE MATTER.   Unsaponifiable matter in fats and oils, and to a lesser extent in fatty acids and derivatives, consists of sterols, methyl sterols, natural hydrocarbons such as squalene, long-chain alcohols, triterpene alcohols, pigments, and other trace materials. Their analysis and assay depend on the fact that they are soluble in the common low-boiling "fat solvents" and they are determined by methods based on extraction of them from the unsaponifiable matter left as residue in the saponification of the fat products. There are six AOCS standard methods for the determination of unsaponifiable matter that use various solvents depending on the content and the nature of the unsaponifiable matter.

American Oil Chemists' Society Method Tk 1a-64 (Reapp. 73) or ASTM Method D 1965, intended originally for tallow and grease fatty acids, is recommended for all acids, drying oils and polymerized fatty acids but, in practice, has not been applied generally to all members of the fatty acid class. In this method seven successive extractions of the alcoholic aqueous saponification medium are carried out with petroleum ether (AOCS Specification H 2-41) or until such time as extractions afford less than a 0.005-gram residue. A correction for any free fatty acid content of the combined residues is made.

Although developed for fats and oils, especially marine oils, AOCS

Method Ca 6b-53 (Reapp. 77) is applicable to fats and oils that "usually contain more unsaponifiable matter than found in tallows and greases." It is applicable to vegetable oil deodorizer distillates and sludges, but is not useful with feed grade fats. The method employs diethyl ether as the extractive solvent and the extractions are repeated until 0.001 gram or less of unsaponifiables is in the last extract.

American Society for Testing and Materials Standard Method D 803 includes a determination of the unsaponifiables in tall oil fractions based on the use of diethyl ether as extraction solvent, presumably applicable to tall oil fatty acids. Although not identical to AOCS Method Ca 6b-53 (Rev. 76), it is similar, and equally satisfactory results should be obtained with it. Bauer et al. (180) investigated the extraction efficiency of diethyl ether and petroleum ether in the removal of unsaponifiable matter from tall oil distillates using the AOCS and ASTM analytical procedures and showed essentially no difference between them.

Three AOCS test methods for the analysis of raw and acidulated soap-stocks involve extraction of saponified or unsaponified samples with petroleum ether and/or diethyl ether. In AOCS Method G 3-53 (Reapp. 73), a so-called wet extraction method, a saponified alcoholic solution of the soapstock is extracted with petroleum ether after acidulation. The unsaponifiable matter is thus included with the total fatty acids obtained by weighing the residues from the evaporated ether portions. In AOCS G 5-40 (Reapp. 73), neutral oil is extracted with petroleum ether from non-saponified alcoholic soap solutions, and the unsaponifiable matter is included in the neutral oil. In AOCS Method G 4-40 (Reapp. 73), applicable to raw or acidulated soapstocks derived from coconut, palm kernel, or similar oils, a saponified alcoholic solution of the soapstock is extracted with petroleum ether after acidulation. Again, unsaponifiable matter is included with the total fatty acids obtained by evaporating the extracts and weighing the residues.

TRACE METALS.   The natural metal content of fat and oil products arises largely from the soils in which the vegetable plants were grown. If this were the only source of metal contamination, product quality problems would be miniscule. However, traces of iron and copper appear in many fat and oil products, and, irrespective of their source, their presence is objectionable because both play an adverse role in metal-catalyzed autoxidation; cause fat and oil deterioration, odor and off-taste development in food materials; and occasionally cause color development. As little as 0.1 ppm of copper in margarine is said to be deleterious to flavor stability on storage (181), and maximum levels should be below 0.02 ppm for good product stability. Consequently, fat and oil products must be monitored to keep the level of these metals low. Tin, copper, and nickel are components of common catalysts widely used in fat and oil technology and must be removed from the finished products. The detection of environmental contamination and pollution of important food materials by heavy metals such as lead, mercury, cadmium, and chromium is becoming increasingly important. Even cod liver oil contains traces of naturally derived arsenic (182).

In the past many of the standard analytical methods for metals were carried out on the ashes of fats and oils, fatty acids, and other products. Only a few of the many methods can be covered here.

A polarographic method for the determination of iron, copper, and nickel in fats (183) is also applicable to fatty acids. Small amounts of copper and manganese may be determined by the procedure described by Palfrey et al. (184). The use of 1,10-phenanthroline for the detection of iron in biological materials may be extended to its determination in fats, oils, and fatty acids.

For the determination of very small and trace quantities of metals, spectrographic methods are suitable. General treatments of this subject have been given (185) specifically for fats, oils, and fatty acids, and a method for the determination of iron and copper (186) is sensitive for copper down to 0.001 and 0.01 ppm for iron. Recent advances in spectrographic techniques have been made in which savings in time have been achieved in the ashing techniques (187) and in the use of rotating electrode assemblies in which direct ashing on the electrode of the ARL Spectrographic Analyzer is possible (188, 189). Improvements in sensitivity for trace contaminants is also attained through preconcentration of the metals by means of precipitation ion exchange (190, 191).

Recent development in the technique of atomic absorption analysis has been extraordinary. This technique offers accurate, rapid analysis of trace metals without the need for time-consuming ashing steps. The AOCS offers two atomic absorption spectrophotometric methods for trace metals, described in the following two paragraphs.

AOCS Method Ca 15-75 (Corr. 76) is intended for the analysis of chromium, copper, iron, and nickel in vegetable oils (crude or refined) that contain at least 2 ppm of each of the metals. Hardened oils and fats have limited solubility in methyl isobutyl ketone (MIBK); even at reduced solute concentrations they foul the burner. Mixed solvent systems, such as MIBK–chloroform, are helpful for saturated fats if the standards used are prepared in the same solvent systems. The equipment is an atomic absorption spectrophotometer, air–acetylene laminar-flow burner head, and appropriate hollow cathode lamps. The recommended wavelengths are: Cr, 3579; Cu, 3248; Fe, 2483; and Ni, 2320. The observed absorptions are compared with those measured for standard metal solutions of known concentration and plotted on a calibration curve. The following tabulation gives the standard deviations for each metal analysis. Eight collaborators analyzed two samples of soybean oil. Test solutions were retained and absorption measurements were repeated on the following day.

Element	Level (µg/gram)	Standard deviations		
		Day 1	Day 2	Average
Cr	2.5	0.0536	0.1155	0.0845
	8.5	0.2654	0.3007	0.283
Cu	3.5	0.1137	0.1344	0.124
	9.5	0.1574	0.1940	0.176
Fe	2.0	0.0627	0.1640	0.113
	8.0	0.4990	0.2875	0.393
Ni	3.0	0.3599	0.3239	0.342
	9.0	0.3159	0.2797	0.298

AOCS Method Ca 18-79 is a new method intended for assay of chromium, copper, iron, nickel, and manganese using a technique of direct injection into a graphite tube furnace. It is said to be effective for metal contents down to 0.1 ppm. The equipment used is an atomic absorption spectrophotometer with deuterium arc background corrector, graphite tube furnace, and appropriate hollow cathode lamps. In precision determinations, several laboratories participated in collaborative studies of the graphite furnace technique as used to determine copper, iron, nickel, chromium, and manganese at levels below 1 ppm and mostly at 0.2 ppm and under. Results showed a relative standard deviation (RSD) of 30% at levels of 0.2 and 0.1 ppm, respectively. At the 0.05-ppm level, RSD values of 50% and above were often observed. For further details, refer to the publication of Olejko (192).

Recently, Tsai et al. (193) described an improved ashing procedure for the determination of lead, copper, cadmium, manganese, zinc, and iron in edible oils by atomic absorption spectrophotometry said to give improved recovery of the trace metals. A technique recently developed for the rapid and precise analysis of trace metals is "activation analysis," which employs thermal neutrons activated by a nuclear reactor as a means of analysis. Lunde (182) measured arsenic, bromine, sodium, copper, zinc, and iron contents of some marine and vegetable oils and found most of the contaminants in the phosphatide fractions. Trace elements found in several vegetable oils were: Zn, 0.85–1.1 ppm; Cu, 0.005–1.0 ppm; Na, 0.03–5 ppm; and Br, under 0.01–4 ppm.

ASH.   The need for ashing of many fat and oil products has been substantially lowered since the advent of atomic absorption methods of analysis. The AOCS lists seven methods for the determination of ash and various products: acid insoluble oil seed meals (Ba 5a-68), cake, meal, and meats (Ba 5-49), drying oils (Tm 1a-64), fats and oils (Ca 11-55), fatty quaternary ammonium chlorides (Tm 2a-64), glycerin (Ea 2-38), and soya flour (Bc 5-49). AOCS Method Ca 11-55 (Reapp. 73) is an incineration method useful for animal fats and oils and also for marine oils. It is not applicable to heat bodied oils containing lead or zinc. The sample is incinerated in a platinum crucible for 1 hour at 550–680°C in an electric muffle furnace after about 75 grams of it has been carefully burned in increments in the crucible over a small flame. AOCS Method Tm 1a-64 (Reapp. 73) is applicable to drying oils, fatty acids, but not to boiled oils. It uses a porcelain or a high-silica-content crucible for the incineration to avoid damage to platinum by heavy metals that could be present in certain of these products. In AOCS Method Tm 2a-64 (Reapp. 73), a 20-gram sample of fatty quaternary ammonium chloride is ashed similarly, but since the ash is likely to consist of NaCl and NaOH, the weighing must be carried out quickly to avoid water absorption. Whereas ashes of some fats and oils could be as high as 0.3%, those of fatty acids are rarely above 0.1%.

WAXES.   With the exception of sperm oil (about 76% waxes), jojoba oil (about 95% waxes), and a few other fats and oils, most common fats and oils contain only small quantities of waxes, the esters of fatty acids and long-chain alcohols, and occasionally sterols. Sunflower oil contains a relatively high level of waxes; a range of 0.008–0.044% was found in five domestic samples that apparently were partially responsible for the high cloud points observed, since these were substantially decreased by solvent winterization with 85% acetone and 15% hexane (194).

A rapid estimate of wax content in sunflower seed oil may be obtained by heating the oil to 130°C and subsequent cooling in ice for 10–15 min. Microcrystalline waxes are formed that may be quantitatively measured by a turbidimetric technique (195).

PHOSPHORUS.   The presence of phosphorus in fats and oils is generally attributed to phosphatides. AOCS Method Ca 12-55 (Reapp. 73) is a method for the determination of phosphorus or the equivalent phosphatides by ashing the sample in the presence of zinc oxide followed by the colorimetric measurement of phosphorus as molybdenum blue. AOCS Method Ja 5-55 (Reapp. 73) is a method for the determination of phosphorus content in vegetable lecithins. It is an adaptation of the conventional phosphomolybdate method of analysis.

SULFUR.   Except in the cases of high-erucic rapeseed oil which contains traces of sulfur-containing compounds (196), the occurence in *Crambe* of a thioglucoside associated with intense unpalatability (197), and the occurrence in mustard seed of a pungent factor capable of generating objectionable allyl isothiocyanate (198), sulfur-containing compounds are rare in fat and oil products. The AOCS gives two methods for the qualitative presence of sulfur in carbon disulfide-extracted olive oil; (*a*) the coin test (AOCS Ca 8a-35), involving staining of a silver coin immersed in the oil by heating to 210°C in 30 min and (*b*) the silver benzoate test (Ca 8b-35 Reapp. 73), which depends on darkening of the melted oil when solid silver benzoate is added to it at 150°C.

VEGETABLE PIGMENTS.   The pigments that occur in small quantities in fat and oil products include chlorophyll, β-carotene, lutein (xanthophyll), lycopene, gossypol, and chroman-5,6-quinone. The latter are occasionally formed by oxidation of γ-tocopherol, from vegetable oils that contain this tocopherol. The analytical methods used to detect these "color bodies" are spectrophotometric methods dependent on absorption in the ultraviolet range of the spectrum. These pigments are considered objectionable because they must be bleached, heat treated, or hydrogenated to acceptable color levels in fats and oils.

AOCS Method Ce 13d-55 (Reapp. 73) is used to determine chlorophyll (in parts per million) in vegetable oils from spectrophotometric absorption measurements at 630, 670 and 710 nm. The method is not applicable to hydrogenated oils, deodorized oils, and most "finished" products since the 670 nm absorption is missing in most processed oils. Kaufmann and Vogelmann (199) investigated the chlorophyll A and pheophytin A content of linseed oil by means of their respective absorption at 600 and 670 nm. Kurita duplicated this observation by finding the maxima for pheophytin in soybean oil at 670 nm (200). The chlorophyll content of rapeseed oil was examined by a spectrophotometric method by workers at the Gdansk Technical University in Poland (201). Total carotenoids in rapeseed oil are in the range of 25–50 ppm, with the major components said to be neolutein A and neolutein B (202). β-Carotene is conveniently analyzed spectrophotometrically through its absorption maxima at 452 nm. In AOAC 43.001, a method for the chromatographic separation and spectrophotometric analysis of vitamin A and β-carotene from margarine, β-carotene elutes before vitamin A does from the alumina column. It is spectrophotometrically determined at 450 nm. The AOAC describes a method for the separation of β-carotene from xanthophylls (AOAC 14-160 and 43.021 (b)) that is sometimes useful. Gossypol, the principal pigment of cottonseed oil, is described in a later section since it is considered to be a toxic constituent.

STILBENES IN TALL OIL FATTY ACIDS.   The presence of small quantities of stilbene derivatives in tall oil fatty acids is objectionable for several reasons, and processing is usually modified to remove them (203). *trans*-3,5-Dimethoxystilbene (X = CH_3O—, Y = CH_3O—, Z = H) is the principal stibene found in TOFA. Oxidation of tall oil fatty acids during epoxy ester plasticizer production causes formation of a hydroxy derivative (X = CH_3O—, Y = CH_3O—, Z = OH) and a dark red quinone, 2-hydroxy-5-methoxy-3-styryl-*p*-benzoquinone, undesirable because of its intense color (203a). The stilbene molecule has strong ultraviolet absorption maxima at 298 and 305 nm; the derivatives absorb in the same general area and this is the usual basis for analysis. Other spectral properties of the stilbene system make these compounds relatively easy to detect, even in trace quantities (38, 204):

X
Z
H
C
C
Y
H

trans-ISOMERS.   Increased interest in *trans*-unsaturated fatty acids is derived from evidence that *trans* isomers of hydrogenated fats are incorporated into adipose tissue somewhat differently from natural *cis* isomers (205). Isolated (methylene-interrupted) *trans* isomers in long-chain fatty acids, esters, and triglycerides can be measured with an infrared spectrophotometer by AOCS Method Cd 14-61 (Reapp. 73). An absorption band with a maximum at about 10.3 μ arising from a C–H deformation about a *trans* bond is exhibited in the spectra of all compounds with an isolated *trans* double bond. This bond is not observed in the spectra of corresponding *cis* and saturated compounds.

The method is applicable to the determination of isolated *trans* bonds in natural or processed long-chain acids, esters, and triglycerides that contain only small amounts (<5%) of conjugated materials. The spectra of fatty acids exhibit a band of medium intensity at about 10.6 μ arising from a vibration of the carboxyl group. Correction of any contribution of this band to the 10.3 μ band, along with correction of any background absorption at this wavelength, can be made by the baseline technique. However, if the isolated *trans* content of the fatty acid is small, the correction will become the major factor in the measured absorption at

10.3 μ and quantitative analysis is not attainable. Therefore, fatty acids containing less than 15% isolated *trans* bonds must be converted to their methyl esters before anlysis by the infrared spectrophotometric method. The method is not applicable, or is applicable only with specific precautions, to fats, oils, and fatty acids containing large quantities (>5%) of conjugated unsaturated bonds, such as tung oil or its acids, or to acids that contain groups interfering with the 10.3-μ band.

The separation of *cis* and *trans* isomers by chromatographic separations has been reviewed (206). Smith et al. (207) recommend a procedure for the measurement of *trans*- and other isomeric unsaturated fatty acids in butter and margarine. It involves GLC determination of *cis* and *trans* isomers after fractionation of the saturated, monoenoic, dienoic, and polyenoic fatty acid methyl esters by argentation TLC. Although the method is lengthy, it eliminates the interferences of the infrared spectophotometric method.

Anticipating the need for a more rapid and efficient method for *trans* isomer determination over the conventional infrared method, Perkins et al. (208) developed and evaluated packed column gas chromatography as an improvement. Although preliminary, the results obtained with margarine, human milk, blood, and red cell lipid methyl esters were in reasonably good agreement with the infrared method. The column packing was Chromasorb P coated with 15% OV-275.

New developments in silver resin chromatography offer hope for the improved separation of *cis* and *trans* methyl esters. Scholfield and Mounts (209) used silver-treated macroreticular arylsulfonic resins with greater surface area and smaller particle size than had been obtained previously and efficiently separated many methyl ester systems including the unconjugated *cis,trans* and *trans,trans* methyl octdecadienoates, and conjugated *trans*-9,*trans*-11 methyl octadecadienoate, the *cis*-9,*trans*-11 isomer and the *cis*-9,*cis*-11 isomer.

A promising recent development is analytical ^{13}C NMR, a potentially rapid, nondestructive method for the determination of the *cis,trans*-composition of catalytically treated unsaturated lipid mixtures. Pfeffer et al. (210) have investigated the analytical prospects for this technique, called "high-resolution natural-abundance ^{13}C Fourier transform nuclear magnetic resonance (NMR)." This method has already become an effective tool in hydrogenation studies of unsaturated fats and oils.

OXIDIZED FATTY ACIDS.   The term "oxidized fatty acids" is defined as fatty material insoluble in petroleum ether but soluble in diethyl ether under specified conditions. Vegetable soapstocks and very low grade undistilled fatty acids generally contain substantial amounts, but little is found in the distilled commerical grades of fatty acids. Oxidized fatty acid content gives an indication of the yield and the quality of product that might be expected from these raw materials, as well as the number of successive distillations required. AOCS Method G 3-53 (Reapp. 73) in Procedure D is a method for the extraction of oxidized fatty acids from raw or acidulated soapstocks after saponification, acidulation, and petroleum extraction in an overall method for the determination of "total fatty acids." It is applicable to raw and acidulated soapstocks except those from coconut, palm kernel and similar oils.

GLYCEROL. The presence of free glycerol in some esters, α-monoglycerides, and undistilled fatty acids is considered objectionable and analytical methods are required for it. Glycerol, with trihydroxylic functionality, responds to the hydroxyl number determination; specific directions in the analysis with acetic anhydride in pyridine as reagent have been given (211, 212). Several other analytical methods for glycerol determination are available: oxidation with potassium permanganate (213); oxidation with perchloratoceric acid (214); the Zeisel and Fanto method using hydriodic acid (215); and the Bertram–Rutgers method, which involves the formation of a colored copper-glycerol complex in alkaline solution (216).

AOCS Method Ca 14-56 (Reapp. 73), an iodometric periodic acid method, is a determination of total, free, and combined glycerol in fats and oils. The total glycerol is determined after saponification of the sample, the free glycerol directly on the sample as taken, and the combined glycerol by difference. Collaborative studies for determination of precision have shown that the following 95% confidence limits may be expected:

	Total and Combined Glycerol at Level of		Free Glycerol at Level of	
	10.0%	0.2%	0.5%	0.05%
Duplicate determinations made on the same day by an analyst should not differ more than approximately	0.10	0.10	0.01	0.01
Single determinations made in two different laboratories should not differ more than approximately	0.30	0.17	0.17	0.03
Averages of duplicate determinations made in two different laboratories should not differ by more than approximately	0.28	0.14	0.17	0.03

AOCS Method Ea 6-51 (Reapp. 73), a sodium periodate oxidation method, is useful for the determination of glycerol or substances with three adjacent hydroxyl groups. Glycerol reacts with sodium periodate in acid solution forming aldehydes and formic acid, the latter proportional to the glycerol content. Because of its greater specificity and accuracy, this method has replaced older methods for the determination of glycerol. Collaborative studies for determination of precision have shown that the following 95% confidence limits may be expected:

	Glycerol present at levels of		
	70–95%	30–50%	10%
Duplicate determinations on the same day by an analyst should not differ more than approximately	0.20	0.15	0.10
Single determinations in two different laboratories should not differ more than approximately	0.60	0.45	0.30
Averages of duplicate determinations in two different laboratories should not differ more than approximately	0.56	0.42	0.28

SOAP IN OIL.   Determinations of soap content in vegetable and animal fats and oils are required primarily for process control in refining operations. Soap in edible oils is objectionable because it contributes an off flavor and clouds prematurely in the cloud test and other low-temperature performance tests.

AOCS Method Cc 17-79 is "recommended practice" for a rapid titrimetric determination of soap in refined vegetable oils, expressed as sodium oleate. The sample is titrated in acetone-water solution to the bromophenol blue endpoint with standard 0.01 NHCl. AOCS Method Cc 15-60 (Corr. 77) is a conductivity method applicable to normal, refined, water-washed, and neutral clay-bleached glyceride oils. It is not applicable to oils refined with acidic refining reagents or to oils bleached with acid-activated earth or clay. The method involves estimating soap as sodium oleate by a 4-min hot-water extraction of the oil and measurement of the conductivity of the water extract.

The precision is:

	Level (ppm)			
Average, as sodium oleate	2890	162	41	22
Maximum difference between single determinations by a single analyst	590	52	19	8
Maximum difference between single determinations in different laboratories	810	73	27	12

Nelson (217) recommends isopropanol as a solvent for this soap titration.

HEXANE AND OTHER HYDROCARBON SOLVENTS.   Excess extraction solvents remaining in fats and oils are objectionable because of their odor and the depression they cause in flash and fire points, aside from the hazardous aspects. AOCS Method Ca 6c-65 (Reapp. 73) (AOAC 28.124), applicable to the determination of saturated hydrocarbons in triglycerides, involves passing the sample in petroleum ether through an alumina column and recovering the unadsorbed hydrocarbons in the percolate. Collaborative studies for determination of precision have shown that the following 95% confidence limits may be expected:

	Hydrocarbon Content (%) at Levels of			
	0.00	0.05	0.5	5.0
Two single determinations performed in one laboratory should not differ more than	0.06	0.09	0.14	0.33
Single determinations performed in different laboratories should not differ more than	0.06	0.10	0.14	0.40

Hirayama and Imai (218) describe a rapid determination of residual hexane in extracted vegetable oils by means of gas chromatography using a pyrolyzer.

POLYETHYLENE IN ANIMAL FATS.   Collection of animal fats in polyethylene film bags has resulted in introduction of this material into fats, particularly raw materials that are collected from butcher shops and restaurants. Polyethylene causes difficulty in the processing operations of splitting and distillation.

AOCS Method Ca 16-75 is "recommended practice" for polyethylene and other plastic polymers. The method involves (a) preparing a chloroform solution of the sample, (b) removing the polyethylene and other $CHCl_3$-insoluble matter by filtration, (c) dissolving the polyethylene from the solids in hot tetrachloroethylene, (d) filtering, (e) precipitating with cold methanol, and (f) weighing as polyethylene.

DETERGENT CONTAMINANTS IN FATS AND OILS.   Inadvertent contamination of fats with detergent materials used in cleaning operations has necessitated analytical methods for the detection of these substances. The methods depend on the kind of detergent material present. Rek and Appel (219) suggest a method for the analysis of anionic detergents, mainly sodium lauryl sulfate, involving extracting the detergent from the oil, forming a methylene blue, chloroform-soluble complex, and determining it spectrophotometrically. The method is suitable for oils and fats containing 0.05–5.0 mg of sodium lauryl sulfate per kilogram. Only fully refined oils are applicable.

ROSIN IN TALL OIL FATTY ACIDS.   Rosin in tall oil fatty acids is objectionable because of its more sluggish chemical activity and its tendency to develop color in esterification reactions. Although a number of methods for the determination of rosin at the 1.5% level are available, only the Herrlinger–Compeau method (220), as outlined by AOCS Ts 1a-64 (Reapp. 73) and ASTM D 1240, appears to combine the fairly good precision and the reproducibility at the 0.2–5% levels with ease and rapidity of analysis. It is based on the titration of free unreacted rosin acids in the sample after the fatty acids have been esterified. Precision is as follows:

1. Agreement within laboratories. Two single determinations performed in one laboratory shall not differ by more than 0.20%.
2. Agreement between laboratories. Single determinations performed in two different laboratories shall not differ by more than 0.34%.

For the analysis of rosin in high-rosin content fatty acid mixtures, neither the old Wolff method (221, 222) as specified by ASTM D 803, nor the McNicoll method (223) as specified by AOCS Method Da 12-48 (Corr. 76) are considered acceptable today. Instead, the Linder–Persson method (36), officially approved only by ASTM D 1469 (for rosin acids content of 0–5% in coating vehicles), is finding widespread use and is rapidly supplanting both the Wolf and McNicoll methods in the tall oil fractionating industry. It involves a more complete esterification of the fatty acids by azeotropic removal of water of esterification, but the method requires a 1.5-hour esterification period. Apparently, it has been possible to reduce the reflux time in the method to 20 mins.

Extensive chemical changes in rosin isomers as a result of new technology in the tall oil fractionation industry have resulted in products that are negative to the old qualitative spot tests for rosin. Thus both the Liebermann–Storch (ASTM D 803) and the Halphen–Hicks tests give questionable results on these acids. A polarization method (232) for "rosin oil," recommended in Lewkowitsch's 1921 compilation (233), is also infrequently used ("rosin oil" polarizes in a 200–mm tube at $+30$ to $+40°$, whereas most oils read between $+1$ and $-1°$).

Holmbom (234), using an improved gas chromatography technique based on a 30–40-cm-long, 0.3-in inner-diameter glass capillary column coated with 1,4-butanediol succinate (BDS), separated all the main fatty and resin acids in tall oil. Obviously, further extensions of this method, especially to analysis of tall oil fatty acids, can be anticipated.

MINERAL OIL IN FATS.   The AOAC lists a qualitative method (225) for mineral oil content of fats and oils involving examination for turbidity after saponification with alcoholic alkali. A quantitative method (226) involves saponification, extraction with ether, and recovery by weighing. Saturated hydrocarbons are determined by AOAC Cd 6c-65 (AOAC 28.124–28.126 (1980)).

CHOLESTEROL IN ANIMAL FATS.   Implicated in the incidence of high blood pressure, heart attacks, and other maladies, cholesterol in fat has attracted considerable attention. Since cholesterol occurs primarily in animal fats and only in traces in common vegetable oils, its direct assay has been used for the determination of animal fat in vegetable oils. AOAC Method 28.099–28.106 is such a method, involving saponification of the fat, extraction of the unsaponifiables, TLC, and then GLC of the mixed sterols. The AOAC lists another method (227) for the determination of vegetable fat in butterfat that depends on examination of mixed cholesterol–phytosterol crystals under the microscope, which are visibly different from cholesterol or phytosterol alone.

The Sperry–Webb method (30) is the classic colorimetric method for the determination of cholesterol. It includes saponification, neutralization with acetic acid, color development with digitonin, and measurement of the ultraviolet maxima at 625 nm. Other colorimetric methods for cholesterol are available; Norcia and Rosenthal (39) adapted the Liebermann–Burchard color method for both cholesterol and vegetable oil sterols. The AOAC has an official method (224) based on the digitonin colorimetric procedure, as well as a bromination (229) and a fluorimetric method (230), which are all used for total sterols expressed as cholesterol. The AOAC also has a method for cholesterol acetate (231) applicable to the assay of cholesterol and β-sitosterol mixtures. The separation is by means of dimethylsilyl derivatives using cholestane as a standard.

Analytical Methods for Detection of Constituents that Are Suspected or Proven Toxic but Infrequently Found in Fats and Oils.   Table 7.11 lists contaminants that are suspected or proven to be toxic and infrequently find their way into fat and oil products.

TOXIC METABOLITES OF MOLD ORIGIN.   Mycotoxins, or mold-derived toxins, can infest vegetable oil source materials and fat and oil products, and their determination is of considerable importance (9). The methods used

Table 7.11   Contaminants suspected or proved toxic and infrequently found in fat and oil products

	General Reference	Method or Reference
Toxic metabolites of microbiological origin		
Aflatoxins	9	235–241
Trichothecenes	242	243
Aromatic heating unit fluids	—	244, 245
Isothiocyanates	—	246, 247
Trace heavy metals		
Cu, Cd, Pb, etc.	182, 192	248, 249
		AOCS Ca 18-79
Hg	—	251
Nitrosamines	10	11, 252–255
Gossypol	322	AOCS Ca 13-56
Polychlorinated Biphenyls	—	257–259
Chick edema factor	—	261, 266
1,4-Dioxane	14a	14a
Cyclopropenoid fatty acids	—	270, 271, 273
Chlorinated solvents	—	274

by the AOAC and the AOCS are adaptations of thin layer chromatographic separations in the case of aflatoxins B_1, B_2, G_1, and G_2. The AOCS lists three methods: for copra and copra meal (Ah 1-72), cottonseed products (Aa 8-71), and peanuts (Ab 6-68). Comparable methods by the AOAC include cocoa beans (235), coconut (236), copra and copra meal (236), corn (237), cottonseed products (238), and soybeans (239). Romer et al. (240) have reviewed the minicolumn screening methods for detecting aflatoxin and have compared them with the TLC methods. A rapid method for the detection of aflatoxin M_1 in dairy products involving purification of chloroform extracts with a silica-gel column, TLC, and densitometry has been described by Stubblefield (241). The occurrence and potential hazard of trichothecenes, toxins characterized by a tetracyclic 12,13-epoxytrichothec-9-ene skeleton, have been reported by Pathre and Mirocha (242). Analysis of these mycotoxins has been reviewed by Eppley (243).

AROMATIC HEATING UNIT FLUIDS.   Chlorinated biphenyl compounds, such as the Arochlors, are prohibited as heat transfer media in the processing of edible products in the United States. Therminol 66, a heating fluid used in certain vegetable oil deodorizers, is a mixture of *ortho* and *para* terphenyls. A method for its detection in palm and rapeseed oils down to a level of 100 ppm (1–2% in the unsaponifiables) involves infrared absorption at 700 and 760 cm^{-1} (244). Dowtherm® A (26.5% diphenyl and 73.5% diphenyl oxide) and KSK oil (mixed hydrogenated substituted naphthalenes) may be determined, according to Imai et al. (245), by steam distillation from the contaminated product followed by application of gas chromatography.

ISOTHIOCYANATES.   Glucosinolates in rapeseed and mustard seed are sources of goitrogens in both meal and oil. As such, they are relatively nontoxic, but their hydrolysis products include isothiocyanates and oxazolidinethiones, which are powerful goitrogens, producing varying manifestations of toxicity in nonruminants. Mustard seed contains glucosinolates that hydrolyze to allyl isothiocyanate, an objectionable lachrymator.

Maheshwari et al. (246) used an HPLC method to determine both isothiocyanates and oxazolidinethiones in tyrosinase digests of rapeseed meal. Daun and Hougen (247) identified four sulfur-containing components in rapeseed oil (butenyl, pentenyl and phenethyl isothiocyanates, and 5-vinyl-2-oxazolidinethione) by a GLC method. The structures were proved by a combination of GLC, TLC, and mass spectroscopy.

TRACE HEAVY METALS.   Traces of mercury, cadmium, zinc, tin, lead, chromium, and several other metals may contaminate fats and oils through various pollution avenues, and although their level is generally low, their toxic effect is dangerous because of the cumulative effect of toxic metal buildup in the human body. The analytical method that is used primarily is atomic absorption spectroscopy.

Elsen et al. (248) used anodic stripping voltammetry for the determination of zinc, lead, cadmium, and copper in a series of rapeseeds and meals and atomic absorption spectrophotometry for the oils from them. Partitioning of metals heavily favored the metals, except for lead, where the distribution was only 2:1. These workers also used a series of methods for the analysis of trace metals in Canadian herring oil (249): zinc, differential pulse anodic stripping voltammetry; selenium, differential pulse cathodic stripping voltammetry (250); copper, cadmium and lead, atomic absorption spectrophotometry; and mercury, (EPA method) (251). Following digestion, the Teflon bomb was rinsed with water, and the entire sample plus 1 ml of HNO_3 was introduced into the stripping cell, subsequently reduced by $SnCl_2$, and swept into the "analyzer" by a stream of N_2.

NITROSAMINES.   Amine-fatty acids soaps in cosmetics and fatty nitrogen chemicals as pesticide emulsifiers are potential sources of nitrosamines, which are powerful carcinogens in animals. The recent introduction of the thermal energy analyzer (TEA) (252) is probably the most important development which has occurred in this area. This device employs a technique in which the N–NO bond is catalytically cleaved. The released nitrosyl (NO) radical is reacted with ozone to give an electronically excited nitrogen dioxide. The excited NO_2 decays back to its ground state, with the emission of light energy. The light emitted is detected with a photomultiplier system. The intensity of the energy emitted is proportional to the N-nitrosyl radical concentration. The TEA is usually interfaced with a gas chromatograph. In this way, specific nitrosamines can be detected at extremely low levels (in parts per billion).

Hedler et al. (253) concentrated nitrosamines from vegetable oils and margarine by steam distillation and determined them by GLC using an alkali flame ionizing detector. A method used by Rhodes and Johnson (254), involving the pyrolysis of N-nitrosamines after gas chromatographic separation and applied to the analysis of neutral extracts of cigarette tars, is probably applicable to fats and oils. White et al. (255) isolated volatile N-nitrosamines from cooked bacon fat and efficiently recovered those purposely added to sample vegetable oils (no N-nitrosamines were found in four soybean oils) by extraction with ethylene chloride and silica gel chromatography, followed by GLC with a thermionic detector.

GOSSYPOL.   The level of gossypol in refined cottonseed oil is insufficient to confer toxicity on the oil, but the presence of gossypol must be considered disadvantageous because of its intense pigmentation and potential toxicity. Curiously, gossypol is known to be toxic to rabbits and swine, whereas cattle can consume large quantities with no apparent detrimental effects whatsoever.

AOCS Tentative Method Ca 13-56 (Rev. 74), (now displaced) applicable to cottonseed oils, is an ultraviolet spectrophotometric method for the determination of gossypol and gossypol-like pigments. The method includes a chart for selection of sample size consistent with the expected gossypol content. In the analysis, three mixtures are measured for absorption at 460 nm: p-anisidine, sample and solvent; acetic acid, sample and solvent; and blank solvent (isopropanol-hexane). A calibration curve is constructed using pure gossypol under the same conditions. The calibrations should be run prior to each analysis.

In 1979, AOCS displaced the "Gossypol in Oils" method with a new "Total Gossypol in Oils" method: Tentative Method Ca 13-56 (Rev. '79). This method, applicable to crude cottonseed oil and cottonseed soapstocks, depends upon the reaction of gossypol and derivatives with 3-

amino-1-propanol in dimethylformamide solution to yield a diaminopropanol complex which then reacts with aniline to form dianilinogossypol under conditions of the method, spectrophotometrically measured at 440 nm. The method is an extension of the AOCS Tentative Method Ba 8-78 said to be applicable to decorticated, glanded and glandless cottonseed, cottonseed flour, cooked cottonseed meats, cottonseed press cake, cottonseed meal, crude cottonseed oil and cottonseed soapstock. However, since the precision becomes poorer with decreasing gossypol content levels (coefficient of variation, 0.5–0.10% range was estimated at 3.5–7.0%, and for the 0.06–0.01% range at 10.0–17.0%, respectively), it is apparent that this method is unsatisfactory for the newer glandless cottonseed products, specifically those that have a total gossypol content corresponding to the National Cottonseed Producers' Association proposed glandless cottonseed grades of AA (less than 100 ppm) or AAA (less than 10 ppm). New spectrophotometric development (255a) involving the use of butanol/0.05 N HCl as extraction solvent and TLC as a purification step with a measurement at 238 nm appears to hold promise as does a spectrofluorimetric analysis. The resolution of the various physiological toxic constituents of gossypol appears to be attained through the use of HPLC, and a method based upon this technique might be ultimately preferable to those based on total gossypol.

POLYCHLORINATED BIPHENYLS.   The FDA has established a maximum tolerance of 0.3 ppm of pesticides, such as endrin, dieldrin, heptachlor, lindane and chlordane, in the fat of animals for human consumption. Apparently, vegetable fat suffers less from this contamination because processing tends to eliminate polychlorinated biphenyls (PCBs), but animal fats can have a higher and potentially dangerous content due to the large feed levels consumed by the animals.

The AOAC has a series of general analytical methods (256) for organochlorine pesticides, including two TLC techniques (257), a gas chromatographic method (258), and a paper chromatographic method (259). [The AOAC also recommends a gas chromatographic method (260) for the detection of hexachlorobenzene.]

CHICK EDEMA FACTOR.   This factor is presumably a pesticide residue. It is a mixture of hexa, hepta, and octachlorodibenzo-p-dioxins, and although it has been detected in fatty acids only very infrequently, the consequences are serious in chicken flocks. Before about 1965 the test method employed was a bioassay method still listed today by the AOAC as a surplus method (261). It measures the effect of feeding the fatty acids composited in a standard diet to day-old, white leghorn single comb cockerels. The results of the test were reported in terms of presence or absence of the chick edema factor on the basis of statistical calculations.

For lack of a more definitive and rapid method, the bioassay method was the only analytical control test until about 1966. For this purpose it was obviously unsatisfactory, since a period of 3 weeks was required to complete the test. In an effort for improvement, the AOAC developed a gas chromatograph–microcoulometric method (262), with aldrin as a standard. Finally, the FDA developed additional GLC methodology using an electron capture technique (263) that utilizes TLC (264) and GLC (265) and provides an improved method of analysis (265). The improved method with its clean-up procedure (266) is sensitive to 0.1 ppb.

1,4-DIOXANE. Presumably, 1,4-dioxane can be formed from ethylene oxide by dimerization during production of ethoxylated fatty derivatives such as polysorbate 60, for it has been found from 6.5 to 373 ppm in commercial samples. 1,4-Dioxane is carcinogenic to rats (267, 268).

Birkel et al. (14a) used a closed-system vacuum distillation with water as a "carrier solvent" followed by gas chromatography with a flame ionization detector using Chromasorb 104 to determine 1,4-dioxane. The method is called the *Birkel method.*

Although operating technology can be adjusted to minimize or eliminate the formation of 1,4-dioxane during the production of ethoxylated derivatives, the application of efficient vacuum stripping for removal of any 1,4-dioxane which might be formed in trace quantities is generally satisfactory.

CYCLOPROPENOID FATTY ACIDS. For a discussion of cyclopropenoid fatty acids, see Volume 1, Chapter 4, page 262. The Halphen test (269) [AOCS Method Cb 1-23 (Reapp. 73) (AOAC 28.108)] is a qualitative color test for the presence of malvalic acid or its homologues. It results in a cherry red coloration from heating the sample with a 1% solution of sulfur in carbon disulfide to which has been added in equal volume of amyl alcohol. Cottonseed, kapok, and baobab vegetable oils and fats from animals that have been fed cottonseed feed may give a positive test.

There are three quantitative analytical methods for the determination of cyclopropenoids. A stepwise HBr titrimetric method (270) was developed about 1964; it was used by Bailey et al. (50) on 25 varieties of domestic cottonseed oils. They were found to contain cyclopropenoid acid, calculated as malvalic acid, of 0.56–1.17%. Feuge (270a) prefers an improved modification with glycerides or methyl esters involving titration at 60–65°C to just past the endpoint, cooling to 25°C, and back titrating with aniline in toluene. At a cyclopropenoid content of 1%, reproduction was ±0.003%.

Using a modified Halphen spectrophotometric method (271), Lawhon et al. (272) analyzed 16 varieties of southern United States cottonseed oil and found that cyclopropenoid acids ranged from 0.10 to 0.31%.

Schneider et al. (273) employed a combination of derivatization with silver nitrate and GLC on 15% diethylene succinate columns to separate high cyclopropenoid mixtures and silver nitrate derivatization, alumina chromatography, and GLC for low-content cyclopropenoid oils. Cottonseed and *Sterculia foetida* oil analyses showed good agreement with those carried out with the HBr titrimetric method.

CHLORINATED SOLVENTS.   Because they are assumed to be toxic, some chlorinated hydrocarbons previously used to extract fats and oils from vegetable-source materials have already been displaced by less objectionable hydrocarbons. Since 1935 the AOCS has listed a color flame test, Method Ca 7-35 (Reapp. 73), the classical Beilstein test, which qualitatively determines the presence of chlorine from chlorinated solvents (or other sources). It is the green flame from copper chloride that results from immersing a copper wire in the sample and igniting the flame.

The AOAC lists an infrared spectrophotometric method for the determination of chlorinated hydrocarbons in drugs (274) that could be adapted for the analysis of common chlorinated hydrocarbons. Characteristic infrared absorption maxima for common chlorinated hydrocarbons are: CCl_4, 12.80; $C_2H_3Cl_3$, 10.77; and $C_2H_2Cl_4$, 11.70 μ.

2.4   ANALYTICAL METHODS FOR QUALITATIVE OR QUANTITATIVE DETECTION OF RELATIVELY SMALL QUANTITIES OF NONTOXIC CONSTITUENTS THAT HAVE AN ADVANTAGEOUS OR BENEFICIAL EFFECT ON THE QUALITY OR PERFORMANCE OF FAT AND OIL PRODUCTS

Vitamins.   Although hardly ever used for quality control for edible fat and oil products, a number of bioassay methods for vitamins continue to be listed. The AOAC lists bioassay methods for vitamins B_1 (275) and D (276) and a microbiological method for B_2 (277).

Provitamin A (β-carotene) is frequently assayed for since it generates vitamin A ultimately. In a number of methods the characteristic wavelength of 478 nm and the "shoulders" at 430 and 453 nm provide a reliable means of identification (143, 278). Vitamin A in margarine is determined (278a) by alumina (occasionally MgO) chromatography and then ultraviolet spectrophotometry using 450 nm for β-carotene and 310 nm for vitamin A. In a method for the determination of vitamin A in feeds and food (279), vitamin A is determined in the absence of other vitamins by adsorption, followed by a colorimetric measurement with $SbCl_3$.

Vitamin B_1 (thiamine) may be analyzed by an AOAC method (280) involving extraction, enzyme hydrolysis, purification by column chromatography, oxidation to thiochrome, and a fluorimetric procedure. Vitamin B_2 (riboflavin) is determined by two methods: a fluorimetric method (281), and a procedure (282) for high-potency samples. Vitamin C (ascorbic acid) may be determined by a 2,6-dichloroindophenol method (283) and a microfluorimetric procedure (284). Vitamin D is assayed (285) by saponification, treatment with maleic anhydride, and then colorimetrically with $SbCl_3$ at 500 nm. Vitamin D_3 requires a HPLC method (286). Vitamin E is known also as α-tocopherol, and its analysis is covered in the paragraphs that follow.

Natural Antioxidants.   The Emmerie–Engel method (31a) for total tocopherol assay was developed in 1938 and used for over 30 years. It is based on the determination of Fe^{2+} produced as a result of the oxidation of total tocopherols with $FeCl_3$; its limitations include variation in the rate of oxidation among the varius tocopherols and some inhibition of color in triglycerides. Other methods for the separation and quantitation of individual tocopherols include polarography (286, 287), TLC (288, 289), GLC (290), column liquid chromatography (291, 292), and paper chromatography (293).

Muller-Mulot (40) separated and determined many tocopherols in a series of vegetable oils by silica gel TLC. Gas chromatographic methods are most useful, and the trimethylsilyl derivatives are well suited. Slover (52) used extraction, saponification, TLC, and GLC of trimethylsilyl derivatives on a group of vegetable oils to evaluate and quantify their tocopherol contents. Meijboom and Jongenotter (15a) used TLC and GLC to determine both tocotrienols and tocopherols in palm oil. α-Tocopherol may be conveniently analyzed, according to Hartman (290), by saponification of the oil, extraction of the unsaponifiables, esterification with butyric anhydride, and GLC separation of the butyrate ester. The method eliminates the need for the intermediate TLC step.

The AOCS has an official method (Ce 3-74), applicable to soya sludges and residues, that is based on butyrate (or isovalerate) ester separations by GLC. The method may be used to separate α-tocopherol (vitamin E) from sterols such as campesteryl, stigmasteryl and sitosteryl butyrates, or cholesteryl isovalerate. Precision is as follows:

1. Two single determinations performed in one laboratory shall not differ

by a coefficient of variation more than:

| Total tocopherols | 9.7% |
| Total sterols | 14.7% |

2. Two single determinations performed in different laboratories shall not differ by a coefficient of variations more than:

| Total tocopherols | 11.0% |
| Total sterols | 26.0% |

The AOAC methods for α-tocopherol (vitamin E) include a number of special methods. One method (294) is designed for the determination of α-tocopherol or α-tocopheryl acetate. It requires extraction, saponification and reextraction, TLC, and colorimetry using bathophenanthroline reagent. To report vitamin E in IU units, it is necessary to analyze for all-*rac*-α-tocopherol by a procedure (295) requiring a measurement of optical rotation and an additional oxidation step. Another method (296) is intended for α-tocopheryl acetate in feeds and foods and still another (297), for fortified pharmaceuticals.

Carpenter (298) has developed an HPLC analytical method with UV absorbance detection for the determination of tocopherols that is rapid (10 min) and selective. Comparison of data for soybean, olive, cottonseed, peanut, and corn oils with the HPLC method (using a detection wavelength of 295 nm) against the Emmerie–Engel method, as modified by Stern and Baxter (299), showed higher values for α-tocopherol by the HPLC method.

The natural antioxidants (sesamol, sesamoline, and sesamine) present in sesame oil respond to a number of specific color tests that make their qualitative identification relatively simple. One of these, the Villavechia test, as described in AOCS Method Cb-2-40 (Reapp. 73) (AOAC 28.118) is well known. The reagent is furfural in ethanol and shaking the sample, acidified with concentrated HCl, gives a crimson or pink color when compounds possessing the 3,4-ethylenedioxyphenyl structure are present. The Baudouin test was first observed prior to 1850 by Camoin, but Baudoin studied the reaction in detail and the test eventually bore his name. Sesame oil gives a red color when shaken with concentrated HCl to which sucrose has been added. For details, refer to Utz (300) and Budowski and Markley (301).

Pavolini and Isidoro (302) devised another color test consisting of the formation of a brilliant red color (turning blue-green) when sesame oil is permitted to react with sulfuric acid, acetic anhydride, and furfural. A standard colorimetric method that has served for many years is the method of Suarez (303), which is a modification of the previously described test. The procedure involves the extraction of sesamol and se-samoline into aqueous alkali, the separate determination of sesamol and sesamolin by color reaction with sulfuric acid and furfural, and the determination of sesamine by spectrophotometric absorption measurements, with a correction for the sesamoline present. Beroza (304) described a procedure for the separation of pure sesamine by silicic acid chromatography followed by UV absorption measurements. He also suggested a reverse-phase paper chromatographic method (305). Finally, a method based on the development of fluorescence and phosphorescence (306) has been devised that is said to be more sensitive than any of the colorimetric procedures.

The natural ester derivatives of ferulic acid are powerful antioxidants that occur in vegetable oils in trace quantities. Although no great effort has yet been expended on analytical methods for ferulates, it is apparent that sensitive microtechniques will eventually be required. Traces of methyl ferulate in rice bran "dark oil" were detected and isolated by Tanaka et al. (307) using silic gel chromatography and eluting with diethyl ether–hexane mixtures.

Synthetic Antioxidants.   Among the natural and synthetic antioxidants for use in foods in the United States are tocopherol, gum guaiac, propyl gallate (PG), BHA (butylated hydroxyanisole), BHT (butylated hydroxytoluene), and THBP (2,4,5-trihydroxyhydroquinone). The last six are synthetic.

Generally, the synthetic phenolic types have characteristic absorption maxima in the ultraviolet that facilitate their identification. Doeden et al. (21) suggest a method for the rapid analysis of BHT, BHA, and TBHQ based on gel permeation chromatography and ultraviolet detection that requires no preliminary isolation of the antioxidants from the matrix. Detection limits with the instruments used in this method are less than 1 ppm for BHA and BHT and about 20 ppm for TBHQ.

Page (308a) described a rapid HPLC method for determination of BHA, TBHQ, 2,6-di-*tert*-butyl-4-hydroxymethylphenol (Ionox 100), BHT, THBP, PG, OG (octyl gallate), DG (dodecyl gallate), and NDGA (nordihydro-guaieretic acid) in vegetable oils, lard, and shortening.

Lactones in Butterfat, Coconut Oil, and Other Oils.   The presence of small quantities of aliphatic lactones (derived from precursor hydroxy acids) are responsible for part of the desirable and characteristic flavor of butter and milk fats, coconut oil, and other fats of the milk and lauric acid groups. The lactones are those derived from γ- and δ-hydroxyalka-noic acids: δ-C_6 to δ-C_{16}, δ-C_7 to δ-C_{15}, δ-C_8 to δ-C_{16}, and γ-C_9 to γ-C_{15} types, inclusive.

Ellis and Wong (32g) used aluminum oxide chromatography with gas chromatography to quantify total free lactones in fresh butter (7 ppm); commercial samples contained 12–30 ppm; butterfat heated for two hours showed 93 ppm. Kinsella et al. (32c) used a silicic acid column and a TLC method to separate γ- and β-hydroxyacid glycerides from butterfat. Many other investigators have employed various other techniques to assay lactone content (see Table 6.11, Chapter 6, Vol. 1, p. 305), such as steam distillation, column chromatography, or other separation methods. The AOAC lists a pH meter titration method (308) for the determination of lactones and free acidity in honey. The titration is carried out on a 10-gram sample with CO_2-free water utilizing $0.05N$ NaOH at the rate of 5 ml/min to a pH of 8.50, then 10 ml of $0.05N$ NaOH is pipetted in, and, without delay, the mixture is back titrated with $0.05N$ HCl to a pH of 8.30. It is assumed that the lactones present saponify instantly.

Antifoam Agents. Dimethylpolysiloxanes are permitted in cooking fats at concentrations of 0.1–1.0 ppm; they have desirable antifoam action and minimize exposure of oil to the air. Doeden et al. (20) have developed a direct-aspiration flame atomic absorption method for the determination of dimethylpolysiloxanes (DMPS) in fats and oils. The method has a detection limit of 1 ppm with a standard deviation of samples containing 1–10 ppm of 0.3. The technique is rapid, requires a minimum of sample preparation, and is applicable to both hydrogenated and unhydrogenated fats and oils.

Emulsifiers. There are 16 natural and "synthetic" generally regarded as safe (GRAS) food emulsifiers approved for use, and their analysis would be somewhat straightforward were it not for the fact that foodstuffs, such as bread, cake, rolls, and many other foods, usually require multiple systems of emulsifiers. Thus the overall complete analysis of emulsifiers in fats and oils or foods is a difficult task. Recent attempts have been made to systematize the analysis of emulsifier combinations in fats and oils.

Dieffenbacher and Bracco (309) have suggested a comprehensive approach to the separation of emulsifier mixtures. Their overall method consists of three steps: (*a*) extraction of total lipids from food with a mixture of chloroform-methanol, (*b*) removal of nonpolar lipids from the lipid extract by column chromatography with an elution mixture based on ethyl ether–petroleum ether mixtures, and (*c*) TLC of the polar lipids extracted from the foods. The technique offers promise in the simplification of a difficult separation problem.

The analysis of common emulsifier mixtures, like mono- and diglycerides, is usually accomplished by α-monoglyceride analysis with the periodic acid method (AOCS Cd 11-57). In the absence of triglycerides, diglycerides are assigned by difference. Riisom and Hoffmeyer (310) have developed an HPLC method for the direct analysis of mono- and diglycerides involving the direct assay of each on a 25-cm, 10-mμ LiChrosorb DIOL-packed column. The acylglycerols are eluted with isooctane–isopropanol (95:5) within 10 min, and the components monitored by absorption at 213 nm. The method shows excellent reproducibility and accuracy with standard deviations of a distilled monoglyceride containing 92.1% monoglyceride and 6.5% diglycerides of 0.70 and 0.42%, respectively. The method is applicable to other types of emulsifier, for instance, acetylated monoglycerides and propylene glycol esters of fatty acids.

The AOAC lists a method (311) for the chromatographic separation of glycerides in monoglycerides. Dissolved in chloroform, the sample is chromatographed through a silica gel column; the triglycerides are eluted with benzene, diglycerides and free fatty acids with 90% benzene–10% ether, and monoglycerides with ether.

Sequestrants. The sequestrants that are commonly used in fats and oils include citric acid and ethylendiamine tetraacetic acid (EDTA) and its sodium salts. They have sequestering action in combining with troublesome trace metals, particularly iron, nickel, and cobalt, the autoxidation catalysts. Citric acid in soybean (and presumably other vegetable oils) can be determined (312) by extracting the oil with water, esterifying with butanol, and subjecting to GLC on a 10% DEGS Chromosorb WAW temperature-programmed column with a flame-ionization detector. Similarly, citraconic, itaconic, and aconitic acids can be determined. Ethylendiamine tetraacetic acid is determined (313) by titration with copper sulfate reagent using 1-(2-pyridylazo)-2-naphthol (PAN) indicator. The method is used for soaps and detergents.

Soap Additives. The presence of trichlorocarbanilide and Triclosan in deodorant soaps may be determined by a rapid and direct analysis (314) using reverse-phase high-performance liquid chromatography (HPLC). The method compares quite favorably with the colorimetric (315) and ultraviolet absorption methods (316) that are also used.

Miscellaneous. Sodium borohydride has been used in the fats and oils industry to reduce impurities that impart color, odor, or instability. As little as 50–300 ppm doses are generally quite effective. A simple direct titration method has been developed by Rudie and Demko (317) for the

Table 7.12 Procedures for the detection of small or trace quantities of components that are ordinarily assumed to have little or no effect on the quality or performance of finished fat and oil products

Component	For Detection in	Method	Reference
Branched-chain, odd-numbered, or unusual unsaturated fatty acids in fat and oil triglycerides	Butterfat	Urea inclusion and GLC	321
Furanoid fatty acids	Cod liver oil	Silver ion chromatography and urea inclusion	26
Sterols	Vegetable oils	TLC (Wakogel B); GLC on unsaponiables	23, 37
	Rapeseed, sunflower, and poppyseed oils	Saponification, TLC, and GLC (silylated derivatives)	318
4-Methylsterols	Vegetable oils	TLC (silica gel or Wakogel B); GLC on unsaponifiables	24, 37
Triterpene alcohols	Vegetable oils	TLC (Wakogel B); GLC on unsaponifiables	22–24, 37
Natural hydrocarbons	Olive oil	Column chromatography (Al_2O_3); TLC on unsaponifiables	37, 54
	Marine oils	Column chromatography; TLC on unsaponifiables, silical gel plus GLC	37, 319
	Butterfat	Saponification, TLC (either Al_2O_3 or silica gel), and GLC (either Chromosorb Q or W)	320

analysis of sodium borohydride in organic media at 1–2000-ppm levels. A purple solution of crystal violet in dimethylformamide is used to titrate the sodium borohydride in dimethylformamide to a purple endpoint.

2.5 TECHNIQUES OF ANALYSIS THAT CAN BE APPLIED TO THE DETERMINATION OF SMALL OR TRACE QUANTITIES OF COMPONENTS THAT ARE ORDINARILY ASSUMED TO HAVE LITTLE OR NO EFFECT ON THE QUALITY OR PERFORMANCE OF THE FINISHED PRODUCTS

Table 7.12 summarizes several components that meet this requirement.

References

1. *Official and Tentative Methods of the American Oil Chemists' Society,* 3rd ed., Including Additions and Revisions 1979, American Oil Chemists' Society, 508 South Sixth Street, Champaign, IL.

2. *Annual Book of ASTM Standards,* Part 20 (Paint, Varnish, Lacquer and Related Products, etc.), April 1980; Part 22, American Society for Testing and Materials, 196 Race Street, Philadelphia, PA.

3. *Official Methods of Analysis,* 13th ed., 1980, Association of Official Analytical Chemists, 1111 North 19th Street, Suite 210, Arlington, VA.

4. Reference 1; Methods Ca 13–56, Ba 7–58 and Ba 8–55.

5. Reference 1; Method Cb 1–25; A. V. Bailey, R. A. Pittman, C. F. Magne, and E. L. Skau, *J. Am. Oil Chemists' Soc.,* **42,** 422–424 (1965); A. V. Bailey, J. A. Harris, E. L. Skau, and T. Kerr, *J. Am. Oil Chemists' Soc.,* **43,** 107–110 (1966).

6. Reference 1; Method Ca 16–75.

7. J. Boldingh and R. J. Taylor, *Nature,* **194,** 909 (1962); J. E. Kinsella, S. Patton, and P. S. Dimick, *J. Am. Oil Chemists' Soc.,* **44,** 449–454 (1967); R. C. Lawrence, *J. Dairy Res.,* **30,** 161 (1963); O. W. Parks, M. Keeney, L. Katz, and D. P. Schwartz, *J. Lipid Res.,* **5,** 232 (1964); B. Van Der Ven, P. H. Begeman, and J. C. Schodt, *J. Lipid Res.,* **4,** 91 (1963); J. E. Langler and E. A. Day, *J. Dairy Sci.,* **47,** 1291 (1964).

8. D. Firestone, J. Ress, N. L. Brown, R. P. Barron, and J. N. Damico, *J. Assocn. Anal. Chem.,* **55,** 85 (1972).

9. K. Sargeant, A. Sheridan, J. O'Kelly, and R. B. A. Carnaghan, *Nature,* **192,** 1096–1097 (1961); W. A. Pons, Chairman, *Aflatoxins in Oilseeds-Problems and Solutions,* 67th AOCS Meeting, New Orleans, LA, April 21–24, 1976; L. A. Goldblatt, *J. Am. Oil Chemists' Soc.,* **54,** 302A–309A (1977); O. L. Shotwell, Chairman, *Mycotoxins I and II,* 69th AOCS Meeting, St. Louis, MO, May 17, 1978.

10. Anon., *Chem. Eng. News,* **58,** (13) 20–27 (March 31, 1980).

10a. D. H. Fine and D. P. Rounbehler, in *Environmental N-Nitroso Compounds—Analysis and Formation,* E. A. Walker, E. Pogovski, and L. Griciute, Eds., Lyon International Agency for Research on Cancer, IARC Scientific Publication No. 14, 1976, p. 117.

11. L. Hedler, C. Schurr, and P. Marquardt, *J. Am. Oil Chemists' Soc.,* **56,** 681–684 (1979).

12. Anon., *Chem. Eng. News,* p. 6 (October 1, 1979).

13. J. E. McGhee, L. D. Kirk, and G. C. Mustakes, *J. Am. Oil Chemists' Soc.,* **42,** 889–891 (1965).

14. G. C. Mustakes, L. D. Kirk, V. E. Sohns, and E. L. Griffen, Jr., *J. Am. Oil Chemists' Soc.,* **42,** 33–37 (1965).

14a. T. J. Birkel, C. R. Warner, and T. Fazio, *J. Assoc. Off. Anal. Chem.,* **62,** 931–936 (1979).

15. W. Lange, *J. Am. Oil Chemists' Soc.,* **27,** 414–422 (1950).

15a. P. W. Meijboom and G. A. Jongenotter, *J. Am. Oil Chemists' Soc.,* **56,** 33–35 (1979).

16. Reference 3; Methods 43.001 and 43.008.

17. T. Tsuchiya, A. Kato, and Y. Endo, *Rep. Gov. Chem. Ind. Res. Inst. Tokyo,* **51,** 359 (1965); T. Tsuchiya and A. Kato, *Tokyo Kogyo Shikensho Hokoku,* **53,** 230 (1959); A. Tanaka, A. Kato, and T. Tsuchiya, *Yukagaku,* **13,** 260 (1964); A. Tanaka, A. Kato, and T. Tsuchiya, *J. Am. Oil Chemists' Soc.,* **48,** 95–97 (1971).

18. Reference 1; Method Cb 1–25; M. Beroza, *Anal. Chem.,* **26,** 1173 (1954); M. C. Bowman and M. Beroza, *Residue Rev.,* **17,** 1 (1967).

19. T. Ramura and T. Matsumoto, *Yukagaku,* **10,** 625 (1961).

20. W. G. Doeden, E. M. Kushibab, and A. C. Ingala, *J. Am. Oil Chemists' Soc.,* **57,** 73–74 (1980).

21. W. G. Doeden, R. H. Bowers, and A. G. Ingala, *J. Am. Oil Chemists' Soc.,* **56,** 12–14 (1979).

22. E. Fedeli, A. Łanzani, P. Capella, and G. Jacini, *J. Am. Oil Chemists' Soc.,* **43,** 254–256 (1966).

23. T. Itoh, T. Tamura, and T. Matsumoto, *J. Am. Oil Chemists' Soc.,* **50,** 122–125 (1973).

24. C. Urakam, S. Oka, and J. S. Han, *J. Am. Oil Chemists' Soc.,* **53,** 525–529 (1976).

25. P. Capella, E. Fedeli, M. Cirimele, and G. Jacini, *Riv. Ital. Sost. Grasse,* **40,** 603 (1963).

26. C. M. Scrimgeour, *J. Am. Oil Chemists' Soc.,* **54,** 210–211 (1977).

27. Reference 1; Method M 1–59.

28. Reference 1; Method M 2–65.

28a. Reference 2; Method D 2245.

28b. W. J. Youden and E. H. Steiner, *Statistical Manual of the AOAC,* Association of Official Analytical Chemists, 1975.

29. Reference 1; Method Cd 14–65.

29a. J. C. Kuck and A. J. St. Angelo, *J. Am. Oil Chemists' Soc.,* **57,** 128–129 (1980).

29b. V. T. Srinivasan, *J. Am. Oil Chemists' Soc.,* **56,** 1000–1003 (1979).

29c. J. A. Robertson and W. H. Morrison III, *J. Am. Oil Chemists' Soc.,* **56,** 961–964 (1979).

30. M. W. Sperry and M. Webb, *J. Biol. Chem.,* **187,** 97–106 (1950).

31. Reference 1; Method Cd 5–10.

31a. A. Emmerie and C. Engel, *Rec. Trav. Chim. Pays-Bas.,* **57,** 1351–1355 (1938).

31b. M. L. Quaife and R. Biehler, *J. Biol. Chem.,* **159,** 663–665 (1945).

31c. M. L. Quaife and P. L. Harris, *Ind. Eng. Chem., Anal. Ed.,* **18,** 707–708 (1946).

31d. M. L. Quaife, *J. Am. Chem. Soc.*, **66**, 308–309 (1944).

31e. L. Weisler, C. D. Roberson, and J. G. Baxter, *Anal. Chem.*, **19**, 906–909 (1947).

32a. J. Boldingh and R. J. Taylor, *Nature*, **144**, 909 (1962).

32b. B. W. Tharp and S. Patton, *J. Dairy Sci.*, **43**, 475 (1960).

32c. J. E. Kinsella, S. Patton, and P. S. Dimick, *J. Am. Oil Chemists' Soc.*, **44**, 202–205 (1967).

32d. G. Jurriens and J. M. Oele, *J. Am. Oil Chemists' Soc.*, **42**, 857–861 (1965).

32e. S. Patton, *J. Agr. Food Chem.*, **6**, 132 (1958).

32f. A. I. Virtenen, *Science*, **153**, 1603 (1958).

32g. R. Ellis and N. P. Wong, *J. Am. Oil Chemists' Soc.*, **52**, 252–255 (1197).

33. Reference 3; Methods 43.001 (UV spectrophotometric) and 43.008 (colorimetric).

34. Reference 1; Method Cb 5–40.

35. P. Colborn, *J. Am. Oil Chemists' Soc.*, **46**, 385–386 (1969).

36. A. Linder and V. Persson, *Svensk Papperstid.*, **52**, 331 (1949); *J. Am. Oil Chemists' Soc.*, **34**, 24–27 (1957).

37. A. Hase, T. Hase, and R. Anderegg, *J. Am. Oil Chemists' Soc.*, **55**, 407–411 (1978).

38. W. E. Hillis and N. Ishikura, *J. Chromatogr.*, **32**, 323–336 (1968).

39. L. N. Norcia and R. E. Rosenthal, *J. Am. Oil Chemists' Soc.*, **43**, 168–170 (1966).

39a. Method 2.403 (Identification and Determination of Sterols by GLC), IUPAC Commission on Oils, Fats and Derivatives, in *Standard Methods for the Analysis of Oils, Fats and Derivatives*, 6th ed., C. Paquot, Ed., Pergamon Press, New York, 1979; also, AOAC Method 28.099–28.106 (1980).

40. W. Muller-Mulot, *J. Am. Oil Chemists' Soc.*, **53**, 732–736 (1976).

41. R. R. Allen, *Chem Ind. (London)*, **1965**, 1560.

42. R. L. Willee and E. S. Lutton, *J. Am. Oil Chemists' Soc.*, **43**, 491–496 (1966).

43. G. M. Chapman, E. E. Akehurst, and W. B. Wright, *J. Am. Oil Chemists' Soc.*, **48**, 824–830 (1971).

44. E. S. Lutton, *J. Am. Chem. Soc.*, **68**, 676–679 (1946).

45. R. D. Plattner, G. F. Spencer, and R. Kleiman, *J. Am. Oil Chemists' Soc.*, **55**, 381–382 (1978).

46. M. H. Coleman, *J. Am. Oil Chemists' Soc.*, **42**, 751–754 (1965).

47. Reference 1; Method Ca 13–56.

48. Reference 1; Method Cb 1–25.

49. A. V. Bailey, R. A. Pittman, F. C. Magne, and E. L. Skau, *J. Am. Oil Chemists' Soc.*, **42**, 422–424 (1965).

50. A. V. Bailey, J. A. Harris, R. L. Skau, and T. Kerr, *J. Am. Oil Chemists' Soc.*, **43**, 107–110 (1966).

51. H. P. Kaufmann, *Fette, Seifen, Anstrichm.* **48**, 51–59 (1941).

52. H. T. Slover, J. Lehmann, and R. J. Valis, *J. Am. Oil Chemists' Soc.*, **46**, 417–420 (1969).

53. H. E. Pattee and A. E. Purcell, *J. Am. Oil Chemists' Soc.*, **44**, 328–331 (1967).

54. K. Taufel, H. Heinisch, and W. Heimann, *Biochem. Z.*, **303**, 324–328 (1940); J. Fitelson, *J. Assoc. Off. Agr. Chem.*, **26**, 499–510 (1943); *J. Assoc. Off. Agr. Chem.*, **28**, 282–289 (1945).

55. Reference 1; Method Cb 4–35.

55a. M. R. Sahasrabudhe, *J. Am. Oil Chemists' Soc.*, **54**, 323–324 (1977).

55b. A. Minguzzi and P. Capella, *Scienza Tec. Almenti*, **2**, 155 (1974); also E. Fedeli in *Progress in the Chemistry of Fats and Other Lipids*, **15**, 63 (1977), Pergamon Press, New York, 1977.

56. D. S. Ingraham, B. A. Knights, I. McEvoy, and P. McKay, *Phytochemistry*, **7**, 1241–1245 (1968).

57. M. H. Coleman, *J. Am. Oil Chemists' Soc.*, **40**, 568–571 (1963); F. E. Luddy, R. A. Barford, S. F. Herb, P. Magidman, and R. W. Riemenschneider, *J. Am. Oil Chemists' Soc.*, **41**, 693–696 (1962).

58. G. Losi and A. Piretti, *Riv. Ital. Sost. Grasse*, **51**, 129 (1974).

59. M. Kofler, *Helv. Chim. Acta*, **28**, 26 (1945); ibid., **30**, 1053 (1947).

60. Reference 3; Method 43.014.

61. R. P. Hansen and F. B. Shorland, *Biochem. J.*, **50**, 207–210 (1951).

62. Reference 1; Method Cb 2–40.

63. M. C. Bowman and M. Beroza, *Residue Rev.*, **17**, 1 (1967).

64. H. Niewiadomski, I. Bratkowska, and E. Mossakowska, *J. Am. Oil Chemists' Soc.*, **42**, 731–734 (1965).

65. T. Tamura, N. Sakaedani, and T. Matsumoto, *Nippon Kagaku Zasshi*, **79**, 1110 (1958).

66. Reference 1; Method Cd 12–57.

67. U. Riiner and R. Ohlson, *J. Am. Oil Chemists' Soc.*, **48**, 860–862 (1971).

68. Reference 1; Method Cb 3–39.

69. T. Tsuchiya, A. Kato, and T. Endo, *Rep. Gov. Chem. Ind. Res. Inst. Tokyo*, **51**, 359 (1965); T. Tsuchiya and A. Kato, *Tokyo Kogyo Shikensho Hokoku*, **53**, 230 (1958).

70. A. Tanaka, A. Kato, and T. Tsuchiya, *Yukagaku*, **13**, 260 (1964).

71. A. Tanaka, A. Kato, and T. Tsuchiya, *J. Am. Oil Chemists' Soc.*, **48**, 95–97 (1971).

72. D. G. Dorrell, *J. Am. Oil Chemists' Soc.*, **48**, 693–696 (1971).

73. Reference 1; Methods Ca 8a–35 and Ca 8b–35.

74. P. Capella and G. Losi, *Ind. Agr.*, **6**, 277–282 (1968).

75. U. Riiner, *J. Am. Oil Chemists' Soc.*, **47**, 129–133 (1970).

76. F. H. Mattson and R. A. Volpenheim, *J. Biol. Chem.*, **236**, 1891 (1961).

77. L. Zeleny and D. A. Coleman, *Oil Soap*, **15**, 253–256 (1936).

78. B. Van Der Ven and K. De Jong, *J. Am. Oil Chemists' Soc.*, **47**, 299–302 (1979).

79. Reference 1; Methods Ca 8a–35 and Ca 8b–35.

80. F. A. Norris and K. F. Mattil, *J. Am. Oil Chemists' Soc.*, **24**, 274–275 (1947).

81. R. W. Eckey (to Procter & Gamble Co.), U.S. Pat. 2,442,531 (June 1, 1948).

82. Reference 1; Method Ce 3–74.

83. Reference 2; Method D 1953.

84. H. T. Slover, L. M. Shelley, and T. L. Burks, *J. Am. Oil Chemists' Soc.*, **44**, 161–166 (1967).

85. T. Tamura, T. Hibino, K. Yokoyama, and T. Matsumoto, *Nippon Kagaku Zasshi*, **80**, 215 (1959).

86. The Browne and Worstall tests are empirical evaluations involving heating that have

been used to assess the overall drying and polymerization qualities of tung oil. ASTM Methods D20, D555, and D1964 suffice for this purpose.

87. Reference 1; Method Cd 4–40.

87a. Reference 2; Method D 1955.

87b. Reference 2; Method D 1953.

88. L. J. Morris, H. Hayes, and R. T. Holman, *J. Am. Oil Chemists' Soc.*, **38**, 316–321 (1961).

89. J. A. Kneeland, D. Kyriacou, and R. H. Purdy, *J. Am. Oil Chemists' Soc.*, **35**, 361–363 (1958).

90. J. P. Varma, S. Dasgupta, B. Nath, and J. S. Aggarwal, *J. Am. Oil Chemists' Soc.*, **34**, 452–454 (1957); D. A. Sutton, *J. Am. Oil Chemists' Soc.*, **32**, 16–22 (1955); J. P. Varma, B. Nath, and J. S. Aggarwal, *Nature*, **175**, 84–85 (1955); *J. Chem. Soc.*, **1956**, 2550; P. K. Faure and J. C. Smith, *J. Chem. Soc.*, **1956**, 1818–1821; D. G. Brooke and J. C. Smith, *Chem. Ind. (London)*, **1967**, 49–50; B. A. Lewis and R. A. Raphael, *Chem. Ind. (London)*, **1957**, 50.

91. K. L. Mikolajczik, C. R. Smith, Jr., and I. A. Wolff, *J. Am. Oil Chemists' Soc.*, **42**, 939–941 (1965).

92. L. Crombie and J. L. Taylor, *J. Chem. Soc.*, **1954**, 2816–2819; S. C. Gupta, S. S. Gupta, and J. S. Aggarwal, *J. Am. Oil Chemists' Soc.*, **31**, 287–289 (1954); J. D. von Mikusch, *Deutsch Farben-Z.*, **8**, 166–169 (1954); S. Das Gupta and J. S. Aggarwal, *J. Am. Oil Chemists' Soc.*, **32**, 501–503 (1955); M. M. Chakrabarty and S. Bhattacharyya, *Naturwissenschaften*, **44**, 91 (1957).

93. A. Steger and J. van Loon, *Rec. Trav. Chim.*, **52**, 593–600 (1933).

94. F. D. Gunstone and W. C. Russell, *J. Chem. Soc.*, **1954**, 1112.

95. M. J. Chisholm and C. Y. Hopkins, *Can. J. Chem.*, **35**, 358–364 (1957).

96. C. F. Krewson, *J. Am. Oil Chemists' Soc.*, **45**, 250–256 (1968).

97. C. Y. Hopkins and M. J. Chisholm, *J. Am. Oil Chemists' Soc.*, **37**, 682–684 (1960).

98. C. F. Krewson and W. E. Scott, *J. Am. Oil Chemists' Soc.*, **43**, 171–174 (1966).

99. R. L. Shriner and R. Adams, *J. Am. Chem. Soc.*, **47**, 2727–2739 (1925).

100. E. W. Eckey, *Vegetable Fats and Oils*, Reinhold, New York, 1954, p. 694.

101. E. Twitchell, *Ind. Eng. Chem.*, **13**, 806–807 (1921).

102. B. L. Madison and R. C. Hill, *J. Am. Oil Chemists' Soc.*, **55**, 328–331 (1978).

102a. H. B. Stevenson, *Oil Soap*, **13**, 18–20 (1936).

103. K. S. Konigsbacher, W. F. Danker, and R. L. Evans, *Drug Cosmetic Ind.*, **70**, 332, 406 (1952).

103a. A. von Hübl, *Dinglers Polytech J.*, **253**, 281–295 (1884); *J. Soc. Chem. Ind. (London)*, **3**, 641 (1884).

104. Reference 3; Method 28.018.

105. J. C. Nevenzel and D. R. Howton, *J. Org. Chem.*, **22**, 319–321 (1957).

106. G. Ponzio and C. Gastaldi, *Gazz. Chim. Ital.*, **42**(II), 92–95 (1912).

107. D. J. Hiscox, *Anal. Chem.*, **20**, 679–680 (1948).

108. J. E. Hashemy-Tonkabony, *J. Am. Oil Chemists' Soc.*, **54**, 233 (1977).

109. N. O. V. Sonntag, in *Fatty Acids*, 2nd ed., K. S. Markley, Ed., Part 2, Interscience, New York, 1961, pp. 1074–1109.

110. Delmar Scientific Laboratories, 317 Madison St., Maywood, IL; H. C. Brown, K. Sivasankaran, and C. A. Brown, *J. Org. Chem.*, **28**, 214 (1963).

111. A. Polgar and J. L. Jungnickel, in *Organic Analysis*, Vol. 3, Wiley, New York, 1956, pp. 256–265.

112. G. M. Chapman, *J. Am. Oil Chemists' Soc.*, **56**, 77–79 (1979).

113. H. P. Kaufmann and E. Schmulling, *Fette, Seifen, Anstrichsm.*, **64**, 319 (1962).

114. B. D. Sully, *Analyst*, **87**, 940 (1962).

115. W. T. Robinson, R. H. Cundiff, and P. C. Markunus, *Anal. Chem.*, **33**, 1030 (1961).

116. J. S. Fritz and G. S. Schenk, *Anal. Chem.*, **31**, 1808 (1959).

117. C. H. Lea, *J. Soc. Chem. Ind.*, **65**, 286–291 (1946).

118. A. Taffel and C. Revis, *J. Soc. Chem. Ind.*, **50**, 87–91T (1931).

119. W. O. Lundberg, Ed., *Autoxidation and Antioxidants*, Interscience, New York, 1961.

120. D. Swern, in *Fatty Acids*, 2nd ed., K. S. Markley, Ed., Part 2, Interscience, New York, 1961, pp. 1387–1443.

120a. R. D. Wood, R. K. Raju and R. Reiser, *J. Am. Oil Chemists' Soc.*, **42**, 161–165 (1965).

121. J. B. Bedale, D. E. Just, R. E. Morgan, and R. A. Reiners, *J. Am. Oil Chemists' Soc.*, **42**, 90–95 (1965).

122. H. Zmachinski, A. Waltking, and J. D. Miller, *J. Am. Oil Chemists' Soc.*, **43**, 425–428 (1966).

123. A. E. Waltking, *Nutr. Rep. Int.*, **5**, 17 (1972).

124. R. B. Alfin-Slater and D. Melnick, *J. Am. Oil Chemists' Soc.*, **41**, 145–150 (1964).

125. W. G. Bickford, P. Krauczunas, and D. H. Wheeler, *Oil Soap*, **19**, 23–27 (1942).

125a. G. F. Spencer, W. F. Kwolek, and L. H. Princen, *J. Am. Oil Chemists' Soc.*, **56**, 972–973 (1979).

126. R. G. Ackman, *Progress in the Chemistry of Fats and Other Lipids*, Vol. 12, R. T. Holman, Ed., Pergamon Press, New York, 1972, pp. 185–284.

127. G. R. Jamieson, in F. D. Gunstone, Ed., *Topics in Lipid Chemistry*, Vol. 1, Wiley-Interscience, New York, 1972, pp. 107–159.

128. W. W. Christie, in *Topics in Lipid Chemistry*, F. D. Gunstone, Ed., Vol. 3, Wiley Interscience, New York, 1972, pp. 171–197.

129. N. Pelick and V. Mahadevan, in *Analysis of Lipids and Lipoproteins*, E. G. Perkins, Ed., American Oil Chemists' Society, Champaign, IL, 1975, pp. 36–62.

130. A. Kuksis, *Sep. Purif. Meth.*, **6**, 353 (1977).

131. W. R. Morrison and L. M. Smith, *J. Lipid Res.*, **5**, 600 (1964).

132. L. D. Metcalfe, A. A. Schmitz, and J. R. Pelka, *Rapid Esterification of Lipids for Gas Chromatography Analysis*, Armour Industrial Chemical Company, Chicago, 1965.

133. N. B. Lorette and J. H. Brown, *J. Org. Chem.*, **24**, 261 (1959).

134. G. R. Jamieson and E. H. Reid, *J. Chromatogr.*, **17**, 230 (1965).

135. H. T. Slover and E. Lanza, *J. Am. Oil Chemists' Soc.*, **56**, 933–943 (1979).

136. J. MacGee and K. G. Allen, *J. Am. Oil Chemists' Soc.*, **54**, 375–379 (1977).

137. R. F. Paschke and D. H. Wheeler, *Oil Soap*, **21**, 52–57 (1944).

138. A. E. King, H. L. Roschen, and W. H. Irwin, *Oil Soap*, **10**, 204–207 (1933); also Reference 1, Method Cd 12–57.

139. N. T. Joyner and J. E. McIntyre, *Oil Soap*, **15**, 184–186 (1938).

140. H. S. Olcott and E. Einset, *J. Am. Oil Chemists' Soc.*, **35**, 161–162 (1958).

141. J. E. Bennett and M. J. Byer, *J. Am. Oil Chemists' Soc.*, **41**, 505–507 (1964).

142. R. W. Beasley and R. L. Sampson, *Modification of the Oxygen Bomb Stability Technique*, presented at Cereal Chemists' Meeting, May 4, 1960.

143. G. J. Marco, *J. Am. Oil Chemists' Soc.*, **45**, 594–598 (1968).

144. H. E. Miller, *J. Am. Oil Chemists' Soc.*, **48**, 91 (1971).

145. A. Banks, *J. Soc. Chem. Ind.*, **63**, 8–13 (1944).

146. D. L. Berner, J. A. Conte, and G. A. Jacobson, *J. Am. Oil Chemists' Soc.*, **51**, 292–296 (1974).

147. M. F. Pool and A. N. Prater, *Oil Soap*, **22**, 215–216 (1845).

148. F. Bernheim, M. L. C. Bernheim, and K. M. Wilbur, *J. Biol. Chem.*, **174**, 257–264 (1948); K. M. Wilbur, F. Bernheim, and O. W. Shapiro, *Arch. Biochem.*, **24**, 305–313 (1949); H. Schmidt, *Fette, Seifen, Anstrichsm.*, **61**, 127–133 (1959).

149. T. C. Yu and R. O. Sinnhuber, *J. Am. Oil Chemists' Soc.*, **41**, 540–543 (1964).

150. T. C. Yu and R. O. Sinnhuber, *J. Am. Oil Chemists' Soc.*, **44**, 256–258 (1967).

151. R. Marcuse and L. Johansson, *J. Am. Oil Chemists' Soc.*, **50**, 387–391 (1973).

152. S. S. Arya, S. Ramanujam, and P. K. Vijayaraghavan, *J. Am. Oil Chemists' Soc.*, **46**, 28–30 (1969).

153. P. K. Jarvis, G. D. Lee, D. R. Eeickson, and E. A. Butkus, *J. Am. Oil Chemists' Soc.*, **48**, 121–124 (1974).

154. W. D. Pohle, R. L. Gregory, and B. Van Giessen, *J. Am. Oil Chemists' Soc.*, **41**, 649–650 (1964).

155. W. S. Singleton, in *Fatty Acids*, K. S. Markley, Ed., Part 1, Wiley-Interscience, New York, 1960, pp. 532–544.

156. E. H. Pryde, Ed., *Fatty Acids*, The American Oil Chemists' Society, Champaign, IL, 1979, pp. 173–176.

157. K. S. Singleton, in K. S. Markley, Ed., *Fatty Acids*, Part 1, Wiley-Interscience, New York, 1960, pp. 582–592.

158. R. T. Hall and W. E. Shaefer, in *Organic Analysis*, Vol. 2, Wiley-Interscience, New York, 1954, pp. 19–70.

159. G. Maerker, *J. Am. Oil Chemists' Soc.*, **42**, 329–332 (1965).

160. N. I. Sax, *Dangerous Properties of Industrial Materials*, 2nd ed., Reinhold, New York, 1963, p. 1202.

161. L. S. Silbert, *J. Am. Oil Chemists' Soc.*, **39**, 480–487 (1962).

162. D. Swern and L. S. Silbert, *Anal. Chem.*, **35**, 880 (1963).

163. E. W. Eckey, *Oil Soap*, **23**, 38 (1946).

164. P. H. Gill and A. H. Gill, *Ind. Eng. Chem., Anal. Ed.*, **12**, 94 (1940).

165. L. D. Metcalfe, *Anal. Chem.*, **34**, 1849 (1962).

166. Unpublished method, courtesy L. D. Metcalfe and A. A. Schmitz, Armak Company, McCook, IL.

167. J. R. Pelka and L. D. Metcalfe, *Anal. Chem.*, **37**, 603–604 (1965).

167a. L. D. Metcalfe, *J. Am. Oil Chemists' Soc.*, **56**, 786A–789A (1979).

167b. L. D. Metcalfe, *J. Am. Oil Chemists' Soc.*, **56**, 819A–822A (1979).

167c. D. O. Hummel, *Identification and Analysis of Surface Active Agents*, Interscience, New York, 1962, 386 pp.

167d. H. A. Boekenoogen, *Analysis and Characterization of Oils, Fats and Fat Products*, Interscience, New York, 1964, 421 pp.

167e. E. G. Perkins, Ed., *Analysis of Lipids and Lipoproteins*, The American Oil Chemists' Society, Champaign, IL, 1975, 299 pp.

167f. M. S. Balsam and E. Sagarin, Eds., *Cosmetic Science and Technology*, 2nd ed., Wiley-Interscience, New York, Vol. 1, 1972; Vol. 2, 1972; Vol. 3, 1974.

167g. M. J. Rosen and H. A. Goldsmith, *Systematic Analysis of Surface-Active Agents*, 2nd ed., Wiley-Interscience, New York 1972.

167h. V. C. Mehlenbacher, *The Analysis of Fats and Oils*, Garrard Press, Champaign, Il., 1960.

168. H. P. Kaufmann and G. Mankel, *Fette, Seifen Amstrichsm.*, **65**, 179 (1963).

169. Reference 3; Method 28.113.

170. Reference 3; Method 28.114.

171. Reference 3; Method 28.121.

172. Reference 3; Method 28.099–28.106.

173. Reference 3; Method 28.085.

174. Reference 3; Method 28.117; Reference 1, Method Cb 3–39.

174a. C. Imai, H. Watanabe, N. Haga and T. Ii, *J. Am. Oil Chemists' Soc.*, **51**, 326–330 (1974).

175. R. D. Plattner, K. Wade, and R. Kleiman, *J. Am. Oil Chemists' Soc.*, **55**, 381–382 (1978).

176. Reference 1; Method Cb 5–40. Reference 3; Method 28.119.

176a. Method 2.210, IUPAC Commission on Oils, Fats, and Derivatives, in C. Paquot, Ed., *Standard Methods for Analysis of Oils, Fats and Derivatives*, 6th ed., Pergamon, New York, 1979; L. El Sayed and A. El Dashlauty, *Rev. Ital. Delle Sost. Grasse*, **56**, 52(1979).

177. A. W. Ralston and C. W. Hoerr, *J. Org. Chem.*, **7**, 546 (1942).

178. D. N. Eggenberger, F. K. Broome, A. W. Ralston, and H. J. Harwood, *J. Org. Chem.*, **14**, 1108 (1949).

179. Reference 2; Method E 203–64.

180. S. T. Bauer, R. E. Price, and P. R. Gill, *J. Am. Oil Chemists' Soc.*, **35**, 120–121 (1958).

181. W. G. Mertens, C. E. Swindells, and B. F. Teasdale, *J. Am. Oil Chemists' Soc.*, **48**, 544–546 (1971).

182. G. Lunde, *J. Am. Oil Chemists' Soc.*, **48**, 517–522 (1971).

183. J. M. Lupton, J. Mitchell, A. W. Oemler, and L. B. Woolaver, *J. Am. Oil Chemists' Soc.*, **25**, 216–218 (1948).

184. G. F. Palfrey, R. H. Hobert, A. F. Benning, and I. W. Dobratz, *Ind. Eng. Chem., Anal. Ed.*, **12**, 94–96 (1940).

185. R. T. O'Connor, D. C. Heinzelman, and M. E. Jefferson, *J. Am. Oil Chemists' Soc.*, **24**, 185–189 (1947).

186. E. H. Melvin and J. E. Hawley, *J. Am. Oil Chemists' Soc.*, **28**, 346–347 (1951).

187. D. Hoggan, C. E. Marquart, and W. R. Battles, *Anal. Chem.*, **36**, 1955 (1964).

188. D. L. Nash, *Appl. Spectrosc.*, **19** (3), 89 (1965).

189. L. J. Meyers, in *Spectrographer's Newsletter,* Applied Research Laboratories, Inc., Glendale, CA, Vol. 18, No. 4, 1965.

190. F. Tera, R. R. Ruch, and G. H. Morrison, *Anal. Chem.*, **37**, 358 (1965).

191. R. R. Ruch, F. Tern, and G. H. Morrison, *Anal. Chem.*, **37**, 1565 (1965).

192. J. T. Olejko, *J. Am. Oil Chemists' Soc.*, **53**, 480–484 (1976).

193. W.-C. Tsai, C.-P. Lin, L.-J. Shiau and S.-D. Pan, *J. Am. Oil Chemists' Soc.*, **55**, 695–698 (1978).

194. W. H. Morrison III and J. A. Robertson, *J. Am. Oil Chemists' Soc.*, **52**, 148–150 (1975).

195. U. I. Brimberg and I. C. Wretensjö, *J. Am. Oil Chemists' Soc.*, **56**, 857–860 (1979).

196. T. von Fellenberg, *Mitt. Lebens. Hyg.*, **36**, 355–359 (1945).

197. J. E. McGhee, L. D. Kirk, and G. C. Mustakas, *J. Am. Oil Chemists' Soc.*, **42**, 889–891 (1965).

198. G. C. Mustakas, L. D. Kirk, V. E. Sohns, and E. L. Griffen, *J. Am. Oil Chemists' Soc.*, **43**, 33–37 (1965).

199. H. P. Kaufmann and M. Vogelmann, *Farbchemiker,* **61**, 6–10 (1959).

200. K. Kurita, *Abura Kagaku,* **5**, 347–349 (1956).

201. H. Niewiedomski, L. Bratkowska, and E. Mossakowska, *J. Am. Oil Chemists' Soc.*, **42**, 731–734 (1965).

202. L. A. Appelqvist, *J. Am. Oil Chemists' Soc.*, **48**, 851–859 (1971).

203. A. F. Wicke, H. E. McLaughlin, and J. H. Stump, Jr. (to Tenneco Chemicals, Inc.), U.S. Pat. 3,257,438 (June 21, 1966).

203a. R. P. Scharrer and M. Epstein, Paper no. 487,ISF-AOCS World Congress, New York, May 1, 1980.

204. W. W. Simons, ed., *The Sadtler Handbook of Ultraviolet Spectra,* Sadtler Research Laboratories, Philadelphia, Pa., 1979, Spectrum 159 (*trans*–stilbene).

205. P. V. Johnston, O. C. Johnson, and F. A. Kummerow, *J. Nutr.*, **65**, 13 (1958).

206. H. B. S. Conacher, *J. Chromatogr. Sci.*, **14**, 405 (1976).

207. L. M. Smith, W. L. Dunkley, A. Franke, and T. Dairiki, *J. Am. Oil Chemists' Soc.*, **55**, 257–261 (1978).

208. E. G. Perkins, T. P. McCarthy, M. A. O'Brien, and F. A. Kummerow, *J. Am. Oil Chemists' Soc.*, **54**, 279–281 (1977).

209. C. R. Scholfield and T. L. Mounts, *J. Am. Oil Chemists' Soc.*, **54**, 319–321 (1977).

210. P. E. Pfeffer, F. E. Luddy, and J. Unruh, *J. Am. Oil Chemists' Soc.*, **54**, 380–386 (1977).

211. W. D. Pohle and V. C. Mehlenbacher, *Oil Soap,* **23**, 48 (1946).

212. W. D. Pohle and V. C. Mehlenbacher, *J. Am. Oil Chemists' Soc.*, **24**, 155–159 (1947).

213. A. Revenna, *Zymologica,* **3**, 174, 176 (1928).

214. G. F. Smith and F. R. Duke, *Ind. Eng. Chem., Anal. Ed.*, **13**, 558 (1941); *ibid.*, **15**, 120 (1943).

215. R. Neuman, *Z. Angew. Chem.*, **30**, 234 (1917).

216. J. T. R. Andrews, *Oil Soap*, **16**, 19–20 (1939); **18**, 14–18 (1941).
217. R. M. Nelson, *J. Am. Oil Chemists' Soc.*, **50**, 207–209 (1973).
218. S. Hirayama and C. Imai, *J. Am. Oil Chemists Soc.*, **54**, 190–192 (1977).
219. J. H. M. Rek and A. C. M. Appel, *J. Am. Oil Chemists' Soc.*, **56**, 861–863 (1979).
220. R. Herrlinger and G. M. Compeau, *J. Am. Oil Chemists' Soc.*, **29**, 342–344 (1952).
221. H. Wolff and E. Scholze, *Chem. Ztg.*, **38**, 369, 370, 382, 430 (1914).
222. R. Hastings and A. Pollak, *Oil Soap*, **16**, 101 (1939).
223. D. McNicoll, *J. Soc. Chem. Ind.*, **40**, 124T (1921).
224. Reference 3; Method 14.148.
225. Reference 3; Method 28.122.
226. Reference 3; Method 28.123.
227. Reference 3; Method 28.085.
228. B. M. Milwidsky, *Practical Detergent Analyses*, MacNair Dorland, New York, 1970.
229. Reference 3; Method 14.149.
230. Reference 3; Method 14.150.
231. Reference 3; Method 28.089.
232. Reference 3; Method 28.107.
233. J. Lewkowitsch, *Chem. Technol. Anal. Oils, Fats Waxes*, 6th Ed., MacMillan, New York, 1921.
234. B. Holmbom, *J. Am. Oil Chemists' Soc.*, **54**, 289–293 (1977).
235. Reference 3; Method 26.037.
236. Reference 3; Method 26.044.
237. Reference 3; Method 20.014, 26.020, and 26.049.
238. Reference 3; Method 26.052.
239. Reference 3; Method 26.069.
240. T. R. Romer, N. Ghouri, and T. M. Boling, *J. Am. Oil Chemists' Soc.*, **56**, 795–797 (1979).
241. R. D. Stubblefield, *J. Am. Oil Chemists' Soc.*, **56**, 800–802 (1979).
242. S. V. Pathre and C. J. Mirocha, *J. Am. Oil Chemists' Soc.*, **56**, 820–823 (1979).
243. R. M. Eppley, *J. Am. Oil Chemists' Soc.*, **56**, 824–829 (1979).
244. J. G. Marcus, *J. Am. Oil Chemists' Soc.*, **51**, 472–474 (1974).
245. C. Imai, H. Watanabe, N. Haga, and T. Ii, *J. Am. Oil Chemists' Soc.*, **51**, 495–498 (1974).
246. P. N. Maheshwari, D. W. Stanley, J. I. Gray, and F. R. Van de Voort, *J. Am. Oil Chemists' Soc.*, **56**, 837–841 (1979).
247. J. K. Daun and F. W. Hougen, *J. Am. Oil Chemists' Soc.*, **54**, 351–354 (1977).
248. C. M. Elson, D. L. Hynes, and P. A. MacNeil, *J. Am. Oil Chemists' Soc.*, **56**, 998–999 (1979).
249. C. M. Elson and R. G. Ackman, *J. Am. Oil Chemists' Soc.*, **55**, 616–618 (1978).
250. M. W. Blades, J. A. Dalziel, and C. M. Elson, *J. Assoc. Off. Anal. Chem.*, **59**, 1234 (1976).
251. U.S. Environmental Protection Agency, *Manual of Methods of Chemical Analysis of Water and Waste*, Pub. No. EPA 625/66-74-003, 1974, p. 118.

252. J. N. Schoolery, *Varian Application Note* NMR-75-3.

253. L. Hedler, C. Schurr, and P. Marquardt, *J. Am. Oil Chemists' Soc.*, **56**, 681–684 (1979).

254. J. W. Rhodes and D. E. Johnson, *J. Chromatogr. Sci.*, 616–617 (1970).

255. R. H. White, D. C. Havery, E. L. Roseboro, and T. Fazio, *J. Assoc. Off. Anal. Chem.*, **57** (6), 1380–1382 (1974).

255a. Food Protein R & D Center, Texas Engineering Experiment Station, *Annual Progress Report*, Sept. 1, 1978–August 31, 1979; E. W. Lusas, Director, Texas A & M University, College Station, Texas, pp. 83–106.

256. Reference 3; Method 29.001.

257. Reference 3; Method 29.019.

258. Reference 3; Method 29.018.

259. Reference 3; Method 29.028.

260. Reference 3; Method 29.123.

261. Reference 3; Method 28.127.

262. Association of Official Agricultural Chemists, *Official Methods of Analysis,* 10th ed., Washington, D.C., 1965, p. 440.

263. T. C. Campbell and L. Friedman, *J. Assoc. Off. Agr. Chem.*, **49**, 824 (1966).

264. A. Huang, D. Firestone, and A. D. Campbell, *J. Assoc. Off. Anal. Chem.*, **50**, 16–21 (1967).

265. D. Firestone, J. Ress, N. L. Brown, R. P. Barron, and J. N. Damico, *J. Assoc. Off. Anal. Chem.*, **55**, 85 (1972).

266. Reference 3; Method 28.128.

267. *Monographs on the Evaluation of Carcinogenic Risk of Chemicals to Man* Vol. 11 (1976), International Agency for Research on Cancer, Lyons, France, pp. 247–256.

268. *Carcinogenesis Technical Report* Series No. 80 (1978), National Cancer Institute, Bethesda, MD.

269. G. Halphen, *J. Pharm.*, **6**, 390 (1897); *Analyst*, **22**, 326 (1897).

270. J. A. Harris, F. C. Magne, and E. L. Skau, *J. Am. Oil Chemists' Soc.*, **41**, 309–311 (1964).

270a. R. O. Feuge, L. P. Codifer, and J. H. Zeringue, Jr., Paper No. 383, ISF-AOCS World Congress, New York, April 27–May 1, 1980.

271. Reference 3; Method 28.109.

272. J. T. Lawhon, C. M. Cater, and K. F. Mattil, *J. Am. Oil Chemists' Soc.*, **54**, 75–80 (1977).

273. L. Schneider, S. P. Loke, and D. T. Hopkins, *J. Am. Oil Chemists' Soc.*, **45**, 585–590 (1968).

274. Reference 3; Method 36.013.

275. Reference 3; Method 43.194.

276. Reference 3; Methods 43.195 and 43.209.

277. Reference 3; Methods 43.126 and 43.141.

278. L. E. Khoo, F. Morsingh, and K. Y. Liew, *J. Am. Oil Chemists' Soc.*, **56**, 672–675 (1979).

278a. Reference 3; Method 43.001.

279. Reference 3; Method 43.008.

280. Reference 3; Methods 43.024–43.030.

281. Reference 3; Method 43.039.

282. Reference 3; Method 43.043.

283. Reference 3; Method 43.056.

284. Reference 3; Method 43.061.

285. Reference 3; Method 43.068.

286. A. E. Waltking, M. Kiernau, and G. W. Bleffort, *J. Assoc. Off. Anal. Chem.*, **60**, (4) 890 (1977).

287. O. Podlaha, A. Erickson, and B. Toregard, *J. Am. Oil Chemists' Soc.*, **55**, 530–532 (1978).

288. H. R. Bollinger, in E. Stahl, Ed., *Thin Layer Chromatography*, Academic, New York, 1957, pp. 229–232.

289. A. Seher, *Nehrung*, **4**, 466 (1960).

290. K. T. Hartman, *J. Am. Oil Chemists' Soc.*, **54**, 421–423 (1977).

291. P. J. Van Niekerk, *Anal. Biochem.*, **52**, 533 (1973).

292. M. Matsuo and Y. Tahara *Chem. Pharm. Bull.*, **25** (12), 3381 (1977).

293. A. Mariani and C. Vicari, *Rend, 1st. Super Sanita*, **21** 160–166 (1958); *Pharm. Acta Helv.*, **33**, 120–125 (1958).

294. Reference 3; Method 43.088.

295. Reference 3; Method 43.097.

296. Reference 3; Method 43.106.

297. Reference 3; Method 43.111.

298. A. P. Carpenter, Jr., *J. Am. Oil Chemists' Soc.*, **56**, 668–671 (1979).

299. M. H. Stern and J. G. Baxter, *Anal. Chem.*, **19**, 902 (1947).

300. F. Utz, *Apoth. Ztg.*, **15**, 28 (1900).

301. P. Budowski and K. S. Markley, *Chem. Rev.*, **48**, 125–154 (1951).

302. L. Pavolini and R. Isidoro, *Olii Miner. Grasse Saponi, Colori Vernici*, **29**, 33 (1952).

303. C. C. Suarez, R. T. O'Connor, E. T. Field, and W. G. Bickford, *Anal. Chem.*, **24**, 668 (1952).

304. M. Beroza, *Anal. Chem.*, **26**, 1173 (1954).

305. M. Beroza, *Anal. Chem.*, **28**, 1550 (1956).

306. M. C. Bowman and M. Beroza, *Residue Rev.*, **17**, 1 (9167).

307. M. Tanaka, A. Kato, and T. Tsuchiya, *J. Am. Oil Chemists' Soc.*, **48**, 95–97 (1971).

308. Reference 3; Method 31.160.

308a. B. D. Page, *J. Assoc. Off. Anal. Chem.*, **62**, 1239 (1979).

309. A. Dieffenbacher and U. Bracco, *J. Am. Oil Chemists' Soc.*, **55**, 642–646 (1978).

310. T. Riisom and L. Hoffmeyer, *J. Am. Oil Chemists' Soc.*, **55**, 649–652 (1978).

311. Reference 3; Method 28.133.

312. K. Miyakoshi and M. Komoda, *J. Am. Oil Chemists' Soc.*, **54**, 331–332 (1977).

313. Reference 2; Method D 1767.

314. E. D. George, E. J. Hillier, and R. Krishnan, *J. Am. Oil Chemists' Soc.*, **57**, 131–134 (1980).

315. E. Jungermann and E. Beck, *J. Am. Oil Chemists' Soc.*, **38**, 513–515 (1961).

316. Board of Standards, Toilet Goods Association, Inc., TGA Specification No. 76.

317. C. N. Rudie and P. R. Demko, *J. Am. Oil Chemists' Soc.*, **56**, 520–521 (1979).

318. A. Johannson and L. A. Appelqvist, *J. Am. Oil Chemists' Soc.*, **56**, 995–997 (1979).

319. L. J. Gershbein and E. J. Singh, *J. Am. Oil Chemists' Soc.*, **46**, 554–557 (1969).

320. H. Rodenbourg and S. Kuzdzal-Savoie, *J. Am. Oil Chemists' Soc.*, **56**, 485–488 (1979).

321. J. L. Iverson, J. Eisner, and D. Firestone, *J. Am. Oil Chemists' Soc.*, **42**, 1063–1068 (1965).

322. A. E. Bailey, *Cottonseed and Cottonseed Products, Their Chemistry and Chemical Technology*, Interscience, New York, 1948, pp. 213–362.

323. P. J. Elving and I. M. Kolthoff, Eds., *Chemical Analysis*, Wiley-Interscience, New York (a series of monographs on analytical chemistry and its applications).

324. W. Slavin, Ed., *Atomic Absorption Spectroscopy*, Vol. 25 in *Chemical Analysis* series, Wiley-Interscience, New York, 1968.

325. F. T. Weiss, *Methods and Procedures*, Vol. 32, in *Chemical Analysis* series, Wiley-Interscience, New York, 1970.

326. G. F. Longman, *The Analysis of Detergents and Detergent Products*, Wiley-Interscience, New York, 1975.

327. F. D. Snell and B. F. Biffen, *Commercial Methods of Analysis*, rev. ed., Chemical Publishing Co., New York, 1964.

328. *Standard Methods for the Examination of Water and Wastewater*, 14th ed. (Am. Public Health Assoc., Am. Waterworks Assoc., and Water Pollution Control Assoc.), ed. board, M. J. Taras (AWWA), A. E. Greenberg (APHA), R. D. Hoak, and M. C. Rand (WPCF), 1976.

329. O'D. L. Kurtz and K. L. Harris, *Microanalytical Entomology for Food Sanitation Control*, AOAC, Arlington VA, 1962.

330. *FDA Training Manual for Analytical Entomology in the Food Industry*, AOAC, Arlington, VA, 1975.

8

Environmental Aspects of Animal and Vegetable Oil Processing

Introduction

This chapter is an assemblage of information on the impacts that processing of animal and vegetable oils have on water quality, air quality, and land pollution through manufacturing process wastewaters, gaseous and particulate emissions to the air, and waste materials. The manufacturing processes covered are oil-seed extraction, edible oil refining, and soap and fatty acid manufacture. Manufacturing operations for production of derivatives of fatty acids or of compounded products such as cosmetics are not discussed, nor is animal fat rendering. The scope does not include environmental aspects of the work space (occupational health) such as noise or dust exposure. The environmental aspects of the generation of steam or electric power usually found in conjunction with the fat and oil processing operations are not embraced in the scope. Each phase of the environment—water, air, and land—is addressed in separate sections. Odor is included in the section involving air emissions.

One intent is to provide information sufficient for an inexperienced engineer or technician to be informed of the environmental control needs of the industry. A second is to give a new member of production management a basic understanding of the environmental control aspects of the processes.

For each phase of the environment, the sections discuss the sources of the potential environmental insult, pollution impacts, regulations, and control methods. The material presented on regulations is limited because the regulations needed for implementing relatively recent laws have not been developed to their full state at the time of this writing.

Water Pollution

1.1   Sources of Wastewater

General Comment.   Water and chemicals diluted in water are important process materials in refining raw oil and fat and production of soap and fatty acids. The water used becomes a wastewater stream conveying away the chemicals, impurities, and heat. Undesired constituents of the oil are eliminated from the products in this manner. Some valued oil is inevitably lost in these wastewaters. In addition to the process wastewaters themselves, the wastewaters include oil and chemicals lost to flows and surfaces through leaks and spills. The latter become part of housekeeping wastes from such operations as washing floors and paved areas. Liquid wastes also originate from sanitation and cleaning measures in which water supplemented with cleaning agents is used in sanitizing pipes, tanks, and equipment.

Water is used for temperature control in cooling and condensing operations. Wastewaters are generated from indirect heat exchangers. Such water, used in a once-through-and-to-waste fashion, will not have any changes in its characteristics other than increased temperature, provided, of course, that there are no leaks of process streams into the cooling water. Indirect cooling water is commonly passed over cooling towers and recycled. In such cases some discharge to waste must be practiced to avoid an intolerable increase in the mineral content of the recirculating water.

Large volumes of wastewaters also originate in systems for creating low pressures for evaporation, distillation, or drying. Water vapor in these systems is condensed by sprays or streams of cold water. The condensing water is discharged from the system through a pipe that terminates below a water surface and thus provides a seal to maintain a vacuum. Volatile matter from the materials being processed will be part of the vapors reaching the condensing water sprays. Also, some droplets of the material being processed may carry over or be entrained in the stream of vapor to the condensing water. Thus wastewater from such condensing systems will contain condensed and entrained volatile materials. Such condensing waters may be passed over cooling towers and recycled. Some discharge to waste must be practiced from such recirculating systems to avoid an intolerable increase in the minerals in the water supply and an excessive concentration of contaminants from the process.

Steam is used in heating tanks of oil to maintain fluidity and for many process steps. This steam condenses and thus yields its heat. The condensate is returned to the boiler feed where it is economic to do so and where the possibilities of contamination are acceptable. At many locations the condensate enters the sewer at the point of generation. This condensate is unpolluted except in those instances where there is a leak into the steam line from the process stream.

Sources of Wastewaters from Oil Extraction.   Mechanical or solvent extraction processes for separating oil from the oilseed or other source do not cause the discharge of major amounts of wastewaters or water-borne pollutional materials. Oil extraction plants are commonly at different geographical locations than are oil refining plants. Consequently oil extraction plants are considered separately in this discussion.

Plants that use solvents to extract the oil use large volumes of cooling water in indirect condensing systems. The water is used to condense the solvent in the solvent recovery system. This water is recirculated over cooling towers, and only that necessary to limit the buildup of dissolved solids enters the sewer. In preparation for extraction of soybean and other seeds, the beans are pressed into thin sections. This operation creates appreciable dust. Usually exhaust ducts convey the dust to scrubbers or filters to control the emissions of particulate matter. Wastewaters from such wet scrubbers constitute a considerable part of the routine wastes from extraction operations. In addition to the bean preparation, the remaining seed after extraction is usually processed in some way for marketing as animal feed. Such operations may include drying and packaging and other steps from which the dust must be controlled.

Mechanical extraction plants have minor quantities of wastewaters. Both methods share wastewaters from cleaning and housekeeping operations. Small volumes of steam condensate used in heating the oils to maintain them in liquid state will be discarded.

A small fraction of soybean oil is subjected to a prerefining process referred to as *degumming*. This can be a part of the extraction plant or a preliminary step at an oil refining plant. The degumming process is employed to remove lecithin, a mixture of several phosphatides. This commercial product, including the chemical lecithin, is used as a food additive and for industrial purposes. Degumming consists of contacting the oil with a small volume of water, a process referred to as *hydration*. Lecithin and other water-soluble materials are subsequently removed in the water stream by centrifugation. The lecithin is then dried. The oil is filtered using cloth or cloth and filter aid before lecithin removal. The market for lecithin is limited; therefore, the degumming process is carried out on only a fraction of soybean oil produced, about 30%. The market value of the raw oil is the same regardless of whether the oil is degummed. Wastewater production from these operations is low and of little importance. Degumming is important to the quantity of pollutional matter discharged because if lecithin is not recovered, it becomes part of the wastewaters from subsequent processing.

Sources of Wastewaters from Oil Refining. The objective of the process of refining edible oils is to separate undesired nontriglycerides from the triglycerides. Some vegetable oils contain about 2–4% of organic compounds that are not triglycerides or free fatty acids. The raw oil will usually contain a few percent of fatty acids, mono- and diglycerides, glycerine, lecithin, phosphatides, sterols, sugar, a variety of color- and flavor-producing compounds, particles of the oil seed, and low concentrations of other nontryglyceride consituents. The oil supply to which the refining process is applied is either raw fat or oil or soybean oil that has been degummed.

The refining process used practically exclusively is the extraction of the oil with a solution of caustic or a combination of caustic and soda ash. The alkaline materials convert the free fatty acids to sodium soaps and dissolve other materials from the raw oil. The alkaline extracts, referred to as *soapstock*, is separated from the oil by centrifugation. The centrifugation produces two streams, the extracted oil stream and the alkaline extract. The alkaline extract contains practically all the fatty acids and water-soluble nontriglycerides.

A few plants in the United States use steam distillation for refining instead of the process just described. This distillation process is similar to the process for oil deodorization which is described later. The U.S. Department of Agriculture also reports that about 2% of oil product is refined by solvent extraction processes.

The alkaline extract from alkaline refining is acidified (acidulated) for recovery of the fatty acids and any emulsified oil. The agent used to neutralize the alkaline material is sulfuric acid. The soap stock may be heated to help release the emulsified oil. Separation of the fatty acid and oil from the extracting solution may be accomplished either by centrifugation or gravity sedimentation or by the two operations in series. The water and the materials it contains are discarded. The fatty acids and oil separated in the acidulation plant are used in animal feed or for raw materials for a fatty acid processing plant. The soap stock may be shipped to a central or independent plant for the acidulation process.

Immediately following the extraction of the oil with caustic, the oil stream is extracted with water to remove any fatty acid soaps remaining with the oil stream or any water-soluble materials, including any traces of caustic. The water stream, referred to as *soapy water*, is separated from the oil by centrifugation. In the majority of plants the water extract (the soapy water) is added to the alkaline extract and passed through the acidulation process just described. In some plants, particularly those which ship the soap stock to other locations for acidulation, the soapy water is discarded.

The alkaline-extracted oil must be further processed to remove trace components that contribute flavor and odor, to create the bland, colorless product desired by edible oil consumers.

The introduction of this discussion of refining mentioned plants that do not use alkaline extraction for refining but instead use steam stripping at low pressures for the essential refining step. This process is similar to the deodorization process described later in this section, under "Sources of Wastewater from Deodorizing." No information is available in the literature specifically describing the wastewater sources and characteristics of the operation. Statements have been made that the wastewaters are reduced in magnitude compared to alkaline refining. This could refer to the absence of the neutralized alkaline minerals in the waste.

Bleaching.   The removal of color is accomplished by an adsorption process using natural clays whose adsorptive properties are suitable. A typical clay is a diatomaceous earth with the desired adsorptive property. The clay is added continuously to the oil stream, the slurry is mixed in a contact chamber, and finally the clay is separated from the oil by filtration. Filtration equipment is typically plate and frame or septum type filters. Filtration is continued until pressure limits are reached, or the flow rate is reduced to a limit, or the filter has a full load of clay. The oil remaining in the lines, the filter spaces, and the filter clay is purged into the product flow by use of air or nitrogen. The routine followed at different plants in further separation of oil from the clay differs from plant to plant. A hot water wash may follow the air purge. This oil-laden water flush is discharged to a gravity oil separator. The oil is skimmed from this separator and is marketed for animal feed or returned to process. The small volume of water enters the sewer.

These purging operations produce a spent filter clay containing at least 35% oil. This spent clay from decolorizing constitutes the major solid waste material of the industry and is discussed in a subsequent section on waste materials. However, in addition to the wastewater from the purging operation previously discussed, other wastewaters enter the sewer from the spent clay processing operation at some plants. The oil-laden spent clay must be cleaned from the filtration equipment. At many plants this is done by manual labor. The cake is knocked or scraped from the leaves and falls to a trough below. A screw conveyor in the trough moves the plastic clay mass to a bin or truck tank. Labor time in such operations plus problems with the pyrophoric property of such spent clay has led some plants to the practice of hosing the cake from the filters and receiving trough with water. The resultant slurry then flows to a truck tank. There are at least four ways in which such a slurry may be further processed. Three of these involve the sewering of liquid wastes. In former years this entire slurry was discharged to the sewers. In circumstances where the slurry was transported and treated satisfactorily, this was a technically feasible practice. Sewage treatment has become more expensive and the practice may not be economical today.

Another alternative is to dewater the slurry with a rotary drum filter. The relatively dry cake is discharged to a bin or a truck. The water is discharged to the sewer. This water is referred to as "spent-filter-clay-transport wastewater."

Still another alternative is to process the slurry for oil separation and recovery. The operation consists of adding limited quantities of caustic, which helps to separate some oil from the clay. The process is carried out batchwise; each caustic treated batch is heated and then settled. The oil in representative operations is reduced from an original content of 35% by weight of solids to 20% or lower. The batch is settled, the oil layer skimmed, the water layer drawn to the sewer, and the clay slurry either hauled as a slurry or dewatered as described in the previous paragraph. The decanted water or filtrate is another form of the spent filter cake transport water.

Sources of Wastewaters from Deodorizing. Volatile organics not removed by previous processes are removed by steam stripping at low pressure and high temperature. A significant percent of the oil and of volatile oil-soluble constituents are removed from the oil. Much of the material is the more volatile, short chain length fatty acids. Modern deodorizers are continuous, multistage systems. The quantity of steam used requires tremendous quantities of cooling water for its condensation. A number of systems may be employed to remove material from the water vapor before it is condensed. Initially, these may include an operation for removing entrained material in the form of droplets. This is a gravity separation system in a chamber providing reduced velocities of the vapor, typically followed by a partial condensation system to capture the majority of fatty acids. One such partial condensation system utilizes a vegetable oil or a fatty acid that is not volatile at temperatures maintained in the partial condenser. This oil at low temperature is circulated to a vapor contactor, where it cools the vapor stream to just below the condensation temperature of the volatile organics that are to be removed from the vapor stream. Not all the organics can be condensed without condensing water vapor also, which is not desired in such a partial condenser. It is obvious that the vapor temperature, the condensing oil temperature, and the pressures must be delicately balanced to achieve the optimum removal.

Following the partial condenser, the vapors pass to a system for condensation of the water vapor as well as any remaining volatile organics. Such condensation may be accomplished with either direct or indirect condensers. The direct condensers use large flows of water, thousands of gallons per minute. This water is typically recirculated over cooling towers. The organics carrying over to the condensing water build up to high concentrations. Some insoluble organics such as fatty acids can be removed by gravity settling from this stream. A significant blowdown stream is typically discarded to the sewer.

Some systems operate without a partial condenser. The steam and the vaporized material are condensed together in a spray of water. The cooling water is recirculated over a cooling tower. A settling and surge tank is included in the system. The organic material which floats to the surface in the surge tank is periodically removed. Makeup to these systems is minimal, if any is necessary at all, because of the addition of condensed steam to the systems. A significant diversion of the recirculating water to the sewer is practiced to control the buildup of soluble matter.

In the winter the amount of heat dissipated by convection rather than evaporation is so great that more condensate enters such systems than evaporates in the cooling tower. When such conditions prevail, there is a need to continuously or periodically discharge the excess.

The production and characteristics of the wastewaters from these vapor condensing systems was materially changed at a number of plants in recent years because of local odor problems with the recirculating systems. Some odorous materials in the recirculated cooling water with direct condensing systems are released to the air. One solution is to use surface condensers so that the condensate is not mixed with the cooling water. The cooling water is uncontaminated and may be recirculated without significant odor production. The indirect condensers condense the water vapor as well as any material condensable at the temperatures and pressures involved. The stream of condensed steam, fatty acids, and other organics is relatively small, making effective gravity separation of floatable material possible. The skimmed water is discharged to waste. A second waste comes from cleaning the indirect condenser of oil and fat sticking on the surfaces. This oil is removed by isolating the condenser and circulating hot water through the vapor side. This flush water is also passed through the gravity settling tank to waste.

Sources of Wastewaters from Hardening. Following processes previously described, a portion or all of the oil processed may be subject to another process referred to as *hardening* if the objective is to increase the congealing temperature of the product. The addition of hydrogen to the unsaturated linkages lowers the temperature of congealing. The property of being semisolid or plastic at ambient temperatures or slightly higher is important for products such as margarine and the solid form of shortening. The addition of hydrogen is accomplished by reacting the oil with hydrogen in the presence of a nickel catalyst. The nickel catalyst is a compound of nickel deposited on diatomaceous earth. The reactants and the catalyst are intimately mixed at the desired temperature and pressure, and when the reaction is complete, the catalyst is removed by filtration. The filtration equipment employed is of the plate and frame type. The catalyst has a limited active life and must eventually be discarded or regenerated. This problem is discussed in the section of this chapter on solid wastes.

The generation of wastewaters from hardening has not been significant in the past. Any wastewaters that result consist of water overflow from hydrogen gas tank seals and housekeeping wastes. Since the mid-1970s, information on the significance of nickel in the environment has caused much concern, as summarized in a subsequent section on pollutional properties, toxic pollutants. At times of repair of pumps, tanks and piping systems, small losses of the oil and nickel slurry to floor drains may occur. Also, the cleanup of filters and filter cloths where used results in wastewater. The practice of dry or no water use cleanup eliminates these wastewaters. In such practice the material is adsorbed in granular commercial cleaning material and disposed of in the manner of a waste material and not discarded in the sewer.

Sources of Wastewater from Winterizing. For the production of an edible oil end product that will remain clear and unclouded at temperatures of household storage and use, a process is employed for removing those triglycerides whose solidification temperature is in the range of storage and use. This is accomplished by cooling the oil to low temperature and removal of the solidified material. No routine wastewaters are produced from the oil processes, so there is no need to discuss the operation further. Auxiliary to this process is usually one of a variety of refrigeration systems for cooling the oil. These involve indirect heat exchanges with which are associated recirculated cooling water systems. The usual discharge from such recirculation systems to control buildup of minerals is employed. An alternative is the use of large volumes of once-through cooling water. A concern of such recirculation systems can be ammonia leaks or spills if such material is involved in the refrigerant system.

Sources of Wastewaters from Drying. For some products it is desirable to remove all traces of water from the oil so that shelf life is not affected by trace amounts of water. This is done by drying the oil under vacuum. Direct condensation of the vaporized water or other volatile material by water sprays is practiced. Ordinarily the quantity of entrained oil or other volatile material is insufficient to produce a significant concentration in the condensing water. However, misoperations do occur, and oil will appear in significant concentrations in condensing water recirculating systems. The wastewater stream from such recirculating systems may occasionally have a low concentration of dispersed oil.

Sources of Wastewaters from Filtration. To ensure that no sediment is present in a final product and to assure clarity, the refined and purified oil may be subjected to a final filtration prior to transfer to its container or as a part of such a packaging operation. Filtration is done by plate and frame systems using as filter media diatomaceous earth similar to or the same as is used in decolorizing. There are no routine liquid wastewaters from this system. Filtration system of earlier years utilized filter cloths that were periodically washed similarly to a clothing laundering operation. Filters using cloths have been replaced by metallic or plastic materials. The removal of oil from such cake may involve similar wastewaters to those discussed in decolorizing.

Sources of Wastewaters from Packing. The filling, packaging, and warehousing of edible oil products does not involve the release of routine process wastewaters. During processing, the oil is maintained at temperatures that destroy bacteria or other life forms. The oil is handled in closed systems and blanketed with nitrogen so that atmospheric sources of any organism are minimal and oxidation is minor. The system is essentially sterile at all times so that the usual periodic cleanouts for sanitizing purposes, such as those found in a milk-bottling plant, are not needed. At times of maintenance or weekend shutdowns, steam flushing of pipes, tanks, and equipment may result in some condensed steam and flushing water being added to the sewer. Water with detergents or caustic may be used occasionally in cleanup.

The packing line and even the warehouse may experience breaking of bottles. Clean-up procedures may include flushing to the sewers.

A peculiar wastewater may be associated with warehousing. The wastewater comes from the service area for the carts. These carts, if of the internal combustion engine type, may be washed down and the washwater discarded in the sewer. In one instance, drainage from this area, which likely included some used crankcase oil, contributed a small amount of mineral oil.

Sources of Wastewaters from Storage and Handling. Numerous sources of wastes originate in the unloading and storing operations. One is leaks and spills from tank car and tank truck unloading and loading. Small losses can occur at each uncoupling of a line and loss of the line contents will take place unless precautions are taken. Glands of pumps used to transfer the oil may leak.

A major source of oil-bearing wastewaters can take place from the washouts of tanks, trucks, and railway cars. At one plant this loss of vegetable oil was the greatest single source of oil to waste. Furthermore, during winter months this loss was twice the quantity of such losses in the summer months. This likely reflected the effect of cold tank walls retaining more oil or slowing drainage.

Major quantities of waste materials may be in wastewaters from cleaning storage tanks. Storage tanks of raw oil can accumulate major amounts of sediment, pieces of beans or nuts, or, in the case of animal fats, of hair, bone, and flesh. Some plants practice removal of much of such accumulations and mere disposal in the manner of solid waste material. Smaller residues are washed from the tanks in the wastewaters from cleaning.

Much of the wastewater that enters the sewers serving the storage tanks and unloading areas that are not under roof is rainfall runoff. The drainage from such areas may be slightly polluted from leaks and drips and constitutes a significant volume of wastewaters.

Sources of Wastewaters from Soapmaking.   Wastewater volume and characteristics from soapmaking are markedly affected by which of two processes are employed. The manufacture of soap by the batch kettle process has a very low discharge of wastewaters. In the kettle, a purified fat and sodium hydroxide are reacted to form soap and glycerine. The aqueous caustic stream is used countercurrent to the oil in a series of batch reactors so that the caustic is essentially exhausted. After the final contact, the soap is drawn off the top. The lower layer is water containing the color bodies and other impurities from the fat or the oil. This material, called *niger*, is recirculated to fresh fat reactors until the alkali content is low. The niger can be processed to recover a low grade dark soap. The water residual after separation of the dark soap is eventually neutralized and discarded to the sewer when it becomes particularly loaded with color bodies. Salt and sulfuric acid may be used in the process of recovering the dark soap, and these will be present in the wastewaters.

To prepare a high-quality soap by the batch process, the fat must be refined before use. One of the most frequently used methods employs activated clay as the extraction agent. No liquid wastes are produced except for washouts, provided that the clay is handled dry or as a waste material and not discarded to the sewer. Other ways in which fats are refined include caustic extraction, steam stripping, and proprietary treatments. Wastewaters from caustic extraction have been discussed in the section on edible oil refining. Steam stripping produces a waste of condensate and condensing water that is discarded to the sewer after separation of the fatty acids by gravity settling.

The alternative process of making soap by fatty acid neutralization produces a small quantity of wastewaters. The fat splitting operation preliminary to soap making has significant volumes of wastewaters. The equipment for mixing of the fatty acid and caustic and other materials that may be used is only occasionally washed out. A small volume of water containing unreacted alkali may be separated from the product and discarded to the sewer. For some products it is necessary to remove water from the product of the neutralization step. This may be accomplished by subjecting a thin layer of the product to a vacuum. The vacuum is created by use of steam and a barometric condenser. Very small amounts of volatile fatty acids may be drawn over to the condensing water. The wastewaters from this operation will be the once-through condensing water or the discard stream from a recirculating cooling water system.

Fatty acid manufacture by fat splitting produces several liquid waste streams. The hot fat at high pressure is contacted with water countercurrently, and hydrolysis takes place. The acids are released to atmospheric pressure, and the vapors are condensed by water sprays and discarded to the sewer following gravity separation of the floatable fatty acids. The fatty acids are purified by distillation at low pressures. The low pressure is achieved commonly by a barometric condensing system using steam ejectors and direct water sprays for condensing. The condensing water may be recirculated over cooling towers. Wastewaters consist of a discard stream from this system. The recirculating water is passed through a gravity settling system for recovery of floatable fatty acids and oil. The unreacted oil and impurities will accumulate in the bottoms of the fatty acid still. These still bottoms are treated with acid to break emulsions, floatable oil recovered, and they are then neutralized and discarded to the sewer. Typically, this is a batch operation.

The other product stream from acid splitting is that containing the glycerine. Glycerine streams from the kettle process are neutralized by the addition of mineral acid. The glycerine is evaporated to a low water content usually by a two-stage process. Barometric condensers are used in the two stills, and the vapors are condensed by sprays of cold water. Although mist eliminators are commonly employed, there will be some carryover of crude glycerine to the condensing water. The major volume of wastewater is the condensing water or the discard stream from a recirculating condensing water system. The still bottoms are typically dark brown amorphous solids when cooled. This material is commonly disposed of in the manner of a solid waste.

The process of forming bars or other form of soap seldom involves a process waste, although reworking of scrap can produce a water drawoff. There may also be waste from washout of equipment.

1.2 POLLUTIONAL PROPERTIES

Oxygen Depression. Much organic matter in wastewaters will be degraded by bacteria ever present in the environment. An initial population of bacteria is omnipresent in rivers and lakes. In degrading the material, the bacteria utilize dissolved oxygen present in the water. This oxygen is also used by gill-breathing fish in their life processes. Therefore, biodegradable organic matter in a stream or lake may result in a lowering of the dissolved oxygen content so that the fish are harmed or killed. The common measurement of biodegradable organic matter is the biochemical oxygen demand (BOD) test. This test measures the amount of oxygen used in degradation of the organic matter under test conditions simulating to a degree those in a stream.

The standard BOD test period is 5 days and the temperature 20°C. The BOD is the amount of oxygen used expressed as oxygen and varies from about one-half to two or more times the weight of the organic matter oxidized. During this 5-day test, about 68% of the carbon and hydrogen in the wastewater compounds is oxidized. All the organic matter in wastewaters from animal and vegetable oil processing are readily biodegradable, so the wastewaters have a significant BOD, that is, use of oxygen in degradation. The ratio of BOD to biodegradable organic matter for triglycerides is about 1.8:1. Therefore, the oil is eliminated in biological treatment.

Suspended Solids. Another common parameter used in measuring pollutional characteristics is the suspended solids test. This measures materials that can be filtered using specific filter media. In the wastewaters under discussion there will be some suspended solids consisting of fat particles and perhaps filtering clay. Suspended solids are of interest because of their effect on appearance of the water as well as their potential of settling out on the river or lake bottom.

Acidity–Basicity. The presence of acids producing too low a pH or of alkaline materials producing too high a pH will adversely affect aquatic life. A pH in the range of 6–8 is usually specified in discharges to public waters. Both acids and alkaline materials are used in these industrial processes, and there is a potential of either high or low pH.

Oil. Unfortunately for the processors of animal and vegetable oils, in early water pollution control activity, oil of all kinds is generally lumped together as one measurement and controlled as one parameter. The standard analytical procedure measures any material extractable by a specified solvent under acidic conditions. The substances measured include petroleum oil, animal–vegetable fat or oil, fatty acids, soaps, and any other compound that is extracted. A water quality standard based on this "total" oil measurement has been adopted by two or three states, and limitations on total oil have been specified in effluent standards for a few industry categories. The most recent technical policy document for water quality standards of the U.S. Environmental Protection Agency (EPA) discusses differences in the pollutional properties of two major kinds of oil—petroleum and animal–vegetable. Both kinds of oil have a wide variety of compounds of differing properties; nevertheless, broad categorical differences exist between the two kinds mentioned. The potential pollutional properties of petroleum oil include:

Low rate of aerobic biodegradation in the environment and thus ability to persist for extensive periods in the water.

Complete lack of biodegradability in the absence of oxygen; therefore indefinite persistence when agglomerated with silt and settled in bottom muds.

Impairment of the flavor of fish flesh.

Resultant taste and odor in drinking water.

Creation of fire hazards or explosive atmospherics in sewers and treatment plants.

The pollutional properties of animal–vegetable oils include the consumption of oxygen in biological degradation as described for the BOD test previously at rates comparable to most biodegradable organics. It is more dramatic to list the pollutional properties that animal–vegetable oil do *not* exhibit that petroleum oil does:

Animal–vegetable oil triglycerides are not toxic to fish or aquatic life; this oil is a part of commercial fish food.

Animal–vegetable oil does not persist for great lengths of time in a river flow.

Animal–vegetable oil does not persist in bottom deposits where oxygen is absent.

The two kinds share some properties, such as that of spreading to thin layers on water surfaces, for example, causing oil slicks and other unaesthetic effects. Therefore, there is logic in common restrictions on oil spills or discharge of free floating oil of any kind. However, with the universal application of biological treatment, the animal–vegetable oil remaining in treated effluents will be in solution, finely dispersed or associated with suspended particulates; therefore, such effluents are not involved in slicks or aesthetic problems.

The most meaningful pollutional measurement for animal–vegetable oil is the BOD test. Biochemical oxygen demand effluent standards are adequate for control. One state, Illinois, has recognized in its effluent standards for oil the two different kinds of oil and permits separate measurement and reporting. A standard (approved) analytical measurement for petroleum oil (hydrocarbons) is available.

Toxic Pollutants: Zinc, Nickel, and Phenols.   Three substances currently listed as toxic pollutants in federal water pollution control law have the potential of being present at detectable levels in wastewaters of some plants of the industry. The three substances are nickel, zinc, and phenols.

One or more of three properties form the basis of listing these compounds, namely, direct adverse effects at relatively low concentrations on saltwater aquatic life, on freshwater aquatic life, and on human health. The criteria (standards) for the protection of the aquatic environment for these pollutants are in the process of being established by the U.S. EPA as provided under federal law. Technology-based limitations on these substances in industrial waste effluents containing significant concentrations are in the process of being established by the U.S. EPA. The current limitations for the wastewaters of the soap manufacturing industry do not contain limitations on any of the compounds listed as toxic because the compounds have not been found to be present in significant concentrations. Effluent limitations for the edible-oil-refining industry have not been proposed at this time. The preliminary document for the effluent limitations for the edible-oil-refining industry does not propose limitations on any of the compounds.

Nickel has the potential of being present in wastewaters of edible-oil plants that carry out hydrogenation of oils with a nickel catalyst. Not all edible oil refining plants perform this operation. Zinc may potentially be present in wastewaters from fat splitting (the hydrolyzer). Many plants do not use zinc in the hydrolysis process. Substances that analyze as phenols have been found in low concentrations in the effluents of two edible-oil-refining plants. Phenols are defined as materials that react with the reagent, 4AAP, in a specified analytical procedure. At the time of this writing, the nature of the phenols in animal and vegetable oil wastewaters has not been defined.

The incentives of loss prevention and water quality protection, coupled with the treatment necessary to remove the biodegradable organic matter, will probably produce wastewaters from these industries that do not contain concentrations of phenols of significance to the aquatic environment or to human health.

To provide the reader with some information on the basis of concern for these substances, the texts of certain pertinent releases of the U.S. EPA and the National Academy of Sciences are quoted verbatim in the outline that follows. As previously mentioned, there is no evidence on wastewater characteristics of those industries of interest that would indicate that any limitations are needed. The information is presented to inform the reader of the necessity of maintaining control. The quoted material is presented without editing or comment, without endorsement of the selection of data, and without interpretation.

On July 25, 1979, the U.S. Environmental Protection Agency announced in the Federal Register, the availability of water quality criteria and asked for comments on the criteria as included in the announcement. The announcement stated that the water quality criteria may form the basis for enforceable standards for stream water quality as may be adopted by agencies involved. These criteria are as follows:

Phenol

CRITERIA SUMMARY

Freshwater Aquatic Life.   For phenol the criterion to protect freshwater aquatic life as derived using the Guidelines is 600 μg/l as 24-hour average, and the concentration should not exceed 3,400 μg/l at any time.

Saltwater Aquatic Life.   For saltwater aquatic life, no criterion for phenol can be derived using the Guidelines, and there are insufficient data to estimate a criterion using other procedures.

Human Health.   For the protection of human health from phenol ingested through water and through contaminated aquatic organisms the concentration in water should not exceed 3.4 mg/l.

BASIS FOR THE CRITERIA

Freshwater Aquatic Life.   The maximum concentration of phenol is the Final Acute Value of 3,400 μg/l and the 24 hour average concentration is the Final

Chronic Value of 600 μg/l. No important adverse effects on freshwater aquatic organisms have been reported to be caused by concentrations lower than the 24-hour average concentration.

For phenol the criterion to protect freshwater aquatic life, derived using the Guidelines is 600 μg/l as a 24-hour average, and the concentration should not exceed 3,400 μg/l at any time.

Summary of Available Data.   The concentrations below have been rounded to two significant figures.

Final fish acute value	4,000 μg/l
Final invertebrate acute value	3,400 μg/l
Final acute value	3,400 μg/l
Final fish chronic value	Not available
Final invertebrate chronic value	600 μg/l
Final plant value	20,000 μg/l
Residue limited toxicant concentration	Not available
Final chronic value	600 μg/l
*0.44 × final acute value	1,500 μg/l

Saltwater Aquatic Life.   No saltwater criterion can be derived for phenol using the Guidelines because no Final Chronic Value for either fish or invertebrate species or a good substitute for either value is available, and there are insufficient data to estimate a criterion using other procedures.

Summary of Available Data.   The concentrations below have been rounded to two significant figures.

Final fish acute value	620 μg/l
Final invertebrate acute value	960 μg/l
Final acute value	620 μg/l
Final fish chronic value	Not available
Final invertebrate chronic value	Not available
Final plant value	Not available
Residue limited toxicant concentration	Not available
Final chronic value	Not available
*0.44 × final acute value	270 μg/l

Human Health.   Heller and Pursell (1938) reported no significant effects in a multi-generation feeding study in rats at 100, 500, 1000 mg/l of phenol in drinking water for five generations and 3000 and 5000 mg/l for three generations. Assuming a daily water intake of 30 ml and an average bodyweight of 300 grams, these rats would have received daily doses of 10, 50, 100, 300, and 500 mg/kg/

---

* The 0.44 multiplier is a factor aquatic toxicologists use as an application factor to take into account uncertainties to arrive at a criterion.

day. The upper range approaches a single I.D. dose per day. Deichmann and Oesper (1940) reported no significant effects in rats receiving approximately 70, 100 or 163 mg/kg/day in their drinking water for 12 months. However, both of these studies did not report detailed pathological or biochemical studies but relied mostly on the weights and the general appearance of the animals for evaluation. In a more recent study (Dow Chemical Co., 1976), 135 dosings by gavage over six months at 100 mg/kg/dose resulted in some liver and kidney damage. At 50 mg/kg/dose the exposure resulted in only slight kidney damage. It must be borne in mind that in the first two studies the phenol is incorporated into the drinking water so that the daily dose is taken gradually. In the Dow study the phenol is administered in a single dose. A 500-fold uncertainty factor applied to the 50 mg/kg exposure in the Dow study would provide an estimated acceptable level of 0.1 mg/kg/day for man. In the case of phenol a great deal of information on human exposure exists. Long term animal data are available as well; however, the detail in these studies is very incomplete. Shorter term studies of sufficient detail provide the lowest dose level in animal studies for which an adverse effect was seen. It was judged that the existing data did not fully satisfy the requirements for the use of a $100\times$ uncertainty factor but were better than the requirements for a $1000\times$ uncertainty factor (Table 1). Consequently, an intermediate $500\times$ uncertainty factor was selected.

When one examines the quantity of phenol absorbed through inhalation near the TLV of 20 mg/m^3 for occupational exposures by using the Stokinger and Woodward model (1958), then at a breathing rate of 10 m^3 for an eight hour day with 75 percent absorption and a body weight of 70 kg, a man would absorb approximately 2.14 mg/kg/working day, assuming no skin absorption. The use of the Stokinger–Woodward model may be applicable to estimate acceptable intake from water.

Table 1—Guidelines for using uncertainty factors [NAS *Drinking Water and Human Health*, 1977]

Uncertainty factor	10	Valid experimental results from studies on prolonged ingestion by man, with no indication of carcinogenicity.
Uncertainty factor	100	Experimental results of studies of human ingestion not available or scanty (e.g., acute exposure only). Valid results of long-term feeding studies on experimental animals or in the absence of human studies, valid animal studies on one or more species. No indication of carcinogenicity.
Uncertainty factor	1000	No long term or acute human data. Scanty results on experimental animals. No indication of carcinogenicity.

It has been established that phenol is absorbed rapidly by all routes and subsequently is distributed rapidly. If a tenfold safety factor is applied to the projected doses absorbed from inhalation at the TLV (which already incorporates some safety factors), then the projected acceptable level would be 0.2 mg/kg/day. The estimate from animal data is 0.1 mg/kg/day. On the basis of chronic toxicity data in animals and man, an estimated acceptable daily intake for phenol in man should be 0.1 mg/kg/day or 7.0 mg/man, assuming a 70 kg body weight. Therefore, consumption of 2 liters of water daily and 18.7 grams of contaminated fish having a bioconcentration factor of 2.3, would result in a maximum permissible concentration of 3.4 mg/l for the ingested water (assuming 100% gastrointestinal absorption of phenol):

$$\frac{7.0 \text{ mg/day}}{(2 \text{ liters} + (2.3 \times 0.0187) \times 1.0} = 3.4 \text{ mg/l}$$

This water quality criterion is in the range of reported taste and odor threshold values for phenol which have been reported. It is recognized that when ambient water containing this concentration of phenol is chlorinated, various chlorinated phenols may be produced in sufficient quantities to produce objectional taste and odors. However, the ambient water quality criterion for phenol is based on phenol alone. For the criteria of 2-chlorophenol, 2,4-dichlorophenol and other chlorophenols, reference should be made to their specific criterion documents.

Nickel

CRITERIA SUMMARY

Freshwater Aquatic Life.* For nickel the criterion to protect freshwater aquatic life is derived using the Guidelines "$e^{(1.01 \ln (\text{hardness}) - 1.02)}$" as a 24-hour average and the concentration should not exceed "$e^{(0.47 \ln (\text{hardness}) + 4.19)}$" at any time.

Saltwater Aquatic Life. For nickel the criterion to protect saltwater aquatic life, as derived using procedures other than the Guidelines, is 220 μg/l as a 24-hour average, and the concentration should not exceed 510 μg/l at any time.

Human Health. For nickel the criterion to protect human health is 50 μg/l.

---

* Water body characteristics affect the concentrations at which toxicity is registered. One such characteristic is the hardness of the water. Hardness is a measure of calcium, magnesium, and a few other elements. The various criteria are computed by application of the hardness of the water to the formulas.

BASIS FOR THE CRITERIA

Freshwater Aquatic Life. * The maximum concentration of nickel is equal to the Final Acute Value as given by $e^{(0.47 \ln (hardness) + 4.19)}$ and the 24-hour average concentration is the Final Chronic Value as given by $e^{(1.01 \ln (hardness) - 1.02)}$. No important adverse effects on freshwater organisms have been reported to be caused by concentrations lower than the 24-hour average concentration other than the possible changes in diatom diversity discussed in the Criterion Document.

Summary of Available Data. The concentrations below have been rounded to two significant figures. All concentrations herein are expressed in terms of nickel.

Final fish acute value	$e^{(0.74 \ln (hardness) + 4.82)}$
Final invertebrate acute value	$e^{(0.47 \ln (hardness) + 4.19)}$
Final acute value	$e^{(0.47 \ln (hardness) + 4.19)}$
Final fish chronic value	$e^{(1.01 \ln (hardness) - 1.02)}$
Final invertebrate chronic value	$e^{(1.01 \ln (hardness) + 0.13)}$
Final plant value	100 µg/l
Residue limited toxicant concentration	not available
Final chronic value	$e^{(1.01 \ln (hardness)}$

Saltwater Aquatic Life. Criterion: for nickel the criterion to protect salt water aquatic life, as derived using procedures other than the Guidelines, is 220 µg/l as a 24-hour average, and the concentration should not exceed 510 µg/l at any time.

Summary of Available Data. The concentrations below have been rounded to two significant figures. All concentrations herein are expressed in terms of nickel.

Final fish acute value	14,000 µg/l
Final invertebrate acute value	510 µg/l
Final acute value	510 µg/l

Human Health. In arriving at a criterion for nickel, several factors must be taken into account. There is little evidence for accumulation of nickel in various tissues. Absorption through the gastrointestinal tract is minimal. Acute exposure of man to nickel is chiefly of concern in workplaces where nickel carbonyl or nickel dust are present at high levels. In these situations inhalation is the main route of entry and the lung is the critical organ although, in some instances of high exposure, the central nervous system may also be involved.

---

* Water body characteristics affect the concentrations at which toxicity is registered. One such characteristic is the hardness of the water. Hardness is a measure of calcium, magnesium, and a few other elements. The various criteria are computed by application of the hardness of the water to the formulas.

The major problem posed by nickel for the U.S. population at large is nickel hypersensitivity, mainly via contact with many nickel-containing commodities. Nickel could play a role in altering defense mechanisms against xenobiotic agents in the respiratory tract, leading to enhanced risk for respiratory tract infections.

While nickel has a possible role as a cocarcinogen in producing respiratory cancer, as suggested by animal studies, this remains to be confirmed. There is not evidence for carcinogenicity due to the presence of nickel in water. The role of nickel as an essential element is a confounding factor in any risk estimate.

To develop a risk assessment based on toxicological effects other than carcinogenicity, dose-response data would be most helpful. However, while the frequency or extent of various effects of nickel are related to the level or frequency of nickel exposure in man, the relevant data do not permit any quantitative estimation of dose-response relationships. The lowest levels of nickel associated with adverse health effects, therefore, must be used in establishing a criterion level for nickel in drinking water.

To arrive at a risk estimate for nickel, a modification of the approach used for non-stochastic effects (Fed. Register 44(52):15980, March 15, 1979) has been adopted.

The studies cited in this document have not demonstrated a "no observable" effect level (NOEL). Therefore, the study demonstrating the lowest observable effect level (LOEL) for nickel in drinking water has been used to arrive at a non-stochastic risk estimate.

In the study of Schroeder and Mitchener, adverse effects in rats were demonstrated at a level of 5 ppm in drinking water. Three generations of rats were continuously exposed to 5 ppm of nickel in drinking water. In each of the generations, increased numbers of runts and enhanced neonatal mortality were seen. A significant reduction in litter size and a reduction in the proportion of males in the third generation was also observed.

To adapt the LOEL into an Acceptable Daily Intake (ADI) for man, the LOEL is divided by an uncertainty factor of 100, as detailed in a recent National Academy of Sciences report and adopted by the U.S. Environmental Protection Agency (Fed. Register 44(52):15980, March 15, 1979). The choice of this factor is based on the absence of long-term or acute human data, scanty results on experimental animals, and an absence of evidence for carcinogenicity.

When the uncertainty factor of 100 is applied to 5 ppm, the lowest level at which adverse effects may occur is then 0.05 ppm. It can be concluded that levels in water below this concentration would not result in adverse health effects.

Zinc

CRITERIA SUMMARY

Freshwater Aquatic Life. For zinc the criterion to protect freshwater aquatic life, as derived using the Guidelines, is "$e^{(0.67 \ln (\text{hardness}) + 0.67)}$" as a 24-hour average, and the concentration should not exceed $``e^{0.64 \ln \text{(hardness)} + 2.46)}\text{''}$ at any time.

Saltwater Aquatic Life. For saltwater aquatic life, no criterion for zinc can be derived using the Guidelines, and there are insufficient data to estimate a criterion using other procedures.

Human Health. For the prevention of adverse effects due to the organoleptic properties of zinc, the current standard for drinking water of 5 mg/l was adopted for ambient water criterion.

BASIS FOR THE CRITERIA

Freshwater Aquatic Life. The maximum concentration of zinc is the Final Acute Value of $e^{(0.64 \ln \text{(hardness)} + 2.46)}$ and the 24-hour average concentrations is the Final Chronic Value of $e^{(0.67 \ln \text{(hardness)} + 0.67)}$. No important adverse effects on freshwater aquatic organisms have been reported to be caused by concentrations lower than the 24-hour average concentration.

For zinc the criterion to protect freshwater aquatic life, as derived using the Guidelines, is $``e^{(0.67 \ln \text{(hardness)} + 0.67)}\text{''}$ as a 24-hour average and the concentration should not exceed $``e^{(0.64 \ln \text{(hardness)} + 2.46)}\text{''}$ at any time.

Summary of Available Data. All concentrations herein are expressed in terms of zinc.

Final fish acute value	$e^{(0.67 \ln \text{(hardness)} + 3.63)}$
Final invertebrate acute value	$e^{(0.64 \ln \text{(hardness)} + 2.46)}$
Final acute value	$e^{(0.64 \ln \text{(hardness)} + 2.46)}$
Final fish chronic value	$e^{(0.67 \ln \text{(hardness)} - 0.67)}$
Final invertebrate chronic value	$e^{(0.64 \ln \text{(hardness)} + 2.00)}$
Final plant value	30 µg/l
Residue limited toxicant concentration	Not available
Final chronic value	$e^{(0.67 \ln \text{(hardness)} + 0.67)}$

Saltwater Aquatic Life. No saltwater criterion can be derived for zinc using the Guidelines because no Final Chronic Value for either fish or invertebrate species or a good substitute for either value is available, and there are insufficient data to estimate a criterion using other procedures.

Summary of Available Data. The concentrations below have been rounded to two significant figures. All concentrations herein are expressed in terms of zinc.

Final fish acute value	9,000 µg/l
Final invertebrate acute value	41 µg/l
Final acute value	41 µg/l
Final fish chronic value	Not available
Final invertebrate chronic value	Not available
Final plant value	50 µg/l

Residue limited toxicant concentration   Not available
Final chronic value   50 µg/l
*0.44 × final acute value   18 µg/l

Human Health.   Zinc is an essential element and is not a carcinogen. Studies on experimental animals and on human beings given zinc for therapeutic purposes together with observations of occupationally exposed persons show that large doses of zinc can be tolerated for a long period provided that the copper status is normal.

Daily ingestion of about 150 mg of zinc as the sulphate has not resulted in adverse effects in most patients even after several months of treatment. A reduction of copper levels has been reported in patients with diseases such as sickle cell anemia and coeliac disease. A reduction of the dose of zinc and copper supplementation corrected the copper deficiency.

Laboratory animals have been shown to tolerate zinc concentrations in the range of 100 to 300 mg/kg food and even higher for long periods when the intake of copper has been adequate. Copper deficient animals have been shown to be more susceptible. In many animal experiments zinc concentration in the diet of 1000 to 2000 mg/kg have been reported to be without effect. These concentrations should be compared to the average zinc content of human food, which is about 10 mg/kg.

The water quality criterion for zinc in water based on available data on effects of ingested zinc would be about 10 mg/l for the adult U.S. population. Assuming a water intake of 2 liters per day, this exposure would not cause more than an additional intake of 200 mg which can be well tolerated. This concentration is above the present standard for drinking water which is 5 mg/l based on organoleptic effects.

There are some indications that infants and small children may have a high intake of water and an additional intake of 10 to 20 mg might have an influence on copper metabolism in children with low copper intakes or with copper deficiency due to intestinal diseases, for example. However, because of insufficient information available for this special group at risk, derivation of criteria lower than the current standard would be difficult to justify. Therefore, it is recommended that the current level be maintained for water quality criteria purposes (5 mg/l). As additional information becomes available reconsiderations of appropriateness of the current standard should be performed.

The Safe Drinking Water Act (Pub. L. 93-523) required recommendations on drinking water and health be obtained from the National Academy of Sciences. A summary of the report was published in the Federal Register, V 42, N132 P 35764. The summaries concerning nickel and zinc follow:

---

* The 0.44 multiplier is a factor aquatic toxicologists use as an application factor to take into account uncertainties to arrive at a criterion.

Nickel

Nickel may occur in water from trace concentrations of a few micrograms/liter to a maximum of 100 µg/liter. At these levels the daily intake of nickel from water ranges from less than 10 µg/day to a maximum of 200 µg/day, as compared to a normal food intake of 300–600 µg/day. Available information indicates that nickel does not pose a toxicity problem because absorption from food or water is low. The principal reason for considering nickel stems from epidemiological evidence that occupational exposure to nickel compounds through the respiratory tract increases the risk of lung cancer and nasal-cavity cancer. There is difficulty in separating the effect of nickel from the effects of simultaneous inhalation of other carcinogens including arsenic, chromium and cobalt.

Because of the generally low concentration of nickel in drinking water and its reported low oral toxicity, there is no present need to set primary health effect limits for nickel in water. WHO and the U.S.S.R. have set no standards for nickel in drinking water.

Zinc

Concentrations of zinc in surface water are correlated with man's activities and with urban and industrial runoff. The solubility of zinc depends upon the pH of the water. Concentrations ranging from 2–1200 µg/liter were detected in 77% of 1577 surface water samples and 3–2000 µg/liter in 380 drinking waters.

Zinc is relatively nontoxic and is an essential trace element. Recommended minimum intake levels are 15 mg/day for adults and 10 mg/day for children over one year of age. A wide margin of safety exists between normal intake from the diet and doses likely to cause oral toxicity. Concentrations of 30 mg/liter or more impart a strong astringent taste and a milky appearance to water. Some acute adverse effects have been reported from consumption of water containing zinc at 40–50 mg/liter. There are no known chronic adverse effects of low-level zinc intake in the diet, but human zinc deficiency has been identified.

The proposed EPA secondary maximum contaminant level is 5 mg/liter.

1.3   REGULATION OF WASTEWATER EFFLUENTS

The water pollution control laws are structured for two groups of the industry's plants, namely, those that discharge wastewaters directly to the nearby lake, river, or stream and those that discharge their process wastes to publicly owned treatment works. These are known as the direct and indirect discharges. Plant management usually interfaces with municipal or state authorities in dealing with legislated control, but the requirements originate for the most part in federal law and regulations emanating from that law.

Regulation of Wastewater Discharges to Publicly Owned Treatment Works. Dischargers to publicly owned treatment works are required to comply with local sewer use regulations. These regulations must be as stringent at least as the U.S. EPA General Pretreatment Regulations and with pretreatment standards, if any, for the edible oil refining, soap manufacturing, or other appropriate industry category. The particular industry category limitations applicable will be either pretreatment standards for the existing sources or pretreatment standards for new sources. There also may be pretreatment standards for discharges to marine (deep ocean) waters. The new source standards will apply to new plants whose construction commenced after a specified date. This date will be fixed by the date of EPA's issuance of the standards.

The pretreatment standards specific to the industry categories of edible oil refining and soapmaking are not expected to be completed and promulgated by EPA until about 1981. Such categorical pretreatment requirements are intended for control of materials that are incompatible with municipal wastewater treatment processes. Materials would be incompatible if they are toxic or inhibitory to the treatment process or pass through the process unaltered to the extent that the effluent limitations for the treatment works are violated. The toxic materials that may be present in the wastes are discussed in the section on pollutional properties. The heavy metals used as catalysts are employed at relatively few plants, or the processes are so isolated from wastewater production that the metals, if present in the wastewaters at all, will likely be at concentrations well below that capable of being achieved by best available technology. Therefore, no categorical pretreatment limits are expected for the heavy metals.

Sewer use ordinances may contain a restriction on the oil content. This subject is rather complicated and is important to a subsequent section on the treatment of wastewaters from the industry in municipal systems. A full discussion is found there.

Regulation of Industrial Discharges to Streams, Rivers, or Lakes. The basic requirements for direct dischargers originate in the Clean Water Act of 1977 and the retained measures of the 1972 amendments to the Water Pollution Control Act. These laws require each industrial discharger to receive a discharge permit from the control agency with jurisdiction, either a state or federal EPA branch. The industry's permit is to reflect limitations established by the EPA for the particular industry category involved or such limitations as are necessary to meet water quality standards adopted for the particular public waters involved. The limitations established for the industry will be the minimum treatment requirement in any case. Additional treatment will be required if water quality standards are not met. A state may set effluent or water quality standards more severe than the U.S. EPA, and such will prevail over any less severe federal effluent limitation or water quality standard.

The control agencies in complying with the law were forced to issue discharge permits to the individual direct dischargers long before either the industry categorical limitations or the water quality standards had been developed and adopted. The industrial plants that are direct dischargers have been complying with discharge permit limitations in the absence of their industry category limitations. The national category limitations likely will not be more severe than the present permit limitations when they are issued. The limitations that will be promulgated and proposed for direct industry eventually will include:

Best Conventional Technology (applies to conventional pollutants such as BOD, suspended solids, pH, etc.).

Best Available Technology (applies to toxic pollutants—if any industry is found to have one or more).

New Source Performance Standards.

These have not all been promulgated for the industries involved in this discussion as of this writing. Some of the industry categories involved are expected to be promulgated in 1980 or 1981. Compliance dates are July 1984, or three years after final regulations are issued.

New source standards will apply to new sources, the construction of which is started after the regulations become final.

Controversy surrounds the best conventional technology standards that have been issued for other industries. The differences of opinion will likely be litigated so these standards may not be made final for some time.

In those cases where plants are located on small streams or where large pollution sources contribute to low water quality, the national categorical standards will not be adequate treatment. The U.S. EPA estimates that over half the miles of streams in this country will need better discharge quality than the industrial categorical standards permit, to meet water quality standards in the streams. Such streams include reaches of streams as large as the Ohio River. In such cases extreme wastewater control and treatment will have to be practiced. Just what this control and treatment will be is still undecided. If standards based on application of Best Available Technology are not enough, one may ask what is the meaning of "best available" if there is something more effective. Water quality standards have not been adopted for many parameters or by all states. When these are adopted and permits modified to reflect compliance, plants on small streams or critical reaches of rivers will likely be faced with extreme control and treatment measures.

1.4 WASTEWATER VOLUMES AND CHARACTERISTICS

For the purposes of developing limitations for the wastewaters of the edible oil refining and soapmaking industries, much information has been gathered on the subject in recent years by the EPA. The limitations for the soap industry are currently being reviewed. The limitations for the edible-oil industry have not been completed as of this writing. Therefore, there will be refinements and additions to the information presented here within a few months or a year or two when the EPA work is expected to be available.

The wastewaters are characterized by three principal parameters, which were discussed in a previous section on pollutional properties: BOD, suspended solids, and oil. The values are expressed in terms of a production unit, that is, as pounds per 1000 pounds (μg per 1000 μg) of initial raw oil processed. The volume figures are averages over the operating hours of the process; they are average flows during that period per 1000 pounds of oil processed in the period. The Commerce Department lists a common capacity of an edible oil refinery as 1 million pounds per day. The information for wastewaters from edible-oil refining is listed in Table 8.1. Data for wastewaters from soapmaking are listed in Table 8.2.

1.5 WASTEWATER CONTROL AND TREATMENT METHODS

Introduction and Scope. The intent of coverage of this subject is to assist the industrial employee in formulating and carrying out a waste control program. Discussion of wastewater treatment methods is limited to brief presentation of principles. There are two reasons for this. First, only a small percent of the industries involved provide their own wastewater treatment facilities for direct discharge to nearby public waters. The vast majority discharge their pretreated wastewaters to publicly owned treatment works. So there is limited need to know wastewater treatment processes in much detail. Second, the art and science of wastewater treatment is too complex to be adequately presented here. The industry is advised to utilize the services of specialists who are educated and experienced in industrial wastewater treatment for the design, construction, and operational guidance of such facilities.

However, one treatment operation is covered in detail, namely, gravity separation of floatable oil. The process is so important to subsequent

Table 8.1 Characteristics of wastewaters from edible-oil processing[a]

Category[b]	Industrial Process(es)	Volume (gal/1000 lb Raw Oil)	BOD (lb/1000 lb Raw Oil)	Suspended Matter (lb/1000 lb Raw Oil)	Oil (lb/1000 lb Raw Oil)
1	Solvent extraction and degumming	22	0.06	0.04	—
2	Caustic refining, bleaching, and deodorization	130	4.67	1.69	1.62
3	Caustic refining, bleaching, deodorizing, and acidulation	185	9.36	3.35	2.82
4	Tank car cleaning	27	0.49	0.19	0.20
5	Margarine from refined oil	180	1.93	1.34	2.86

[a] These characteristics apply to wastewaters that have been treated by gravity sedimentation for removal of floatable oil, in other words, floatable oil has been removed.

[b] The operations included in categories 2 and 3 constitute all the processes employed at more than half of the edible oil plants in the United States. Processes such as winterizing and hardening do not ordinarily involve significant wastes and are, therefore, not included in the processes listed. Those plants not acidulating the caustic extract may ship the extract to other plants so that the acidulation process wastewaters may be for larger amounts of oil processed than for other processes at a plant.

Table 8.2 **Characteristics of untreated wastewaters from soapmaking**[a]

	Pollutant		
Process	BOD per 1000[b] Units of Product	Suspended Solids per 1000 Units of Product	Oil and Grease per 1000 Units of Product
Soap manufacture from kettle boil	6	4	0.9
Fatty acid manufacture by fat splitting	12	22	2.5
Soap from fatty acid neutralization	0.1	0.2	0.5
Glycerine concentration	15	2	1
Glycerine distillation	5	2	1
Soap flakes and powders	0.1	0.1	0.1
Bar soap	3.4	5.8	0.4
Liquid soap	1.5	1.0	1.0
Storage and handling	Variable	Variable	Variable

[a] These data are mass quantities of pollutant in wastewater following removal of floatable oil by passing through gravity settling tanks.

[b] Units are unit mass per 1000 units of anhydrous product of process, for example, lb/1000 pounds, μg/1000 μg.

treatment steps that it warrants emphasis. The animal–vegetable oil industry has some factors in such treatment that are unique to the industry.

The final portion of the section on treatment deals with the appropriate regulations and pretreatment for the practice of discharge to public sewers for treatment with other community waters in publicly owned treatment plants.

In-Plant Control. Water pollution control by operator attention to preventing losses to sewers and the use of equipment that inherently has less loss of materials to wastewaters are major elements in a control program. A quantity of process material that is negligible in terms of production is very important and significant as a percent of materials in wastewaters.

A complete water pollution control program includes:

Minimization of materials entering the wastewater by keeping valued materials in the product stream.

Separation of materials for marketing as byproducts that otherwise would be associated with wastewaters.

Recovery of materials from wastewaters for reuse or recycle.

Treatment of the minimized wastewaters.

The cost of wastewater treatment to meet the more severe limitations is such that prevention and recovery incentives are greatly supplemented by costs of treatment. Some representative values of the cost of wastewater treatment are as follows:

Constituent	Costs Associated with Constituent
BOD	8 cents per pound
Suspended solids	6 cents per pound
Volume of wastewater	15 cents per 1000 gallons

A pound of oil contributes about 1.5–2 lb of BOD, say, 1.8. Oil is measured as a suspended solid to some extent in the range of 0–100% depending on the physical form and temperature, say, 25% for this illustration. Thus each pound of oil lost is a total cost of:

BOD 8 × 1.8	14.4 cents
Suspended solids 6 × 0.25	1.5 cents
Value of original oil	30.0 cents
	45.9 cents

The cost of the water conveying the oil in and through the sewers is not included in this estimate. Pollution prevention measures, referred to as "good housekeeping," include such maintenance steps as never allowing a leaky pump or stirrer gland to continue leaking. "Pollution prevention" means using seals or equipment that minimize leaks and the immediate repair of those that leak. It means never draining oil or process material from a line or tank or pump to the sewer. The material should be returned to process or disposed of in the manner of a solid waste. "Good housekeeping" means scooping up spilled material and picking up remnants on absorbent material rather than flushing the surfaces to the sewers. Pollution prevention means not allowing tanks to overflow and by using alarms, alert operators, and catchment systems. Most importantly, it involves not operating equipment at over its rated capacity if it is at the cost of producing a high quantity of material in wastewaters. It also means avoiding misoperations and correcting those that occur without delay.

Gravity Settling.   The art and science of design of sedimentation and flotation facilities is found largely in engineering disciplines involving municipal wastewater and water supply treatment. A technical principle for design of sedimentation and flotation tanks has been in existence for decades. Chemical engineers have utilized sedimentation processes for liquid streams in chemical processes. The volume of such process material streams compared to water and wastewater treatment streams makes for great differences in sophistication of design. In chemical processing, many systems of feeding the inflow to a tank and taking out an effluent with settleable matter removed have been employed satisfactorily. However, more thought to scientific principles should be helpful in large flow rate situations.

The objectives of a gravity sedimentation or flotation facility is obviously to achieve a slow, smooth, tranquil, and uniform passage of the liquid stream horizontally from the inlet end to the effluent end. To achieve this ideally, the inflow would have to be distributed evenly over a cross section of the tank at the inlet and collected evenly across a cross section of the tank at the outlet. During the period when the water is moving through the tank, the oil particles that rise at sufficient velocity and are near enough the surface to reach it will be removed. Thus it appears that a shallow tank would be advantageous because an oil particle in a shallower tank would, on the average, have less distance to rise to the surface than in a deeper tank. On the other hand, for the same surface area, the deeper the tank, the longer the water would reside in the tank and hence the time available for a particle to reach the surface would be longer. A mathematical analysis shows that the theoretical removal should be independent of depth and be proportional to the surface area divided by the inflow rate. This criterion used in design is referred to as the *surface overflow rate* and is usually expressed as gallons per day per square foot. This parameter can also be expressed as a rise rate or settling rate such as feet per day. This parameter conforms to the theory that removal should be independent of depth. A shallow tank will thus have economic advantage. However, there is the obvious limit of keeping the velocity compatible with nonturbulent flow and low enough not to drag floating oil back into the current. Also depth is needed for oil accumulations at the surface and sludge on the bottom.

The concept of separation of an oil particle in a gravity-settling system may be applied to a test program to determine the time necessary for the oil particles to rise to the water surface. In the absence of reasons for expecting otherwise, the oil particles dispersed in water should be distributed uniformly from large particles that reach the surface rapidly, down to very small particles that are subject to other actions and will not rise at appreciable rates. Theoretically, assuming a uniform distribution of particle sizes by weight and that the particles remain discrete, the removal per unit of time would be maximum at first and would drop off with time such that the removal per unit of time would soon be low. A removal time plot would theoretically be a curve of parabolic form. In laboratory tests, such removal phenomena data often fail to follow the theoretical curve closely. Tests to obtain such data are called column settling tests. The test is performed by first placing fresh oil-bearing waste-water samples in a column of about the depth to be used in the full scale settling system. Samples are collected from ports at several points in the column at selected intervals of time. The samples are analyzed for oil and a plot made of oil concentration versus time. From such a plot, a designer can theoretically choose a residence time for a wastewater flow in a settling tank that will achieve removal of the preponderance of floatable oil or some chosen portion so as to achieve all removal possible that it is cost-effective to remove.

As previously mentioned, work experience with wastewaters from edible-oil refining and soapmaking have generally not followed the theoretical prediction. The data typically show a rapid initial removal and very little additional removal after 15–30 min. A number of factors could contribute to the accuracy of the sampling and measurement as to make it difficult to predict plant performance accurately. These wastewaters are typically hot and inevitably there will arise strong convection currents in the test column. These will flow vertically down at the walls and up in the center and keep oil in suspension that otherwise might gravitate to the surface. Measures can be taken to avoid such convection currents, but in a practical sense they are unavoidable.

The employment of such a column test has been advocated for waste-water and water treatment plant design for several decades. In practice, useful data similar to the theoretical are seldom obtained. The test is more an exercise in the training of technicians and engineers than one of design utility. Thus use of such a sophisticated test to design a flotation system is not considered a worthwhile activity. To satisfy a logical need for evidence that a sedimentation system will remove a significant amount of floatable oil a simple bench top test using a beaker or graduated cylinder suffices. Accurate prediction of how much oil will be removed is difficult because if there is a layer or globs of free oil floating on the surface in the sewer, it is impossible to obtain a representative sample.

Thus in the absence of a technical basis for sizing a gravity settling facility, an alternative is to rely on past practices. An EPA-sponsored survey of wastewaters of the edible-oil-refining industry reported that the design loads advocated by industry technicians were as follows:

Surface overflow rate	600 gal/(day · ft^2)
Residence time	60 min
Minimum liquid depth	5 ft

Such a system meets in a practical degree an objective of an effective size of facility; a slightly larger facility would cost more than the additional removal would be worth. Nevertheless, some plants employ gravity separation tanks with residence times of 6 or 12 hours. The incremental additional removal accomplished by the larger facilities are relatively small; therefore, costs per unit are high.

The details of the design of oil separation facilities are theoretically important. However, tests for floatable oil remaining in the effluents of a variety of odd and incongruous designs seldom have revealed a significant concentration of oil escaping. This is attributed to the fact that a liberal theoretical residence time was present in the systems tested. With liberal residence time involved, poor performance has been rare.

Ideal design achieves distribution of the flow coming in, which is typically in a pipe of 12-in diameter or less, evenly over a cross section of a tank typically 50 ft^2 or more. (Here the discussion involves a rectangular sedimentation tank; circular tanks are discussed subsequently.) Traditionally in large settling tanks of municipal sewage systems a number of ports or slots and a baffle ahead of the ports or slots is the practice for distributing the inflow. Such an adoption to a small tank may consist of a small baffle immediately in front of the inlet pipe. With floating oil in the system, it is important that the inlet pipe does not enter below the water surface because oil will accumulate in the pipe upstream wherever the free surface is and may congeal in the sewer and plug it. Also, the pipe should not come in above the water surface because the plunging water plume will cause currents in the tank.

Complex devices to achieve the distribution have been used. A circular baffle with a curved surface just ahead of the inlet directing the inlet velocity back against the wall is on the market. The American Petroleum Institute recommends a design of vertical slots spaced across the tank width. This design is difficult to adopt to the smaller tanks for the smaller flows of the edible-oil refinery and other industries involved. The most definitive work on inlet devices was reported by Mau (1). His laboratory investigations showed that to achieve even distribution across a tank section, the slots in a distribution device must be sufficiently narrow to cause a significant difference in the water surface level upstream and downstream of the slots. However, with such design he found that the velocity of water through the slots was such that the jets extended well into the downstream tank and caused undesirable mixing.

This jet problem was overcome by placing target baffles a few inches downstream of the slots. Also, with the use of the target baffles a lesser velocity in the slots did not sacrifice much in the distribution pattern. Larger openings corresponding to lower velocities are better because of lessened potential for plugging.

Designers have also used a distribution device consisting of a wall across the entire tank, just downstream of the entrance, in which many small holes, say, 1-in diameter, are spaced. Slots are much easier to clean in place than holes. The placing of target baffles opposite each hole such as with the slots discussed previously would be possible but would entail an expensive design.

At the downstream end of the tank there is a need to collect the flow uniformly over the cross section. The usual arrangement is a baffle extending below the water surface to hold the oil layer and following this, an overflow weir. Flow is under the baffle and up over the weir. This flow pattern leaves opportunity for some stagnant areas of the tank. Mau (1) found these in his observations and suggested the use of a wall with slots spaced in it to achieve the full cross-sectional distribution at the outlet as is achieved at the inlet. This eliminates the need of an underflow baffle and overflow weir if a submerged pipe outlet is used. Theoretically, such a design would enable a smaller tank to do the job.

Circular sedimentation tanks are the most popular form for sewage treatment because they are more economical to construct. The most common flow pattern is from a distribution chamber, located in the center, radially to an overflow weir around the periphery of the tank. Floating scum is captured behind a baffle slightly interior of the weir. Many designs of circular tanks with different flow patterns exist, including peripheral feed and central weir, peripheral feed at low depth with a peripheral weir also, and spiral flow from wall to center. The technical considerations are basically the same as described for a rectangular tank. Some circular tanks are in use in the industries involved in this discussion. Commercial units of circular form and of rectangular form are available.

An important adjunct to the system is a means of removing the oil layer accumulating on the surface. The most sophisticated system is a continuous mechanical collector. For a rectangular tank, such a commercial device uses shallow moving collecting baffles extending a couple of inches or so below the water surface. These are mounted on endless chains and are moved slowly along the tank. At the end of the tank the baffle enters a wide trough with the floor sloping upward and ending a few inches above the high water surface. The oil trapped in this trough behind a collecting baffle is pushed up the trough as the baffle moves through it and the oil falls into a sump. This system operates satisfactorily in some applications. If the oil is very thin (nonviscous), there is difficulty in the seal between the moving baffle and the trough. These devices are also expensive, in the order of $25,000 installed. Such a device for a circular tank consists of a moving baffle at the end of a rotating arm pivoted at the center. The collecting baffle extends from the peripheral baffle inward just a few feet.

An alternative to the continuous mechanical skimmer is a stationary collector. The more sophisticated form is a cylinder about 12 in. in diameter and positioned across the water surface and mounted at the ends in slip bearings. The pipe has a slot cut in it lengthwise. The pipe is equipped with a means of rotating it so that the slot may be positioned at any elevation desired. The slot is left at the top normally. When it is desired to skim oil, the pipe is turned until the slot is below the oil surface. The oil flows into the pipe and to a sump at the end. This device enables taking a thin layer the full length of the slot so that the trap can be skimmed quickly.

The simplest skimmer is a pipe system with swing connections. It may be fitted with a funnel inlet. The end of the pipe is raised or lowered by a hand-cranked ratcheted axle and cable or chain.

Within the last 10 years skimming devices have come on the market that make use of the property of some plastic surfaces for having an affinity for oil in preference to water. A belt or a hose of such material is drawn through the oil layer on the basin. The adhering oil is wiped off at a point above and flows to collectors. One advantage is that the oil removed is relatively free of water. Some such devices have experienced plugging problems from pieces of plastic trash floating in the water. The device is not operable when not kept at temperatures that prevent congealing. Stationary baffles in tanks and use of multiple tanks in series add to the skimming work because of the need to skim at several points.

The collection of the oil is not possible unless it remains fluid. In severe climates the oil will congeal on the surface. This may happen only following plant shutdowns as on weekends. Discharge of live steam directly in the tank to liquefy the congealed material will disturb the flow and cause loss of oil. To overcome this, a system of heating pipes placed just below the surface can be used. Steam or hot water in the pipes will keep the oil liquid. The convection currents from such practices are limited because the high-temperature material tends to stay where it is created, that is, on top. Heating has also been practiced by providing a cover for the tank and admitting steam to the air space.

The accumulations at the surface in a gravity flotation tank will commonly consist of an oil layer or an oil-and-fatty-acid layer and a second layer of water-in-oil emulsion—typically about 10–30% oil. The emulsion layer may be skimmed with the oil or the emulsion layer may be left for a suitable time to break the emulsion and release the oil.

Usually some water is drawn with the oil or oil-and-emulsion skimmings. The mix of skimmings is pumped to an accumulation tank, where the emulsions are broken by the use of heat. The operation and decanting steps may be automatic or by hand. The pumping of the skimmings must be with pumps that do not create emulsions.

In gravity flotation system design and operation, thought has to be given to any problems from settleable matter accumulating in the tank. A representative plant has a minimum of such materials in the wastewaters. Such matter is usually limited to a little decolorizing clay and perhaps some bean fragments in the raw oil. Of course, the settling tank can be equipped with mechanical devices for continuous or frequent removal of the contaminant. In plants of the industries involved here this is rarely needed. The material that settles out is suitable for transport and treatment in municipal systems. Thus contaminated material can be drawn from the tank to the downstream sewer periodically if the quantity involved does not cause a slug load at the municipal plant. If the economics favor disposal in the manner of a solid waste, this may be the method of choice. Of importance is that the sludge not be allowed to accumulate to such extent that it occupies so much of the volume of the tank as to cause poor efficiency of the system in removing oil. There is a human tendency to allow a layer of oil, emulsion and sludge to accumulate so that the system is not as effective as it could be.

The effectiveness of gravity-settling systems is significantly improved in typical cases if the wastewaters are highly acidified. A pH of about 2 is most effective. The acid conditions break emulsions and also convert fatty acid soaps to free fatty acids that are insoluble and will float. Such acid conditions can be maintained in some plants if wastes from the acidulation process are continuously discharged to the sewers without neutralization. The acid has to be neutralized, of course, downstream from the settling tank before discharge.

This leads to a caution about materials of construction of gravity-settling tanks and sewers leading to them. Typically, the wastewaters of the plants involved contain sufficient concentrations of fatty acids that they cause disintegration of common concrete in time. Disintegration of the concrete is not corrosion because of acidity, but a property of fatty acids. There-fore, unprotected concrete tanks, concrete sewers, and manholes are not recommended. This is particularly true if the acidic conditions recom-mended in the previous paragraph are utilized. Tanks of stainless steel and of reinforced fiberglass have been employed. Also, concrete tanks have been provided with troweled-on liners.

Logic would indicate that pumping wastewaters to such a settling sys-tem can lower removal effectiveness by emulsifying oil as a result of the shearing action of the pump. For this reason, construction of the tank in the ground so that flow to it is by gravity is recommended. If economics dictate pumping, pumps can be selected with low peripheral speeds so as to minimize emulsification.

In some circumstances there may be an advantage in improved flotation effectiveness if continuous or frequent removal of the oil as it accumulates is practiced. If there are periodic losses of highly alkaline wastes or wastes containing emulsifiers, an accumulated oil layer can be emulsified and pass out in the effluent. In the absence of such events, the accumulation of a depth of oil and its automatic collection by an overflow set at a selected elevation are possible.

Before leaving the subject of gravity flotation, it is of interest to mention European design of such systems. European practice and even recent European literature describes practices in which three circular tanks are used in series. Inlets are baffled and outlets are from below the surface via elbows or tees. Such a design would be similar to a rectangular basin which was compartmentalized by baffles. Such baffles add to dead (un-used) space in the tank and, hence, are not recommended.

Treatment with Community Wastewaters.   At least 90% of the indus-trial establishments that process animal or vegetable oil discharge their wastewaters to public sewers for treatment with residential sewage and other industrial wastewaters of the community. This practice has eco-nomic advantages for wastewaters compatible with such joint treatment. The lower costs per unit capacity of larger treatment works is shared among residential, commercial, and industrial users. For many urban industries, this means of treatment is the only practical way because of limited space for industry to take care of its wastewaters itself. Unfor-tunately, municipalities have traditionally placed limitations on concen-trations of oil in wastewaters that are accepted in the public sewers. At one time most municipalities had a limitation of 100 mg/liter of total oil in the wastewaters. Many of these ordinances had provisions for variances so that industries with high concentrations of dispersed animal–vegetable oil were able to continue to discharge their wastewaters with concentra-tions of dispersed oil of 500 mg/liter or more. Some municipalities have formal variance provisions, but few actually enforce the limit on a case-by-case basis when they realize it does not serve a useful purpose. At present, only a few communities are known to enforce a 100-mg/liter limitation on wastewaters from animal–vegetable oil processors.

With ever-increasing interest in environmental control and with stepped-up emphasis of the EPA on pretreatment, renewed and rigorous defense of this practical way of treatment of dispersed animal–vegetable oil is required on the part of industry. This is best done by determining the reality of problems alleged to occur from higher concentrations of oil, examining the reasons for these problems, and, finally, comparing the economics of removing the oil before discharge.

The problems encountered in community wastewater systems in transporting and treating oil-bearing wastewaters, although rather numerous, have solutions that allow continuation of this practice. With a modest degree of control, the evidence is that treatment of *certain kinds* of oil in the physical form referred to as the *dispersed* state, and at the concentrations encountered in normal industrial practice, is a practical and advantageous means of water pollution control.

One of the problems in municipal systems is caused by the congealing of animal–vegetable oil in the sewers and the subsequent obstruction of sewers and control mechanisms. This problem is largely avoided if oil, which is of a droplet size sufficient to float to the surface in the sewers and pump stations, is removed at the plant before discharge. The removal of free-floating oil by gravity settling is rather easy and has previously been described in this section. Municipal system managers have reported that obstructions in small sewers serving restaurants and foodservice establishments have been a problem. These have been largely solved by traps or settling tanks in the drains for removing free-floating oil.

One major municipality has reported the most frequent sewer blockages occur in small sewers serving apartment buildings equipped with garbage grinders. The blockages are composed of fibrous material, garbage pieces, and congealed fat. It is not surprising that congealed oil or fat is found. Residential wastewaters contain appreciable amounts of oil and fat. In fact, the normal 30–50 mg/liter in strictly domestic sewage makes up to 15% or more of all the organic matter present. Furthermore, these household wastewaters receive no pretreatment, so part of this oil is in floatable form. Gravity settling of domestic sewage will remove appreciable oil; about 30–40% is a representative figure.

There can be problems with the handling of oil-containing skimmings from the treatment of the sewage as it will congeal and coat surfaces. Pipes can become completely plugged. Even pumps can plug and systems can be ruptured by the pressures such obstructions cause. But satisfactory systems for removal and transport of such skimmings are a necessity for a good design. Any design that does not make this possible is faulty and should be laid to inadequate knowledge of sewage because all household sewage has oil and grease. Designers have learned to use larger pipe sizes, to use pipe with very smooth surfaces, to avoid elbows of short radius, to provide steam or other methods of heating the lines, to include cleanout plugs at appropriate locations, and to provide adequate access to cleanout locations. Thus systems have been created that work without intolerable problems.

In about half the municipal plants in the country the skimmings (with their content of oil) and other residues (sludges) are processed by anaerobic digestion. In this process the sludge is held in large tanks for prolonged periods, 15–30 days or more. Anaerobic bacteria convert some organic matter in the skimmings to methane. The remaining residue is less offensive and more easily disposed of. Prior to about 1950, these digesters were not heated nor were the contents mixed or stirred regularly to a significant extent. Consequently, oil entering the digester would rise to the surface and accumulate there. Layers of congealed oil and other floating material would solidify to depths of several feet. Periodically the digester would have to be emptied and this material laboriously removed. The scum layers were real problems, and cleaning was a disagreeable task.

In designs of the last 30 years or so digesters have been provided with means for keeping them mixed and heated to speed up the process. The scum-layer-forming tendency in such digesters is much reduced and is manageable. Coincident with this practice came the knowledge that some of the oil when kept fairly dispersed was converted to methane in the digester. Triglycerides and fatty acids, if kept dispersed in the digester, are degraded to methane by bacterial action. In fact, these materials are converted at a greater rate and to a greater extent than are other classes of organics, such as carbohydrates. In dispersed form these materials are valuable sources of methane. Petroleum oils (hydrocarbons), on the other hand, are not degraded in the anaerobic digesters. The undegraded liquid oil remains in the liquid separated from the stabilized solid matter. At the plant where this liquid material is recycled to the incoming sewage, the oil is recycled through the system over and over again.

Another supposed problem with oil in sewage treatment plants concerned the fate of the oil that passed through the initial settling process and was present, therefore, in dispersed form in the main wastewater flow going to subsequent treatment processes. The subsequent treatment process is, practically speaking, almost always biological. Fundamentally, biological treatment consists of maintaining a great mass of bacteria in a system and bringing the wastewaters into close contact with the bacteria. The bacteria oxidize and thus degrade organic matter in the sewage. The masses of bacteria are subsequently separated from the wastewater, and the treated wastewaters flow on to the river or other water course. The most common system maintains the bacteria in a suspended particle state with the particles large enough to be separated by settling from the treated wastewater. This is known as the *activated sludge process*. In other forms of biological treatment the bacteria are maintained attached to the surface of some media such as small rock. One such system is referred to as a *trickling filter*. Early in sewage treatment the belief prevailed that oil was not degraded in these bacteriologic processes. Again this case against oil in wastewaters was brought about through failure to recognize differences in the properties of the different kinds of oil. Evidence slowly built up beginning more than two decades ago that triglycerides and fatty acids of animal and vegetable oils were biodegraded in these processes without any problem. Some evidence was developed that petroleum oil (of the crankcase variety) was not degraded at like rates and that too much would cause agglomerates of oil and bacteria that would not settle from the wastewater. Consequently, these particles would be present in the treated effluent.

Very recently information has been published showing data from 55 publicly owned treatment works (2). Recently the U.S. EPA has made data available on two municipal plants whose treatment effectiveness was evaluated in a research study (3). The data from these 57 plants is on total oil, that is, as measured by the common analytical method for wastewater analysis. The data show that more than 90% of these 57 plants achieved a greater percent removal of oil than their percent removal of biodegradable organic matter as measured by the BOD test. The concentrations of total oil in the wastewaters entering the treatment plants ranged up to an average of over 100 mg/liter. Since domestic sewage typically contains only 30–50 mg/liter of oil, the contribution from industrial wastewaters was much in excess of 100 mg/liter. The average oil content of the effluents was in the range of 3–8 mg/liter for those plants that were in compliance with the national BOD limitations for biological treatment. These effluent concentrations were for total oil; there were no data available on these plants on the kinds of oil present. These removals were obviously achieved without limitations on the concentration of oil in the discharges of industrial users.

Evidently, municipal treatment can be effective in removing some petroleum-type oil even though the degradation rate is slow. This could be accomplished in two ways. The oil could become agglomerated with the clumps of bacteria and be removed as part of the excess bacterial sludge as it is separated from the system. Or, if the oil were held long enough in the presence of the bacteria, it could be degraded even though the rate would be slower than for other materials. The average period for bacterial clumps to be kept in the system is 5–10 days, sufficient time for even a slowly degraded material to be degraded to an appreciable extent.

Thus it appears that there is no reason to limit concentrations on dispersed animal–vegetable oil on the basis of their not being biodegraded in the system. These materials are degraded at about the same rate and to about the same extent as is the broad spectrum of biodegradable organic compounds found in domestic sewage.

Animal–vegetable oil in municipal effluents can logically be controlled as components of the organics that are measured in the test used for biodegradable organic matter, namely, the BOD test. Animal–vegetable oils in well-treated effluents of biological treatment systems do not exhibit any surface film problems. Animal–vegetable oil is not toxic, does not impair the flavor of fish flesh, does not cause taste and odors in the water, and does not persist long in the natural waters because it is biodegradable at a nominal rate. Therefore, no effluent limitation or water quality criteria for animal–vegetable oil is needed.

Treatment for Direct Discharge.   The first step in treatment of wastewaters from this industry is the removal of floatable oil by gravity settling. This process is, of course, applied to wastes that have been minimized by in-plant control as previously discussed. The oil remaining after gravity settling is in a dispersed state or in solution. The removal of the dispersed oil and other biodegradable organics can be accomplished by either of two treatment systems: (*a*) biological treatment or (*b*) chemically enhanced air floatation followed by biological treatment. The choice is an economic one between chemically enhanced air flotation and the additional biological treatment capacity. Practically all the plants in the industry have chosen to use chemically enhanced air flotation followed by biological treatment. Some managers have utilized air flotation because it offered a backup process to gravity settling when a slug or an emulsifying agent loss would cause abnormal passthrough of oil through gravity settlers. Troublesome massive overloads of biological treatment with oil are thus avoided.

The dissolved air flotation process consists of creating minute air bubbles in the wastewater introduced to a settling tank; these air bubbles tend to become attached to surfaces such as the surfaces of oil particles in the wastewater (see Figure 8.1). A few air bubbles on a particle of oil will rapidly buoy it to the surface. The air bubbles seldom become attached in sufficient numbers to smaller oil particles to permit their re-

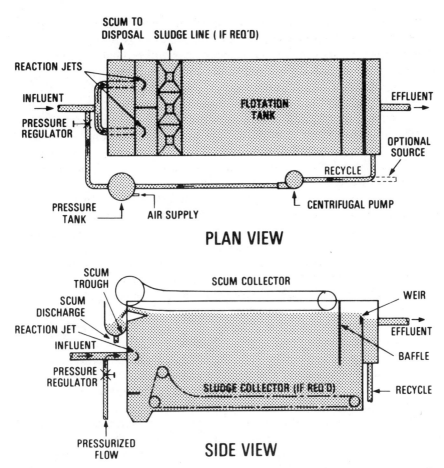

Figure 8.1 Dissolved air flotation process equipment. This model utilizes rectangular tanks and pressurization of a recycle stream.

moval, so air flotation alone will not be very effective. A necessary step is formation of the small oil droplets into larger agglomerates by use of coagulating chemicals, hence the name *chemically enhanced air flotation*. Chemicals, such as aluminum hydroxide (added as aluminum sulfate) or an organic flocculant, form bridges between particles so that an agglomerate of oil particles is formed that is large enough for the air bubbles to buoy them to the surface.

Small air bubbles may form by mechanically creating them or by a process that causes them to be created by air coming out of solution. The latter process, called *dissolved air flotation*, accomplishes the solution of air by pressurizing the flow at about 50 psi and contacting the water with air. In the flotation tank the pressurized water is released to atmospheric pressure and the air comes out of solution in the form of fine bubbles. Various designs are marketed embodying the air flotation process. In principle, they resemble gravity flotation. The surface overflow rates may be four times or more than those used in gravity flotation because the rise rate of the particles is greater with attached air bubbles.

The flotation process effectiveness is very dependent on the chemical coagulation step. Coagulation will not take place optimally outside a certain pH range nor in the presence of higher concentrations of dispersing agents, emulsifiers, and surface-active materials. The latter are likely to be present in the waste in variable concentrations. Dispersing and surface active agents such as caustic, detergents, and phosphates are likely to be present from cleanup operations. Emulsifiers such as mono- and diglycerides are naturally present, and such compounds are used in salad dressings and margarine for their emulsifying properties. Therefore, the process is not 100% reliable. For good results in a majority of cases, the wastewaters must be rendered more uniform in composition by provision of a large equalization basin. Close control of the pH so that it remains in the desired range is also a prerequisite. Even with reasonable application of these precautions, chemically enhanced air flotation cannot be depended on under the technical control and level of attention practical in the field to consistently and continuously produce an effluent of low oil content. Occasionally, the emulsion simply cannot be broken for a variety of reasons, such as variations in acid and alkaline conditions, and in the concentration and kind of emulsifiers. The science or art of coagulation is not at the state that perfect performance can be assured.

The effluent from the gravity-settling step or from chemically enhanced air flotation must be given extensive biological treatment to meet the categorical limitations for direct discharge to be imposed by the U.S. EPA. The agency is expected to promulgate limitations for conventional pollutants based on Best Conventional Technology in 1981, which must be complied with by every direct discharger by mid-1984. The agency will also promulgate limitations based on Best Available Technology if toxic pollutants are found in the effluents.

Biological treatment is described briefly in the subsection on municipal treatment. Two biological treatment processes are presently utilized for biological treatment of wastewater from animal and vegetable oil processing: (a) the *activated sludge process* (see Figure 8.2) and (b) the *aerated lagoon process*. Operating data from existing plants show that the loading parameters used in designing such processes for domestic sewage or other biodegradable industrial wastewaters apply. The bio-

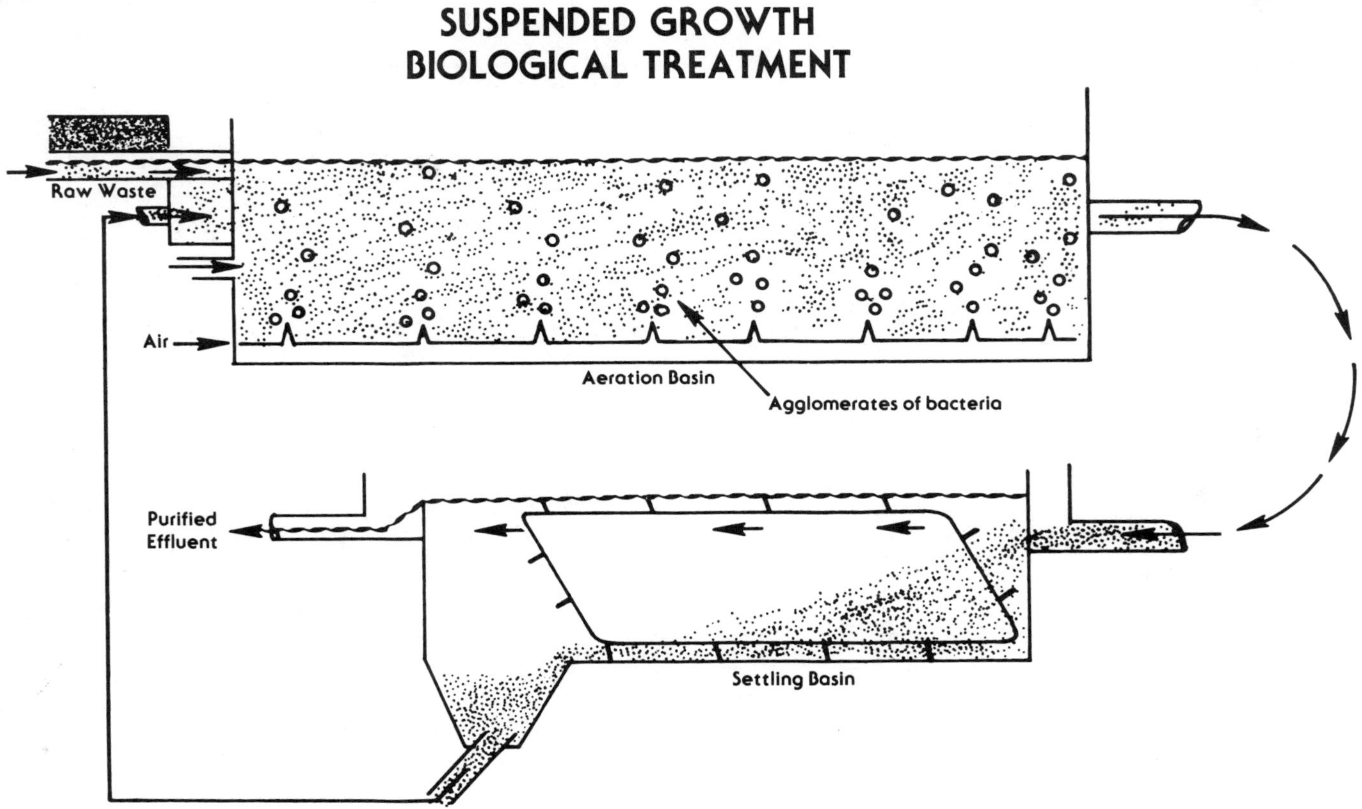

SUSPENDED GROWTH BIOLOGICAL TREATMENT

Raw Waste

Air →

Aeration Basin

Agglomerates of bacteria

Settling Basin

Purified Effluent

Figure 8.2 Activated sludge wastewater. Treatment process consists of contacting wastes with bacteria culture in aerated basin and settling to remove and recycle settleable bacterial agglomerates.

degradable material to microorganism ratio (pounds of BOD per pound of biological mass) is in the same range as used in such designs for municipal sewage. One plant is using a combination of activated sludge and aerated lagoon treatment.

At plant locations on small streams an additional treatment requirement has been applied. This consists of filtration through beds of fine material, including sand. The processes used in one of the more complex treatment systems is shown in Figure 8.3.

As discussed in the section on pollutional effects, one or possibly a few animal and vegetable processing plants practicing joint treatment are or will be subject to limitations on phosphorus. The practice and methodology for removing phosphorus from wastewaters is limited to the precipitation of phosphorus occurring in the form of phosphates. Phosphorus in these industries' wastewaters are in the form of organic phosphorus compounds, and it is doubtful that the present precipitation technology for phosphate will apply. The treatment technology has yet to be worked out.

Air Emissions and Odor

2.1   SOURCES OF EMISSION

Emissions to the air to be addressed here are those from extracting and processing the oil to produce edible products or soap. Emissions from a boiler for steam generation, usually found at a plant, are not included. There is nothing unique about the boilers of the fat and oil industry. Other sources of information on air pollution control of boilers are readily available and are not included in this chapter.

The raw fats and oils involved are of low volatility at ambient temperatures and pressures. Also, they are not prone to form mists or dusts, and their sulfur content is negligible; therefore, it is not surprising that, in general, animal–vegetable oil refining processes have no significant air emissions.

Solvent Extraction.   One segment of the industry, however, uses volatile materials and has received some attention as a potential air pollutant source. The process involved is the solvent extraction of oil seeds. For the extraction of soybeans and cottonseed, for example, hexane is commonly used. The solvent is volatilized from the extracted oil, condensed and recycled. Solvent losses from the extract are not large. The solvent remaining with the oil seed is the potential problem. The oil-seed residue

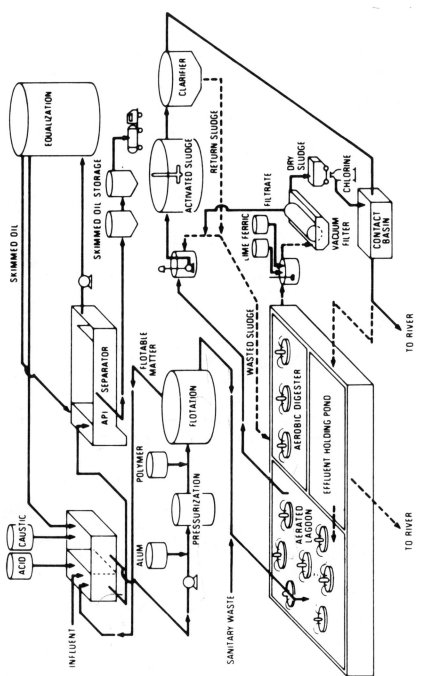

Figure 8.3 Complete complex treatment train.

571

is processed to remove as much solvent as practical in equipment referred to as the *desolventizer–toaster,* which is defined in further detail at the end of this paragraph. Substantial volumes of offgases (air) are involved. A representative airflow is approximately 30,000 scfm* for a 1000-ton/ day plant. In June 1978 the U.S. EPA published a Control Techniques Guideline (CTG) for the manufacture of vegetable oils entitled "Control of Volatile Organic Emissions from the Manufacture of Vegetable Oils" (4). The guidance provided by this document was admitted to be in need of verification; hence regulatory action has been modest pending better data. That document described quite well common soybean processing. Significant differences exist for processing other oil seeds (corn, cottonseed, and peanuts) and for special desolventizing processes, such as human food protein production from soybeans.

The common process consists of first screening the seeds to remove sticks, stones, leaves, and soil. The seeds are then cracked, hulled, and conditioned by steam cooking. The seeds are then flaked, that is, pressed into thin pieces, so as to achieve a high extraction rate. The flakes are then extracted with hexane in a system providing intimate contact in countercurrent flow of solvent and flakes. The oil–solvent mixture is pumped to the separation operation where the solvent is removed from the oil. This is accomplished in film evaporators and distillation columns using steam as a stripping medium; virtually all solvent is recovered. The consumption of solvent commonly experienced in the industry is about 1 gal per ton of beans extracted. The extracted seed (meal) is conveyed to a solvent stripping device referred to as a *desolventizer-toaster* in which steam and air are forced through and across the meal to volatilize the solvent. The offgases are discharged to the atmosphere through tall stacks.

Refining Operations. In the refining processes of the industry, not involving a solvent, materials are volatilized in process, but in every case the volatilized material is condensed for recovery or the vapor material is condensed along with water vapor to maintain low pressures in the systems. Therefore, significant atmospheric emissions are rarely, if ever, created. Organic materials in the vapors in some of these systems are condensed directly in the condensing water sprays. This condensing water is typically recirculated over cooling towers. The exhaust air stream from the cooling tower may contain some organic material from droplet entrainment or volatilizations. Only one local community with extreme emission limitations has found such a source of organic matter to be significant.

* Cubic feet per minute of gas flow at specified standard conditions of temperature and pressure.

A small quantity of soap production for laundry purposes is marketed in the form of small particles of soap (flakes). These are produced in drying towers. The soap in the liquid state is sprayed into the tower at the top and falls countercurrently to a stream of hot air. Water evaporates from the drops, and soap flakes are produced. The offgases from the drying towers contain a small quantity of dust. Significant volatile organics are not present. The air velocity in the tower is so low that a very small amount of dust is carried into the offgases. The offgases are typically cleaned using cyclones.

Vents from storage tanks of liquid raw materials and products that have a certain vapor pressure can be a source of vapors. This emission occurs principally on the filling of the tanks. The vapor pressure of animal and vegetable fats and oils is such that the contribution is not significant.

2.2   AIR POLLUTION IMPACTS

As discussed in the previous section on air emission sources, animal–vegetable oil processors need have limited concerns. The single source of attention to date has been the solvent, hexane, used in extraction of the oilseed. The volatile organics are controlled in air quality programs because they are involved in the formation of photochemical smog. The smog is a result mainly of the interaction of hydrocarbons from automobile emissions and other sources with nitrogen oxides under the influence of sunlight to form ozone. Ozone then reacts further with the substances in polluted air to form a variety of substances. One of the products of reaction between ozone and hydrocarbons is free radicals that accelerate smog formations. Photochemical smog is linked to eye and bronchial irritation, plant and crop damage, and reduced visibility.

To date, the other air pollutants for which control requirements have been or are being set include particulate matter, oxides of nitrogen, sulfur dioxide, photochemical oxidants, and carbon monoxide. These, plus volatile hydrocarbons, are specific pollutants identified by the U.S. EPA as pollutants that have an adverse effect on public health or welfare (Clean Air Act, 42 USC, Paragraph 1857-C-3 and 40 CFR, Part 50, 1976). Except for volatile hydrocarbons from solvent extraction, none of these is likely to be emitted from animal–vegetable oil processing operations.

2.3   REGULATIONS ON AIR EMISSIONS

Knowledge of the constituents involved in irritating and hazy atmospheres is not in a mature status. Scientific debate is continuing on the levels of pollutants that can be tolerated with acceptable health risk and on the selection of harmful pollutants and their measurement. The animal–vegetable oil processing industries appear to be impacted only slightly by air pollution laws and regulations.

The Clean Air Act Amendments of 1977 required each state in which there is an area not attaining the ambient air standards involving volatile organic components to adopt a plan to achieve those standards. The plan is required to contain regulations reflecting the application of "reasonably available control technology" (RACT). The RACT program is currently being developed for the manufacture of vegetable oils by the EPA and is directed at the *desolventizer–toaster* operation of oilseed extraction. Although not final, the required control technology likely will be one of the following: (*a*) a mineral oil scrubber system, (*b*) a carbon adsorption system, or (*c*) other means. Whatever methods are chosen, removal of 90 or even 95% of volatile organic carbon being vented without the controls will probably be required.

If the meal is not well desolventized in the *desolventizer–toaster*, there may be appreciable loss of solvent in subsequent meal cooling and drying; thus RACT technology for control of the latter may be imposed.

2.4 ODOR SOURCES

Odor appears to be a logical subject to associate with air pollution discussions but on close examination is quite different. Odor is not subject to mass limitations because in most cases no specific substance can be quantitated. Odor is also not directly health related, so the basis for regulation is absent. However, both odor and air pollution involve the atmosphere, and control methods may be similar.

Animal and vegetable oils as they are separated from fresh animal and oilseed sources do not have a significant or strong odor. However, on storage, with time and the growth of bacteria, and on autoxidation the triglycerides will be broken down to short-chain fatty acids and other species that have an odor that most people find highly objectionable. Those materials are also fairly volatile. Therefore, on storage and transport of oil care is taken to keep bacterial degradation and autoxidation from occurring.

Odorous materials naturally present in the oil or developing during processing are removed from the finished products. In soapmaking processes the odorous materials are transferred to water, mainly to the condenser water used in sprays or in the steam condensate itself if surface condensers are used. Specifically, the fatty acid still condensate may contain some objectionable odorous material. If a direct condensing system is used, the odor will be in the water. If this water is recirculated over a cooling tower, odors will be present in the air emissions of the cooling tower.

Similarly, in edible-oil refining, odorous materials are contained in the vapors from the deodorizer operation. These vapors will be contained in the condensate from surface condensers or in the condensing water of direct condensers. These odors will be transferred to the exhaust of the cooling tower if the direct condensing water is cooled and recycled.

Miscellaneous odor sources include any vents of steam to the atmosphere. Also, if a direct-to-atmosphere steam eductor is used for removing noncondensibles from the distillation and deodorizing operations, the process may be objectionable.

The acidulation operation for recovering triglycerides or fatty acids from soapstock, soapy water, or fatty acid still bottoms may also be a significant odor source. The source of these streams is discussed in the preceding section on water pollution. The acidulation process, if done batchwise in open tanks, usually includes boiling the tank contents. The vapors arising from this can be highly odorous.

In addition to odors entering condensing water of recirculated systems, the oil and fatty acids and other material in the recirculated water can under certain conditions be degraded by bacterial action and autoxidation and thus odors can be created in the system. The odorous materials are most likely produced in the more static sections of the basin under the cooling tower or in the hot well in the skimming basin.

2.5   REGULATIONS ON ODOR

There are no federal regulations on odor; therefore, no description of universal abatement regulations is possible. State and local regulations must be addressed, with all their opportunities for diversity, in relating the regulation picture. Federal regulations have been recently considered by Congress. Section 403(b) of the 1977 Amendment to the Clean Air Act directed the EPA to study "the effects on public health and welfare of odors or odorous emissions, the sources of such emissions, the technology or other measures available for control of such emissions and the costs of such technology of measures, and the costs and benefits of alternative measures or strategies to abate such emissions."

To assist the EPA in this effort, the National Research Council of the National Academy of Sciences conducted a study. They concluded that the establishment of federal ambient air quality or emission standards for odors would confront conceptual and technical difficulties. First, the adverse effects of odors on people are variable, and our knowlege is very incomplete. It would be difficult to define standards that will be widely accepted. Odor measurement methods are costly and time consuming. With the assistance of this report, the EPA has recommended that no federal law for control of odor be passed.

The abatement of nuisance or objectionable odor has and continues to be pursued under local and state laws. Lack of federal participation will probably not interrupt or change these programs. Some local programs are based on establishing the existence of an annoying or nuisance odor level, although regulations can state that *no* odor is to be detected beyond the property line. Odor regulations can also be stack emission type standards as opposed to ambient odor type standards.

Abatement of odor can be pursued through state nuisance laws under two fields of liability, termed "public nuisance" and "private nuisance." A public nuisance is an act resulting in the public being denied a right common to the public such as enjoyment of a park. A private nuisance interferes with an individual person using his or her land and property. These remain valid avenues for seeking abatement even if odor law and regulation programs are not in effect. In a private nuisance action, the complainant must establish interference with use of land by disagreeable odors for which a defendant can be held responsible. To be at nuisance level, the odors must be judged a substantial annoyance by a reasonable person. The establishment of a nuisance odor in an industrialized neighborhood may be impossible if the odor is characteristic of an industrialized neighborhood or considered harmless by residents. The conduct of the producer of the odor must be established as unreasonable.

Public nuisance charges can be pursued if the rights invaded are genuine public rights and not just a number of individuals, however large. This action can be taken by public authorities. The case hinges mainly on the number and the reliability of witnesses who testify to the existence of the nuisance. Traditional nuisance law has serious shortcomings of costs, time, risks, management, and other factors. Its chief merit is its more convincing evidence of annoyance by complaining witnesses than measurements, surveys, comparison with standards, and so on.

The problems with nuisance law being a timely and truly practical tool for the public has led many states and local authorities to adopt specific odor control regulations. Such laws allow a simpler administrative decision that a nuisance exists, administrative enforcement rather than court interference, and enable public prosecution of cases that otherwise would have been left to private individuals. Various ways have been followed in determining what constitutes an "acceptable lack of odorous emissions." These programs utilize one or a combination of the following criteria for abatement action:

Number of complaints received of disagreeable odors.

Community census or survey views.

Measurement of ambient odor by some accepted technical method.

Specific odor limits on emissions from stacks, either substance or odor units.

Specific numerical ambient standards for specific odorous substances.

Applications of specified odor control technology such as incineration of offgases at certain temperatures and dwell time minimums.

Each of these has its deficiencies and can be arbitrary, unfair, or unreasonably expensive.

Corrective actions on the basis of complaints alone are questionable on the basis that many such complaints are typically found invalid for such reasons as being upwind; registered when operations were down; or registered by supersensitive people, cranks, or acts of harassment.

The census approach or representative survey is costly, time consuming, and itself may generate complaints.

Ambient air standards based on measurement of odor are challengeable on the basis that no absolute methodology exists, that the relationship between measurements and a nuisance or annoyance level has not usually been established, and difficulty in obtaining ambient air samples at the required time and place. The use of ambient air standards for specific substances has been in question because it is unlikely that many odorants can be measured accurately by analytical techniques that are practical.

Emission limits based on odor measurements suffer from the methodology challenge, by lack of correlation between emission standards and annoyance thresholds in each community, and the fact that fugitive emissions are not included.

Universal application of specified control technology could result in unneeded expenditures of energy, capital, and labor unjustified by annoyance odor cases. They could also deter technologic innovation.

The previous discussions have pointed out the use of measurements of odor intensity in connection with odor regulation. Odors are sensations and must be assessed by measuring human response to them. Several sensory methods assisted by instruments have been developed and applied to sources of odor and to ambient air measurements. The most common measurements use the limit of detection, known as the *odor threshold,* in which the odorous air or gases are mixed with odor-free air until no longer detected. By use of a trained panel reasonable reproducible values can be obtained. The methods require careful attention to the acquisition and preservation of a representative sample and the selection of appropriate human judges.

One dilution method is the ASTM syringe dilution technique (5). This incorporates a batch dilution technique; a sample is diluted to a desired fixed volume. The original method has been improved to eliminate the trial-and-error feature (6, 7). A more sophisticated instrument uses a dynamic dilution; a flow of the odorous sample is mixed with a flow of the odorless air in a desired ratio. One such instrument is the Scentometer (8). Other dynamic olfactometers are in the development stage.

When complaints form the basis of control, there is little need for process measurement, and control programs can function without specific odor measurements.

A major criticism is that all of these are applicable to after-the-fact corrections and leave no procedure for evaluating the likelihood of odor problems and the control needed for new sources. One element of the new source problem is that atmospheric dispersion models are not well perfected.

The odor control program that the animal and vegetable processing industry is likely to encounter will be based on odor complaints in the plant locality and validation of complaints by agency personnel using quantitative methods at property line limits. Air emission standards or specific substance ambient air standards are not likely.

2.6   Control of Air Emissions and Odor

As discussed in previous subsections, the significant air emission from industrial plants involved in animal–vegetable oil processing is primarily from some solvent extraction operations. Only a small segment of the industry utilizes solvent extraction, namely, those plants utilizing solvent extraction for separating the raw oil from oilseeds. The losses of the solvent from oilseed extraction has been the subject of investigation by the U.S. EPA (4). That agency is charged with establishing best control technology. The EPA is scheduled to promulgate the various control technology-based standards for oilseed extraction plants within the next few years. It would be inappropriate at this time to speculate on what control technology is practical and economically feasible. Managers of these plants are apprehensive about the costs and the energy needs likely to be involved. All plants have solvent recovery and recycle systems; the degree of recovery is a question of economics.

The industry has faced isolated instances of odor control problems. Air emissions and odor are discussed together in this control section because it is estimated that 50% or more of the complaints about air pollution deal with odors. Also, the control of odors in most of the cases involves air emissions released with the odors.

As discussed in a preceding subsection on odor sources, the major source is from condensing water recirculating systems. Odors from such recirculated systems of oil-bearing water can originate in materials condensed from the process vapor streams or from anaerobic conditions of bacterial growth in portions of the water system.

The correction of problems from bacterial growth is relatively easy. The addition of a bactericidal agent, such as chlorine or chlorine dioxide, has been reported to be successful. The dose is low and the residual chlorine dissipates readily in the discard stream. Manufacturers of chlorine dioxide and chlorine will advise potential customers including dose requirements. Many recirculating contaminated cooling water systems do not experience odor problems. An ample discard stream helps to maintain low concentrations of contaminants.

Reduction of odors in recirculated water streams is being accomplished by treatment of the recirculated water and by treatment of the exhaust air from the cooling tower, or both. The treatment of the water stream with chlorine dioxide has been reported to achieve a significant odor reduction as reflected in panel observations of the plant atmosphere and the number of complaints. Further reduction in odor has been reported to be achieved by mist sprays of dilute hypochlorite solutions into the offgases of the cooling tower. In assessing potential odor problems from recirculating contaminated water in cooling towers and the effectiveness of control, it has been found that threshold odor measurements on the recirculating water are an effective tool and are easier to run than are odor threshold measurements on the air emissions.

Additional measures have been taken, but these and the ones discussed earlier have not been described, nor have data on effectiveness been presented for peer review in technical literature. One such measure is the addition of powdered activated carbon to the recirculating water. This carbon is discarded as a component of the discarded water stream from the process. It seems improbable that this measure is effective because the high concentration of other organics in the wastewater would take up much of the adsorptive capacity of the carbon.

At a number of locations in the United States the control measures described in the preceding paragraphs have not obtained the odor control necessary to satisfy complaints or odor limitations, as made by agency personnel or private complaints. In several cases the control measure taken was to abandon the direct vapor condensing systems and substitute surface condensers. The condensed water vapors and organic vapors from the surface condensers are discharged to gravity-settling systems. After removal of the flotable matter, the wastes are discharged to the waste-water treatment system. The condensing surfaces become coated with congealed material. For this reason the surface condensers must be installed in multiple units. One unit or more must be off stream for cleaning. Each such installation is a multi-million-dollar investment and requires much labor for operation.

Waste Materials

3.1 INTRODUCTION

The words "waste materials" or "solid wastes" in environmental control vernacular generally mean those materials that are discarded other than as pollutants in the wastewaters or in emissions to the atmosphere. Waste materials may be solids, liquids, semisolids, slurries, or even contained gases. Much of the solid waste from industry is similar to household garbage and refuse as far as environmental significance is concerned. Some waste materials from industrial plants could be hazardous, that is, harmful to humans or other organisms if not properly disposed of. Federal legislation of the late 1970s severely regulated the processing, transporting, storage, and disposal of waste materials defined as hazardous in the law. Notification of the appropriate state or federal agency of the generation of a hazardous waste is required. Permits from such agencies are required for processing, storage, and disposal of hazardous wastes. Certain rules are applied also to transport of the waste. The disposal of wastes that are not hazardous is also controlled by federal legislation, but to a lesser degree. The disposal facilities must be approved and meet certain criteria.

3.2 SOURCES OF WASTE MATERIALS IN THE OIL PROCESSING INDUSTRIES

Waste materials from industrial operations addressed in this chapter include:

Spent bleaching earth.
Spent metal catalysts.
Oilseed residues accumulating in raw-oil storage tanks.
Animal hide scrap, bone, hair, and other residue accumulating in raw animal fat storage tanks.
Diatomaceous filter earth and powdered activated carbon from filtering and decolorizing processes.

Bottom residues from some distilling operations.

Oil and coagulating chemical slurry from chemically enhanced air flotation of wastewaters.

Sludges of biological matrix created in biological treatment.

Impurities separated from the raw oil for which there is no byproduct use.

Sludge from gravity settling tanks used to recover flotable oil.

Off-quality or contaminated products or raw materials.

Residues of unreacted raw materials and process chemicals from batch soapmaking operations.

3.3   SOLID WASTE IMPACTS

Solid wastes of the industry, if not properly disposed of, could have significant environmental impact. For the most part, the solid wastes of the industry are similar to household garbage and refuse with respect to environmental significance. One interest, of course, is simply aesthetic. Also, if left open and unprotected, such material will breed flies, rats and odors. In a dump or a landfill, a certain amount of liquid will ordinarily drain from the material. This drainage, termed *leachate*, can seriously degrade the quality of groundwater or surface waters it might reach. Degradation of the organic matter may produce slightly acidic conditions in the leachate that will tend to dissolve metals that may be present.

Improper incineration of solid waste can cause excessive emissions of particulate matter. Volatile organic materials might enter the air in damaging amounts.

3.4   SOLID WASTE REGULATION

For regulatory purposes, solid wastes are divided into two categories: "hazardous" waste and waste similar to household garbage and refuse in disposal requirements.

The regulations of the federal government provide minimum requirements. Each of the states is expected to have its own regulations that are at least as stringent as the federal. The diversity among state regulations will prevent providing a summary of each in this chapter.

Federal regulations are new as of this writing and are still evolving. The current EPA regulations provide several ways by which a waste may be classified as hazardous. The regulations name a large number of elements and compounds as hazardous if discarded. The amount of these wastes triggering control is also specified. A second way wastes are classified as hazardous is by naming specific wastes as identified by industry category and process source. Still a third way is by certain characteristics. One such is corrosivity, which is simply pH limits. Another is ignitability; the test for liquids is a flash point of 140°F and for solids is a judgment based on whether or not when ignited the material "burns so vigorously and persistently as to create a hazard." The stated intent is to control thermally unstable solids liable to spontaneous chemical changes. Another property is reactivity, which covers explosiveness and/or generation of toxic fumes. Another test is an extraction procedure and analysis for 13 specific elements or compounds. The compounds are those named in the primary drinking water standards of the the U.S. EPA. Concentration limits are specified. None of these compounds would be expected to be components of waste materials from the industries embraced in this chapter.

A few waste materials of the fat and oil industry are possible candidates for eventual designation as hazardous wastes under present definitions. Spent nickel catalysts used in hydrogenation of vegetable oils and containing such oil may be considered hazardous based on their testing for reactivity or in interpretation of the language of "burning vigorously and persistently." No decision has been made as of this writing (1981). Much of the spent catalyst will be processed for regeneration. Regulations covering recovery and regeneration of hazardous wastes have not been issued at this time. Nickel and zinc catalysts may also be designated as hazardous wastes based on their toxicity to aquatic life if the proposed water quality criteria for nickel and zinc are established. (See sections of this chapter on water pollution effects).

Another possible candidate for designation as a hazardous waste is spent bleaching earth of high oil content from purification of edible oils. Bleaching earth with high ($\geq$20%) oil content, when exposed to the atmosphere at warm ambient temperatures, will spontaneously reach ignition temperatures. Spent bleaching earth of lower oil content, which may be achieved by oil recovery or by mixing with earth or other matter, is not pyrogenic. Likewise, spent bleaching earth if wet or in slurry form is not pyrogenic. Even the dry oily earth when spread in a thin layer will not ignite spontaneously.

Other materials that the fat and oil industry uses are potentially hazardous because they are in the range of pH defined to make it a hazardous waste, namely, pH less than 2.5 or greater than 12.5. However, the sodium hydroxide and sulfuric acid streams commonly used in the industry are part of the process and not wastes. These materials are neutralized before discharge in wastewaters. As of this writing, it has not been decided whether these routinely neutralized materials will be regulated as hazardous wastes. However, if there is any off-quality acid or caustic that is discarded and hauled off site for treatment and disposal, the plant involved would be a hazardous waste generator.

Whether or not a waste material is in the hazardous category makes a great difference in the regulations that apply and also in the costs of disposal. If a plant stores, processes, or disposes of hazardous wastes, a permit will be required from the agency with jurisdiction. Shipments must be accompanied by a manifest containing specified information.

Regulation of the nonharzardous wastes are limited to the hauler and the disposer. As a minimum, the hauling regulations will require that the wastes be covered to prevent windblown loss and to control odors. The disposal must be to approved facilities. The types of approved facility include landfills, incinerators, composting plants, or land farming. "Land farming" refers to the spreading of appropriate organic wastes, such as organic sludges, on land. The organic material is stabilized by bacterial action and later tilled into the soil. Slurries of spent bleaching earth have been disposed of in this manner. The use of land spreading is subject to limitations based on the amount of toxic metals such as cadmium in the sludge and the crop to be grown on the land. The productivity of poor soil may be enhanced by the material added. Not all metals are objectionable in controlled amounts; for example, zinc is a necessary trace element for plant growth. It is also needed in the human diet so that plant uptake is not a problem per se. A major portion of the soil in the United States is deficient in zinc, and thousands of pounds of zinc compounds are spread on farms yearly, particularly on orchard lands.

Because of the contribution to the volume of leachate that might be produced in a landfill, the disposal of water slurries of waste materials of any kind to a garbage–refuse-type landfill may be prohibited.

A special waste material control is that required for PCBs. These compounds are used for heat transfer purposes in transformers, capacitors, and heat-exchange systems. This material has been found to be very persistent in the environment and to harm beneficial organisms. Regulations have been proposed to eliminate its use in food and food package manufacturing plants. Disposal of the liquid must be to incineration facilities approved by the EPA. Even the electrical transformer bodies or other equipment in which it was contained, if disposed of, have to be delivered to approved facilities. Other fluids used in transformers may have become contaminated with PCB, and any such fluid with 50 ppm or more of PCB is regulated as to disposal and handling in the same manner as the pure material.

The EPA intends to issue separate regulations governing the disposal of waste oil. Waste petroleum oil is persistent in the ground, so disposal to landfills will be prohibited. (Animal and vegetable oils are degradable in landfills.) Hopefully, the differences in the kinds of oil will be recognized in the regulations, and the regulations should be appropriate to each.

3.5   Waste Material Recovery and Disposal Methods

The federal law and regulations encourage reuse of materials or secondary uses rather than their disposal. Many citizens support such efforts because ever-increasing usage and a limited supply of resources make that position rational. The dictates of cost and economic reality or even technical problems limit the recovery route.

Recovery of the heavy nickel catalyst used in hardening is being practiced by at least one manufacturer of catalyst in the United States. Other users of nickel catalysts currently have no companies offering recovery and regeneration services. The trade associations of the involved industries are joining in efforts to interest companies in operating regeneration facilities.

The economics of scale are such that shipping all spent catalysts generated in a region, or perhaps the whole country, is needed to justify a minimum-sized facility. Shipping costs are a large handicap. If regeneration is not practical, disposal of the spent catalyst at landfill facilities approved for hazardous wastes may be the only approved disposal.

The continuous hydrolysis of fats and oils, which is the cornerstone process of the soap and fatty acid industry, is accomplished at some plants by using medium pressures and catalysts, such as zinc oxides, calcium, or magnesium oxides. The spent zinc catalysts may be recovered and converted to the oxide form for reuse. Thus this recovered material is not involved in waste material disposal.

Spent bleaching earth has two components with recovery potential, the oil and the earth. Unfortunately, the earth has special properties such as its bleaching (adsorptive) property, which is not easily restored. At two locations in the country, the deoiled earth was processed in multihearth furnaces, and the organic-free product was used for various purposes. The recovery of the oil or portions of the oil is likewise possible. One highly effective method is extraction with solvent.

Individual companies practice partial recovery of the oil, as discussed in the section on water pollution. The dewatered bleaching earth with oil content less than the pyrophoric level can be disposed of in a common landfill approved for household garbage and refuse.

Trials have been carried out for the use of the fuel value of the oil in spent bleaching earth. The quantity of material does not justify a facility solely for this purpose. The trial was to test the addition to pulverized coal being fed the boilers of a public utility. The object was more for disposal than use of the fuel value; the test was successful. However, a number of problems arose: (*a*) there was the pyrophoric problem during storing and transfer, if the oil content was high; (*b*) the cold material conceivably could foul up the pulverizer and conveyors by clinging to them; and (*c*) there was also the potential of corrosion or scale problems in the boiler, and even erosion by the clay particles. The utility decided that there was not enough spent clay to justify further investigation. One concern was that the clay would add to the particulates in the stack emissions. It was reported that the clay was removed in the fly ash settling chamber. Particulate emission was below normal during the run.

Another clever secondary use method that accomplishes disposal is the utilization of the value of the oil content of the spent bleaching earth as an animal food additive. Fat and oil are normally added to animal and chicken feed. Because of the content of clay and other materials, this source can constitute only a small percent of the feed content. This use is being practiced to a limited extent.

As to actual disposal practices, the disposal of slurries of spent bleaching earth will quite likely become unfeasible because the disposal of liquid wastes at landfills for garbage and household refuse will be prohibited. Ease of filter cleaning and transport of the spent earth is enhanced by making it into a water slurry. Disposal of a waste slurry will probably be permitted only at facilities approved for liquid hazardous material disposal. Use of these will be expensive. Thus the spent bleaching earth will have to be dewatered or sufficient dry material added to take up the water so that no water will drain from the cake. Dewatering can reduce hauling and disposal costs.

Special arrangements with a commercial landfill have enabled one plant to dispose of the spent bleaching earth in a special landfill separate from the general facility. The clay is covered daily soon after deposit at the site. Because of its isolation from other combustible materials and coverage soon after deposit, there is no concern over the pyrophoric nature of the material. Leachate from the special fill is monitored; so far there has been none.

Recently, one plant has begun disposal of the spent bleaching earth slurry by land farming. The slurry is spread from a truck as a thin layer on land. After spreading, the slurry-laden area is disked, using common farm equipment to incorporate the material in the soil. No problems have been encountered to date, and the agricultural use properties of the soil are expected to be enhanced.

References

1. G. E. Mau, "A Study of Vertical-Slotted Inlet Baffles," *J. Water Pollution Control Fed.*, **31** (12), 1349–1372 (1959).

2. J. C. Young, "Removal of Grease and Oil by Biological Treatment Processes," *J. Water Pollution Control Fed.*, **51** (8), 2070–2087 (1979).

3. A. Cywin and R. C. Loehr, "EPA Pretreatment Requirements for Oil and Grease. Treatability of Oil and Grease Discharged to Publicly Owned Treatment Works," U.S. EPA, April 1975.

4. EPA, "Background Information for Establishing of National Standards of Performance for New Sources, Vegetable Oil Industry," Industrial Standards Branch, Division of Applied Technology, Office of Air Programs Environmental Protection Agency, Raleigh, NC, July 15, 1978.

5. American Society for Testing and Materials, ASTM D1391 (1978), Standard Test Method for Measurement of Odor in Atmospheres (dilution method), ASTM, Philadelphia, PA, 1978.

6. J. L. Mills, R. T. Walsh, K. D. Luedtke, and L. K. Smith, "Quantitative Odor Measurement," *J. Air Pollut. Control Assoc.*, **13**, 467–475 (1963).

7. D. M. Benferado, W. J. Rotella, and D. L. Horton, "Development of an Odor Panel for Evaluation of Odor Control Equipment," *J. Air Pollut. Control Assoc.*, **19**, 101–105 (1969).

8. *Barneby-Cheney Co. Scentometer: An Instrument for Field Odor Measurements.* (manual), Barneby-Cheney Co., Columbus, OH., 1970.

Index